Food Process Engineering and Technology

Junaid Ahmad Malik • Megh R. Goyal •
Anu Kumari
Editors

Food Process Engineering and Technology

Safety, Packaging, Nanotechnologies and Human Health

Springer

Editors
Junaid Ahmad Malik
Department of Zoology
Government Degree College Kulgam
Kulgam, Jammu and Kashmir, India

Megh R. Goyal
College of Engineering
(Retired) University of Puerto Rico
Mayagüez, Puerto Rico

Anu Kumari
Warner College of Dairy Technology
Sam Higginbottom University of
Agriculture, Technology & Sciences
(SHUATS)
Prayagraj, Uttar Pradesh, India

ISBN 978-981-99-6830-5 ISBN 978-981-99-6831-2 (eBook)
https://doi.org/10.1007/978-981-99-6831-2

This Springer imprint is published by the registered company Springer Nature Singapore Pte Ltd.
The registered company address is: 152 Beach Road, #21-01/04 Gateway East, Singapore 189721, Singapore

Paper in this product is recyclable.

Preface

For its growth, innovation, and ability to maintain its competitiveness, the food industry places a high value on the fields of food science and food process engineering. Globalization has altered the boundaries of what was formerly a relatively local activity, which is the reality of the current food industry.

The industry imports the raw materials it uses from all around the world. On the one hand, it gets harder to guarantee that the raw materials going into the production process will be accurately specified in terms of their processing properties, like composition, color, and flavor. The qualities of the raw materials have an impact on the final product. To manage and control such variation in their raw materials, food plants must be equipped. On the other hand, food safety has evolved significantly. It was and still is one of the fundamental principles of food production. There are numerous systems in place to ensure the safety of the products, but because processing lines must be flexible in order to meet consumer demands, new contaminants, allergens, or pathogens, as well as other spoilage, have the potential to cause problems, organisms must be constantly monitored and kept under control.

As a result, advancements in engineering and technology must be made continuously. Food engineering examines and applies engineering, science, and math principles to the production, handling, storage, conservation, control, packaging, and distribution of food products. It is a scientific, academic, and professional field. As a result, its users—engineers, technicians, food scientists, and any other staff members who are anyway involved in the quality of the food product—must be well-versed in the techniques necessary to master food processing activities.

This book has six parts: First part of this book covers the novel technologies in food processing including emerging technologies for food drying, significance and applications of colligative properties on foods and food products. It also includes 3D printing for foods, microwave processing, and renewable energy sources for food and dairy processing. Part II of this book emphasizes on the area of food quality and management. Part III of this book gives information on food packaging including new-age packaging for foods and food products, biodegradable, packaging, and extension of shelf life of packaged foods through herbal extracts and essential oils.

Part IV of this book covers the application and safety of nanomaterials in food packaging. Part V of this book emphasizes on the scope of nanotechnology and nanoencapsulation in nutritional and functional foods. Part VI of this book covers the food science and nutritional research for human health. This part covers the information on bioaccessibility and bioavailability of phenolics in grapes and grape products, formulation of infant foods, fermented *vigna mungo* and carrot pomace cookies using *Lactobacillus casei*, probiotics in gut microflora, and metabolomics applications in food science and nutritional research.

A really good addition to the collection of works on food processing is this book. It will be helpful for teaching undergraduate and graduate students, as well as research scholars, food scientists, and industry professionals.

Kulgam, India — Junaid Ahmad Malik
Mayagüez, Puerto Rico — Megh R. Goyal
Prayagraj, India — Anu Kumari

Contents

Editors and Contributors

About the Editors

Junaid Ahmad Malik holds a diverse educational background, having received his B.Sc. in Science from the University of Kashmir, Srinagar, J&K in 2008. He pursued his M.Sc. in Zoology from Barkatullah University, Bhopal, Madhya Pradesh in 2010, and later achieved his Ph.D. in Zoology from the same university in 2015. To further enhance his academic qualifications, he successfully completed a B.Ed. program in 2017 from the University of Kashmir, Srinagar, J&K.

After completing his education, he embarked on his career as a Lecturer in the School Education Department, Govt. of J&K, where he dedicated 2 years to educating and inspiring students. Subsequently, in 2017, he joined the Department of Zoology at Govt. Degree College Bijbehara, J&K, and contributed his expertise until 2020. At present, he serves as a Lecturer in the Department of Zoology at Govt. Degree College, Kulgam, Kashmir (J&K), where he continues to be actively involved in teaching and conducting research activities.

With over 10 years of research experience, he has demonstrated a profound interest in several areas, including ecology, environmental science, wildlife biology, and conservation biology.

His contributions to the scientific community are notable. He has published more than 20 research papers in various prestigious national and international peer-reviewed journals. Moreover, his dedication to academic authorship and editorship extends to 26 books, 38 book chapters, and over 10 popular editorial articles, published with renowned publishers such as Springer Nature, Elsevier Inc., Taylor and Francis Group, and IGI Global.

He is the founder and CEO of World Biologica Publishers. Apart from being an accomplished author, he holds an esteemed role as the Editor-in-Chief of "Inventum Biologicum," an International Journal of Biological Research published by World Biologica, India. Additionally, he actively contributes to the academic world as an editor and reviewer for various reputable journals.

His involvement in the scientific community extends beyond publishing. He has actively participated in numerous state, national, and international conferences, seminars, workshops, and symposia, presenting over 20 conference papers that further enrich the field of Zoology.

As a testament to his commitment to the advancement of biological sciences, he is a life member of SBBS (Society for Bioinformatics and Biological Sciences) with membership id LMJ-243.

For those interested in connecting with Dr. Malik, he can be reached at editor@worldbiologica.com or malik.junaidahmad@gmail.com.

Megh R. Goyal received his B.Sc. degree in Engineering from Punjab Agricultural University, Ludhiana, India; his M.Sc. and Ph.D. degrees from the Ohio State University, Columbus; his Master of Divinity degree from Puerto Rico Evangelical Seminary, Hato Rey, Puerto Rico, USA.

Since 1971, he has worked as Soil Conservation Inspector (1971); Research Assistant at Haryana Agricultural University (1972–1975) and the Ohio State University (1975–1979); Research Agricultural Engineer/Professor at Department of Agricultural Engineering of UPRM (1979–1997); and Professor in Agricultural and Biomedical Engineering at General Engineering Department of UPRM (1997–2012). He spent one-year sabbatical leave in 2002–2003 at Biomedical Engineering Department, Florida International University, Miami, USA.

He was first agricultural engineer to receive the professional license in Agricultural Engineering in 1986 from College of Engineers and Surveyors of Puerto Rico. On 16 September 2005, he was proclaimed as "Father of Irrigation Engineering in Puerto Rico for the twentieth century" by the ASABE, Puerto Rico Section, for his pioneer work on micro-irrigation, evapotranspiration, agroclimatology, and soil and water engineering.

During his professional career of 53 years, he has received awards such as Scientist of the Year, Blue Ribbon Extension Award, Research Paper Award, Nolan Mitchell Young Extension Worker Award, Agricultural Engineer of the Year, Citations by Mayors of Juana Diaz and Ponce, Membership Grand Prize for ASAE Campaign, Felix Castro Rodriguez Academic Excellence, Rashtrya Ratan Award and Bharat Excellence Award and Gold Medal, Domingo Marrero Navarro Prize, Adopted son of Moca, Irrigation Protagonist of UPRM, Man of Drip Irrigation by Mayor of Municipalities of Mayaguez/Caguas/Ponce and Senate/ Secretary of Agriculture of ELA, Puerto Rico.

The Water Technology Centre of Tamil Nadu Agricultural University in Coimbatore, India recognized Dr. Goyal as one of the experts "who rendered meritorious service for the development of micro-irrigation sector in India" by bestowing "Award of Outstanding Contribution in Micro-Irrigation." This award was presented to Dr. Goyal during the inaugural session of the National Congress on "New Challenges and Advances in Sustainable Micro-Irrigation on 1 March 2017, held at Tamil Nadu Agricultural University.

At Annual Meeting on 1 August 2018 in Detroit, MI, American Society of Agricultural and Biological Engineers (ASABE) bestowed on him Netafim Microirrigation Award for his unselfish contribution. VDGOOD Professional Association of India awarded Lifetime Achievement Award at 12th Annual Meeting on Engineering, Science and Medicine that was held on 20–21 of November of 2020 in Visakhapatnam, India.

He has authored more than 200 journal articles and more than 120 books including: "Elements of Agroclimatology (Spanish) by UNISARC, Colombia"; two "Bibliographies on Drip Irrigation." Apple Academic Press Inc. (AAP) has published his books, namely: "Management of Drip/Trickle or Micro-Irrigation," and "Evapotranspiration: Principles and Applications for Water Management," ten-volume set on "Research Advances in Sustainable Micro-Irrigation" and 14-volume set on "Challenges in Micro-irrigation." During 2016–2022, he has published several book volumes on emerging technologies/issues/challenges under many book series.

Anu Kumari is working as Assistant Professor (Dairy Engineering) in Department of Dairy Engineering, Warner College of Dairy Technology, Sam Higginbottom University of Agriculture, Technology and Sciences, Prayagraj, U.P., India.

She did her graduation (2011) in Dairy Technology from Sam Higginbottom University of Agriculture, Technology and Sciences (SHUATS), Prayagraj, India; M.Tech. (2013) in Dairy Engineering from ICAR-National Dairy Research Institute, Karnal, India. She is recipient of Gold Medal (2011) awarded by SHUATS for excellence performance in academics during graduate program, NDRI fellowship (2011–2013). She secured second position in dairy product judging contest organized by Anand Agriculture University (AAU), Anand.

She is currently working in the area of development of biocomposites for food and dairy products, process mechanization of dairy equipment, mathematical modeling, image analysis technique, rheology of food and dairy products, by-product utilization of agricultural commodity, baking of food and dairy products. She has served Dairy Science College, KVAFSU, Hebbal, Bengaluru as Assistant Professor. She has guided 12 postgraduate students and co-guided several master scholars in dairy and Food technology. She has participated in several national and international conferences and seminars. She has served Birsa Agriculture University, Ranchi; Baba Sahab Dr Bhim Rao Ambedkar College of Agriculture Engineering and Technology, Chandra Shekhar Azad University of Agriculture and Technology, Kanpur, Etawah Campus; UPPSC (Uttar Pradesh Public Service Commission) as an external examiner and paper setter for dairy engineering courses.

She has received Best Research Paper Award (2016) by SHUATS, Prayagraj; Best Paper Award (2019) by ICAR-NDRI, Bengaluru; Young Scientist Award (2019) by Agricultural Technology Development Society (ATDS), Ghaziabad. She has successfully completed training of 15 days at Feeder Balancing Dairy, Patna Dairy Project (Sudha Dairy), Bihar in all the activities related to Microbiological and Chemical Analysis and Production and Storage of Milk and Milk Products; one month at Experimental Dairy of National Dairy Research Institute, Karnal, Haryana in all the activities related to Market Milk Section, Cheese and Butter Section, Condensing and Drying Section and Quality Control Laboratory; 4 months at Mother Dairy Fruits and Vegetable Pvt. Ltd., Patparganj, New Delhi in all the activities related to Q.A lab, Dock lab and Process and Utilities and 21 days CAFT Training on Dairy and Food Process Engineering: Equipment, Processing, and Value Addition organized by Dairy Engineering Division at ICAR-NDRI, Karnal.

She has more than 50 publications including 12 papers, 3 books, 8 book chapters, 3 popular articles, 4 articles, and 26 abstracts in compendium of various seminars/conferences. She is Life member of IDEA (Indian Dairy Engineering Association); Agricultural Technology Development Society (ATDS), Ghaziabad.

Contributors

Farhana Mehraj Allai Faculty of Agricultural Sciences, Department of Post-Harvest Engineering and Technology, Aligarh Muslim University, Aligarh, Uttar Pradesh, India

Sartaj Ahmad Allayie Zoology Research Laboratory, Department of Zoology, Government Degree College, Anantnag, Jammu & Kashmir, India

Anuj Department of Dairy Technology, College of Dairy Science and Technology, Lala Lajpat Rai University of Veterinary and Animal Sciences, Hisar, Haryana, India

Jobil J. Arakal Department of Food Technology, Saintgits College of Engineering, Kottayam, Kerala, India

Shalini Arora Department of Dairy Technology, College of Dairy Science and Technology, Lala Lajpat Rai University of Veterinary and Animal Sciences, Hisar, Haryana, India

Z. R. Azaz Ahmad Azad Faculty of Agricultural Sciences, Department of Post-Harvest Engineering and Technology, Aligarh Muslim University, Aligarh, Uttar Pradesh, India

Neriman Bağdatlıoğlu Department of Food Engineering, Faculty of Engineering, Manisa Celal University, Manisa, Turkey

Tarun Kanti Bandyopadhyay Chemical Engineering Department, National Institute of Technology Agartala, Jirania, Tripura, India

Mainak Barman Department of Genetics and Plant Breeding, Bidhan Chandra Krishi Viswavidyalaya, Nadia, West Bengal, India

Vibhuti Batra Suryan Enterprises, Baddi, Himachal Pradesh, India

Biswanath Bhunia Bioengineering Department, National Institute of Technology Agartala, Jirania, Tripura, India

Sabzar Ahmad Dar Zoology Research Laboratory, Department of Zoology, Government Degree College, Anantnag, Jammu and Kashmir, India

Anamika Das Warner College of Dairy Technology (WCDT), Sam Higginbottom University of Agriculture, Technology and Sciences (SHUATS), Prayagraj, Uttar Pradesh, India

Mamoni Das Department of Food Science and Nutrition, College of Community Science, Assam Agriculture University, Jorhat, Assam, India

Nabendu Debnath Centre for Molecular Biology, Central University of Jammu, Jammu, Jammu and Kashmir, India

Arpan Dubey Agricultural and Food Engineering Department, Indian Institute of Technology Kharagpur, Kharagpur, West Bengal, India

Abhijit Ganguli SSD Conglomerate Projects DST (Government of India), Science & Technology, Government of Haryana, PHFI, Gurgaon, Haryana, India

Nidhi Gaur School of Basic and Applied Sciences, K.R. Mangalam University, Gurgaon, Haryana, India

Rajni Gautam School of Basic and Applied Sciences, K.R. Mangalam University, Gurgaon, Haryana, India

Soumitra Goswami Department of Food Science and Technology Programme, College of Agriculture, Assam Agriculture University, Jorhat, Assam, India

Khalid Gul Department of Food Process Engineering, National Institute of Technology, Rourkela, Odisha, India

Pinar Gumus Faculty of Health Sciences, Department of Nutrition and Dietetics, Kilis 7 Aralik University, Kilis, Turkey

Annika Durve Gupta Department of Biotechnology, B.K. Birla College of Arts, Science and Commerce (Autonomous), University of Mumbai, Kalyan, Maharashtra, India

V. Harsha Agricultural and Food Engineering Department, Indian Institute of Technology, Kharagpur, West Bengal, India

Faisal Sualeh Hayyat Department of Physical Education and Sports, Govt. Degree College, Pulwama, Jammu and Kashmir, India

Pawan Jalwal Faculty of Pharmacy, Baba Mastnath University, Rohtak, Haryana, India

Jayasree T. Joshi Agricultural and Food Engineering Department, Indian Institute of Technology, Kharagpur, West Bengal, India

Vijayasri Kadirvel Department of Food Technology, Rajalakshmi Engineering College, Chennai, Tamil Nadu, India

Müzeyyen Berkel Kaşıkçı Faculty of Engineering, Department of Food Engineering, Manisa Celal University, Manisa, Turkey

Amit Kotiyal Department of Horticulture, Lovely Professional University (LPU), Phagwara, Punjab, India

Arya S. Krishnan Department of Food Technology, Amal Jyothi College of Engineering, Koovappally, Kerala, India

R. Kurup Krishna Department of Food Technology, Bhaskaracharya College of Applied Sciences, University of Delhi, Dwarka, Delhi, India

Amit Kumar Agricultural and Food Engineering Department, Indian Institute of Technology Kharagpur, Kharagpur, West Bengal, India

Anu Kumari Warner College of Dairy Technology, Sam Higginbottom University of Agriculture, Technology & Sciences (SHUATS), Prayagraj, Uttar Pradesh, India

Prakash Kumar Agricultural and Food Engineering Department, Indian Institute of Technology Kharagpur, Kharagpur, West Bengal, India

Rimpa Kundu Department of Genetics and Plant Breeding, Bidhan Chandra Krishi Viswavidyalaya, Nadia, West Bengal, India

Deeptimayee Mahapatra Department of Food Science and Nutrition, College of Community Science, Assam Agriculture University, Jorhat, Assam, India

Junaid Ahmad Malik Department of Zoology, Government Degree College Kulgam, Kulgam, Jammu and Kashmir, India

Shweta Mall Animal Genetics and Breeding Section, ICAR-National Dairy Research Institute, Southern Regional Station (SRS), Bengaluru, Karnataka, India

Arshied Manzoor Department of Food Technology, Institute of Engineering and Technology, Bundelkhand University, Jhansi, Uttar Pradesh, India

Rita Mehla Dairy Chemistry Department, National Dairy Research Institute, Karnal, Haryana, India

Sonia Mor Warner College of Dairy Technology (WCDT), Sam Higginbottom University of Agriculture, Technology and Sciences (SHUATS), Prayagraj, Uttar Pradesh, India

Navdeep Nain University Institute of Engineering & Technology, Kurukshetra University, Kurukshetra, Haryana, India

Pinku Chandra Nath Bioengineering Department, National Institute of Technology Agartala, Jirania, Tripura, India

Sadaf Nazir Faculty of Agricultural Sciences, Department of Post-Harvest Engineering and Technology, Aligarh Muslim University, Aligarh, Uttar Pradesh, India

Amiya Ojha Bioengineering Department, National Institute of Technology Agartala, Jirania, Tripura, India

Kaushik Pramanik Department of Agricultural Entomology, Bidhan Chandra Krishi Viswavidyalaya, Nadia, West Bengal, India

Neha Rai Department of Pharmaceutical Sciences, Kurukshetra University, Kurukshetra, Haryana, India

Sakshi Rai Centre for Molecular Biology, Central University of Jammu, Jammu, Jammu and Kashmir, India

Kamalesh Raja Department of Biotechnology, Rajalakshmi Engineering College, Chennai, Tamil Nadu, India

Poonam Rani Agricultural and Food Engineering Department, Indian Institute of Technology Kharagpur, Kharagpur, West Bengal, India
Teagasc Food Research Centre, Cork, Ireland

Lakshita Rao Department of Pharmaceutical Sciences, Gurugram University, Gurugram, Haryana, India

Ameeta Salaria Discipline of Dairy Technology, Sher-e-Kashmir University of Agriculture Sciences and Technology, Jammu, Jammu and Kashmir, India

Shalini Sehgal Department of Food Technology, Bhaskaracharya College of Applied Sciences, University of Delhi, Dwarka, Delhi, India

Ramesh Sharma Bioengineering Department, National Institute of Technology Agartala, Jirania, Tripura, India

Pooja Singh Department of Horticulture, Lovely Professional University (LPU), Phagwara, Punjab, India

Satish Kumar Singh Department Plant Breeding and Genetics, Dr. Rajendra Prasad Central Agricultural University, Pusa, Bihar, India

Tarun Pal Singh Goat Products Technology Laboratory, ICAR-Central Institute for Research on Goats, Makhdoom, Uttar Pradesh, India

Harshita Sonarthi Dairy Technology Section, ICAR-National Dairy Research Institute, Southern Regional Station (SRS), Bengaluru, Karnataka, India

Thiruvengadam Subramaniyan Department of Biotechnology, Rajalakshmi Engineering College, Chennai, Tamil Nadu, India

S. Supreetha Dairy Technology Section, ICAR-National Dairy Research Institute, Southern Regional Station (SRS), Bengaluru, Karnataka, India

Vini Swarup Warner College of Dairy Technology (WCDT), Sam Higginbottom University of Agriculture, Technology and Sciences (SHUATS), Prayagraj, Uttar Pradesh, India

Ashok K. Yadav Centre for Molecular Biology, Central University of Jammu, Jammu, Jammu and Kashmir, India

Deepika Yadav Aravali College of Pharmacy, Rewari, Haryana, India

Pooja Yadav Centre for Molecular Biology, Central University of Jammu, Jammu, Jammu and Kashmir, India

Ritika B. Yadav Department of Food Technology, Maharshi Dayanand University, Rohtak, Haryana, India

A. R. Yeswanth Department of Food Technology, Bhaskaracharya College of Applied Sciences, University of Delhi, Dwarka, Delhi, India

Insha Zahoor Faculty of Agricultural Sciences, Department of Post-Harvest Engineering and Technology, Aligarh Muslim University, Aligarh, Uttar Pradesh, India

List of Abbreviations and Symbols

μm	Micrometer
1D	One-dimensional
2D	Two-dimensional
3D	Three-dimensional
A	Swept area
AAP	American Academy of Pediatrics
ABTS	2,2′-azino-bis(3-ethylbenzothiazoline-6-sulfonic acid) diammonium salt
AC	Alternating current
A_c	Surface area of collector
AFM	Atomic force microscopy
Ag	Silver
AgNMs	Silver nanomaterials
AI	Artificial intelligence
ALA	5- aminolaevulinic acid/alpha-linolenic acid
ALE	Amaranthus leaf extract
ANN	Artificial neural network
Antho	Anthocyanins
AOAC	Association of Official Agricultural Chemists
ARA	Arachidonic acid
ARGs	Antibiotic resistant genes
Ast	Astilbin
ATP	Adenosine triphosphate
a_w	Water activity
B:C ratio	Benefit–cost ratio
BASF	Baden Aniline and Soda Factory
BC	Bacterial cellulose
BHA	Butylated hydroxyanisole
BHT	Butylated hydroxytoluene
BNC	Bacterial nanocellulose
BPA	Bisphenol A

BPE	Boiling point elevation
BSA	Bovine serum albumin
BSG	Bovine skin gelatin
Bt	Bacillus thuringiensis
BV	Baculoviruses
C.V.	Coefficient of variation
CA	Cellulose acetate/coutaric acid
CAD	Computer-aided design
CAGR	Compounded annual growth rate
CARK	Controlled additive-manufacturing robotic kit
CAT	Catalase
Cbl	Cobalamin
CDAD	*Clostridium difficile* associated diarrhea
CE	Capillary electrophoresis
Ce	Cerium
CFs	Cell-free supernatant
CFU	Colony forming units
CL	Crosslinking
CMA	Cow milk allergy
CMC	Carboxymethyl cellulose
CMP-eHF	Cow's milk protein-based extensively hydrolyzed formulas
CNC	Cellulose nanocrystals
CNcbl	Cyanocobalamin
CNF	Cellulose nanofibers
CNT	Carbon nanotubes
CO_2	Carbon dioxide
CoQ10	2,3-dimethoxy, 5-methyl, 6-decaprenyl benzoquinone
C_p	Power coefficient
CPLA	Polylactide aliphatic copolymer
CPS	Canadian pediatric society
CSP	Concentrator solar power
CSS	Corn syrup-based solids
Cu	Copper
CuO	Copper oxide
CUPRAC	Cupric reducing antioxidant capacity
CVD	Chemical vapor deposition
CYP450	Cytochrome-450
DBP	Dibutyl phthalate
DC	Direct current
DEHP	Bis(2-ethylhexyl)phthalate
DEP	Diethyl phthalate
DHA	Docosahexaenoic acid
DLS	Dynamic light scattering
DLT	Distributed ledger technology

DMBI	5,6-dimethylbenzimidazole
DMP	Dimethyl phthalate
DNA	Deoxyribonucleic acid
DNBS	Dinitrobenzene sulfate
DPPH	2,2-diphenyl-1-picrylhydrazyl
DSS	Dextran sodium sulfate
DT	Decision tree
EA	Ethyl gallate
EBABs	Expanded-bed adsorption bioreactors
EDTA	Ethylenediamine tetraacetic acid
EJ	Exajoules
ELISA	Enzyme-linked immunosorbent assays
EM	Electron microscopy
ENMs	Engineered nanomaterials
EO	Essential oil
EPA	Environmental Protection Agency/eicosapentaenoic acid
EPF	Entomopathogenic fungus
Epi	Epicatechin
EPN	Entomopathogenic nematodes
EPS	Exopolysaccharides
EPV	Entomopathogenic virus
ESEM	Environmental scanning electron microscopy
EWNS	Engineered water nanostructures
FAO	Food and Agriculture Organization
FDA	US Food and Drug Administration
Fe	Iron
Fe_3O_4	Ferrous oxide
FESEM	Field emission scanning electron microscopy
FFQ	Food frequency questionnaire
FL	Fuzzy logic
FMN	Flavin mononucleotides
FOS	Fructooligosaccharides
FP	Freezing point
FPD	Freezing point depression
FRAP	Ferric reducing antioxidant power
FTIR	Fourier Transform infrared
FW	Fresh weight
GABA	Gamma amino butyric acid
GAD	Glutamic acid decarboxylase
GC	Gas chromatography
GC-MS	Gas chromatography-mass spectroscopy
GERD	Gastro-esophageal reflux disease
GHz	Gigahertz
GI	Gastrointestinal

GIT	Gastrointestinal tract
GMC	Monocaprylin glycerate
GML	Glycerol monolaurate
GNCs	Graphene nanocomposites
GOS	Galacto oligosaccharides
GPCRs	G protein-coupled receptor
GPD	Glycerol-3-phosphate dehydrogenase
GRAS	Generally recognized as safe
GSH	Glutathione
GTF	Glycosyltransferases
HACCP	Hazard analysis and critical control point
HDC	Hydrodynamic chromatography
HDPE	High density polyethylene
HEF	High electric field
HEPS	Heteropolysaccharides
HHP	High-pressure pasteurization
HMF	Hydroxy methyl furfural
HMO	Human milk oligosaccharides
H-NMR	Proton-nuclear magnetic resonance
HOPS	Heteropolysaccharides
HP	Himachal Pradesh
HPC	Hydroxypropyl cellulose
HPHP	High-pressure homogenization processing
HPLC	High performance liquid chromatography
HPMC	Hydroxypropyl methyl cellulose
HV	Hydroxy-valerate
Hz	Hertz
I	Solar energy radiation incident on the collector
IaaS	Infrastructure as Service
IBD	Inflammatory bowel disease
IBS	Irritable bowel syndrome
I_c	Current
ICA	Independent component analysis
ICAR	Indian Council of Agricultural Research
ICP	Intrahepatic cholestasis of pregnancy
IF	Infant formula
Igs	Immunoglobulins
IIR	International institute of refrigeration
IMF	Infant milk formula
IoT	Internet of Things
IPM	Integrated pest management
IQ	Intelligent quotient
JAM	Junctional adhesion molecules
K	Potassium

KE	Kinetic energy
LA	Lauric acid
LAB	Lactic acid bacteria
LC-MS	Liquid chromatography-mass spectroscopy
LCPUFA	Long chain polyunsaturated fatty acids
LDPE	Low-density polyethylene
LI	Lactose intolerance
LLDPE	Linear low-density polyethylene
LM	Lactose malabsorption
LNMS	Lipid nanomaterials
LOS	Lactulose
LPA	Lipoteichoic acid
$\dot{m}$	Air mass flow rate
MALDI	Matrix-assisted laser desorption ionization time
MALS	Multiangle light scattering
Malv-3-acglc	Malvidin-3-acetylglucoside
Malv-3-glc	Malvidin-3-glucoside
MBA	Microbiological assay
MBC	Minimum bactericidal concentration
MC	Methyl cellulose
MFGMs	Bovine milk fat globule membranes
Mg	Magnesium
mg	Milligram
MgO	Magnesium oxide
MHz	Megahertz
MIC	Minimum inhibition concentration
miRNAs	MicroRNAs
ML	Machine learning
MMT	Montmorillonite
Mn	Manganese
Mo	*Molybdenum*
MPs	Microplastics
MRGs	Metal resistant genes
MRS	de Man Rogosa and Sharpe
MS	Mass spectroscopy/mechanical strength
MSN	*Mesoporous silica nanoparticles*
Mub	Mucin-binding proteins
MucBP	MUCin-binding protein
MVR	Mechanical vapor recompression
MW	Microwave
N	Nitrogen
NaCl	Sodium chloride
NAM	Nicotinamide
NASA	National Aeronautics and Space Administration

NBAIR	National Bureau of Agricultural Insect Resources
NC	Not calculated
NEC	Necrotizing enterocolitis
NLCs	Nanostructured lipid carriers
NM	Nanomaterial
Nm/nm	Nanometer
NMR	Nuclear magnetic resonance
NP	Nanoparticle
NSMs	Nanostructured materials
NT	Nanotechnology
NTA	Nanoparticle tracking analysis
O/W	Oil/water
OA	Oxygen absorber
OP	Osmotic pressure
OTA	Ochratoxin A
P	Phosphorus/electrical power output from solar cell
PA	Phenolic acids/propionic acid
PaaS	Platform as service
PACA	Polyalkylcyanoacrylate
PAHs	Polycyclic aromatic hydrocarbons
PBS	Phosphate buffered saline/polybutylene succinate
PC	Personal computer
PCA	Principal component analysis
PCBs	Polychlorinated biphenyls
PCL	Polycaprolactone
PDI	Polydispersity index
PDX	Polydextrose
PEG	Polyethylene glycol
PET	Polyethylene terephthalate
PF	*Propionibacterium freudenreichii*
PFAs	Perfluoroalkyl
PG	Propyl gallate
PGA	Polyglycolide
PGL	Pyrogallol
pH	Potential of hydrogen ions
PHA	Polyhydroxyalkanoate
PHB	Poly(3-hydroxy butyrate)/polyhydroxy butyrate
PHBV	Poly(3-hydroxy butyrate-co-3-hydroxy valerate)/polyhydroxy-butyrate-valerate
PLA	Polylactic acid
PLGA	Poly(lactic-co-glycolic acid)
PLS-DA	Partial least squares-discriminant analysis
PNC	Polymer nanocomposites
PNE	Pine needle extract

POD	Peroxidase
PP	Polypropylene
Ppb	Parts per billion
PPIs	Proton pump inhibitors
PPO	Polyphenol oxidase
Prl	Prolactin
PROBIT	Promotion of breastfeeding intervention trial
PS	Potassium sorbate
P_T	Turbine power output
PUFA	Polyunsaturated fatty acid
PV	Photovoltaics
PVA	Polyvinyl alcohol/polyvinyl acetate
PVC	Polyvinyl chloride
PVD	Physical vapor deposition
PVOH	Polyvinyl alcohol
PV-T	Photovoltaic-thermal
P_w	Wind power
Q	Quercetin
Q-3-glcr	Quercetin-3-glucuronide
QD	Quantum dots
QPS	Qualified presumption and safety
QR	Quick response
Q_u	Useful energy output from a collector or heat gained by the fluid
RDA	Recommended daily allowance/recommended dietary allowance
rDNA	Recombinant DNA
Ref	Reference
RES	Renewable energy sources
RF	Riboflavin
RFID	Radio frequency identification
RH	Related humidity
RNA	Ribonucleic acid
ROP	Ring-opening polymerization
ROS	Reactive oxygen species
RWD	Refractance window drying
SaaS	Software as service
SAM	S-adenosyl-L-methionine
SANS	Small angle neutron scattering
SCFA	Short-chain fatty acids
SDGs	Sustainable development goals
SEC	Size exclusion chromatography
SEM	Scanning electron microscopy
SHUATS	Sam Higginbottom University of Agriculture, Technology and Sciences
Si	Silicon

SLAPs	Surface layer associated proteins
SLS	Static light scattering
SNE	Safflower oil nanoemulsion
SNP	Single nucleotide polymorphism
SOD	Superoxide dismutase
SOS	Sulfite-based oxygen scavenger
SPI	Soy protein isolate
SPM	Semi-permeable membrane
SRS	Southern regional station
SVM	Support vector machines
SWCNT	Single walled carbon nanotubes
Syringetin-3-glc	Syringetin-3-glucoside
T80	Thermal treatment at 80 °C for 30 min
T90	Thermal treatment at 90 °C for 30 s
TEM	Transmission electron microscopy
TF	Total flavonoids
Tg	Glass transition temperature
Ti	Titanium
TiO_2 NMs	Titanium dioxide nanomaterials
TiO_2	Titanium dioxide
TiPED	Tiered protocol for endocrine disruption
Tm	Melting temperature
TNF-α	Tumor necrosis factor
TP	Total phenolics
TPGS	Tocopheryl polyethylene glycol succinate
TPS	Thermoplastic starch
TS	Tensile strength
TTI	Time-temperature indicator
TTP	Polyanion tripolyphosphate
TVBN	Total volatile basic nitrogen
UC	Ulcerative colitis
UDP-Glu	Uridine diphosphate glucose
UDP-GPPR	Uridine diphosphate glucose pyrophosphorylase
UHPLC	Ultra-high performance liquid chromatography
UHT	Ultra-high temperature
USD	United States dollar
v	Mean wind speed for an appropriate time period
V	Voltage
VAR	Vapor absorption refrigeration
VFA	Variable fatty acids
Vit B12	Vitamin B12
Vit.	Vitamin
VP	Vapor pressure
VPE	Vapor pressure osmometer

WCDT	Warner College of Dairy Technology
WHO	World Health Organization
WPI	Whey-protein isolate
WPS	Water-soluble polysaccharides
WSN	Wireless sensor technology
WVP	Water vapor permeability
WVTR	Water vapor transmission rate
Zn	Zinc
ZnO NMs	Zinc oxide nanomaterials
ZnO	Zinc oxide
ZO	Zonula occludens
η_{cell}	Solar cell efficiency
η_{th}	Thermal efficiency of solar collector
ρ	Air density

Part I
Novel Technologies in Food Processing

Chapter 1
Emerging Novel Technologies for Food Drying

Harshita Sonarthi, S. Supreetha, and Shweta Mall

Contents

Abstract Food is vital for the survival of human beings. However, due to its perishable nature, it will spoil rapidly; therefore, it must be preserved. Moisture content plays a critical role among others factors involved in the spoilage of food as it is a favorable for the growth of microorganisms. There are two main techniques to prevent spoilage by moisture content, viz., freezing and drying. Drying nowadays is a prevalent technique to preserve any food product. There are various techniques that

H. Sonarthi (✉) · S. Supreetha
Dairy Technology Section, ICAR-National Dairy Research Institute, Southern Regional Station (SRS), Bengaluru, Karnataka, India

S. Mall
Animal Genetics and Breeding Section, ICAR-National Dairy Research Institute, Southern Regional Station (SRS), Bengaluru, Karnataka, India

J. A. Malik et al. (eds.), *Food Process Engineering and Technology*,
https://doi.org/10.1007/978-981-99-6831-2_1

are already available in the industry at commercial scale for drying food (such as: solar drying, tray drying, roller drying, spray drying, etc.); however, researchers are focusing more on novel techniques for drying for enhancing the efficacy and efficiency of drying process. Ultimately, it will lessen the energy consumption and it also improves the quality of final dried product. Examples of these novel techniques are microwave drying, refractance window drying, freeze drying, infrared drying, high electric field drying, vacuum drying, and heat pump drying.

Keywords Heat pump drying · High electric field drying · Microwave drying · Novel drying technology · Refractance window drying · Vacuum drying

1.1 Introduction

Drying is an important technique in food industry for preservation of foods. Novel drying techniques are now-a-days emerging as good alternatives to conventional drying. Refractance window drying, vacuum drying, heat pump drying, infrared drying, freeze drying and high electric drying are examples of novel drying techniques. These new concepts are helpful in preserving the nutrients and sensory characteristics of the final product with low energy and time requirement. This chapter focuses on the novel approaches for drying of food and their advantages over conventional drying methods.

1.2 Importance of Drying in Food Industry

Drying is a process in which water is removed from the food product by evaporation under controlled conditions for the production of a solid food product. The uppermost aim of drying is to increase the shelf-life of the product by minimizing the water activity. When sufficient water is not available in food product, there is minimum growth of microorganisms that are responsible for spoilage of food product, and enzymes that induces chemical changes. Drying is done for almost all food products, such as: dairy products, coffee, processed cereal-based foods, potatoes, sugar beet pulp, vegetable, fruits, spices, etc. (Goodenough et al. 2007). Other than preserving foods and increasing their shelf-life, it reduces the need for refrigeration system for transport and storage, as the volume of food is reduced, the space for storage and transport is also reduced, and it expands the supply of foods by providing different flavor and texture; hence, consumers will have more choices while buying the foods (Chen and Mujumdar 2009; Isleroglu and Turker 2019).

1.3 Conventionally Available Drying Techniques

1.3.1 Solar Drying

Sun drying has been used for generations as a traditional method of food preservation. Farmers with small quantities of agricultural produce can benefit from solar radiation, which is a renewable resource that is readily available. The solar drying was developed to address issues of sun drying and to utilize the potential of solar radiations. Solar drying reduces the consumption of fossil fuels and energy in the drying process. This method improves moisture removal rate, resulting in faster drying, increased manufacturing capacity, and reduced contamination risks. The main aim of solar drying is to collect solar energy and heat the air volume, then transmit the hot air from the collector to an attached enclosure, where food is held for drying (drying chamber), and the passage of hot air above the food aids in drying.

1.3.2 Tray Drying

This drying technique is a batch method for drying of solid foods with a capacity of 2000–20,000 kg/day. These dryers are equipped with closed compartments in which shelves are fixed at proper distance, so that there is free movement of hot air. On these shelves, trays are placed with a food product for drying, and hot air then flows through them. Also, for appropriate movement of hot air, sometimes trays are perforated. Product, which is near to the entrance of the hot air, has less moisture content than the rest. Therefore for uniform drying, baffles are placed or shelves that can be rotated.

1.3.3 Roller Drying

The roller drying method includes a metal drum that is heated from inside with steam, and a thin layer of product for drying is applied on the drum with different techniques. As the layer of product gets dried, it is scraped off from the other side with a knife, which is placed at the opposite side of the liquid product application point. Steam pressure inside the drum ranges from 4 to 8 bars according to the product. This drying method comes under conduction drying (Chen and Mujumdar 2009). The way wet product is put on the drum surface is the basis of different type of drum dryers. And the most common of these is dip feeding. In this method, the drum is partially dipped in the feed solution, which is in the tray. As the drum rotates the film of the feed solution in the dipped part of the drum. Fresh feed solution is constantly delivered to the tray. Drum dryer types according to the number of drums are of single or double drum. In double drum dryers, 2-drums are used, which are rotating in opposite directions to each other, and have narrow and adjustable gap between them. For double drum dryer, nip feed method for delivering feed is suitable. In this method, there is a pool, which is formed between the drums and the feed is supplied into the pool. The narrow gap between the drums is adjusted according to the thickness of the film.

1.3.4 Spray Drying

This technique deals with spraying of the feed into the hot drying medium and resulting dried product is collected. For preparing milk, coffee and fruit juice powders, it is mostly used. The fundamental principle of spray drying is to atomize the feed with the help of an atomizer, after that the droplets of the feed come in contact with the hot air stream, which makes the liquid to evaporate and only dried product is left, and at powder is collected. For producing droplets from the feed, rotary (wheel) or nozzle atomizers are used. There are controlled temperature and airflow conditions for the evaporation of moisture from the feed droplets. There are different dryer designs and operating conditions, which are based on the attributes of the product desired and specifications of product (Samantha et al. 2015).

Generally single type of dryers is used for spray drying. In this single source of drying, air is added from the top of drying chamber and this is called co-current air flow and the feed droplets flows in the same direction. However, production of fine powder with poor powder flowability and production of more dust are some drawbacks. To overcome this, multiple spray dryers are used to dry the liquid is in two steps. The first step is the same procedure as in single effect dryer. The second step is an integrated static bed that is used at the bottom of the drying chamber having a humid environment to help in clumping of the small powder particles and forming more uniform particle sizes of 100–300 μm. This increases the free flowability of the powder because of increased size of particles. Then, these particles enter in the drying chamber either from top or from bottom of the drying chamber. For final drying of the particles, fluidized bed dryer is installed at the end (Straatsma et al. 1999).

1.4 Novel Drying Techniques

1.4.1 Refractance Window Drying

The refractance window drying technique is an innovative drying technique that is helpful in transferring different liquid- and semi-liquid food products into powder, dried sheets and flakes. In this technique, a thin film of product is applied on a plastic conveyor, which receives thermal energy from hot water (30–95 °C). Evaporation of water takes place from the belt and is removed by an exhaust system. The product is scraped-off from the conveyor after passing through a cold-water bath. The key benefits of this unique technology include: the avoidance of cross-contamination because there is no straight contact between the product and the heating medium; and the operation is done under vacuum thus reducing the overall cost of drying (Moses et al. 2014). There is high level of retention of bioactive compounds in the food product. Overall, refractance window drying offers favorable food quality with ability to retain high nutrients.

This drying process is completely reliant on Mylar film, which is at its core. The film aids in the transmission of radiant energy to the product for drying and transferring it from one end to the other. For the movement of the film during the drying process, two-end pulleys are used. A shallow water bath is present beneath this film, with hot water flowing from the water heating unit; in addition to this hot water bath, there is a cold water bath to cool the product before it leaves the dryer. With the help of a heating unit, the water temperature in the hot water bath is kept constant at around 99 to 92 °C. A single pump transports hot water from the heating unit to the hot water bath. There is a valve between the heating unit and the hot water bath that assists in pumping hot water from the heating unit to the water bath, if the water bath temperature falls below the desired value. Puree with a high moisture content is evenly applied to one side of the film and then removed after drying. A stainless steel hood is installed at the top of the film to vent out the vapors that are vanished from the product.

1.4.2 Microwave Drying

Now-a-days electromagnetic waves are used in most of the food processing operations. Microwaves varying between 1 mm to 1m and frequency from 0.3 GHz and 3 GHz. Frequency of radio waves is lesser as compared to the microwaves; therefore, microwaves can concentrate more tightly. These waves propagate through space and air with a speed of light. Use of microwaves for the drying of the food product can be useful against the drawbacks of conventional drying methods. This drying technique helps in preserving the quality of food materials. Also, there is no loss in heating by conduction or convection, as the radiations emitted by the microwave are confined to the cavity and most of the energy is taken by the wet food material, which is kept in cavity. Furthermore, water molecules in the wet food product take up this energy and help in raising the temperature of the food product, which helps in the evaporation of water and reduction of moisture level.

Microwave oven works at two frequencies 2450 MHz and 915 MHz. For domestic work, microwave with 2450 MHz frequency is used; and for industrial work, microwave with 915 MHz is used. Microwave radiations with these frequencies pass through the food and the molecules like water, fat and sugar in the food absorb energy from microwave radiation and produce dielectric heating. As these molecules have electric dipole, therefore, they have both positive and negative charge on each end. This helps these to rotate themselves as they try to orient themselves in response to alternative electric field, which is produced by the microwave radiation (Feng et al. 2012). This hasty movement of these molecules helps in production of friction and ultimately generates heat in the food materials (Guo et al. 2017).

1.4.3 Infrared Drying

Infrared radiations are chief energy sources for food processing applications, such as: drying, pasteurization, roasting, blanching, peeling, and elimination of antinutrients from legumes. By uniting this technique with supplementary drying techniques (such as: microwave, hot air vacuum, and freeze drying) helps to speed the process with improved results. Far, near and medium are different types of infrared drying methods in food industry. Wide variety of agricultural produce can be dried using this technique in food industry (Adak et al. 2017). Compared to the traditional drying method, this method has advantages, such as: efficient product heating, exceptional efficiency of energy and good quality of dried product; since this method preserves the chemical and organoleptic characteristics of the original product. Many studies indicate that far-infrared drying is a good choice for drying of agricultural produce for efficiency, because whatever heat is produced in this method that helps in the drying up of the product and preserve the quality of the product too. This technique of drying can be used alone as well as in supplement to other techniques (Liu et al. 2019). Because of its simplicity, steady heat distribution and its uniform heat distribution, rapid drying speed makes it fairly practical. Amid various infrared technologies, far-infrared drying is most appropriate for food processing, because many food components showed more absorption of radioactive energy in the far-infrared region.

Surface properties of material, radiation type and emitter's and receiver's shape are crucial factors that affect the infrared drying. Infrared radiation heating depends on surface property of material, type of radiation and shape of emitter and receiver. Infrared drying is now becoming prevalent in the food drying because of its unique heating properties and surface drying of food in less time. Rehydrating property is also good with infrared-treated drying compared to the conventional drying system. Infrared drying is an alternate energy source and offers several benefits, such as: lesser drying time, maximize energy efficiency, while drying constant temperature is maintained in the product, finished product quality is better, lesser need of air flow through the product and needs less space (Sakare et al. 2020).

1.4.4 Freeze Drying

Freeze drying is used for preservation of high-value food products, and it is also known as lyophilization. The basic principle behind freeze drying is sublimation.

Any solvent, which is present in food product, gets crystallized at low temperature and switches from solid to vapor state under vacuum. The main focus in freeze drying is drying of product at low temperature, which is helpful in retaining the quality of food product while also preventing thermolabile chemical destruction. When compared to traditional drying methods, freeze drying has various advantages, such as: biochemical, morphological and unique characteristics of food product are preserved and recovery of volatiles is also high (Martinez 2017).

Freeze drying removes solvent from a liquid composition. Freeze drying process is completed in three stages: freezing, primary drying and secondary drying. Freezing, sublimation, desorption, vacuum pumping, and vapor condensation are five key operations that make up the freeze-drying process. Firstly, the liquid is cooled at a low temperature, and any remaining water is fully frozen. Then the solvent, which is frozen, is heated and it converts to vapor phase directly from solid phase without changing to liquid phase; therefore, this process is known as sublimation drying (primary drying). After that, the unfrozen solvent is removed using a desorption process (secondary drying). Therefore, freeze drying is a process in which both freezing and drying have equal significance (Waghmare et al. 2021).

1.4.5 Vacuum Drying

For drying of heat-sensitive food products, vacuum drying is safest and highly efficient technique. In this technique, drying is done at low temperature as compared to other drying methods. For creating vacuum, the ambient pressure is reduced to lower the flashpoint of liquids significantly. The vacuum drying is the leading technique for heat sensitive product, hygroscopic, toxic powders and granules. Eluding extreme heat while drying powder will be beneficial for product quality and safety. Heat and vacuum together is an effective way, which can dry food at relatively low temperatures. Vacuum drying gives final dried product with low moisture content than other normal drying methods.

Vacuum pump creates vacuum around the product to be dried in a vacuum chamber. This causes reduction in boiling point of the water thus increases the rate of evaporation considerably, ultimately the rate of drying is increased. Generally, 0.0296–0.059 atmospheres of pressure and 25–30 °C boiling temperature for water are maintained (Parikh 2015). Vacuum drying has many benefits compared to other techniques. Most important is its energy conservation, less drying time, less damage to the product and preserve the integrity of the original product (Ravichandran and Upadhyay 2019).

1.4.6 High Electric Field Drying

High electric field drying is a non-thermal drying technique, it is fast as well as gentle and is acquiring interest in industrial drying operations (Atungulu 2007). In this technique, the moisture is removed by high intensity alternating current (AC) or direct current (DC) with normal frequency (nearly 60 Hz). Sample drying rate is directly affected by the potential difference produced between the electrodes. Rapid evaporation of moisture is achieved by the exothermic interaction of the electric field within the dielectric food material. Overall, high electric drying technique comes under convective mode of drying, which efficiently reduces the moisture content of food product and maintains the level of nutrients like ascorbic acid. High electric field drying was used for spinach drying and was compared with other drying method. Results show more evaporation of moisture in high electric field drying compared to other methods. The color of spinach was preserved; also ascorbic acid content was higher after six weeks (Bajgai and Hashinaga 2001).

1.4.7 Heat Pump Drying

For drying of agricultural products, heat pump technology is effectively used. In this technique, an oxygen-free atmosphere and low temperature are used for drying of products at low energy consumption compared to other techniques. Generally, a heat pump dryer includes a drying chamber and a heat pump. Compressor, condenser, expansion valve and evaporator are main components of the heat pump. Heat pump and dryer are connected by an air cycling circuit.

The fundamental principle behind the close heat pump drying is whatever exhausted air from the dryer enters the evaporator of the heat pump, and cooling takes place in this part and moisture in the air gets condensed and discarded. The final air, which is dried and cooled in an evaporator, will go to the heat pump's condenser and heating takes place here. Then, the hot and dried air invades the drying chamber, where it removes the moisture from the product and gets out from the dryer as an exhausted air, and again the cycle is repeated. This method is highly efficient for drying of biological materials and oxygen sensitive materials.

In open heat pump drying, there is no recirculation of exhausted air and ambient air is used for the drying process. Heat pump drying has several advantages, such as: high energy efficiency, low energy consumption, high coefficient of performance, high drying efficiency, low drying temperature, less time required, low-cost and food quality is good compared to other methods (Salehi 2021).

1.5 Advantages of Novel Drying Techniques as Compared with Conventional Drying

Although drying is a method to enhance the shelf-life of food product, yet it is also helpful in storage and transit of food product by removing the requirement of costly cooling systems, the quality of dehydrated food is usually significantly worse than the original product. Therefore, it is necessary to reduce the chemical changes like enzymatic and non-enzymatic browning and on the other hand it increases the nutrient retention including macronutrients, micronutrients or bioactive compounds. Shrinkage is one of the common issue for many food products, and it has a significant impact on their structure and texture (Guiné 2015). To overcome all these drawbacks, novel drying techniques were developed. The main purpose was to retain the quality of the food product.

Advantages of novel drying techniques include: better product quality, shorter drying time, better cost economics, benefits of hybrid drying, non-polluting, better process control and operational safety.

1.6 Future Aspects

Drying is an earliest and most effective physical technique of food preservation. Currently in industries, over 85% of dryers are convective, and use combustion gases or hot gases as their medium for transfer of heat. Drying is an energy-intensive process that involves simultaneous heat and mass transfer. And it takes almost 12–20% of the energy utilized in any food industry. Other than energy consumption, overall product qualities like nutritional, sensory and functional qualities are major issues. As traditional drying processes depends on convective and conductive manner of heat transfer, therefore the end product may be of low standard, and the product contamination risk is more. This chapter addresses the growing demand for better drying procedures to retain the end product's quality. These innovative approaches take advantage of many physical processes to improve on existing commercial drying techniques. While there has been a substantial amount of scientific research in the subject of innovative drying, commercialization of these systems has been limited due to a lack of understanding of the capital costs and energy efficiencies.

1.7 Conclusion

From ancient times, drying is well-known food preservation method. And presently, it is a very significant industrial process for the treatment of assorted food products. As drying is useful in many food products, therefore more innovation and technological advancements are possible for improvement in drying process in terms of efficiency, power requirement and nutritional and organoleptic qualities. As this method is applicable to wide range of food products with various characteristics, therefore this makes drying as a prominent processing operation in food industry. The conventional drying methods have their own disadvantages; therefore, switching to novel drying technologies is favorable to overcome all these drawbacks.

References

Adak N, Heybeli K, Ertekin C (2017) Infrared drying of strawberry. Food Chem 219:109–116

Atungulu GG (2007) High electric field technology in postharvest drying. Fresh Produce 1(1): 23–31

Bajgai TR, Hashinaga F (2001) Drying of spinach with a high electric field. Dry Technol 19(9): 2331–2341

Chen XD, Mujumdar AS (eds) (2009) Drying technologies in food processing. New York, Wiley–Blackwell, p 352

Feng H, Yin Y, Tang J (2012) Microwave drying of food and agricultural materials: basics and heat and mass transfer modeling. Food Eng Rev 4(2):89–106

Goodenough TI, Goodenough PW, Goodenough SM (2007) The efficiency of corona wind drying and its application to the food industry. J Food Eng 80(4):1233–1238

Guiné R (2015) Food drying and dehydration: technology and effect on food properties. Lambert Academic Publishing, Saarbruecken

Guo Q, Sun DW, Cheng JH, Han Z (2017) Microwave processing techniques and their recent applications in the food industry. Trends Food Sci Technol 67:236–247

Isleroglu H, Turker I (2019) Thermal inactivation kinetics of microencapsulated microbial transglutaminase by ultrasonic spray-freeze drying. LWT 101:653–662

Liu Y, Zeng Y, Hu R, Sun X (2019) Effect of contact ultrasonic power on moisture migration during far-infrared radiation drying of kiwifruit. Food Bioprocess Technol 13:430–441

Martinez N (2017) Biopolymers and freeze-drying shelf temperature have an impact on the quality of a Mandarin snack. LWT 99:57–61

Moses JA, Norton T, Alagusundaram K, Tiwari BK (2014) Novel drying techniques for the food industry. Food Eng Rev 6(3):43–55

Parikh DM (2015) Vacuum drying: basics and application. Chem Eng 122(4):48–54

Ravichandran C, Upadhyay A (2019) Use of vacuum technology in processing of fruits and vegetables. In: Processing of fruits and vegetables. Apple Academic Press, Burlington, pp 139–174

Sakare P, Prasad N, Thombare N, Singh R, Sharma SC (2020) Infrared drying of food materials: recent advances. Food Eng Rev 12(3):381–398

Salehi F (2021) Recent applications of heat pump dryer for drying of fruit crops: a review. Int J Fruit Sci 21(1):546–555

Samantha SC, Bruna ASM, Adriana RM, Fabio B, Sandro AR, Aline RCA (2015) Drying by spray drying in the food industry: micro-encapsulation, process parameters and main carriers used. Afr J Food Sci 9(9):462–470

Straatsma J, Van Houwelingen G, Steenbergen AE, De Jong P (1999) Spray drying of food products: 1. Simulation model. J Food Eng 42(2):67–72

Waghmare RB, Perumal AB, Moses JA, Anandharamakrishnan C (2021) Recent developments in freeze drying of foods. In: Knoerzer K (ed) Innovative food processing technologies: a comprehensive review. Elsevier, New York, pp 82–99

Chapter 2
Foods and Food Products: Significance and Applications of Colligative Properties

Ramesh Sharma

Contents

Abstract The colligative properties include boiling point elevation (BPE), vapor pressure (VP), freezing point depression (FPD), and osmotic pressure (OP) of solutions, which are used to determine various parameters of food materials. The water activity (a_w) of the food sample is determined by estimating the FPD of the liquid or by measuring the equilibrium relative humidity of a liquid or solid sample. The water vapor pressure in the food sample is correlated to a_w of the food sample. The vapor pressure osmometer is used for the measurement of osmolality in solution by sensing the vapor pressure in the isolated chamber. The other approaches for measuring the a_w in food sample includes the dew point method, electric

R. Sharma (✉)
Bioengineering Department, National Institute of Technology Agartala, Jirania, Tripura, India

J. A. Malik et al. (eds.), *Food Process Engineering and Technology*,
https://doi.org/10.1007/978-981-99-6831-2_2

hygrometer, hair hygrometer, vapor pressure manometer, and psychrometer. Considering the wider application of colligative properties in food products this review provided the principles for the measurement of food parameters using colligative properties. The milk samples containing the added water can be evaluated by a vapor pressure osmometer and a thermistor cryoscope. Similarly, the molar mass of solute in the milk sample can be obtained with the measurement of freezing point depression after determining the osmolality value. Boiling point elevation temperature of fruit juices is widely applied for the concentration of fruit juice and used to evaporate the juice as well it is fruitful to design the process and equipment. Similarly, osmotic pressure is a useful term in the membrane separation technology to produce concentrated juice with more retention of nutrients than other heat-sensitive methods.

Keywords Colligative properties · Cryoscope · Freezing point depression · Water activity

2.1 Introduction

Colligative properties are considered the intensive features of solutions that mainly depend on the solute concentration and are not affected by solute chemical identity (Bogsanyi and Weeden 1968). Although, structurally the mass of salt (sodium chloride) and sugar (sucrose) are distinct the intensity of these solutes is related to some solute molecules more important than the individual molecular part. The influence of colligative properties including boiling point elevation (BPE) and freezing point depression (FPD), vapor pressure (VP) and osmotic pressure (OP) over the addition of solutes changes these properties in the solutions (Eckhoff and Okos 1986). The VP signifies the equilibrium pressure of the food substances at a particular temperature of a closed system. The addition of solute lowers the vapor pressure. The phenomena occurred due to the occupying the space at the surface of the solutions thereby lowering the quantity of solvent to be liberated.

The boiling temperature at which substances change from liquid phase to gas phase. At a particular temperature, the solutions retaining the vapor pressure are lower than comparing the vapor pressure of the solvent. Hence, absorb sufficient heat to get the same VP for boiling to occur. This results in elevation of the boiling point that is considered the physical property of a solution and it has a major impact on the sensorial characteristics of food materials. Similarly, the freezing point is considered the temperature after which substances change from liquid phase to solid phase. Addition of solute decreases the freezing point. The freezing point lowering occurred due to the interference of solute on the crystallization process of solvent thereby reducing the temperature required to alter the phase from water to ice (Jenkins 2008).

The applied pressure to the solution to interfere with the process of osmosis is called the osmotic pressure (OP) (Wickware et al. 2017). Hence, the fluctuation of colligative properties interferes with the motion of the solvent in permeable membrane with the system trying to get the equilibrium. Interestingly, solute concentration increases the OP as we know salt and sugar are the main ingredients in food products hence OP plays a significant role to stabilize the microorganism in food products.

The control over the amount of water present in the foods is one primitive approach to preserving the food materials. This method restricts the water availability for the growth of microbes and the biochemical reaction. Commonly the water is removed with drying, solidification of water by freezing, or with the addition of electrolyte including NaCI or non-electrolyte including sugar (sucrose). The addition or removal of solutes from the water altered the colligative properties. As for the case of vapor pressure and freezing point, the value is decreased while the osmotic pressure is increased. The change resulted in decreases in the a_w value.

The growth of microbes and biochemical reactions is estimated by the water availability in a food sample and that is expressed by water activity (a_w). The a_w is the ratio of VP of the sample at equilibrium (P) to VP of pure solvent under equilibrium condition (P_o) at a particular temperature (Scott 1957). Hence, the value of "a_w [P/P_o]" ranges from 0 to 1. The a_w varies with the change in temperature that facilitates the growth of microorganisms (Prior 1979). The relationship of water with spoilage microorganisms in food has been reported and the concept of a_w was introduced and it has been followed by all food technologists (Scott 1957). However, the use of these methods is finite due to the lack of a reliable method for the determination of a_w (Prior 1979).

This review chapter provides information regarding the application of colligative properties in various food products with the application of various principles and methods. The various colligative properties are discussed in this chapter.

2.2 Vapor Pressure (VP)

The a_w is the ratio of water's partial pressure (which is available in the food sample) to the vapor pressure of pure water at the same temperature as the food sample. Thus, the measurement of the VP of the water in the food sample directly helps to obtain a_w of food. The food sample undergoes to achieve equilibrium conditions and the manometer or transducer is used to measure the partial pressure. The effectiveness of the measurement is affected by the size of the sample, time taken for equilibrium, volume, and temperature. The demerits of the method are that it is not applied to the biological sample with an active rate of respiration or with containing the volatile compound.

Recent studies have suggested the application of a thermistor cryoscope for measuring the added water in the milk sample (Boyer 1969; Pensiripun et al. 1975). The instrument called vapor pressure osmometer has been applied for diagnosis in the medical profession (Pensiripun et al. 1975). The depression value of dew point temperature is determined with the instrument and is known as a thermocouple hygrometer. The VP is determined by measuring the temperature change in the chamber of the instrument. The osmometer was found more effective than the cryoscope while comparing for measurement in the serum sample. The major advantages of the application are reported to the dairy industry and in food research laboratories to determine a_w in intermediate food. The potential application of vapor pressure osmometry and other instruments to determine added water in milk and a_w of food sample is summarized below.

2.2.1 Vapor Pressure Osmometer (VPO)

This apparatus consists of a filter paper disc of a diameter of 0.64 mm in diameter where milk samples are inserted using forceps. Then, the milk sample is allowed to saturate in the disc through the capillaries. Once the disc is saturated with milk sample it is allowed to overflow with milk sample before the insertion of the disc into the instrument chamber. Then, the automatic sequence of time is initiated after insertion. The chamber containing the sample is closed with a plastic knob. After closing the sample, the high-temperature transient (A) will occur after the creation of high chamber pressure. Then, the chamber is allowed to reach the ambient temperature of the instrument.

The instrument is adjusted to achieve zero reference points during the thermal and vapor equilibration process. The control flow of the direct current is allowed to pass through the thermocouple junction which causes the temperature to drop at the dew point temperature (T_d). The time required for cooling depends on the sample's osmolality. The flow of electric current is turned off, once the equilibrium has been obtained and the current of the junction is dependent on the T_d value. The final reading is collected from the digital meter. After the opening of the chamber, the ambient temperature reading is established after ensuring lagging due to the temperature of wet-bulb depression with the evaporation of water from the junction. The water gets evaporated immediately after opening the chamber and the temperature of the thermocouple reaches ambient temperature. Thus, the instrument operates precisely as a thermocouple hygrometer.

2.2.2 Measurement of Water Activity

2.2.2.1 The Dew Point Method

The measurement of a_w is performed by placing the sample in the chamber which is contained with a mirror, sample holder which is equipped with detection of condensate in the mirror, and equilibration is allowed to the air of the surrounding space. The formation of vapor occurs in the mirror and the mirror is cooled to measure the a_w with measurement of temperature. The mirror is mainly cooled by using Peltier coolants (Anagnostopoulos 1973; Leistner and Rodel 2012) with the use of cooling agents including petroleum (Ayerst 1965a) or acetone (Blackmore 1960). The temperature is called the dew point temperature and is associated with a_w of the measured sample (Anagnostopoulos 1973). The formation of dew can be observed visually or by the photoelectric cell (Ayerst 1965a). Similarly, the a_w can be estimated at the dew point with the measurement of its vapor pressure in the chamber (Blackmore 1960) but Anagnostopoulos (1973) reported that the a_w estimation with the measurement of temperature is the easier process. The a_w values ranging from 0 (Blackmore 1960) to 1 (Ayerst 1965a; Leistner and Rodel 2012) were determined with the dew point method. The precision with measurement of standard deviation and method was 0.003 a_w unit (Northolt 1972), while the variation ranged from 0.003 to 0.005a_w.

The time required to achieve the equilibration was between 2 h and 3 h for the solid food sample to measure the a_w (Leistner and Rodel 2012). However, the time of equilibration required to measure the a_w for liquid food is less than 10 min (Blackmore 1960). Additionally, the determination of a_w can be performed at a wide temperature range for equilibration. Many authors have demonstrated their approach to measuring the dew point (Anagnostopoulos 1973; Blackmore 1960; Northolt 1972). The commercial dew point hygrometer has been used to determine the a_w (Rodel and Scheuer 1999). The a_w of foods has been estimated by using the dew point method: including syrups (Blackmore 1960) meats products (Leistner and Rodel 2012), bakery products (Blackmore 1960), and wheat, blackseed pepper, coffee beans, and sorghums. kernels, groundnut meal (Ayerst 1965a).

2.2.2.2 Electronic Hygrometer

This instrument can estimate the a_w of food samples despite its expensive cost. The instrument performs with the benefits of precision, convenience, and accuracy; and various types are available in the market. The components of the instrument include the LiCl (hygroscopic nature), potentiometer, and chamber.

The hygroscopic materials show their conductivity, which changes with changing the relative humidity of the chamber containing the sample (Leistner and Rodel 2012; Troller 1977; Wolf 1970). The Sina-Equi-hygroscope instrument can measure a_w in the range of 0.02–0.99 by switching the sensor. The precision of 0.27% (Karan-Djurdjic and Leistner 1970) to 0.53% (Troller 1977) was obtained while accuracy was from 0.002 (Troller and Stinson 1975) to 0.02 unit of a_w (Labuza et al. 1976). The precision of measurement depends on the process of calibration against the saturated salt solution and it is used to plot the standard curve (Leistner and Rodel 2012).

The time required to equilibrate ranges from 30 min (Leistner and Rodel 2012) to 24 h (Karan-Djurdjic and Leistner 1970) for the measurement of a_w in a food sample. The longer equilibration time may result to alter the a_w values as microbial growth will occur that may damage the food product.

The hydrodynamics hygrometer performs similarly to that of Sina instrument and can determine a_w in the range of 0.05–0.99 using different sensors. The precision data has been obtained with a coefficient of variation (C.V.) from 3.6 to 4.8% (Labuza et al. 1976) with an accuracy of 0.005 (Baur and Ensminger 1997) to 0.11 (Labuza et al. 1976).

An electric hygrometer measures the a_w of protein-rich food by measuring the sensor resistivity of sulfonated polystyrene over the variation of relative humidity (Hagerdal and Lofqvist 1973). The instrument can measure the a_w from 0.11 to 0.92, with a maximum error of 0.016 a_w at an equilibration time of 1 h. The additional advantage of the instrument is that it is precisely determined with small sample size. Some researchers have reported that the hygrometer is inefficient to measure a_w above 0.90 (Fett 1973; Vos and Labuza 1974) and the sensors may lose their effectiveness with its storage period (Gal 1975; Juven and Gertshovki 1976; Vos and Labuza 1974). The volatile absorption resulted to cause an error in the instrument and the volatile included the glycerol. The precision of Sina-equi-hygroscope instrument can help to improve values of a_w (Troller 1977).

The duration and degree of exposure factors will decide the extent of the contamination. The contaminants (such as glycerol, propylene glycol, and water) can be removed by evacuating the sensor in the desiccator (Labuza et al. 1976). The manufacturing company of Sina-equi-hygroscope provided the filter to screen out the volatile from the filter. Similarly, the electric hygrometer was used to estimate the wide range of a_w in food products (such as chocolate syrup and jelly (Troller 1977) meats, sausage (Leistner and Rodel 2012) intermediate moisture foods, cheese, bread, and dry soup mix (Labuza et al. 1976).

2.2.2.3 Hair Hygrometer

This instrument is used to estimate the a_w of meat and its products (Leistner and Rodel 2012; Rodel and Scheuer 2002). The a_w measurement was performed by changing the hair length for the change of relative humidity in a closed chamber (Northolt 1972). The Lufft-Metall-barometer-fabriek instrument has been used to obtain the a_w value of meat products varying from 0.85 to 1 and the manufacturer also claimed that it can be used for a_w below 0.4. The calibration of the instrument is required at least weekly (Rodel and Scheuer 2002) and it is preferred to calibrate at the same temperature as that of the sample.

The equilibration time of 3 h is required to reach a constant temperature of the sample (Leistner and Rodel 2012) though, the longer time of equilibration of higher a_w values has also been reported (Labuza et al. 1977). The manufacturer advises the equilibration at 20 °C. In case of inconvenience for the measurement of temperature, the a_w value is corrected for temperature but the accuracy of measurement is higher while keeping at constant temperature in the incubator. The precision with C.V. was 0.26% and 0.36% for dual sets of the experiments in the case of sausage (Rodel and Scheuer 2002); although the lower precision with C.V. of 2.18% was observed (Labuza et al. 1977) for Parmesan cheese of 0.73 a_w. Measurements with C.V. of

<2% were achieved by Sina-equi-hygroscope for the meat samples (Rodel and Scheuer 2002). Estimation of the a_w of the food sample using a hair hygrometer ranged from 0.3 to 0.8 (Labuza 1974). The presence of glycerol as a volatile with a longer period of equilibration resulted in a large variation in the sensitivity of the instrument (Labuza et al. 1977).

2.2.2.4 Vapor Pressure Manometer

Manometers are helpful to estimate the a_w of the food sample (Acott and Labuza 1975; Karel and Nickerson 1964; Vos and Labuza 1974). This method is mostly accurate for liquid food with a high a_w range and is not applicable for the solid food sample (Barbosa-Cánovas and Vega 1996). The flask with a sample that is connected to the manometer gets evacuated with minimal moisture loss through the sample. The operation of the system takes place at a fixed temperature and the time of equilibration for the sample is about 1 h (Labuza 1974). The value of VP is recorded. The water is collected after evaporation and is taken away and the VP of the volatiles and gases is recorded. The a_w was calculated by finding the difference obtained from two readings with the division of pure water VP at a temperature of the same condition (Labuza et al. 1976).

The method is efficient to determine the a_w ranging from 0 to 0.9 (Karel and Nickerson 1964; Vos and Labuza 1974). At a_w of value >0.9, the method is less precise due to problems while controlling the temperature (Vos and Labuza 1974). With the a_w of 0.85, the precision was ±0.005; but with >0.85, the precision expanded to ±0.02 (Vos and Labuza 1974). The a_w of 0.9 resulted in poor control with shorter equilibration time over-temperature, thus causing inaccurate observations (Labuza 1974). The precision measurement with C.V. of 1.2% was obtained with measurement in a saturated solution of Li_2S0_4 (Labuza et al. 1976) 0.62% with cheese sample, and 1.02% for animal food (Labuza et al. 1977).

Labuza et al. (1977) used this method for the estimation of a_w in cheese, pancake batter, bread, egg products, soup mix, and pet foods (Labuza et al. 1976; Sloan and Labuza 1976). The method is inaccurate to predict the results, in case of fermented food or sample associated with the growth of microorganisms due to the evolution of gases. The presence of volatiles contributes to the creation of vapor pressure and alters the value of a_w. However, the food containing propylene glycol and glycerol has been used to determine a_w (Labuza et al. 1976).

Osmotic pressure is mainly measured by using an osmometer. The VPO was used for the measurement of a_w and obtained value was greater than 0.98 (Mozumder et al. 1970). However, this instrument restricts its application at the higher range of a_w and is considered an expensive instrument. Therefore, it has limited use as an a_w meter.

An instrument was developed to measure the a_w in confectionery syrups (Norrish 1966) and it was based on changing the resistance of a ceramic pellet with the change of relative humidity. The measuring cell with the sample containing the pellets is equilibrated at 25 °C for 1 h. The instrument is effective in the a_w range of 0.5–1 a_w.

With the a_w range of 0.9 and 0.95, the accuracy of prediction was ±0.001 a_w while at lower a_w of 0.5 to 0.55 the accuracy was ±0.02 a_w. Similarly, The pressure cell has been developed to estimate the a_w of flour samples with higher moisture value (Gur-Arieh et al. 1965). The sample was equilibrated for 36 h with separation of water from the sample with the pressure creation through porous membranes. Once, the equilibrium was achieved at constant temperature the sample water content was analyzed. The relationship is shown below:

$$\Delta P = -\left[\frac{RT}{V}\right] \tag{2.1}$$

The a_w calculation can be obtained with applied pressure and equilibration of moisture content of the sample. The reported range of a_w with this apparatus is 0.67 to 0.96 a_w.

2.2.2.5 Psychrometer

The water availability in plant material is also measured by psychrometer (Bookin et al. 2008). Prior et al. (1977) measured the a_w of bread, cheese, and meat using a psychrometer. The instrument contains the thermocouple in the chamber. The sample is kept in the chamber with an equilibration time of at least 10 min (for liquid food sample) and 1 h for solid food sample. Peltier cooling was used to cool the water vapor and a thermocouple condenses the water vapor. The evaporation rate of vapors in thermocouples is related to the reading of psychrometers and the a_w value was estimated by using the standard curve, which is plotted against the standard known value of a_w. The thermocouple of the psychrometers is cleaned to remove the contamination.

Additionally, the Wescor psychrometer is featured with a_w measurement ranging from 0.935 to 1 (Van Zyl and Prior 1990) and another electric hygrometer was inefficient in measuring the a_w for this range (Fett 1973; Vos and Labuza 1974) thus, the psychrometer is a potential option to measure the a_w of moist foods. The Wescor psychrometer showed its accuracy with C.V. ranging from 0.18% to 0.35% a_w, while testing in a food sample (Prior et al. 1977).

2.3 Freezing Point Depression (FPD)

The FPD is the difference in the freezing temperature of pure solvent and the solution. The significance of FPD is mainly in the juice industry for the concentration purpose of providing the equilibrium freezing curve (EFC), which is used to determine the solute molecular weight of the sample. Though the molecular weight value estimated by FPD is not exact, it provides the equivalent molecular weight of the dissolved products. The freezing point determination with the conventional method is more time-consuming (Novo et al. 2007).

The Fiske osmometer is a widely popular and cheap instrument that measures the freezing point depression with an easy and faster approach. The time taken for the measurement is the 90s, where a small quantity is required. Fiske osmometers provided the value of osmolality (Osm), which signifies the osmotically-active particles' concentration in the solution, which is equivalent to molalities of the dissolved particles. The value of one osm of the aqueous solution is equal to 1.86 °C of freezing point depression. Hence, the calculated value of osmolality is directly related to the freezing point depression (Eq. (2.2)).

$$T_f = \left[\left(T_f{}^* - 1.86\left(\frac{°\mathrm{C}}{Mol\ kg^{-1}}\right)\left(x\ Mol\ kg^{-1}\right)\right]\right. \tag{2.2}$$

where, T_f = solution's freezing point; T_f^* = pure solvent freezing point (and for pure water, it is equal to zero); and X = the osmolality.

Another important significance is for the preservation of fruit juice because preservation of juice is achieved through the combination of the processes including solid concentration, freezing, and pasteurization; hence, the freezing characteristics are needed in the juice industry for calculating the requirement of a refrigerant. The other application of FPD is to predict the rest of the colligative properties, such as OP, a_w, and the rise in boiling point, which are also required in food processing.

The accuracy for the prediction of FPD is dependent on the composition and concentration of juice. The estimation of molecular weight was more accurately calculated for non-acidic foods (Northolt 1972). For the sample with higher solute concentration, the theoretical prediction by FPD was not accurate and the measurement of the data has to be performed experimentally to obtain the predicted model (Northolt 1972). In the reported literature, the effort was to extend the theory of ideal solutions to characterize the real solutions and to determine characteristic constants for theoretical and empirical equations.

2.3.1 *Factors Influencing the Milk Freezing Point*

The vacuum thermal pasteurizer has been used to control the flavor of milk (Lazar and Bellamy 1957). It was found that the vacuum thermal pasteurizer was efficient in separating the onion flavor through the milk sample. Operating conditions including standardization, clarification, pasteurization, and homogenization did not affect the value of the milk freezing point (Demott 1967; Pensiripun et al. 1975; Smith 1964; Smith et al. 1962); therefore, it was concluded that milk was adulterated with water from the farmhouse at the time of reception and during the time of processing and handling in the milk plant. It has been reported that 0.25–2% of increased water occurred during handling and mainly during the processing of milk (Pensiripun 1975).

The factors affecting the changes in freezing point are vacuum pasteurization, sterilization, homogenization, and freezing storage (Henderson 1963). The pasteurization of the herd milk at 161 °F and 169 °F for 16 s could raise the freezing point while decreasing the conductivity, and no clear effects were seen on the lactose content (Pinkerton and Peters 1958).

Ultra-High Temperature (UHT) treatment did not affect the milk freezing point (Aschaffenburg et al. 1958), and it was concluded that little fluctuation in the chemical composition of the milk sample resulted in difficulty to achieve uniformity over the process of sampling. The freezing point of the milk was 0.004 °C has been reported, which was higher than the raw milk (Beuchat 1974). The raised freezing point of 0.005 °C by vacuum pasteurization was able to measure 1% added to water in milk (Demott 1967; Henningson and Lazar 1959). Similarly, the increase of the freezing point was by 0.001 °C by pasteurization and further homogenization could raise it to 0.0016 °C (Demott 1967; Henderson 1963; Smith et al. 1962). The differences in the freezing point between a fresh sample and homogenized milk after pasteurization were found to be 0.0026 °C.

The effectiveness of vacuum treatment over the milk freezing point was performed by removing the CO_2 and H_2O (Henderson 1963; Henningson and Lazar 1959; Shipe 1964). In the case of removing the CO_2, it lowered the freezing point; but in case of removing the H_20 could raise the freezing point. The elevation of the freezing point after vacuum treatment was found by 0.008 °C. The CO_2 addition could reduce the milk freezing point but the application of vacuum elevated the freezing point (Henderson 1963). The elevation of the freezing point has been recorded after removing the 25% of CO_2 from the milk (Moore and Smith 1961; Moore et al. 1961). In the processing unit, the location of the vacuum also added the difference in values. While keeping the processing unit in the diverse flow direction, the increment was 0.003 °C but keeping the unit below the generator could increase the value to 0.005 °C.

Similarly, all double chambers of the vacuum units elevated the freezing point by 0.008 °C; while for the single-chamber, the observed value was 0.003 °C. The difference in the temperature of pasteurization of milk sample and recorded temperature of vacuum chamber exceeded 20 °F and water was evaporated from the sample and the total solid content of milk was exceeded by 0.05% after pasteurizing under vacuum and the freezing point was lowered by 0.006 °C which was 1% of the added water in the milk (Lazar and Henningson 1960). The change in the value occurred due to the removal of the gas after pasteurization with the vacuum. The sterilization of milk with the application of steam under pressure could increase the freezing point of milk serum by 0.01 °C, which is equivalent to 1.84% of apparent dilution (Pensiripun 1975).

Also, the milk freezing point was changed by boiling the sample in a beaker and was gradually cooled down and the obtained value was changed by 0.004–0.008 °C; while compared to the raw milk sample, boiling was performed at 95 °C for a time interval of 5–20 min. The overall changes of FPD were improved by 0.028–0.04 °C after 5 min and 20 min of simmering which resulted in to increase by 0.129–0.140 °C. Similarly, boiling under reflux conditions for 30 min and sterilization at 15 lbs of pressure for 15 min resulted in no or little change over the freezing point of the sample. Thermal treatment at 60 °C for 20 min resulted in an increase of the freezing point by 0.002 °C (Shipe 1959), and it was concluded that thermal treatment does not alter the freezing point depression unless the moisture and dissolve gas are driven out from the sample.

2.3.2 *Estimation of Effective Molar Mass of Milk*

The molar mass of the milk powder was determined by dissolving the powder in water (Novo et al. 2007). The milk powder in the range of 0.010 and 0.100 g is added to a milliliter of water. The solution is properly homogenized by shaking for a few minutes. Then, the osmolality (mmol/kg) measurement is performed by repeating the experimental run. The plots obtained for osmolality vs weight ratio (powdered milk/water in solutions) are generated for skim and whole milk (Fig. 2.1).

The variation occurs linearly for both types of milk (Fig. 2.1). The expected relationship between the weight ratio to the FPD concerning the obtained value for solute and solvent is given in Eq. (2.3).

$$\Delta T_f = -\left[\frac{K_f}{M_B}\frac{W_B}{W_A}\right] \tag{2.3}$$

where, M_B = solute molar mass; W_B = solute weight; and w_A = solvent weight. The solute molar mass is calculated using the slope from Fig. 2.2. The Eq. (2.3) is mainly

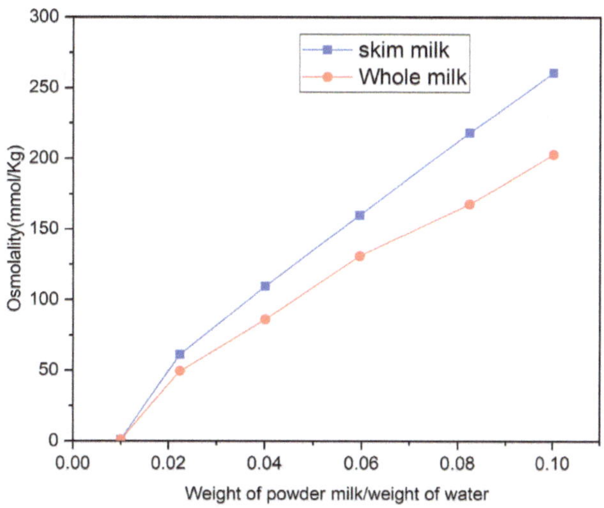

Fig. 2.1 The plot representing the values of osmolality after adding different concentrations of milk powder in water

Fig. 2.2 The estimated effective molar mass value obtained from data in Fig. 2.1

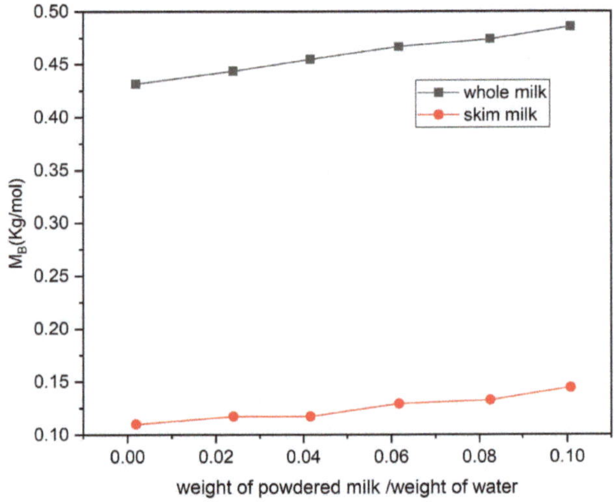

applicable for the dilute solution and needs to modify to a higher-order for the concentrated solution (Garland et al. 2003). Hence, the molar mass is more accurately estimated by extrapolation to zero concentration of solute from the generated molar mass plot values for the solution corresponding to the weight ratio of solute and solvent molecules. Hence, the value of M_B was calculated using experimental data using Eq. (2.4) given below.

$$M_B = -\left[\frac{K_f}{\Delta T_f}\frac{W_B}{W_A}\right] \tag{2.4}$$

The molar mass was decreased with decreasing the solute concentration. Linear extrapolations were performed to calculate the values. In the solution containing the mixture of solute, The M_B represented the average molar mass of solute (i.e., the ratio of total solute weight to the mole particles of solution). However, for the suspension of particles in the milk, the average molar mass is considered an effective molar mass and these particles do not contribute to the FPD, whereas they affect the overall weight of powder. Therefore, the effective molar mass is truly dependent on the suspended particles of the solution. As we can see the obtained effective molar mass which is lower for whole milk in comparison to skim milk. The effect is explained due to the high percentage of fat in whole milk when compared to skim milk.

2.3.3 *Cryoscopy of Milk*

One of the major problems associated with milk quality is the addition of water to milk products. The relevant methodology to detect the adulterants with references

Fig. 2.3 Representing the dependency of the Cryoscopic index with added water in the milk sample

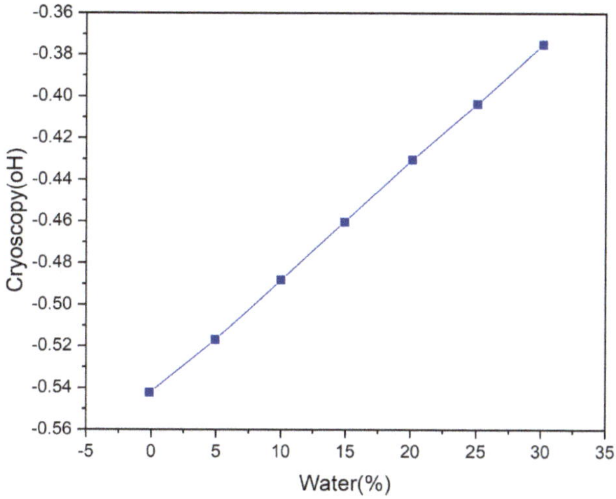

from the (ISO-IDF 2009) (Bagnall and Smith 1955) is the determination of freezing point (FP) (Fig. 2.3).

Cryoscope device is portable, quite expensive, and constantly requires the process of calibration. The adulterants in milk include water and some chemicals like NaCl which are difficult to detect and vary as per the composition of milk. The official approaches have been recommended to determine the adulterated cow milk with water (Pensiripun 1975), which emphasized using standardized equipment and techniques to detect the water in milk. It involves the estimation of the difference of FP of the standard (sucrose 7%) and milk sample. The choice of sucrose as the standard has been unfortunate because of the difference in the freezing characteristics of sucrose and milk (Foust et al. 2008). Additionally, there is difficulty in obtaining reproducible results using the sucrose solution. Similarly, the addition of the agar in the sucrose could provide a similar freezing behavior to the milk (Pensiripun 1975). The FP of milk was presented through a graph (Fig. 2.3), where the linear equation of freezing temperature and extra added water was observed with the equation, $C = -0.5446 + 0.00567\ W$: with C signifying the cryoscopic index (Hortvet) and W is the added water % (Nascimento et al. 2013). For the concentration of sodium chloride added to the milk, a similar procedure was used as shown in Fig. 2.4.

The FP was lowered after adding salt to the milk. The correlation was linearly plotted and the negative coefficient was obtained as shown by the equation: $C = -0.54232 - 0.06346 \times S$; where, S is the concentration (g/L) of salt in milk. For a cryoscope with a refrigerated bath for the cooling system, Horvat's ether was used (Shipe 1959). However, a stirring of the sample is still required and manual tapping of the thermometer is also required. A semi-automatic cyroscope was developed with the solutions of glycol-water and it was fitted with mechanical stirring with a refrigerated bath and was adjusted mechanically for sample stirring and tapping of thermometer (Shipe et al. 1953). The tapping with mechanical force reduced the

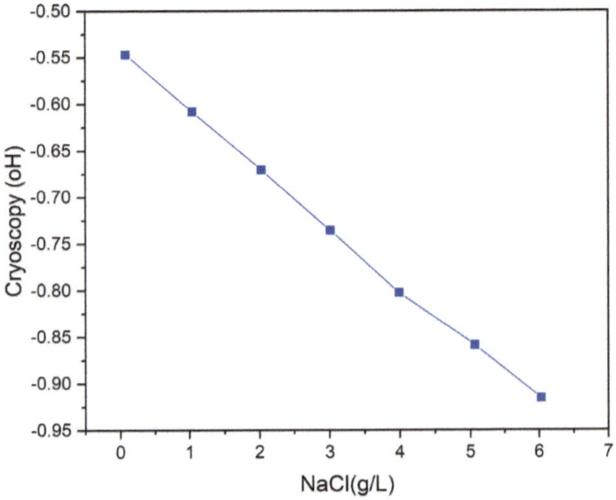

Fig. 2.4 Representing the dependency of the cryoscopic index with added NaCl in the milk sample

variation for the analysis. The taping manner of the thermometer is considered an important aspect of the variation (Dubin 1954).

The cryoscope was also designed and developed with a stir-tapping system by mechanical force (Green 1956). A cryoscope was designed having four freezing tubes with four sample determinations that can be made in 15 min and finally made the water cooler operated by electricity to provide a bath in the cryoscope (Shipe 1959); and it was modified using thermometers as thermal sensing elements replacing the mercury in the thermometer with additional beneficial features (such as a faster response to fluctuation of temperature, a minimum amount of milk is required, with a faster rate of cooling). Fiske's cryoscope for the measurement with temperature gave an accuracy of 0.001 °C (Blackmore 1959). The use of thermistor by AOAC for estimation of FP in milk samples has been reported (Henningson 1969). The variation in results of cryoscope among the 18 experiments using the same sample has been observed (England and Neff 1963), due to fault in the operation of the cryoscope, inaccuracy for calibration of standard, dialing the cryoscope non-linearly, and the pre-freezing of sample in solid carbon dioxide which restricted the dissolving of carbon dioxide in milk.

The factors contributing to the difference while calculating the FP by cryoscope using thermistor (Henningson 1966) were: refrigerated bath level, the temperature of the sample, stirring rate, the extent of supercooling, size of the sample, and the freezing conditions of the sample. The statistics predicted the variation with these variables with an average range of 0.001–0.011 °C in the value of the FP. The specific protocol for measurement in thermistor cryoscope was considered (Henningson 1967), which included the uniform degree of cooling and supercooling, non-uniform in seeding, and non-uniform clear procedure for taking the readings. The Associate referee dealing with milk cryoscopy addressed the recommendation to add the "specific directions" on the official method to determine the FP of milk by using a thermistor cryoscope.

The FP standard deviation of the milk sample was 0.0015 °C and 0.004 °C for standard (Henningson 1968), which was estimated by thermistor cryoscope. It was recommended to add the concept for estimation of FP of milk with AOAC to make part of the official protocol for the plan of action (Henningson 1969). Milk has been recommended as the added water-free with an FP value of −0.525 °C or below (Henningson 1970). In the case of higher FP than −0.525 °C, the milk was considered as "presumptive added water conforming to the added water in milk". The automatic cryoscope (called Kry-0-mat) was introduced to estimate the FP of milk (Ligugnana 1970). The audible sound will come once the milk has been supercooled so that the operator can operate the instrument to initiate the crystallization process while taking the reading. Of course, added water has been detected in the milk sample (Ystgaard et al. 1951). Cryoscopic measurement was not followed at that time. Certain customers used a lactometer to detect the milk total solids to determine the excess water and the added water was estimated using Eq. (2.5).

$$Water\ added\ in\ milk\ (\%) = \frac{SS_1 - SS_2}{SS_2} \times 100 \tag{2.5}$$

where, SS_1 = the minimum content of serum solids (8.5%); and SS_2 = the content of solids measured by a lactometer.

$$SS_1 = 0.4F + 7.07 \tag{2.6}$$

The limited formula was modified (Pensiripun 1975) as indicated in Eq. (2.8).

$$\%added\ water = 110.89 - 111.11\frac{SS_2}{SS_1} \tag{2.7}$$

$$\%added\ water = \left[\frac{T - T_1}{T}\right] \tag{2.8}$$

where, T = the average FPD for normal milk; and T_1 = the FPD of the sample.

Lactometer can only detect the water content in the milk but were inefficient to determine the less than 10% of the water added to milk (Pensiripun 1975). The equation reported by the AOAC method is given below:

$$\%added\ water = \left[\frac{T - T_1}{T}\right] \tag{2.9}$$

The AOAC considered the milk FP as −0.55 °C as the standard value. Similarly, a value of −0.530 °C as the milk FP (Blackmore 1959). In the calculation, it was not compensated for the solid content of milk. Equation (2.9) was modified to the following Eq. (2.10).

$$\%added\ water = \frac{T - T_1}{T}\ (100 - \mathrm{T.S})\tag{2.10}$$

where, T. S = total solids % available in the milk; and −0.544 °C was used as the average value of FPD. Therefore Eq. (2.10) reduces to the following equation (Freeman 1957).

$$\%added\ water = \frac{-0.544 - T}{-0.544}\ [0.92 \times 100]\tag{2.11}$$

The limiting FPD value of milk has been reported as −0.530 °C (Ystgaard et al. 1951). Similarly, the minimum added water percentage in milk (Henningson 1967) is as follows:

$$W = \left[100 - \frac{T.S.(T - T_1)}{T}\right]\tag{2.12}$$

where, T = 0. 530 °C; T_1 = the FP of the sample; and T.S. = total solid %.

Equation (2.12) is currently being considered by the AOAC. The cryoscopy method to estimate the excess water in the milk (Lythgoe 1952) has been recommended, and the method is only applicable to the sample with not having a titrable acidity of more than 0.18%. It was concluded that the decrease in the milk FP was 0.003 °C for every 0.01% increment of titrable acidity.

2.3.4 *Freezing Point Depression (FPD) Method for Water Activity*

This method is the most accurate to determine the a_w of the solution at higher than 0.8 a_w and has been used for experimental purposes (Juven and Gertshovki 1976; Kang et al. 1969; Limsong and Frazier 1966; Strong et al. 1970). The FP measurement is done by a thermometer, which is calibrated either with mercury or an electric thermometer. The instrument shows the sensitivity of 0.1 °C; hence, the a_w is calculated with three decimal points. Mainly the cooling of the sample is performed in an alcohol bath at below 0 °C and freezing is accelerated by adding ice crystal to the supercool solution.

The estimation of a_w with FP follows Raoult's Law according to which, "FPD in solution is correlated to the VP lowering above the solution while comparing to the pure water at the same temperature and pressure". The VP of the solution is measured from the value of FPD by following the "Standard Table in the Handbook of Chemistry and Physics (Ystgaard et al. 1951)" and divided with the value of VP of pure solvent providing the a_w. Also, as per Raoult's Law the ratio of solvent VP (P) to pure solvent VP (P_0), is equivalent to the solvent mole fraction of the solution (Strong et al. 1970).

$$\frac{P}{P_o} = N = \left[\frac{n_1}{n_1 + n_2}\right] = a_w \tag{2.13}$$

where, n_1 = the number of solvent moles in the solution; n_2 = the number of moles of the effective solute; and n_2 is estimated from FPD values using Eq. (2.14).

$$n_2 = \left[\frac{G\Delta T_f}{100kf}\right] \tag{2.14}$$

where, G = solvent weight(g) in solution; ΔT_f = FPD (°C); K_f = molal FPD constant (the value is 1.86 for H_2O).

The precision of using this method has not been discussed previously, but the error of the experiment was calculated with ±0.002 a_w unit (Kang et al. 1969).

2.3.5 *Equation for Prediction of FPD*

For an ideal solution (low concentrations), As per Raoult's law, the difference of FP of pure solvent and solution is given by Eq. (2.15).

$$\Delta T = \left[\frac{k_o X}{(1 - X).M}\right] \tag{2.15}$$

where, X = solute concentration; M = solute molecular weight; and K_o = Van Hoff constant (the value for water: 1860 kg/kg mol).

For high concentration, two semi-empirical models for the FPD were proposed to describe the deviation from the ideal solution. The bound water theory assumes a definite amount of unfreeable water at any temperature. The first model is represented by Eq. (2.16).

$$\Delta T = \left[\frac{k_o X}{(1 - b.X).M}\right] \tag{2.16}$$

The second semi-empirical FPD model is based on the solute-solvent interaction theory as shown in Eq. (2.17).

$$\Delta T = \left[\frac{k_o X(1 - C.X)}{(1 - X).M}\right] \tag{2.17}$$

where, C = solute-solvent interacting coefficients. The empirical model of Riedel suggested that most of the juices are frozen at −60 °C.

2.4 Boiling Point Elevation (BPE)

The knowledge of temperature at the boiling point of fruit juices is widely applied for the concentration of fruit juice and used to evaporate the juice as well it is fruitful to design the process and equipment. A previous study has reported on the temperature effect and boiling point of the extract obtained from coffee (Telis-Romero et al. 2002). The rise over the boiling point of fruit juice over the different concentrations was previously reported by (Jacob 2006) for the juice of thai tangerine, by (Crapiste and Lozano 1988) and (McKenna 1984) for the juice of apple, by (Moresi and Spinosi 1980) for the juice of orange, and by (Varshney and Barhate 1978) for mango, lemon juice, and pineapple. The variation in the intensity of sensory flavor in the ice cream was observed under different concentrations of fat and flavoring (Frøst et al. 2005). It was concluded that variation of flavor perception was directly related to the boiling point and hygroscopic natures of the flavor compound. Similarly, the measurement of the boiling point of different samples was performed for refrigerants (Rogdakis and Lolos 2006), for biodiesel from vegetable oils (Goodrum 2002), for methyl jasmonate (Acevedo et al. 2003), and for carbohydrates (Gaida et al. 2006). The VP exerted by the liquid at the temperature is linearly correlated to the VP exerted by the reference liquid at the temperature (Alvarado 1831). Hence for constant concentration, Eq. (2.18) is written as

$$T_A = m_o + m_1 T_{A0} \tag{2.18}$$

where T_A and T_{A0} = boiling temperatures at the same pressure for fruit juice and water. The constant (m_0 and m_1) is calculated using regression analysis linearly for the concentration of juice along with the coefficient of determination and r^2. The correlation over rising in boiling point is described in Eq. (2.19), ΔT_B, as follows:

$$\Delta T_B = T_A - T_{A0} \tag{2.19}$$

With the substation to Eq. (2.20) for the condition of $m_1 = 1$, $\Delta T_B = m_0$, which indicates that the elevation in boiling point only occurs over changing the juice concentration and is not dependent on pressure.

Secondly, the boiling point data of the aqueous solution is expressed by using the suitable equation which describes the temperature dependency of the VP of pure substance using the Antoine equation (Gabas et al. 2008). The empirical equation is shown in Eq. (2.20).

$$ln\ P = A - \frac{B}{T_A + C} \tag{2.20}$$

where, P = pressure (Pa); T_A = boiling temperature (K) of fruit juice; and A, B, & C = empirical constants showing dependency on concentration.

The alternative method to estimate the FPD in aqueous solutions (Cross 1989) was suggested by using the emphatical model and was dependent on the solid concentration and pressure which is presented in Eq. (2.21).

$$AT_B = \alpha W^{\beta} expexp\ (\gamma\ W)\ P^{\delta} \tag{2.21}$$

where, $\Delta T_B = T_A - T_{A0}$, the difference is the rise over boiling point (°C); W = concentration of mass (TSS in °Brix), and the constant parameters including α, β, γ, and δ are computed nonlinearly with regression analysis.

2.5 Osmotic Pressures (OP)

Osmosis involves the transfer of water (solvent) through a membrane that is semi-permeable in nature and movement occurred from the high concentration of solvent to the lower solvent concentration. The difference in the concentration of solute is the driving force in which the semi-permeable membrane (SPM) only permits the passage of water by trapping the molecules of solute or ions (Crapiste and Lozano 1988). Hence the movement of water molecules occurred from a solution to other regions to get maximize mixing with achieving equilibrium conditions. In the thermodynamics approach, the extent of mixing is measured with the osmotic potential or called the OP.

The osmotic potential is always higher for the concentrated solution region while it is lower for the dilute solution and proportional molar concentration of solutes. It is defined as the supplied pressure required to apply in the solution to avoid the net transfer of solvent through an SPM (Hameed 2013). After applying pressure exceeding the OP the pure solvent start flowing from the higher solute to the lower solute concentration by passing through an SPM where the solvent can be separated from the solution. The reverse of the normal process of osmosis is called reverse osmosis (RO) (Jacob 2006; Wang et al. 2011).

For the treatment of wastewater, the solvent and the contaminants are allowed to pass through an SPM and solvent are pass-through and contaminants are not allowed to pass through the membrane. In the wastewater plant, the feed is allowed to flow through the membrane on one side whereas the osmotic agent is passed to another side. The solute can be used by the osmotic agent, and create the OP which is higher as compared to feed and the solute gets a rejection from the surface of the membrane (Flynn et al. 2012).

Similarly, forward osmosis (FO) is another innovative separation technique based on the membrane that is potentially considered the sustainable and economically feasible alternative to RO and is an electrodialysis reversal approach because it can utilize green energy from the natural system (Duranceau 2012). Conventionally, the juice is concentrated with thermal treatment. Such a method reduced the color and flavor of the juice. The membrane separation techniques were applied to produce the concentrated juice with the same flavor as fresh juice with having the shelf life of commercial food products (Cross 1989). Equation (2.23) presented the relationship between the flux occurring crosswise of the membrane surface and the hydrostatic and OP difference across the surface of the membrane (Hameed 2013). Equation (2.22) is most suitably described the RO process as follows.

$$F_w = A_c(\Delta P - \Delta\pi) \tag{2.22}$$

where, F_w = overall available water flux across the membrane (l/m^2 h); A_C = resistant constant for membrane flux (l/m^2 h kPa); the ΔP = hydrostatic pressure (kPa); $\Delta\pi$ = reverse OP (kPa) [equivalent to the difference of the OP of the draw and feed solution].

In the case of FO, the, ΔP is equal to zero, and the Eq. (2.22) is reordered to Eq. (2.23).

$$F_w = A_c\pi \tag{2.23}$$

As we can see Eq. (2.23) is sustainably fitted in the membrane where hydrostatic pressure does not exist across the membrane surface hence no support of the pressure is required in the system. This facilitates the operation of the membrane in soft bags with water walls. Mostly the hydrostatic pressure exists due to the flow of fluid across the membrane. The flow should require on both sides of the membrane and balance should achieve with ΔP equal to 0. Once the balanced flow is achieved the hydrostatic pressure may occur in a forward or reverse direction as per the direction of water flux but in both cases, it is negligible as compared to OP (Gormly et al. 2010).

2.5.1 Water Activity Measurement from Osmotic Pressure (OP)

The OP is the pressure that is supplied mechanically to stop the overall flow of solvent through an SPM. The a_w is estimated concerning equilibrium relative humidity (ERH) and OP and identical instruments are involved in the measurement of these parameters. ERH is related to a_w is which is expressed in percentage and is 100 times the a_w value. The a_w unit is widely preferred by scientists to define the a_w of solid and liquid food, while ERH is the atmosphere of the surrounding (Scott 1957). The OP is inversely proportional to the a_w and is represented by Eq. (2.24).

$$OP\ (\pi) = \frac{-RT}{Vl\ln a_w} \tag{2.24}$$

where, V = water partial molal volume; T = absolute temperature; and R = gas constant (Ayerst 1965b). The OP is denoted with the unit including bar, atmosphere, or ergs cm^{-3} (Bookin et al. 2008).

The OP is classified as the solute OP with the inclusion of the matrix effect which resulted from the water-solid interaction on the surface of the colloid (Nobel 1970). The equation where the value of water activity correlated to the OP is represented by Eq. (2.25).

$$OP = \frac{RT}{V_w \ln(a_w)} \tag{2.25}$$

where, V_w = water's partial molar volume of the solution; R = universal gas constant; and T = absolute temperature.

2.6 Conclusion

The colligative properties have a significant role in the measurement of food parameters. The concept of VP is used in a VPO and a thermistor cryoscope to determine the adulterated water in the milk sample. The VPO has more accuracy over the measurement and performs with additional advantages than the thermistor cryoscope for a_w in food containing intermediate moisture. Based on the concept of VP the a_w of the food sample is measured using the instrument including the dew point method, electric hygrometer, hair hygrometer, vapor pressure manometer, and psychrometer. The thermistor cryoscope method has been widely used to measure the FPD of aqueous solution with a large range of solute concentrations. A cryoscopy method is an efficient approach to detecting the added water and concentration of sodium chloride in milk. However, the method is inaccurate for detection in the case of a mixture of water and NaCl in milk. The significance of BPE temperature is much more in the beverage industry especially for juice concentration and for the design and development of evaporators. The OP is a very important term for the development of membrane separation as the non-thermal approach for the concentration of juice.

Acknowledgement Author is thankful for the guidance and suggestions from his professors Biswanath Bhunia and Tarun Kanti Bandhyopadhyay.

References

Acevedo C, Sánchez E, Young M, Simpson R (2003) Prediction correlation of vapor pressure for methyl jasmonate. J Food Eng 59(4):431–433

Acott K, Labuza T (1975) Inhibition of Aspergillus niger in an intermediate moisture food system. J Food Sci 40(1):137–139

Alvarado JD (1831) Specific heat of dehydrated pulps of fruits. J Food Process Eng 14(3):189–196

Anagnostopoulos G (1973) Water activity in biological systems: a dew-point method for its determination. Microbiology 77(1):233–235

Aschaffenburg R, Burton H, Rowland S, Thiel C (1958) An investigation of the effect of ultra-high-temperature treatment on the freezing-point and composition of milk. Int J Dairy Technol 11(2): 93–95

Ayerst G (1965a) Determination of the water activity of some hygroscopic food materials by a dew-point method. J Sci Food Agric 16(2):71–78

Ayerst G (1965b) Water activity: its measurement and significance in biology. Int Biodeterior Bull 1(2):13–26

Bagnall D, Smith A (1955) Abnormally small freezing-point depressions of genuine milk. Analyst 80(953):623–625

Barbosa-Cánovas GV, Vega H (eds) (1996) Dehydration of foods. Springer, Cham, p 340

Baur FJ, Ensminger LG (1997) Newest analytical methods. The Association of Official Analytical Chemists (AOAC), Washington. https://doi.org/10.1007/BF02670789

Beuchat L (1974) Combined effects of water activity, solute, and temperature on the growth of Vibrio parahaemolyticus. Appl Microbiol 27(6):1075–1080

Blackmore R (1959) One year survey of Ohio raw and pasteurized milk. Milk Dealer 49(3):43–48

Blackmore R (1960) Added water present in Ohio milk during 1959. J Milk Food Technol 23(5): 150–151

Bogsanyi D, Weeden D (1968) A simple apparatus for water vapor pressure-moisture studies by a dew point method. Chem Ind 23:741–745

Bookin D, Ebbing DD, Gammon SD (eds) (2008) General chemistry – students' solution manual. Florence, Cengage Learning Inc, p 481

Boyer J (1969) Measurement of the water status of plants. Annu Rev Plant Physiol 20:351–364

Crapiste GH, Lozano J (1988) Effect of concentration and pressure on the boiling point rise of apple juice and related sugar solutions. J Food Sci 53(3):865–868

Cross S (1989) Membrane concentration of orange juice. In: Proceedings ASME citrus engineering symposium. ASME, Washington, pp 46–63

Demott B (1967) The influence of vacuum pasteurization upon the freezing point and specific gravity of milk. J Milk Food Technol 30(8):253–255

Dubin T (1954) Report on cryoscopy of milk. J Assoc Off Agric Chem 37(2):232–234

Duranceau SJ (2012) Emergence of forward osmosis and pressure-retarded osmotic processes for drinking water treatment. Florida Water Resour J 1:32–36

Eckhoff SR, Okos MR (1986) Kinetics of sulfite reaction in corn grain. J Agric Food Chem 34(2): 239–245

England C, Neff M (1963) The accuracy of cryoscopy methods. J Assoc Off Agric Chem 46(6): 1043–1049

Fett HM (1973) Water activity determination in foods in the range 0.80 to 0.99. J Food Sci 38(6): 1097–1098

Flynn M, Soler M, Shull S, Broyan J, Chambliss J, Howe AS, Gormly S, Hammoudeh M, Shaw H, Howard K (2012) Forward: osmosis cargo transfer bag. In: 42nd International conference on environmental systems. NASA Ames Research Centre, California, pp 35–39

Foust AS, Wenzel LA, Clump CW, Maus L, Andersen LB (eds) (2008) Principles of unit operations, 2nd edn. John Wiley & Sons, New York, p 784

Freeman T (1957) Cryoscopic estimations of added water in milk. J Dairy Sci 10:1392–1393

Frøst MB, Heymann H, Bredie WL, Dijksterhuis GB, Martens M (2005) Sensory measurement of dynamic flavor intensity in ice cream with different fat levels and flavorings. Food Qual Prefer 16(4):305–314

Gabas A, Sobral P, Cardona-Alzate C, Telis V, Telis-Romero J (2008) Influence of fluid concentration on the elevation of boiling point of blackberry juice. Int J Food Prop 11(4):865–875

Gaida LB, Dussap C, Gros J (2006) Variable hydration of small carbohydrates for predicting equilibrium properties in diluted and concentrated solutions. Food Chem 96(3):387–401
Gal S (1975) Recent advances in techniques for the determination of sorption isotherms. In: Duckworth RB (ed.), Water relations of foods: proceedings of an international symposium held in Glasgow, pp 139–154
Garland C, Nibler J, Shoemaker D (eds) (2003) Experiments in physical chemistry. New York, McGraw-Hill Book Co, p 757
Goodrum J (2002) Volatility and boiling points of biodiesel from vegetable oils and tallow. Biomass Bioenergy 22(3):205–211
Gormly S, Flynn M, Polonsky A (2010) Membrane based habitat wall architectures for life support and evolving structures. In: Presented at 40th international conference on environmental systems. NASA Ames Research Centre, California, pp 22–25
Green W (1956) An economical method for determining the freezing point of milk. J Milk Food Technol 19(8):279–281
Gur-Arieh C, Nelson A, Steinberg M, Wei L (1965) Water activity of flour at high moisture contents as measured with a pressure membrane cell. J Food Sci 30(2):188–191
Hagerdal B, Lofqvist B (1973) Screening method based on electric hygrometer for obtaining water sorption isotherms. J Agric Food Chem 21(3):445–451
Hameed KW (2013) Concentration of orange juice using forward osmosis membrane process. Iraqi J Chem Petrol Eng 14(4):71–79
Henderson JL (1963) The effect of handling and processing on the freezing point of milk. J Assoc Off Agric Chem 46(6):1030–1035
Henningson R (1966) Cryoscopy of milk: effect of variations in the method. J Assoc Off Anal Chem 49(3):511–515
Henningson R (1967) Determination of the freezing point of milk by thermistor cryoscopy. J Assoc Off Anal Chem 50(3):533–537
Henningson R (1968) Collaborative study of the thermistor cryoscopy method for the determination of the freezing point value of milk. J Assoc Off Anal Chem 51(4):816–821
Henningson R (1969) Thermistor cryoscopic determination of the freezing point value of milk produced in North America. J Assoc Off Anal Chem 52(1):142–151
Henningson R (1970) Regulatory agency acceptance of the interpretation of the freezing point value of milk as part of the official thermistor cryoscopic method. J Assoc Off Anal Chem 53(3): 539–542
Henningson R, Lazar J (1959) Effect of vacuum pasteurization on the freezing point of milk. J Dairy Sci 42(2):417–419
Jacob A (2006) Critical review of the history, development and future prospects for forward osmosis. Desalination 391:16–29
Jenkins HDB (ed) (2008) Chemical thermodynamics at a glance. Wiley, New York, p 208
Juven B, Gertshovki R (1976) The effect of salting on the microbiology of poultry meat. J Milk Food Technol 39(1):13–17
Kang CK, Woodburn M, Pagenkopf A, Cheney R (1969) Growth, sporulation, and germination of Clostridium perfringens in media of controlled water activity. Appl Microbiol 18(5):798–805
Karan-Djurdjic S, Leistner L (1970) Messung der Wasseraktivität von Fleisch und Fleischwaren mit dem Sina-Gerät (Measuring the water activity of meat and meat products with the Sina device). Fleischwirtschaft 50:1104–1106
Karel M, Nickerson J (1964) Effects of relative humidity air+ vacuum on browning of dehydrated orange juice. Food Technol 18(8):1214–1217
Labuza TP (1974) Interpretation of sorption data in relation to the state of constituent water. In: Duckworth RB (ed) Water relations of foods: proceedings of an international symposium, Glasgow, UK, pp 155–172
Labuza TP, Acott K, Tatini S, Lee R, Flink J, McCall W (1976) Water activity determination: a collaborative study of different methods. J Food Sci 41(4):910–917

Labuza TP, Kreisman L, Heinz C, Lewicki P (1977) Evaluation of the Abbeon cup analyzer compared to the VPM and Fett-Vos methods for water activity measurement. J Food Process Preserv 1(1):31–41
Lazar J, Bellamy W (1957) Use of the vacu-therm pasteurizer for milk flavor control. J Dairy Sci 40: 1393–1393
Lazar J, Henningson R (1960) Influence of vacuum pasteurization upon the freezing point value, total solids, and concentration of fluid milk. J Dairy Sci 43(1):42–47
Leistner L, Rodel W (2012) The significance of water activity for micro-organisms in meats. In: Duckworth RB (ed) Water relations of foods: proceedings of an international symposium held in Glasgow, pp 309–323
Ligugnana R (1970) Automatic cryoscopy of milk. Latte 44(9):655–657
Limsong S, Frazier W (1966) Adaptation of Pseudomonas fluorescens to low levels of water activity produced by different solutes. Appl Microbiol 14(6):899–901
Lythgoe HC (1952) Freezing points of sour milks. J Assoc Off Agric Chem 35(4):855–859
McKenna BM (1984) Engineering and food. In: Proceedings on engineering sciences in the food industry, vol 1. Elsevier, London, pp 21–25
Moore J, Smith A (1961) Observations on the freezing point of vacuum treated milk. J Milk Food Technol 24(4):115–118
Moore J, Smith A, Gosslee D (1961) The effect of carbon dioxide removal upon the freezing point of vacuum treated milk. J Milk Food Technol 24(6):176–179
Moresi M, Spinosi M (1980) Engineering factors in the production of concentrated fruit juices, part I: fluid physical properties of orange juices. Int J Food Sci Technol 15(3):265–276
Mozumder BKG, Caroselli N, Albert L (1970) Influence of water activity, temperature, and their interaction on germination of Verticillium albo-atrum conidia. Plant Physiol 46(2):347–349
Nascimento WWG, da Oliveira MAL, Furtado MAM, dos Anjos VC, Bell MJV (2013) Development and optimization of an alternative methodology for detection of milk adulteration by water. J Food Sci Eng 3(7):363–370
Nobel PS (ed) (1970) Introduction to biophysical plant physiology. W. H. Freeman & Co. Ltd., New York, p 488
Norrish RS (1966) An equation for the activity coefficients and equilibrium relative humidity of water in confectionery syrups. Int J Food Sci Technol 1(1):25–39
Northolt M (1972) De microbiologische betekenis en het meten van de wateractiviteit (aω) in voedingsmiddelen (The microbiological significance and measurement of water activity (aω) in foodstuffs). Food Technol 3:151–155
Novo M, Reija B, Al-Soufi W (2007) Freezing point of milk: a natural way to understand colligative properties. J Chem Educ 84(10):1673
Pensiripun K (1975) Evaluation of a vapor pressure osmometer for determination of added water in milk. All Graduate Theses and Dissertations. Utah State University, Logan, p 75
Pensiripun K, Campbell E, Richardson G (1975) A vapor pressure osmometer for determination of added water in milk. J Milk Food Technol 38(4):204–207
Pinkerton F, Peters I (1958) Conductivity, percent lactose, and freezing point of milk. J Dairy Sci 41(3):392–397
Prior B (1979) Measurement of water activity in foods: a review. J Food Prot 42(8):668–674
Prior B, Theron DP, Henning JS, Fouche E (1977) Staphylococcal food poisoning from infected snoek. S Afr Med J 52(22):889–890
Rodel W, Scheuer R (1999) Redox potential of meat and meat products, Part II: typical redox potentials of meat and meat products. Meat Industry Int 3:29–32
Rodel W, Scheuer R (2002) Redox potential of meat and meat products, Part III: control of redox potential during meat processing - effects of pH value, sodium nitrite, sodium ascorbate, sodium lactate and atmospheric oxygen. Meat Industry Int 79(7):78–81
Rogdakis E, Lolos P (2006) Simple generalized vapor pressure-and boiling point correlation for refrigerants. Int J Refrig 29(4):632–644
Scott W (1957) Water relations of food spoilage microorganisms. Adv Food Res 7:83–127

Shipe W (1959) The freezing point of milk: a review. J Dairy Sci 42(11):1745–1762
Shipe W (1964) Effect of vacuum treatment on freezing point of milk. J Assoc Off Agric Chem 47(3):570–572
Shipe W, Dahlberg A, Herrington B (1953) A semi-automatic cryoscope for determining the freezing point of milk. J Dairy Sci 36(9):916–923
Sloan AE, Labuza TP (1976) Prediction of water activity lowering ability of food humectants at high water activity. J Food Sci 41(3):532–535
Smith A (1964) The carbon dioxide content of milk during handling, processing and storage and its effect upon the freezing point. J Milk Food Technol 27(2):38–41
Smith A, Anderson R, Waldo R, Chaffee C, Parry R (1962) Dairy plant precautions to avoid added water in milk. J Milk Food Technol 25(12):379–382
Strong DH, Foster EF, Duncan CL (1970) Influence of water activity on the growth of Clostridium perfringens. Appl Microbiol 19(6):980–987
Telis-Romero J, Cabral R, Kronka G, Telis V (2002) Elevation on boiling point of coffee extract. Braz J Chem Eng 19(1):119–126
Troller JA (1977) Statistical analysis of aw measurements obtained with the Sina Scope. J Food Sci 42(1):86–90
Troller J, Stinson J (1975) Influence of water activity on growth and enterotoxin formation by Staphylococcus aureus in foods. J Food Sci 40(4):802–804
Van Zyl PJ, Prior BA (1990) Water relations of polyol accumulation by Zygosaccharomyces rouxii in continuous culture. Appl Microbiol Biotechnol 33(1):12–17
Varshney N, Barhate V (1978) Effect of concentrations and vacuum on boiling points of fruit juices. Int J Food Sci Technol 13(3):225–233
Vos PT, Labuza TP (1974) Technique for measurement of water activity in the high aw range. J Agric Food Chem 22(2):326–327
Wang LK, Chen JP, Hung YT, Shammas NK (eds) (2011) Handbook of environmental engineering: membrane and desalination technologies, vol 13. New York, Humana Press, p 716
Wickware CL, Day CT, Adams M, Orta-Ramirez A, Snyder AB (2017) The Science of a sundae: using the principle of colligative properties in food science outreach activities for middle and high school students. J Food Sci Educ 16(3):92–98
Wolf A (1970) Concentration properties of aqueous solutions: conversion tables. In: Weast RC (ed) Handbook of chemistry and physics, 50th edn. CRC Press, Philadelphia, pp D181–D226
Ystgaard O, Homeyer P, Bird E (1951) Detection of adulteration of milk by lactometric and freezing point methods. J Dairy Sci 34:680–688

[illegible]

[illegible] products [illegible] Food [illegible]

[illegible] J Food Sci Technol [illegible]

[illegible] J Food Sci [illegible]

Chapter 3
Scope of Three-Dimensional Printing for Fabrication of Foods

Vijayasri Kadirvel, Kamalesh Raja, and Thiruvengadam Subramaniyan

Contents

V. Kadirvel
Department of Food Technology, Rajalakshmi Engineering College, Chennai, Tamil Nadu, India

K. Raja · T. Subramaniyan (✉)
Department of Biotechnology, Rajalakshmi Engineering College, Chennai, Tamil Nadu, India
e-mail: thiruvengadam.s@rajalakshmi.edu.in

J. A. Malik et al. (eds.), *Food Process Engineering and Technology*,
https://doi.org/10.1007/978-981-99-6831-2_3

Abstract Three–dimensional (3D) printing has been receiving prime focus for the production of customized foods by accurately balancing both palatability and nutrition of foods. The 3D printing is a promising alternative to the traditional methods of processing food since it is proven to assist older people with disabilities effectively, print artistically attractive fortified food products for children, provide an opportunity for vegan consumers to taste plant-based meat and produce foods with an enhanced shelf-life for space travel, being more accurate with high precision. The customizable property of this technique proves to satisfy the needs of all age groups. Despite having various advantages, researchers and industrialists face challenges during production, such as: improving the success rate of 3D printed food for a better dining experience and other engineering problems. The main objective of this review chapter is to collect, analyze and conclude the information regarding food processing and production using 3D printers. This chapter also reviews applications of additive manufacturing in the food industry, various materials available for 3D printing, consumer acceptability of 3D printed products.

Keywords 3D food printing · Customized food · Nutrition · Palatability · Printing materials · Protein-rich food

3.1 Introduction

In today's world, establishing a healthy lifestyle has become an important aspect of life. By adopting healthy diet patterns, health promotion is established by the prevention and management of a wide range of diseases (Hayman and Worel 2014). In general, people consult nutrition experts to perform health assessments, whose services are expensive, time-consuming, and not readily available (Salloum and Tekli 2021). The surge in demand for customized food products in the market is currently satisfied by trained artisans; however, the cost for producing a few pieces is relatively high (Wegrzyn et al. 2012).

The growing attention towards functional foods and the importance of tailoring the diet based on the individual's health may be satisfied by three-dimensional printing. The first 3D printer was developed during 1980s; and since then, it has been equipped in different research areas, including medicine, education, aerospace, and jewelry manufacturing. Various research studies revealed that this technology could be applied even in the food industry (Pérez et al. 2019).

3D printing involves a controlled robotic process and 3D computer design program CAD (Computer-aided Design) or 3D platforms installed from available online services (Gibson et al. 2009). The script of these programs is generally based on C^{++} language. The creation and extrusion of materials into desired models are performed by adding and subtracting the primitives (Wijnen et al. 2014). By employing open-source hardware and software technologies, the cost of manufacturing 3D-printed food may be significantly lowered (Hinton et al. 2017). Several 3D printed shapes have been successfully manufactured by employing various food materials, such as: chocolates, cookie dough, cereals, sugar powder, processed cheese, meat gels, fruits, and vegetables (Derossi et al. 2019).

Many raw foods are classified under natively printable material. In contrast, other food materials can be printed by combining hydrocolloids like xanthan gum and gelatin, and other food additives to design 3D printed food (Cohen et al. 2009). The availability of compatible raw materials necessitates the addition of food additives that health-conscious consumer's dislike, which is a major drawback of additive manufacturing. Therefore, this challenge must prevail over in the future to further enhance consumer acceptability. The limited consumer awareness of 3D printing technology may also negatively affect the acceptability (Zargaraan et al. 2013).

Approximately 60–70 million people in the U.S. are affected by dietary diseases, such as: Celiac Disease, Crohn's Disease, GERD, Diverticulitis, and irritable bowel syndrome. As part of the treatment and management of each of these diseases, customized diet is required, which may be achieved by 3D printing of food materials (Lipton 2017)? The development of appealing new foods via additive manufacturing aids in providing proper nutrition and appropriate texture to make life easier for senior citizens with impaired swallowing reflexes. However, 3D printing may improve personalized nutrition or attract the consumer population for commercial profit (Portanguen et al. 2019).

The market potential of 3D-printed food is expected to rise rapidly as complex and customized designs are printed at low cost without employing complicated machines, molding, and other conventional manufacturing procedures (Feinberg and Miller 2017). In the past years, there has been a drastic escalation in the number of researches being conducted in 3D printing, including the study of rheological properties, food encapsulated with alive cells to increase nutrition, bio-mimicking plant tissues, and many more (Derossi et al. 2019). Recently, TastyBeats, a 3D food printer, has been developed to translate physical activity data into food treats to encourage consumers to switch over to a healthier lifestyle. Similarly, a food 3D printer named "FoodFab" focuses on fabricating food based on the hunger level and the daily average calorie intake of the consumer (Lin et al. 2020). Unlike robotics-based food processing technologies, additive manufacturing integrates printing technology and gastronomy techniques to enable artistic design and customized food from the domestic to industrial culinary sector (Sun et al. 2015a, b, c). The future of 3D printing technology may be the enhancement of consumption of non-traditional food materials, such as: insects, high fiber plant-based materials, and animal-based by-products (Payne et al. 2016).

This chapter focuses on the collection and analyzing data on various techniques available for 3D printing of food materials along with optimized processing parameters to obtain 3D printed materials with high accuracy and precision. This chapter aspires to deliver information on the application of additive manufacturing technology in various food sectors, such as, space foods, medicine in the form of candy for children, soft meals for the elderly, fortified foods as a solution to malnutrition, and alternative protein sources (such as: plant-based and insect-based formulations). In addition, the recent advancement of 3D printed functional foods, and their consumer acceptability is also discussed.

3.2 Printing Materials

The materials employed for 3D printing are classified into natively printable and non-printable traditional materials based on their printability. The non-printable materials may be pre-processed or hydrocolloids are incorporated to enhance their printability and stability. The conventional raw materials are pre-processed to make them suitable for 3D printing and improve the thermal stability of the printed materials during post-processing operations. The pre-extrusion procedures include heating, cooling, refrigeration, incubation, size reduction, and constant agitation; and the post-extrusion procedure includes curing the printed gels. The characteristics of various food materials currently employed in the food industry are discussed in this section.

3.2.1 Natively Printable Materials

The natively printable materials include food components, such as: chocolate, icing sugar, hydrogel, pasta, hummus, butter, cake frosting, processed cheese, and jelly. The printed 3D models are stable enough and do not require incorporating food additives. The final 3D structures possess good nutrition, taste, and texture. However, the food printed from non-traditional materials is not served as the main course (Sun et al. 2015a, b, c). A few preliminary processing steps (such as: cooking, mashing, heating, melting, and agitation) are required to convert the raw food material into printer ink. The non-traditional materials are able to be extruded through a syringe tip and can hold the desired shape under gravity post-extrusion. The characteristics of selected natively printable materials are discussed here.

3.2.1.1 Chocolate

Chocolates are the most commonly employed non-traditional material in additive manufacturing due to their proficient popularity in the food market. Chocolates are non-Newtonian fluids known to melt as soon as they come in contact with the receptors in the mouth, and the rate of melting depends on the viscosity of the chocolate ink. If the thickness of chocolate is too low, the formed 3D model becomes unstable and loses its shape. If the density of chocolate is too high, the printed chocolates cannot be de-molded easily. Apart from viscosity, the flow properties of chocolate are affected by fat content, moisture, presence of food additives, couching, and particle size distribution (Beckett 2008; Lanaro et al. 2017). Even though chocolates are natively printable materials, yet hydrocolloids are being incorporated to enhance the 3D printing precision by improving the flow rate and settling time during the printing process. Apart from modifying the composition of the chocolate ink, various printing parameters (such as: nozzle diameter, nozzle height, extrusion rate, printing speed, and cooling rate) may be optimized to improve the stability of the 3D printed model (Mantihal et al. 2017; Rando and Ramaioli 2021).

3.2.1.2 Table Sugar

Table sugar is additive-manufactured in its molten form. A heating device and a temperature sensor are attached to the sugar extruder to obtain molten sugar with desired viscosity and to maintain a constant temperature during the extrusion process. Sugar in molten form is optically transparent; therefore, dyes of synthetic and natural origin are added to enhance visual perception. Sugars do not have a sharp melting point, and they melt over a wide range of temperatures. The melting point of sugar depends on the purity, contamination, and quantity of water added.

Certain sugars can caramelize and can become brown during the melting process, otherwise may decompose even before melting. Therefore, the material is being combined with other sweeteners (such as: corn syrup) to avoid adverse effects of sugar while heating at high temperatures (He et al. 2015).

Sugar 3D structures may also be manufactured by selective sintering technique, which involves the melting and fusing powdered sugar particles layer-by-layer via laser or hot air as a source of heat. During the printing process, the excess sugar powder acts to support the printed structure (Izdebska and Żołek-Tryznowska 2016).

3.2.2 Non-printable Traditional Materials

The non-printable traditional materials are food materials that are part of the everyday diet, e.g.: rice, meat, fish, fruits, and vegetables. Currently, these non-printable materials are being printed successfully by adopting pre-processing operations and incorporating permitted food additives. The various pre-processing techniques and food additives employed to manufacture firm 3D models are discussed here.

3.2.2.1 Fruits and Vegetables

Fresh fruits and vegetables are highly perishable materials, and through additive manufacturing technology, their wastage is being prevented by developing wholesome food products (Gholamipour-Shirazi et al. 2020). Fruits and vegetables lack mechanisms, such as, gelation agglomeration required for 3D printing due to the low fat, carbohydrate, and protein content and the high fiber content. The viscosity of fruit and vegetable puree depends on temperature, the number of soluble solids, pulp, and fiber content.

The physiochemical properties and rheological behavior of fruits vary based on the type, such as: berries, stones, pomes, and citrus fruits. To create 3D printed food from fruits and vegetables, incorporating hydrocolloids becomes mandatory (Diamante and Umemoto 2015; Pulatsu and Lin 2021). Pectin, gelatin, gellan gum, guar gum, and xanthan gum are commonly employed hydrocolloids that provide the capability to maintain the rigidness of the printed structure during post-processing. By optimizing the concentration of hydrocolloids and combining different hydrocolloids at optimized ratios, a wide range of textures to enhance the mouthfeel are accomplished. However, hydrocolloids possess a neutral flavor, and the mouth-feel also depends on the taste of the final product, which may be obtained by adding a flavoring food additive (Liu et al. 2017).

Conversely, peas, avocado, and potatoes belonging to the same group are stable enough for 3D printing due to their high protein, lipid, and carbohydrate composition (Ricci et al. 2019). The 3D-printed fruits and vegetables attract health-conscious consumers and patients, who prefer healthy food for their children. Therefore, printing fruits and vegetable materials into 3D shapes is predicted to have significant market potential (Severini et al. 2018).

3.2.2.2 Meat and Fish

En route towards enhancing the post-extrusion mechanical stability of the meat ink, hydrocolloids from various sources, such as, plants, animals, and microbes are being employed. The frequently incorporated hydrocolloids are basil seed gum, locust bean gum, carboxymethyl cellulose, carrageenan, xantham gum, guar gum, and gelatin. The other hydrocolloid sources (such as: milk, eggs, and soy protein) are not preferred due to the risk of allergies (Dick et al. 2019a). Most of the meat materials, such as, chicken, pork, and fish may be 3D printed post-cooking the meat followed by uniform mincing into a paste. Incorporating a food additive in significant quantities to the meat paste is necessary to avoid a phase separation between solid meat and liquid juice phases of the meat ink.

Transglutaminase and gelatin are potentially contributing to fabricating complex 3D meat structures by reforming a new protein matrix and preserving their rheological properties. The smooth and precise additive manufacturing of meat has become possible by adding food additives to the meat ink. For example, the addition of NaCl allows the extraction of myofibrillar proteins, which is required to enhance the binding and stability of the meat paste. However, to obtain precise 3D structures, NaCl is incorporated in combination with hydrocolloids (Dick et al. 2019b; Lille et al. 2018; Lipton et al. 2010). Apart from animal meat, additive manufacturing may be effectively employed to produce vegetarian meat from high fiber plant-based sources. Vegetarian consumers are willing to try 3D printed vegetarian meat as an alternative to animal meat (Mehak-Jandyal et al. 2021).

3.2.2.3 Rice

Rice is a traditional non-printable material due to its low protein and gluten-free composition. Protein and gluten are essential for forming a cohesive network and to provide required rheological properties (Liu et al. 2020a, b). The stability and printability of rice-based products depends on the chemical composition of the rice and the pre-processing operations. The commonly adopted pre-processing steps for 3D printing rice-based products include cooking and freeze-dying. The former process is performed to gelatinize the starch, while the later is implemented to lower the moisture content to the desired level (Huang et al. 2019).

The physicochemical properties of rice ink also vary based on the size of the grain, for example, long-grain and short-grain variety. The waxy rice flour exhibits good printability with high precision and a rigid 3D model, while steaming the waxy rice leads to unstable 3D structures. The other rice varieties, such as *Indica* and *Japonica* rice, can print natural and healthy 3D foods. The rheological parameters of rice flour dough resembled that of wheat flour dough after incorporating a gluten substitute.

The typical food additives employed for rice-based formulations include xanthan gum, guar gum, locust bean gum, k-carrageenan, hydroxypropyl methylcellulose (HPMC). The mechanism of blending gums to the pre-extruded product is to enhance the mouthfeel by altering the viscosity of the rice ink (Liu et al. 2020a, b; Sivaramakrishnan et al. 2004; Turabi et al. 2008). Apart from food products, the successful additive manufacture of biodegradable packaging from rice husk has become possible by adding gums. The gums are employed to improve the flowability of the ink and smooth the extrusion process (Nida et al. 2021).

3.3 Techniques for 3D Printing Food Material

Among various techniques available for additive manufacturing, the primary methods for 3D printing food materials are extrusion-based printing, selective laser sintering, binder jetting, and inkjet printing. Not all materials may be utilized for all the available techniques. Therefore, only one of these methods is implemented based on the type of food material printed and the requirement to fabricate complex 3D structures with high resolution. Certain advantages and disadvantages are associated with each of these techniques.

It is vital to optimize the processing parameters (such as: temperature at which the material is extruded, diameter, height, and speed of the nozzle). These play crucial role in the quality of the final product obtained. Figure 3.1 illustrates the processing steps involved in 3D printing technology. Figure 3.2 exemplifies different methods of additive manufacturing technology based on the physical state of the feed. Figure 3.3 schematically depicts various 3D printing techniques.

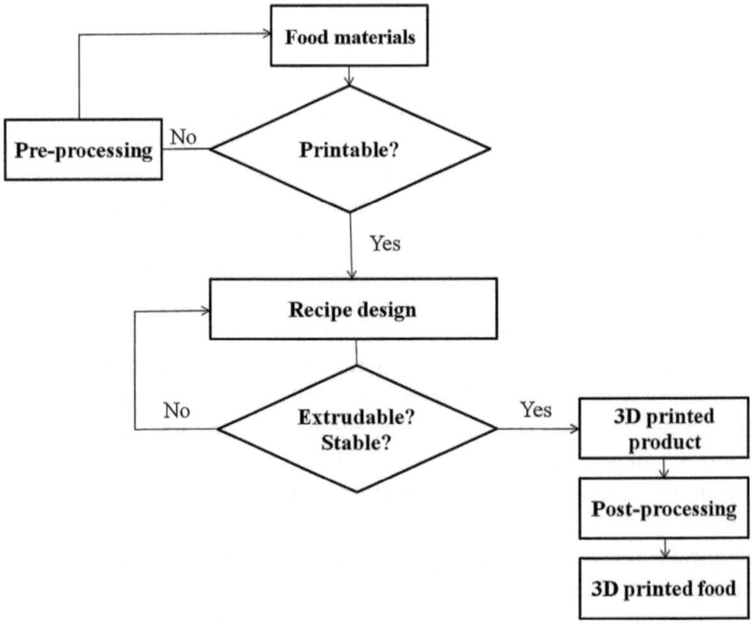

Fig. 3.1 Processing steps involved in 3D printing

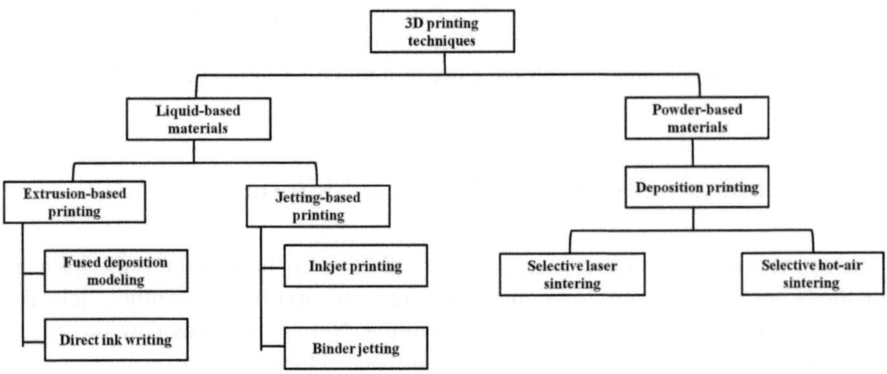

Fig. 3.2 Classification of 3D printing techniques based on the physical state of food material

3.3.1 Extrusion-Based 3D Printing

Extrusion-based 3D printing is commonly employed for printing food materials. The extrusion process starts with loading the food material, and as the extruder moves under controlled conditions, it extrudes the food material into desired geometry and size. The complex structure is deposited layer-by-layer. The extrusion-based printer consists of a digitally controlled multi-axis stage and one or more extrusion units.

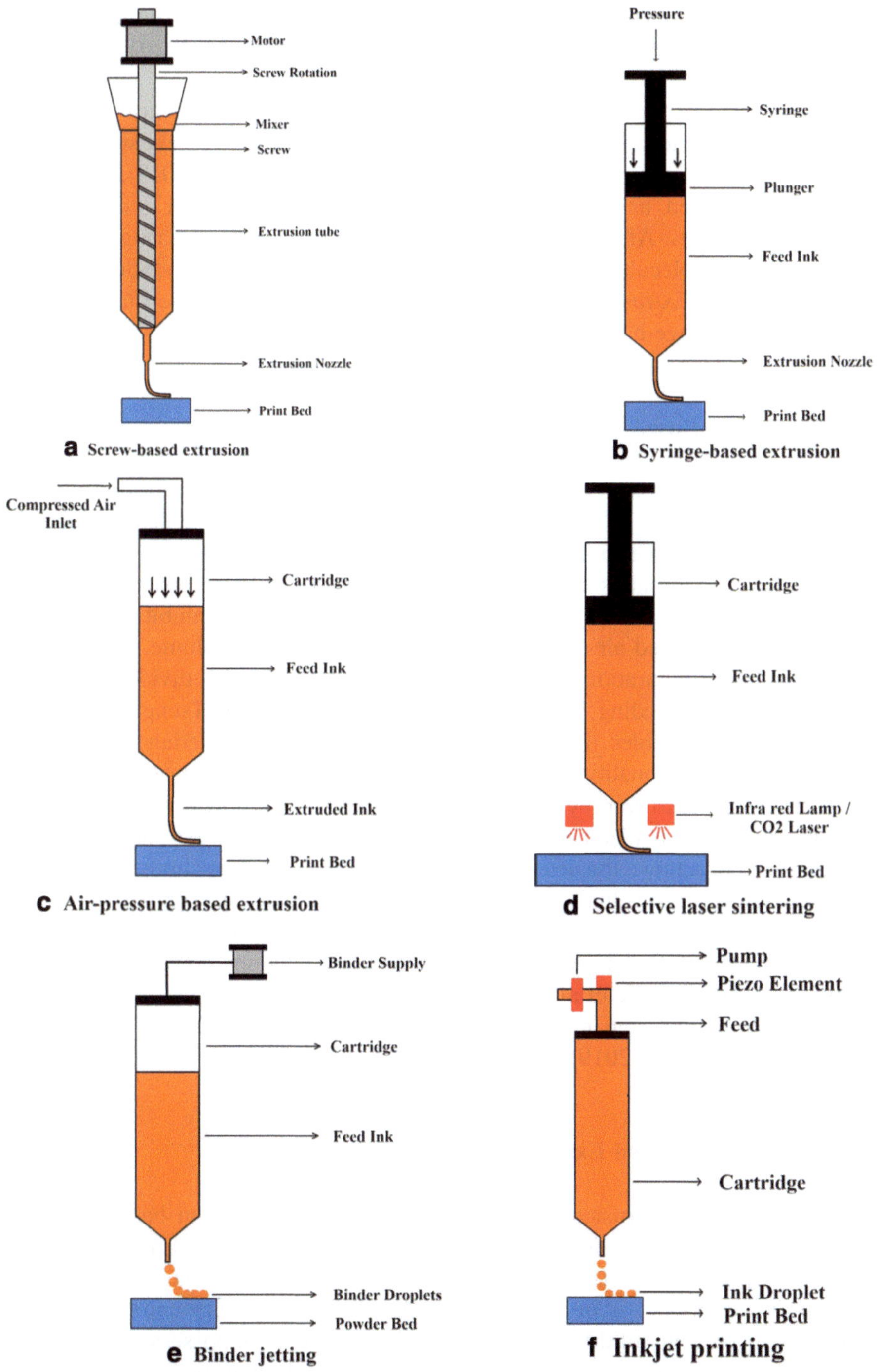

Fig. 3.3 Schematic representation of 3D printing: (**a**) Screw-based extrusion; (**b**) Syringe-based extrusion; (**c**) Air-pressure-based extrusion; (**d**) Selective laser sintering; (**e**) Binder jetting; and (**f**) Inkjet printing

Extrusion technique is traditionally designed to extrude food materials that are tender and slimy, either in the pureed or mashed form (Guo et al. 2019; Sun et al. 2015a, b, c).

The precision of the final product printed via extrusion depends on various processing parameters, including tip-to-collector distance, movement speed, needle diameter, the temperature of the material, and bed (Lanaro et al. 2019). Chocolates are usually printed using extrusion-based methods due to their unique shape-retaining properties. After extrusion, chocolates undergo a phase change where the fat triglyceride molecules melt and re-crystallize after a particular period (Chen and Mackley 2006). Extrusion-based 3D printing techniques may be classified into screw-based, syringe-based, and air-pressure-based (Fig. 3.3).

3.3.1.1 Screw-Based Extrusion

Food paste is fed into the cartridge and is transported to the nozzle for printing. The motor drives a screw to bring down the food paste that is passed through the extrusion nozzle. Since food materials are in direct contact, the screw and cartridge should be of food-grade stainless steel (Sun et al. 2018). The primary focus of the screw-based extrusion technique is to provide high pressure to inhibit the backflow, to purge the entrapped air and to control the extrusion flow volume. Based on the function, the configuration of screw-based extruders may be divided into three sections: material feeding, compression, and metering zone (Tseng et al. 2018). Screw extrusion enables the printing of a wide range of materials, significantly improving the functionality and process ability of high viscous materials.

Screw based extrusion system is a simple, continuous, low-cost process along with good printability (Tian et al. 2021). However, specific process parameters must be optimized to achieve the desired result. For example, there is almost no extrusion when the screw speed is high and continuous extrusion when the screw speed is low. At low printing temperatures, the extruded product tends to become unstable while the higher temperature is beneficial for interlayer bonding however may cause thermal degradation (He et al. 2021). The increase in feed rate reduces the quality of dispersed product, apparent yield stress, mean residence time, and narrowing of distribution (Vergenes 2019).

3.3.1.2 Syringe-Based Extrusion

The syringe-based extrusion employs a simplified pump system based on an air pressure-controlled extrusion of food material through a nozzle of a syringe to 3D print food (Hao et al. 2019). Traditionally in syringe-based extrusion, the food material to be printed is in molten form. A heated jacket surrounds the printing materials to prevent solidifying in the syringe. 3D printers possess a dual needle set-up to improve mixing two feed materials in a spiral mixer before printing (Tan et al. 2018). The application of paste-like materials in food industries presupposes sterile instruments. Standard single-use syringe tubes are employed to clear the demand (Amza et al. 2017). However, the mass of the syringe may be an obstacle during the manufacture and may require compromise in the speed, resolution, and volume. The syringe-based extruders are also susceptible to leakage of food paste (Pusch et al. 2018). The standardization of the manufacturing protocol helped to prevail over the hurdles.

3.3.1.3 Air-Pressure Based Extrusion

Air-pressure based extrusion mechanism employs the pneumatic pump and encapsulated food cartridge. The role of the pneumatic pump is to generate air pressure to extrude the food material via the nozzle. The extruded product obtained via an air-pressure-driven system does not directly contact the mechanical components, which lessens the probability of contamination (Singhal et al. 2020). Air-pressure-driven ejection of feed is confined to batch printing and requires an additional power supply for uninterrupted printing of material. The air-pressure-based system possesses the threat of presence of air bubbles during the deposition of the food material; however, by employing standardized procedures, various feed materials are being printed successfully (Kewuyemi et al. 2021).

3.3.2 Selective Laser Sintering

Selective laser sintering technique performs additive manufacturing, which involves the melting of one or more food materials in the form of powder by using a laser, usually a carbon dioxide laser that aids in binding the powder together due to the fusion mechanism. Once a layer is formed, another layer of powder is spread over it, followed by a laser scan to fuse them. The steps recur until desired shape and size have been achieved.

Selective laser sintering is a quick and low-cost technique as it does not require addition of solvent during processing or post-processing operations. The printed objects possess high resolution, and the sintering printer allows complex designs. Materials produced via selective laser sintering require a settling time of 48 h for the structure to become stable. The sintering technique is one of the first 3D printers designed to be accessible by the consumers; therefore it is equipped with simple parts and is economical (Barakh Ali et al. 2019; Fina et al. 2017; Lanaro et al. 2019).

3.3.3 Binder Jetting

Binder jetting technique involves the deposition of feed powder layer-by-layer and combines selective layers with the aid of a binder (Mostafaei et al. 2017). Before that, the characteristics of the powder should be determined to indicate the particle shape and particle size distribution. These parameters play vital role in the flow ability of the powder. Pre-processing operations, such as, milling and sieving may be included in the analysis. After deposition on the print bed, the layer of powder is exposed to a certain quantity of heat using a heat pump. Binder jetting does not solely depend on the heat for fusion.

For instance, pressure and wet binder may be employed to manufacture 3D printed food. To ensure smooth and even deposition and or drag of new layers, the properties of the binder should be optimized. Binder jetting is designed to operate at high speeds without any error (Holland et al. 2019; Pitayachaval et al. 2018). However, the printed parts are fragile, and to obtain appropriate material densities, tensile strength, and other mechanical parameters, post-processing of printed products becomes mandatory. Post-processing operations involve removing the support chemical and heat treatment (Debroy et al. 2018).

3.3.4 Inkjet Printing

Inkjet printing technology is categorized into drop-on-demand and continuous-jet methods. The former method involves electrical signals that control the extrusion process; while the later ejects the feed under high pressure (Sridhar et al. 2011). The mechanism of inkjet printing technology is to heat the printer head thermally or electrically to generate pressure pulses. Another method involves the creation of a pulse by employing a piezoelectric crystal inside the printer head. Due to the pressure, droplets of feed material are accumulated via a nozzle (Manstan and McSweeney 2019; Murphy and Atala 2014).

Inkjet printing is most suitable for liquid-based food material compared to powdered feed. Inkjet printing is designed to print low viscosity (<0.03 Pa s) food inks and is not employed to print complex structures. Therefore, inkjet printing is considered as 2D printing rather than a 3D printing technique. This limitation is prevailed by printing food materials at elevated temperatures. Cross-linking agents are employed as support systems as the low-viscous inks are not stable enough to form rigid structures. The precision attained by using this method of 3D printing is extremely high compared to other jetting methods (Godoi et al. 2016; Vadodaria and Mills 2020).

3.4 Process Parameters Affecting the Efficiency of 3D Printing Materials

The success rate of 3D printing depends on various process parameters. Out of which, the temperature at which the printing occurs, the rheological and physico-chemical properties of food ingredients, the speed, diameter, and height of the nozzle greatly influence the stability and quality of 3D printed food. Therefore, these parameters are optimized to potentially 3D print food materials.

3.4.1 Temperature

The temperature at which the printing takes place influences the viscosity and flexibility of the fluid material in the nozzle, thus affecting the texture of the end-product post-deposition. The rate of extrusion is affected due to a drastic drop in the viscosity of the food material. There is also a probability for microbial growth due to fluctuations in temperature. Most 3D printing occurs at room temperature to avoid these hindrances (Attalla et al. 2016). Temperature is one of the major factors, which plays a pivotal role in maintaining the expected shape and avoiding de-formation of the food ink post-printing. The time required for the printed material to solidify is based on the ink composition. It is possible when it is left to harden at room temperature without considering humidity and other environmental factors (Zhu et al. 2019).

3.4.2 Nozzle Speed, Diameter and Height

The particle size must be adjusted uniformly according to the diameter of the nozzle to lessen the probability of particles getting clogged in-between. The diameter of the nozzle decides the volume of raw material (follows continuity equation) that has to be considered for printing. Researchers suggest adopting nozzles with small diameters if the food material is less sticky (Wang et al. 2018).

Nozzle height is the distance between the nozzle tip and the printing platform, which determines the accuracy of the final object. The critical nozzle height is calculated based on the volume of food material, diameter, and nozzle speed. The time taken for printing highly depends on the size of the nozzle. In contrast, when a nozzle with a large diameter is used, the precision and resolution are widely affected. Applying a nozzle height higher or lower than the critical nozzle height will lead to inaccurate accuracy. The speed of the nozzle determines the printing speed and extrusion rate per unit length and per unit time. Therefore, the speed of the nozzle has to be optimized to achieve the desired shape and size. Too high nozzle speed results in a dragging effect thus causing breakage of the extruded food filaments and it does not provide sufficient time for the printed food material to deposit; while too low nozzle speed causes flow instabilities of printed food, accumulation of food materials, and the de-formation of the desired shape (Liu and Ciftci 2021; Yang et al. 2018).

3.4.3 *Rheological Properties of Food Ink*

The continuous printing ability of food is achieved by adjusting the rheological properties of feed materials to avoid clogging of food material and obtaining highly-stable structures (Zhu et al. 2019). Based on the measurements, the rheological properties of starter ingredients are improved either by processing before printing or by incorporating a potential additive. The additives used for effective 3D printing vary depending on number of macronutrients present in the food paste. As an alternative to synthetic additives, any raw material that acts synergistically with one another to provide desired rheological properties may be employed (Kaur 2004). The physicochemical properties of all raw materials in 3D printing are evaluated and judged based on their chemical composition, such as: the type of protein, lipid, and carbohydrates present (Dankar et al. 2018a, b).

3.5 Applications of 3D Printing Technology for Foods

The additive manufacturing of food has become a solution to people's health-related problems for various age groups. This technology has application in multiple fields, such as: military food, space food, vegan food, and exceptional food for different age groups.

3.5.1 *Printed Meals for Elderly*

Dysphagia, a swallowing disorder, is common among older people, people with disabilities, patients, who have undergone surgeries and have suffered a stroke. A survey report predicts that around 13% of people between 65 to 70 years, 16% of people with 70–79 years of age, and 33% of people with age ≥80 years. Edentulism, in which aging individuals lose their teeth, causes inappropriate chewing. Around 20–25% of people belonging to the age group of 60 years and above in the U.S. and Canada are diagnosed with edentulism (Aguilera and Park 2016). Health professionals recommend consuming "smooth food" for people, who face difficulty in swallowing to reduce the probability of choking. The risk of aspiration and the reliance on a feeding tube is drastically reduced by optimizing the textural properties of foods.

The soft food is prepared by modifying the texture of the food components by mashing and pureeing operations. Patients refused the thickened "syrup-like" or "yogurt-like" liquid materials with ideal viscosity to enhance smooth swallowing that researchers proposed. The unpalatable appearance and texture may reduce the food intake and can elevate the risk of malnutrition. The production of texture-modified food with satisfactory sensorial properties has become possible by adopting 3D printing technology. The additive manufacturing technology allows tailoring and designing a personalized diet for elder people based on their health conditions. 3D printing also promotes individuals to have better control over food intake (Hemsley et al. 2019; Le-Bail et al. 2020; Pereira et al. 2021; Zargaraan et al. 2013).

3.5.2 *Non-conventional Fish-Meat*

Unlike the traditional methods of culturing meat, 3D printing enables the construction of meat structures from the meat ink consisting of muscle and fat cells. Post-printing, the meat structures can mature in bioreactors for nutrient transportation. The additive manufacturing of cultured meat proficiently balances the nutritional quality and sensory properties without slaughtering animals (Handral et al. 2020). Currently, the scope of clean meat products is limited to hamburgers, sausages, and steaks. This traditional manufacturing and consumption may be notably altered by introducing 3D printing. Modern Meadow focuses on manufacturing cultured meat with additive manufacturing (van der Weele and Driessen 2013). For replicating the conventional fish meat, cultured meat demands spatial resolution and redefinition due to vasculature and intramuscular fat. It is prevailed over by introducing scaffolds in 3D printing technology (Balasubramanian et al. 2021).

The 3D printing of meatless meat necessitates the incorporation of food additives, which is opposed by several health-conscious consumers. Therefore, the future demands the addition of protein-rich materials (such as: plants, algae, mushrooms, and insects) as alternatives to synthetic additives (Hinton et al. 2017). The protein composition of plant-based materials is usually rounded to form ideal fish-meat structures. The plant material must be denatured, aligned, and cross-linked. The 3D printing technology enables the formation of fibrous structures (Kazir and Livney 2021). Apart from cultured meat, 3D printing of plant-based meat is also gravitating. The plant-based materials include tofu, beets, carrots, bell pepper, wheat gluten, nuts, legumes, pulses, cereals, soy, and pea protein. "Impossible Foods," an American pizza company, successfully launched pizza topped with meatless sausages; and a California-based company profitably manufactured and marketed "Impossible Burgers" made out of soy protein and yeast (Boukid and Castellari 2021).

3.5.3 Fortified 3D Printed Food

Fortification is the addition of bioactive compounds to food to prevent the shortfall of nutrient intake and their associated diseases. Although fortification was introduced to reduce the risk of deficiency diseases, yet fortification primarily focuses on enhancing dietary intake. Conversely, the effect of fortification is based on the individual's body conditions and metabolic processes (Dwyer et al. 2015). Recently, the demand for fortified food products integrated with appealing sensorial properties has significantly increased among health-conscious consumers (Hussain et al. 2021). Children have special dietary requirements necessary for proper growth, which is achieved by providing a balanced diet. However, the children refuse to consume healthy but unpalatable foods (Hamilton et al. 2017).

The integration of fortification and 3D printing technology personalizes the nutrient and sensorial parameters according to individual's requirements and manufactures potentially attractive food products. Examples of research work that has integrated additive manufacturing and fortification process are discussed here. Vitamin D is essential for calcium absorption and various body functions. The occurrence of vitamin D in natural food is few, and their acceptability is also low. Azam et al. formulated vitamin D enriched orange concentrates via 3D printed technology. The 3D printed matrix entraps and prevents degradation of vitamin D, thereby reducing the Vitamin D deficiency. Since chocolates are one of the favorite foods for children, therefore fortification of bioactive compounds in 3D printed chocolates can easily attract children to consume healthier snacks (Azam et al. 2018).

Hao et al. (2019) have formulated chocolates incorporated with vitamin C, lutein, and cranberry extract and have checked their printing feasibility.

3.5.4 Space Foods

The food for the astronauts should be nutritious, safe, and palatable, since overall success of the mission entirely depends on the crew's health. The currently adopted food system does not satisfy the nutritional requirements of space travelers during more extended-time. Apart from nutrition, less trash must be generated as there are fewer chances to discard the garbage off the spacecraft. Since refrigeration and freezing systems are not a resource of the space shuttle, the need for a 5-year shelf-stable food has become mandatory. The meal plan for each astronaut varies based on gender, height, weight, and personal preferences. The body utilizes the nutrients differently when in space, which must be considered while planning a diet for space travelers. Therefore, NASA is working on 3D printing food to meet the standards of food in space (Douglas et al. 2016; NASA 2004, 2013).

According to NASA, 3D-printed food may be employed as a part of astronauts' diet on more extended space missions, such as, shuttle to Mars (Lin 2015). The standard nutritional requirements of astronauts, particularly vitamin supplement powders, may be incorporated into the 3D printed foods without altering the sensory parameters. The additive manufacturing technology precisely balances both personalized nutrition for astronauts (Terfansky and Thangavelu 2013).

The prime focus of a NASA-funded project (ICA project) was to analyze the printability of ready-to-eat freeze-dried materials and provide minimum on-board post-processing operations and maximum shelf-life. The researchers successfully printed 3D-foods (such as: "Hummus crunch," "Asian Salad Bar," "Cheese Sandwich," and "Space Jello") to provide accurate nutrition and personalized palatability (NASA 2019). "Systems and Materials Research Consultancy" was awarded $125,000 to additive manufacture highly shelf-life stable foods. The company's 3D printed "Space Pizza" made of cellulose-based food additives and other ingredients was a huge success (Aspler et al. 2016). Table 3.1 summarizes various novel foods formulated via 3D printing technology, the components employed, and the scope of the formulation.

3.6 Consumer Acceptance of 3D Printed Materials

The food industries are actively indulged in creating novel techniques and food products that may impact the health of customers and enhance profit. Food acceptability is intruded by several aspects, which may be associated with the individual, the food, or the environment in which the food is consumed. The lack of knowledge among consumers on the innovative technology may hinder the acceptability of 3D printed food products. Moreover, consumers are not a heterogeneous group and may have varied opinions in accepting the technique of additive manufactured food (Mantihal et al. 2020; Murray and Baxter 2003).

The young aged consumers showed positive attitude compared to the old aged consumers. However, there was no difference in opinion about 3D printed food between male and female consumers (Vidigal et al. 2015). The key reasons for consumers to favor 3D-printed foods are tailored nutrition composition, customization, and the possibility to help in lowering food waste (Nachal et al. 2019). In addition, 3D printing allows consumers to participate in the food production process by choosing specific product characteristics. It may have a positive impact on consumer satisfaction (Caulier et al. 2020).

Ingredients such as algae and insects may be employed as printing materials to manufacture food with relatively high appetizing shapes, potentially improving their acceptability (Burke-Shyne et al. 2020). The majority of 3D printed food requires additives to maintain the stability of the product, which is a shortcoming that must be prevailed in the future to attract health conscious-consumers (Portanguen et al. 2019). If the consumers are unfamiliar with any novel product or technology, they usually tend to dislike it; therefore, intense awareness must be created to elevate the market demand (Manstan and McSweeney 2019). The advantages of the 3D printing technique are illustrated in (Fig. 3.4).

Table 3.1 Summary of novel food materials printed by adopting additive manufacturing technology

3D printed food	Major materials employed	Technique used	Scope	References
3D printed brown rice gels	Brown rice and hydrocolloids	Extrusion based 3D printer	To examine the printing performance by employing various food hydrocolloids	Huang et al. (2020)
3D-printed cordyceps flowers	Cordyceps flowers, vegetable oil, Isomaltose, and food additives		To develop a food product with multifunctional health effects	Teng et al. (2019)
3D-printed potato	Potato, Xanthan gum, and k-carrageenan gum		To analyze the textural properties and microstructure of the printed product	Liu et al. (2018)
3D-printed tablets	Acetaminophen and food polymers		To fabricate tablets with enhanced-extended release properties To achieve improved drug delivery systems	Zhang et al. (2017)
Cereal-based food	Wheat flour, calcium caseinate, and *Lactobacillus Plantarum*		The development of innovative food using functional ingredients To investigate the influence of variation in dough formulation on the printability	Zhang et al. (2018)
Chewable formulation	Isoleucine and food additives		To improve the acceptability of isoleucine supplementation in pediatric patients suffering from Maple syrup urine disease To evaluate isoleucine blood levels after six months of treatment with two types of formulations	Goyanes et al. (2019)
Chicken nuggets	Chicken and refined wheat flour	Controlled additive-manufacturing Robotic Kit (CARK)	To develop a customized fast-food The study investigates the printability of ground chicken	Wilson et al. (2020)

(continued)

Table 3.1 (continued)

3D printed food	Major materials employed	Technique used	Scope	References
			meat with the aid of food additives	
Chocolate	Bitter chocolate, corn syrup, Paracetamol, and ibuprofen	Extrusion based 3D printer	The development of pediatric-friendly oral dosages To examine the *in vitro* digestion studies as well as perform texture profile analysis	Karavasili et al. (2020)
Cookies	Wheat flour, rice flour, tapioca flour, butter, sugar, and milk	3D printer modified with a digital air syringe dispenser	To investigate the factors affecting the printability and post-processing capacity of cookie dough	Pulatsu et al. (2020)
Egg-based formulation	Eggs, rice flour, and maltodextrin	CARK—controlled additive-manufacturing robotic kit	To optimize various extrusion parameters, including printing composition, nozzle height, nozzle diameter, printing speed, extrusion motor speed, and extrusion rate To study the different physical, mechanical and rheological properties of the print ink	Anukiruthika et al. (2020)
Fish surmi	Silver carp fillet and NaCl	Extrusion based 3D printer	The development of 3D printed fish surmi gel To investigate the influence of NaCl on the rheological property, gel strength, water holding capacity of the printed system	Murphy and Atala (2014)
Fruit-based snack	Bananas, mushrooms, canned white beans, and non-fat milk	3D Printer model Delta 2040 embedded with clay kit extruder	Designing ideal snacks for children between the age group of 3–10 years Examining the printing parameters, such as: flow level and print speed	Derossi et al. (2018)

(continued)

Table 3.1 (continued)

3D printed food	Major materials employed	Technique used	Scope	References
Lemon juice gel	Lemon juice and potato starch	Extrusion based 3D printer	To examine the influence of printing parameters on the quality of 3D printed food To investigate the effect of potato starch on rheological properties and mechanical properties of lemon juice gels	Nachal et al. (2019)
Plant-based formulation	Lettuce, pectin and food additives	Extrusion based 3D printer	A novel approach to encapsulate plant cells using the 3D printing technique	Vancauwenberghe et al. (2019)
Plant-based hydrogel	Carrot, sodium alginate, Calcium chloride dihydrate, saccharose, and agar powder	Syringe-extrusion plunger-type 3D printer equipped with lattice scaffolds	To study the feasibility of employing plant cells in 3D printing as the future of real-textured foods	Park et al. (2019)
Printed pavlova	Egg white, sugar, cream, cornflour, and dark chocolate	3D plotter	To optimize the printing of pavlova and chocolate garnish	Kouzani et al. (2016)
Rice-based 3D printed food	Rice flour, jaggery, and water	Extrusion based 3D printer	To determine the best combination for shape retention after printing To analyze the rheological and textural parameters of the optimized composition	Thangalakshmi et al. (2021)
Value-added functional cookies	Broken wheat, grape pomace powder, powdered sugar, and margarine	The extrusion-based 3D food printer—CARK	Utilization of food industry waste to manufacture food Evaluation and optimization of processing and post-processing parameters	Jagadiswaran et al. (2021)

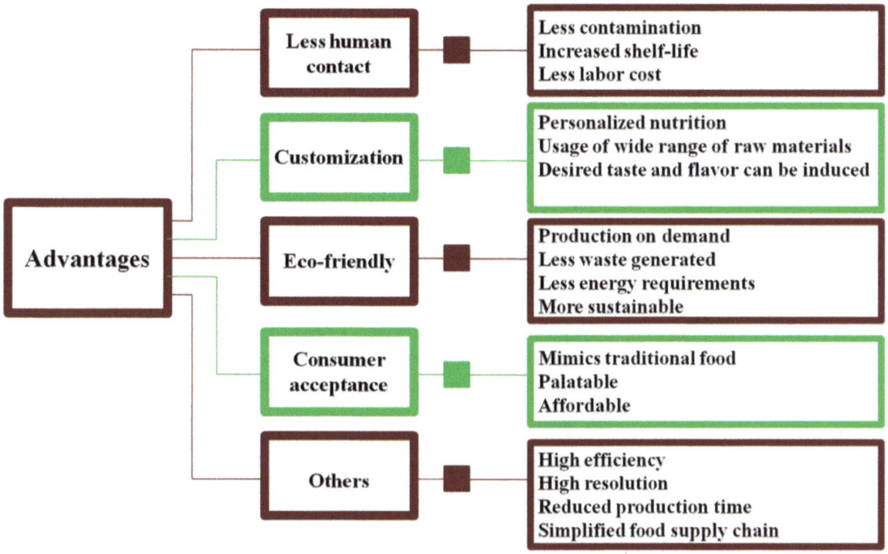

Fig. 3.4 Advantages of 3D printing technology

3.7 Conclusion

3D printing of food materials is a promising alternative for producing customized food at a lower cost, soon expected to replace the conventional cooking system. The chances for food contamination are minimum as there is minimal involvement of human contact. 3D printed food is predicted to possess longer shelf - life. However, several barriers are still unaddressed to make this a flawless technology. These challenges must have prevailed on the road to accomplish the desired results. The process parameters, nature of raw materials, and post-processing treatments must be studied carefully to improve accuracy and high precision. Artificial food additives must be reduced and replaced drastically to attract health-conscious consumers. In recent years, many research studies are being carried out to overcome the challenges faced during mass production, which may pave the way for the broader application of 3D printing of food materials.

References

Aguilera JM, Park DJ (2016) Texture-modified foods for the elderly: status, technology and opportunities. Trends Food Sci Technol 57(1):156–164

Amza C, Zapciu A, Popescu D (2017) Paste extruder: hardware add-on for desktop 3D printers. Technologies 5(3):50

Anukiruthika T, Moses JA, Anandharamakrishnan C (2020) 3D printing of egg yolk and white with rice flour blends. J Food Eng 265(11):109691

Aspler J, Kingsland A, Cormier LM, Zou X (2016) 3D printing: a review of technologies, markets, and opportunities for the forest industry. J Sci Technol Forest Prod Process 5(2):30–37

Attalla R, Ling C, Selvaganapathy P (2016) Fabrication and characterization of gels with integrated channels using 3D printing with microfluidic nozzle for tissue engineering applications. Biomed Microdevices 18(1):17

Azam RSM, Zhang M, Bhandari B, Yang C (2018) Effect of different gums on features of 3D printed object based on vitamin-D enriched orange concentrate. Food Biophys 13(3):250–262

Balasubramanian B, Liu W, Pushparaj K, Park S (2021) The epic of in vitro meat production – a fiction into reality. Foods 10(6):1395–1401

Barakh Ali SF, Mohamed EM, Ozkan T, Kuttolamadom MA, Khan MA, Asadi A, Rahman Z (2019) Understanding the effects of formulation and process variables on the print-lets quality manufactured by selective laser sintering 3D printing. Int J Pharm 570(42):118651

Beckett ST (2008) Chocolate flow properties. In: Beckett ST (ed) Industrial chocolate manufacture and use, 4th edn. Blackwell Publishing Ltd, London, pp 224–246

Boukid F, Castellari M (2021) Veggie burgers in the EU market: a nutritional challenge? Eur Food Res Technol 247(10):2445–2453

Burke-Shyne S, Gallegos D, Williams T (2020) 3D food printing: nutrition opportunities and challenges. Br Food J 123(2):649–663

Caulier S, Doets E, Noort M (2020) An exploratory consumer study of 3D printed food perception in a real-life military setting. Food Qual Prefer 86:104001

Chen YW, Mackley MR (2006) Flexible chocolate. Soft Matter 2(4):304–309

Cohen DL, Lipton JI, Cutler M, Coulter D, Vesco A, Lipson H (2009) Hydrocolloid printing: a novel platform for customized food production. International Solid Freeform Fabrication Symposium, Austin, pp 807–818

Dankar I, Haddarah A, El Omar FEL, Sepulcre F, Pujolà M (2018a) Assessing the microstructural and rheological changes induced by food additives on potato puree. Food Chem 240:304–313

Dankar I, Haddarah A, Omar FEL, Sepulcre F, Pujolà M (2018b) 3D printing technology: the new era for food customization and elaboration. Trends Food Sci Technol 75(5):231–242

Debroy T, Wei HL, Zuback JS, Mukherjee T, Elmer JW, Milewski JO, Beese AM, Wilson-Heid AE, De A et al (2018) Additive manufacturing of metallic components – process, structure and properties. Prog Mater Sci 92:112–224

Derossi A, Caporizzi R, Azzollini D, Severini C (2018) Application of 3D printing for customized food. A case on the development of a fruit-based snack for children. J Food Eng 220:65–75

Derossi A, Paolillo M, Caporizzi R, Severini C (2019) Extending the 3D food printing tests at high speed. Material deposition and effect of non-printing movements on the final quality of printed structures. J Food Eng 275:109865

Diamante L, Umemoto M (2015) Rheological properties of fruits and vegetables: a review. Int J Food Prop 18(6):1191–1210

Dick A, Bhandari B, Prakash S (2019a) 3D printing of meat. Meat Sci 153:35–44

Dick A, Bhandari B, Prakash S (2019b) Post-processing feasibility of composite-layer 3D printed beef. Meat Sci 153:9–18

Douglas G, Cooper M, Bermúdez-Aguirre D, Sirmons T (2016) Risk of performance decrement and crew illness due to an inadequate food system. National Aeronautics and Space Administration (NASA) Lyndon B Johnson Space Center, Houston, pp 1–33

Dwyer JT, Wiemer KL, Dary O, Keen CL, King JC, Miller KB, Philbert MA, Tarasuk V, Taylor CL et al (2015) Fortification and health: challenges and opportunities. Adv Nutr 6(1):124–131

Feinberg AW, Miller JS (2017) Progress in three-dimensional bioprinting. MRS Bull 42(8): 557–562

Fina F, Goyanes A, Gaisford S, Basit AW (2017) Selective laser sintering (SLS) 3D printing of medicines. Int J Pharm 529(1-2):285–293

Gholamipour-Shirazi A, Kamlow MA, Norton T, Mills T (2020) How to formulate for structure and texture via medium of additive manufacturing – a review. Foods 9(4):497–502

Gibson I, Rosen DW, Stucker B (eds) (2009) Additive manufacturing technologies: rapid prototyping to direct digital manufacturing. Springer, New York, p 498

Godoi FC, Prakash S, Bhandari BR (2016) 3D printing technologies applied for food design: status and prospects. J Food Eng 179:44–54

Goyanes A, Madla CM, Umerji A, Piñeiro GD, Montero JMG, Diaz MJL, Basit AW, Taherali F, Sánchez-Pintos P et al (2019) Automated therapy preparation of isoleucine formulations using 3D printing for the treatment of MSUD: first single-center, prospective, crossover study in patients. Int J Pharm 567:118497

Guo CF, Zhang M, Bhandari B (2019) A comparative study between syringe-based and screw-based 3D food printers by computational simulation. Comput Electron Agric 162:397–404

Hamilton CA, Alici G, Panhuis M (2017) 3D printing vegemite and marmite: redefining bread-boards. J Food Eng 220:83–88

Handral KH, Hua TWS, Wan CW, Choudhury D (2020) 3D printing of cultured meat products. Crit Rev Food Sci Nutr 62(1):272–281

Hao L, Li Y, Gong P, Xiong W (2019) Material, process and business development for 3D chocolate printing. In: Godoi FC, Bhandari BR, Prakash S, Zhang M (eds) Fundamentals of 3D food printing and applications, 1st edn. Academic Press, London, pp 207–255

Hayman LL, Worel JN (2014) Healthy lifestyle behaviors: the importance of individual and population approaches. J Cardiovasc Nurs 29(6):477–478

He Y, Qiu J, Fu J, Zhang J, Ren Y, Liu A (2015) Printing 3D microfluidic chips with a 3D sugar printer. Microfluid Nanofluid 19(2):447–456

He Q, Jiang J, Yang X, Zhang L, Zhou Z, Zhong Y, Shen Z (2021) Additive manufacturing of dense zirconia ceramics by fused deposition modeling via screw extrusion. J Eur Ceram Soc 41(1): 1033–1040

Hemsley B, Palmer SR, Kouzani A, Adams S, Balandin S (2019) Review informing the design of 3D food printing for people with swallowing disorders: constructive, conceptual, and empirical problems. In: Paper presentation at 52nd Hawaii international conference on system sciences. University of Hawaii, Hawaii

Hinton TJ, Lee A, Feinberg AW (2017) 3D bioprinting from the micrometer to millimeter length scales: size does matter. Curr Opin Biomed Eng 1:31–37

Holland S, Foster T, Tuck C (2019) Creation of food structures through binder jetting. In: Godoi FC, Bhandari BR, Prakash S, Zhang M (eds) Fundamentals of 3D food printing and applications. Academic Press, London, pp 257–288

Huang M, Zhang M, Bhandari B (2019) Assessing the 3D printing precision and texture properties of brown rice induced by infill levels and printing variables. Food Bioprocess Technol 12(7): 1185–1196

Huang M, Zhang M, Guo C (2020) 3D printability of brown rice gel modified by some food hydrocolloids. J Food Process Preserv 44(7):14502

Hussain S, Arora VK, Malakar S (2021) Formulation of protein-enriched 3D printable food matrix and evaluation of textural, rheological characteristics, and printing stability. J Food Process Preserv 45(2):1–11

Izdebska J, Żołek-Tryznowska Z (2016) 3D food printing – facts and future. Agro Food Ind Hi Tech 27(2):33–36

Jagadiswaran B, Alagarasan V, Palanivelu P, Theagarajan R, Moses JA, Anandharamakrishnan C (2021) Valorization of food industry waste and by-products using 3D printing: a study on the development of value-added functional cookies. Future Foods 4:100036

Karavasili C, Gkaragkounis A, Moschakis T, Ritzoulis C, Fatouros DG (2020) Pediatric-friendly chocolate-based dosage forms for the oral administration of both hydrophilic and lipophilic drugs fabricated with extrusion-based 3D printing. Eur J Pharm Sci 147:105291

Kaur L (2004) Factors influencing the properties of hydroxy propylated potato starches. Carbohydr Polym 55(2):211–223

Kazir M, Livney YD (2021) Plant-based seafood analogs. Molecules 26(6):1559–1565

Kewuyemi YO, Kesa H, Adebo O (2021) Trends in functional food development with three-dimensional food printing technology: prospects for value-added traditionally processed food products. Crit Rev Food Sci Nutr 61:1–13

Kouzani AZ, Adams S, Oliver R, Nguwi YY, Hemsley B, Balandin S (2016) 3D printing of a pavlova. In: 10th IEEE reg conference (TENCON). Singapore, IEEE, pp 2281–2285

Lanaro M, Forrestal DP, Scheurer S, Slinger DJ, Liao S, Powell SK, Woodruff MA (2017) 3D printing complex chocolate objects: platform design, optimization and evaluation. J Food Eng 215:13–22

Lanaro M, Desselle MR, Woodruff MA (2019) 3D printing chocolate. In: Godoi FC, Bhandari BR, Prakash S, Zhang M (eds) Fundamentals of 3D food printing and applications. Academic Press, London, pp 151–173

Le-Bail A, Chieregato MB, Le-Bail P (2020) 3D printing of foods: recent developments, future perspectives and challenges. Curr Opin Food Sci 35:54–64

Lille M, Nurmela A, Nordlund E, Metsä-Kortelainen S, Sozer N (2018) Applicability of protein and fiber-rich food materials in extrusion-based 3D printing. J Food Eng 220:20–27

Lin C (2015) 3D food printing: a taste of the future. J Food Sci Educ 14(3):86–87

Lin YJ, Punpongsanon P, Wen X, Iwai D, Sato K, Obrist M, Mueller S (2020) FoodFab: creating food perception illusions using food 3D printing. In: Proceedings of the conference on human factors in computing systems. ACM (Association for Computing Machinery) Publisher, Honolulu, pp 1–13

Lipton JI (2017) Printable food: the technology and its application in human health. Curr Opin Biotechnol 44:198–201

Lipton JI, Arnold D, Nigl F, Lopez N, Cohen D, Norén N, Lipson H (2010) Multi-material food printing with complex internal structure suitable for conventional post-processing. In: International solid freeform fabrication symposium, Austin, pp SFF, 809–815

Liu L, Ciftci ON (2021) Effects of high oil compositions and printing parameters on food paste properties and printability in 3D printing food processing model. J Food Eng 288:110135

Liu Z, Zhang M, Bhandari B, Wang Y (2017) 3D printing: printing precision and application in food sector. Trends Food Sci Technol 69(A):83–94

Liu Z, Bhandari B, Prakash S, Zhang M (2018) Creation of internal structure of mashed potato construct by 3D printing and its textural properties. Food Res Int 111:534–543

Liu Y, Tang T, Duan S, Qin Z, Li C, Zhang Z, Liu A, Wu D, Chen H et al (2020a) Effects of sodium alginate and rice variety on the physicochemical characteristics and 3D printing feasibility of rice paste. LWT Food Sci Technol 127:109360

Liu Y, Tang T, Duan S, Qin Z, Zhao H, Wang M, Li C, Zhang Z, Liu A et al (2020b) Applicability of rice doughs as promising food materials in extrusion-based 3D printing. Food Bioprocess Technol 13(3):548–563

Manstan T, McSweeney MB (2019) Consumers' attitudes towards and acceptance of 3D printed foods in comparison to conventional food products. Int J Food Sci Technol 55(1):1–9

Mantihal S, Prakash S, Godoi FC, Bhandari B (2017) Optimization of chocolate 3D printing by correlating thermal and flow properties with 3D structure modeling. Innov Food Sci Emerg Technol 44:21-29

Mantihal S, Kobun R, Lee BB (2020) 3D food printing of as the new way of preparing food: a review. Int J Gastron Food Sci 22:100260

Mehak-Jandyal OP, Malav N, Mehta MK, Pavan-Kumar RV, Wagh SK, Tanuj T (2021) 3D printing of meat: a new frontier of food from download to delicious: a review. Int J Curr Microbiol App Sci 10(3):2095–2111

Mostafaei A, Kimes KA, Stevens EL, Toman J, Krimer YL, Ullakko K, Chmielus M (2017) Microstructural evolution and magnetic properties of binder jet additive manufactured Ni-Mn-Ga magnetic shape memory alloy foam. Acta Mater 131:482–490

Murphy SV, Atala A (2014) 3D bioprinting of tissues and organs. Nat Biotechnol 32(8):773–785

Murray JM, Baxter IA (2003) Food acceptability and sensory evaluation. In: Food sciences and nutrition. Academic Press, Cambridge, pp 5130–5136

Nachal N, Moses JA, Karthik P, Anandharamakrishnan C (2019) Applications of 3D printing in food processing. Food Eng Rev 11(3):123–141
NASA (2004) No pizza in space? Available from https://www.nasa.gov/audience/forstudents/k-4/home/F_No_Pizza_in_Space.html
NASA (2013) 3D printing: food in space. Available from https://www.nasa.gov/directorates/spacetech/home/feature_3d_food.html
NASA (2019) 3D printed prototype to enhance the space food system. Available from https://ntrs.nasa.gov/citations/20190033356
Nida S, Anukiruthika T, Moses JA, Anandharamakrishnan C (2021) 3D printing of grinding and milling fractions of rice husk. Waste Biomass Valoriz 12(1):81–90
Park SM, Kim HW, Park HJ (2019) Callus-based 3D printing for food exemplified with carrot tissues and its potential for innovative food production. J Food Eng 271:109781
Payne CLR, Dobermann D, Forkes A, House J, Josephs J, McBride A, Müller A, Quilliam RS, Soares S (2016) Insects as food and feed: European perspectives on recent research and future priorities. J Insects Food Feed 2(4):269–276
Pereira T, Barroso S, Gil MM (2021) Food texture design by 3D printing: a review. Foods 10(2): 1–26
Pérez B, Nykvist H, Brøgger AF, Larsen MB, Falkeborg MF (2019) Impact of macronutrients printability and 3D-printer parameters on 3D-food printing: a review. Food Chem 287:249–257
Pitayachaval P, Sanklong N, Thongrak A (2018) A review of 3D food printing technology. MATEC Web Conf 213:01012
Portanguen S, Tournayre P, Sicard J, Astruc T, Mirade PS (2019) Toward the design of functional foods and biobased products by 3D printing: a review. Trends Food Sci Technol 86:188–198
Pulatsu E, Lin M (2021) A review on customizing edible food materials into 3D printable inks: approaches and strategies. Trends Food Sci Technol 107:68–77
Pulatsu ET, Su JW, Lin J, Lin M (2020) Factors affecting 3D printing and post-processing capacity of cookie dough. Innov Food Sci Emerg Technol 62:102306
Pusch K, Hinton TJ, Feinberg AW (2018) Large volume syringe pump extruder for desktop 3D printers. HardwareX 3(3):49–61
Rando P, Ramaioli M (2021) Food 3D printing: effect of heat transfer on print stability of chocolate. J Food Eng 294:110415
Ricci I, Derossi A, Severini C (2019) 3D printed food from fruits and vegetables. In: Godoi FC, Bhandari BR, Prakash S, Zhang M (eds) Fundamentals of 3D food printing and applications. Academic Press, London, pp 117–149
Salloum G, Tekli J (2021) Automated and personalized nutrition health assessment, recommendation, and progress evaluation using fuzzy reasoning. Int J Hum-Comput Stud 151:102610
Severini C, Derossi A, Ricci I, Caporizzi R, Fiore A (2018) Printing a blend of fruit and vegetables. New advances on critical variables and shelf-life of 3D edible objects. J Food Eng 220:89–100
Singhal S, Rasane P, Kaur S, Garba U, Bankar A, Singh J, Gupta N (2020) 3D food printing: paving way towards novel foods. Ann Braz Sci Acad 92(3):20180737
Sivaramakrishnan HP, Senge B, Chattopadhyay PK (2004) Rheological properties of rice dough for making rice bread. J Food Eng 62(1):37–45
Sridhar A, Blaudeck T, Baumann RR (2011) Inkjet printing as a key enabling technology for printed electronics. Materials 6(1):12–15
Sun J, Peng Z, Yan L, Fuh JY, Hong GS (2015a) 3D food printing an innovative way of mass customization in food fabrication. Int J Bioprint 1(1):27–38
Sun J, Peng Z, Zhou W, Fuh JYH, Hong GS, Chiu A (2015b) A review on 3D printing for customized food fabrication. Proc Manuf 1:308–319
Sun J, Zhou W, Huang D, Fuh JYH, Hong GS, Geok S (2015c) An overview of 3D printing technologies for food fabrication. Food Bioprocess Technol 8(8):1605–1615
Sun J, Zhou W, Yan L, Huang D, Lin L (2018) Extrusion-based food printing for digitalized food design and nutrition control. J Food Eng 220:1–11

Tan C, Toh WY, Wong G, Li L (2018) Extrusion-based 3D food printing – materials and machines. Int J Bioprint 4(2):143
Teng X, Zhang M, Bhandri B (2019) 3D printing of cordyceps flower powder. J Food Eng 107: 2411–2502
Terfansky ML, Thangavelu M (2013) 3D printing of food for space missions. In: Am Inst Aeronaut Astronaut Conference and Exposition, San Diego, California, pp 1–20
Thangalakshmi S, Arora VK, Kaur BP, Malakar S (2021) Investigation on rice flour and jaggery paste as food material for extrusion-based 3D printing. J Food Process Preserv 45(4):1–12
Tian J, Zhang R, Wu Y, Xue P (2021) Additive manufacturing of wood flour/polyhydroxy-alkenoates (PHA) fully bio-based composites based on micro-screw extrusion system. Mater Des 199:109418
Tseng JW, Liu CY, Yen YK, Belkner J, Bremicker T, Liu BH, Sun T, Wang A (2018) Screw extrusion-based additive manufacturing of PEEK. Mater Des 140:209–221
Turabi E, Sumnu G, Sahin S (2008) Rheological properties and quality of rice cakes formulated with different gums and an emulsifier blend. Food Hydrocoll 22(2):305–312
Vadodaria S, Mills T (2020) Jetting-based 3D printing of edible materials. Food Hydrocoll 106: 105857
van der Weele C, Driessen C (2013) Emerging profiles for cultured meat; ethics through and as design. Animals 3(3):647–662
Vancauwenberghe V, Baiye Mfortaw Mbong V, Vanstreels E, Verboven P, Lammertyn J, Nicolai B (2019) 3D printing of plant tissue for innovative food manufacturing: encapsulation of alive plant cells into pectin – based bioink. J Food Eng 263:454–464
Vergenes B (2019) The Effect of processing conditions on the dispersion and microstructure of organoclay/polypropylene nanocomposites in twin-screw extrusion. In: Kenig S (ed) Processing of polymer nanocomposite. Carl Hanser Verlag GmbH & Company, KG, Munich, pp 1–28
Vidigal MCTR, Minim VPR, Simiqueli AA, Souza PHP, Balbino DF, Minim LA (2015) Food technology neophobia and consumer attitudes towards foods produced by new and conventional technologies: a case study in Brazil. LWT Food Sci Technol 60(2):832–840
Wang L, Zhang M, Bhandari B, Yang C (2018) Investigation on fish surimi gel as promising food material for 3D printing. J Food Eng 220:101–108
Wegrzyn TF, Golding M, Archer RH (2012) Food layered manufacture: a new process for constructing solid foods. Trends Food Sci Technol 27(2):66–72
Wijnen B, Hunt EJ, Anzalone GC, Pearce JM (2014) Open-source syringe pump library. PLoS ONE 9(9):e107216
Wilson A, Anukiruthika T, Moses JA, Anandharamakrishnan C (2020) Customized shapes for chicken meat–based products: feasibility study on 3D-printed nuggets. Food Bioprocess Technol 13(11):1968–1983
Yang F, Zhang M, Bhandari B, Liu Y (2018) Investigation on lemon juice gel as food material for 3D printing and optimization of printing parameters. LWT Food Sci Technol 87:67–76
Zargaraan A, Rastmanesh R, Fadavi G, Zayeri F, Mohammadifar MA (2013) Rheological aspects of dysphagia-oriented food products: a mini review. Food Sci Human Wellness 2(3-4):173–178
Zhang J, Feng X, Patil H, Tiwari RV, Repka MA (2017) Coupling 3D printing with hot-melt extrusion to produce controlled-release tablets. Int J Pharm 519(1-2):186–197
Zhang L, Lou Y, Schutyser MAI (2018) 3D printing of cereal-based food structures containing probiotics. Food Struct 18:14–22
Zhu S, Stieger MA, van der Goot AJ, Schutyser MAI (2019) Extrusion-based 3D printing of food pastes: correlating rheological properties with printing behavior. Innov Food Sci Emerg Technol 58:102214

Chapter 4
Sustainable Renewable Energy Sources for Food and Dairy Processing

Poonam Rani, Arpan Dubey, Prakash Kumar, and Amit Kumar

Contents

P. Rani (✉)
Agricultural and Food Engineering Department, Indian Institute of Technology Kharagpur, Kharagpur, West Bengal, India

Food Chemistry and Technology, Teagasc Food Research Centre, Moorepark, Fermoy, Co. Cork, Ireland

A. Dubey · P. Kumar · A. Kumar
Agricultural and Food Engineering Department, Indian Institute of Technology Kharagpur, Kharagpur, West Bengal, India

A. Kumar
Food Chemistry and Technology, Teagasc Food Research Centre, Moorepark, Fermoy, Co. Cork, Ireland

J. A. Malik et al. (eds.), *Food Process Engineering and Technology*,
https://doi.org/10.1007/978-981-99-6831-2_4

Abstract Increasing population, urbanization and rising food demands growing energy needs that cause depletion of conventional fossils fuels at a rapid rate. The switching to renewable energy sources (RES) offers a viable solution to meet the increasing demands of energy and reducing the risks of global food security. Therefore, this chapter deals with application of the different RES (such as: solar energy, wind energy, geothermal energy, biomass energy, etc.) for food processing. Drying of food materials for self-life extension is widely explored application of solar energy and numerous designs of solar dryers have been developed. Heat generated from the solar and geothermal energy can be employed for heating, evaporation, distillation, sterilization and pasteurization of beverages in addition to food drying. Other broad-spectrum utilization of clean energy constitutes solar ovens, solar stoves, cooker, milling and grinding. Different successfully implemented renewable energy-based solutions in food processing sector have been studied to identify future pathways for reducing the dependency of global food processing sector on conventional fossil fuels.

Keywords Geothermal energy · Heating · Refrigeration · Renewable energy · Solar drying · Wind energy

4.1 Introduction

Food processing involves several unit operations to transform the raw agricultural food ingredients into food products, which may be stable for storage and transportable. All food processing segments are highly energy intensive and dependent on the fossil fuels for energy requirement. The agricultural and food processing section accounts for approximately 30% of the global energy. Figure 4.1 shows the different energy intensive food processing operations between farm and end—user of the product (FAO 2017).

The energy is required at every stage of the food sector chain, such as: production of raw materials, processing, transportation, marketing and consumption. About

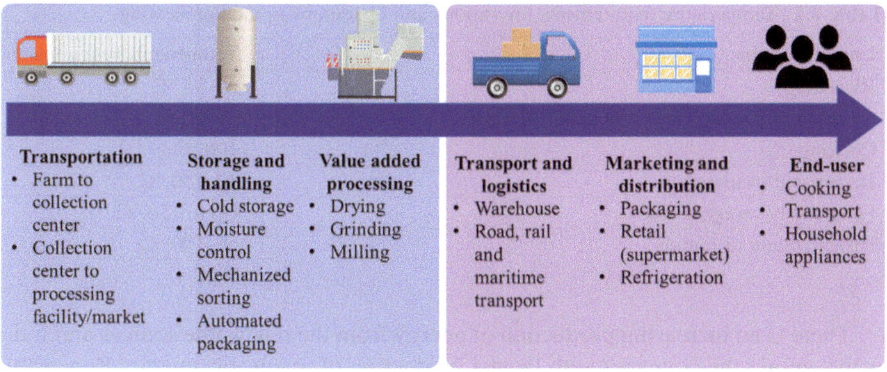

Fig. 4.1 Energy intensive unit—processes in food processing industries

75% of the food industry energy demand is consumed in heating and refrigeration equipments, such as: ovens, freezers, cold storages, furnaces, etc. The mechanical equipments (such as: air compressors, fans, mixers, motors and grinders) are utilized for different food unit operations and these consume 12% for heating and ventilation; and air conditioning and lightning of buildings consume about 8% of the energy in food industry. The demand of energy is dependent on the type and size of food material, equipment, and type of unit operation. It has been observed that heating and cooling of food products are among the most energy intensive processes (Bundschuh et al. 2017).

The transportation of raw materials from the field to processing plants and transportation of final products from the processing unit to the market is widely dependent on trucks, boats and cargo-planes consuming large quantity of diesel and gasoline. The other sources of energy for the food processing unit operations include natural gas, oil, coal, petroleum, electricity, etc. These energy sources are expected to be depleted in the coming decades because of increased energy demands due to accelerated development and increased population. The usage of fossil fuels results in the release of significant volumes of CO_2, volatile organic compounds (VOCs), Sulphur (SOx), and nitrogen oxides (NOx). These additional effects lead to eutrophication, acidification, and ozone depletion, which all contribute to global warming and other major environmental issues. Therefore, the present practices in food processing sector are considered unsustainable (Atuonwu and Tassou 2021).

The dependency of the food processing systems on the non-renewable conventional sources is becoming an alarming concern for the future with increasing population and industrialization. The fossil fuels-based processing systems are not very much operational in the rural areas of developing areas because of lack of availability of electricity; and if available, it is not continuous. Therefore, under these situations, it is important to explore for faster, cheaper and safer methods that are easily viable for poorer, usually rural populations for processing and distribution of food products. Adopting some renewable energy sources for operations of food processing equipment could be an important approach to make the food production sustainable (Bundschuh et al. 2017).

Table 4.1 Temperature requirements for various unit operations in food processing

Unit operations	Temperature requirements
Blanching	65–95 °C
Cleaning of products and production facilities	60–90 °C
Cooking	100 °C
Drying and evaporation	40–130 °C
Food and beverage sector	Up to 150 °C
Pasteurization of liquids	65–100 °C

There is an increasing production of energy from the renewable sources and India is the among the countries with largest production of renewable energy. Renewable energy accounts for 34.6% of the total power capacity in the Indian electricity sector. Remarkably, India has acquired 4th and 5th positions in the wind and solar power installed capacities globally. The country has achieved 29.55 GW of total solar installed capacity, and 34,293 MW of wind power capacity in 2018 (MNRE 2021). Further, Indian location, presence of large sunshine and abundant rain makes it perfect for the biomass production. The renewable energy resources (like solar energy, wind energy, geothermal energy) can be applied at different stages of agri-food chain.

Table 4.1 shows the temperature range generally used for different food processing unit operations. It is possible to use the renewable energy sources for the heating water and other liquids, pasteurization and sterilization of milk, drying grains, fruits and vegetables, refrigeration and cold storages, grain grinding and oil expressing.

The current chapter presents a brief overview of different renewable sources energy technologies and their applications in different food processing unit operations.

4.2 Types of Renewable Energy Sources

Renewable energy sources (such as: solar energy, geothermal energy, wind energy, hydropower, biomass energy, etc.) are abundant and sustainable (Fig. 4.2). These energy sources can be used for generation of electricity, which could be further utilized for running the food processing industry by replacing the conventional energy sources.

4.2.1 Solar Energy

Solar energy is harnessed from the sun's radiant energy. Theoretically, the globe receives approximately 4 million exajoules (1 EJ = 10^{18} J) of solar energy per year,

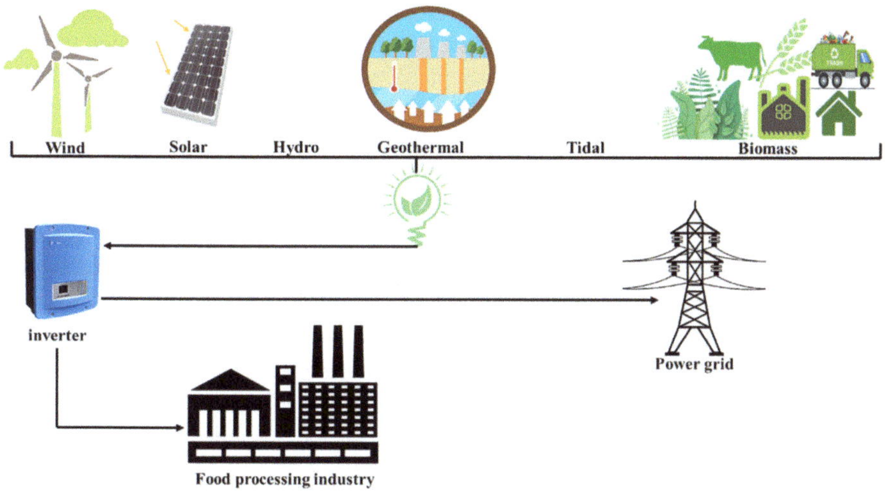

Fig. 4.2 Different sources of renewable energy for utilization in food processing

and 5×10^4 EJ of this amount is thought to constitute usable energy (Devadiga et al. 2022). If the proper harvesting and supplying technologies are available, solar energy has the potential to supply all demands of the world's energy. India is blessed with high solar radiation and theoretical annual solar power reception is approximately 5 Peta Watt-hour per year (5 trillion kWh/year) considering about 300 clear sunny days in a year. The daily average solar radiation incident over the India lies in the range of 4–17 kWh/m^2 depending on the location, with consideration of about 1500–2000 sunshine h/year.

Utilizing certain technology, solar energy is captured and transformed from solar radiation into thermal or electrical energy. The derived energy from the solar energy is used to generate heat, steam and electricity, which can be utilized for various food processing operations (such as: water heating, food drying, cooling and refrigeration, evaporation, cooking, etc.).

The sunlight can be converted directly into the electric power using the photovoltaic systems. The solar thermal systems transfer the solar energy into thermal energy to generate the steam, which can be further utilized for electricity generation using turbines. The solar thermal collectors are of two kinds (Eswara and Ramakrishnarao 2013): (1) non-concentrating (2) concentrating solar collectors.

4.2.1.1 Non-concentrating Solar Collectors

Flat plate and evacuated tube solar collectors are types of non-concentrating solar collectors. They have the absorbing area approximately near to the area to solar energy receiving area.

In a Flat plate solar collector, the solar radiation is received by a transparent cover provided at the top of the enclosure. Most of the solar energy is transmitted through the cover and absorbed by the absorber plate, which is dark colored for maximum absorption. The gap between the cover and absorber plates forms the passageway for the fluid circulating from the collector inlet toward collector outlet and heated by gaining the energy from the heated absorber plate. To prevent heat loss to the environment, the enclosure's sides, edges, and bottom are all well-insulated (Rani and Tripathy 2020).

The evacuated tube solar collector consists of a series of glass tubes, which contain heat pipes that are "plugged-in" to a manifold. The solar radiation falling on the solar collector is received by tube. The inside of the glass is coated, which absorbs the incoming solar energy and gets heated up and further transfers heat energy through an aluminum foil to the copper heat pipe. An aqueous solution is kept in a vacuum inside the copper pipe, where it is permitted to quickly boil and condense in the heat tip. The collector manifold is then used to send the energy produced in the heat tip to the solar circuit. Although evacuated tube solar collectors provide higher efficiency, flat plate collectors are typically more economical. However, evacuated tube collectors operate well in conditions of low solar irradiation and are well-suited to frigid ambient temperatures, producing heat more consistently throughout the year (Bakry et al. 2018).

According to Eq. (4.1), the thermal efficiency (η_{th}) of a solar collector is defined as the ratio of the solar energy received by the collector to the usable energy obtained by the moving fluid inside the channel (Rani and Tripathy 2020).

$$\eta_{th} = \frac{Q_u}{I \times A_c} \tag{4.1}$$

where, Q_u = useful energy output from a collector that is indicated by the heat gained by the fluid; I = solar energy radiation incident on the collector; and A_c = surface area of the collector.

The energy absorbed by the absorber plate is transferred to the useful heat gain by the moving fluid in addition to some heat losses from the collector. The useful energy gain by fluid is an indicative of energy available for useful thermal work output.

4.2.1.2 Concentrating Type Solar Collector

The concentrator solar collector collects solar radiation from large areas and focus the energy to a smaller region resulting to achieve higher temperature. Such type of solar collectors is generally used for the applications when temperature requirement is greater than 100 °C, e.g., electricity generation using the steam. The concentrating thermal solar collector's technology is also denoted as concentrator solar power (CSP) technology. Parabolic trough, linear Fresnel, central receiver/ heliostat (power tower), and parabolic dish/engine systems are four primary categories of CSP technologies (Machrafi 2012), as follows:

- **Parabolic collectors** utilize a parabolic (or ellipse) geometry for concentrating the incoming solar intensity on to a focal point. The fluid passing through the point at the focus collects the thermal energy.
- **Linear Fresnel reflectors** are constructed of long, thin mirror segments that focus solar energy onto a stationary absorber plate, positioned at a shared focal point of reflector. The heat transfer fluid is used to extract the energy from the absorber and the received energy can be further utilized to drive the steam generator using a heat exchanger.
- **Solar Thermal Parabolic Dishes** help to concentrate all incoming sunlight onto a single point rather than a single line as with a parabolic trough; and two-axis tracking parabolic dishes are utilized.
- **Power tower** is accompanied by tracking mirrors (heliostats). The mirror is aligned to concentrate the sun's radiation to the receiver, which is located at top of a tower. When the captured heat is sent to a power plant, resulting in extremely high temperatures.

4.2.1.3 Solar Photo Voltaic System

Photovoltaics (PV) is the phenomenon of the conversion of the photons, i.e., light energy into the electrical energy using suitable semiconducting materials. The conversion of solar energy as light or solar radiation into electric potential is through using photovoltaic cells, which are made of semiconducting materials (e.g.: silicon, germanium, and gallium arsenide, etc.). When a semiconducting material is exposed to sunlight, electrons in the substance's atoms are knocked loose and become free to travel through the material, producing a certain electric current. Due to the semi-conducting properties of the material, the electron flow is a unidirectional flow in nature and results in development of constant potential across the semiconductor. This unidirectional flow of electric charge is known as a direct current (DC), which can be used for direct application, or converted into alternating current, or can be stored for future use.

Generally, the electricity developed from a single PV cell is very small, which has very limited application. Thus to generate an electric potential suitable for conventional electrical appliance, several PV cells are connected in combination of parallel or series arrangements; and these are housed by support structures. The combination of several PV cells is termed as PV module and the collection of several PV modules is called a PV panel or PV array. Thin film, monocrystalline silicon, and polycrystalline silicon are three most prevalent forms of solar modules. The support structure can be rigid, flexible, transparent or opaque based on the application requirements.

Figure 4.3 shows the different components utilized for generation of electricity using the solar PV array. In large segment of PV applications, the PV modules generate the electricity that is stored in suitable storage batteries during sunlight

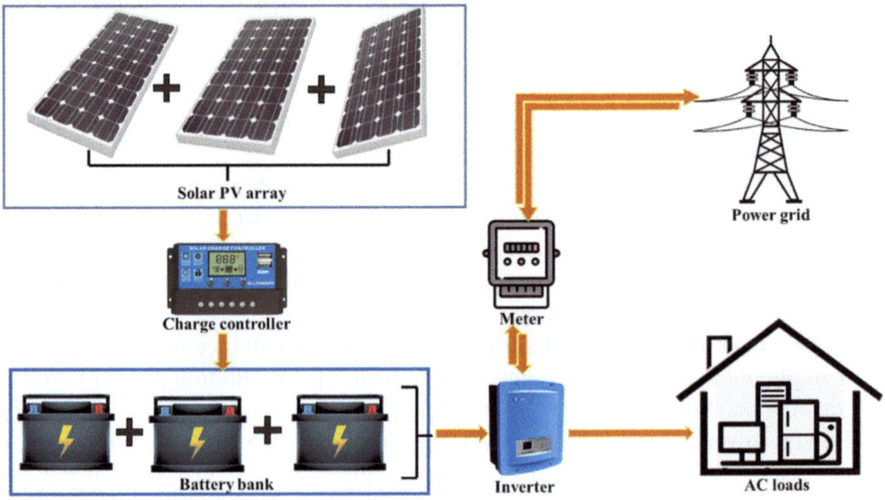

Fig. 4.3 Flowchart showing different components of a solar photovoltaic system

hours. This stored energy powers the DC loads directly or is provided to the alternating current (AC) load through the inverter. In PV systems, batteries are frequently utilized to store the energy generated by the PV array during the day so that it can later be delivered to electrical loads when needed (at night or during cloudy days). Batteries are also employed in PV systems to operate the PV array close to its maximum output, power electrical loads at stable voltages; and supply surge current flow to the inverters and electrical equipments. To safeguard the battery from overcharging and over discharging, these systems typically feature a battery charge controller (Machrafi 2012). The main functions of different components of solar photovoltaic system are:

- **Photovoltaic module** is used to convert the solar radiation into electricity in DC form.
- **Solar charge controller** controls the current and voltage that leave the PV panels and go to the battery. It keeps the battery from being overcharged.
- **Inverter** is used to convert the DC output from the PV panels into AC current to be used by the AC appliances.
- **Battery** conserves energy to provide on-demand power to electrical appliances.

Solar cell efficiency (η_{cell}) is the capacity of a cell to convert the portion of solar energy into electricity via photovoltaics. It represents the amount of solar radiation converted into electricity by a solar cell.

$$\eta_{cell} = \frac{P}{I \times A_c} \tag{4.2}$$

$$P = V \times I_c \tag{4.3}$$

The annual energy output of the PV system is determined by solar cell efficiency, latitude, and climate. For instance, a solar panel with a 20% solar cell efficiency and a surface area of 1 m^2 will generate 200 W under standard testing conditions, but it may generate more power when the sun is high in the sky and less power when the sun is low in the sky or when it is cloudy.

The standard test condition is the international testing criteria for PV panels to test the performance of the PV panels and is defined as being 1 kW/m^2 of irradiance when the cells are at standard atmospheric conditions, *viz.*, temperature 25 °C with sea level air mass of 1.5. The solar cell efficiency is influenced by the its temperature and performance as well as the spectrum, intensity. The available power at the maximum power point divided by the open circuit voltage and the short circuit current is represented by the fill factor that is an indicator to characterize the general behavior of a solar cell (Duffie and Beckman 2013; Jha and Tripathy 2019; Tiwari 2002).

By integrating the PV panels and solar thermal absorber into a single component, photovoltaic-thermal (PV-T) technology aims to improve the quantity of solar energy harvested from a solar energy system. The solar energy in this system will not immediately strike the thermal collector; instead, it will pass through the PV panel and solar thermal collector, where it can be captured as a useful heat. Hybrid photovoltaic thermal (PV-T) collectors can improve the system's overall effectiveness and are specifically created to concurrently produce electrical and thermal energies (Chow et al. 2007).

4.2.2 *Geothermal Energy*

Geothermal energy utilizes heat from the earth to produce energy, making it a clean, sustainable, and renewable resource. According to the estimates, the inner core of the earth is roughly at 5000 °C, while the outer core is at about 4000 °C. When producing geothermal energy, the major source of energy conversion is the temperature difference between the earth's core and its surface. Geothermal energy has the advantage of 24 h availability, whereas a solar energy is available only in day hours while the wind energy is highly dependent upon wind speed.

Figure 4.4 presents different components of a geothermal power plant and process of electricity generation (archive.epa.gov). The geothermal energy provides variable temperature range depending on the depth of the reservoir well of the geothermal power plant. Geothermal power plants require high temperature hydrothermal resources, i.e., water (hydro) and heat (thermal) that can be dry steam or hot water. People can access such hydrothermal sources by boring a well deep into the reservoir and pumping heat from the reservoir to the surface in the form of steam or

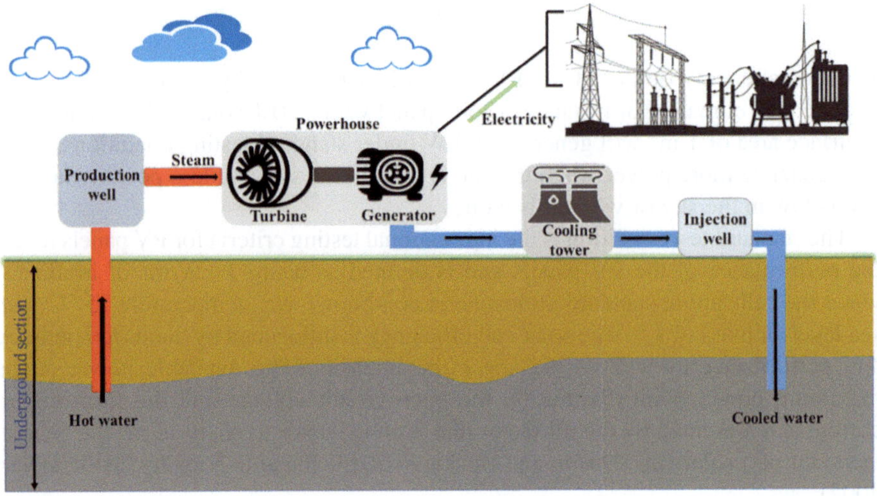

Fig. 4.4 Different components of a geothermal power plant

hot water. There are three kinds of geothermal power plants (Jha and Tripathy 2017): dry steam, flash steam and binary cycle.

4.2.2.1 Dry Steam Geothermal Power Plant

Superheated steam is the heat source for dry steam power plant. Directly from subterranean wells, steam is routed to the power plant, which generates electricity and expands the power grid. Once the water (condensed steam) has cooled, a pipe carries it back into deep wells; where it can be reheated and utilized once more to make steam.

4.2.2.2 Flash Steam Geothermal Power Plant

Geothermal reservoirs of water heated to temperatures above 180 °C and strong geostatic pressures are used in flash steam power plant. The hydrostatic pressure reduces with height as this very hot pressurized water rises through the wells, and some of the hot water boils back into steam. After being extracted from the water, the steam is used to drive a turbine and a generator. This resource is sustainable since any remaining water and condensed steam are fed back into the reservoir. Most geothermal power stations are flash steam plants.

4.2.2.3 Binary Cycle Geothermal Power Plant

Water is used as fuel in binary cycle power plant at temperatures from 110 to 180 °C. In these facilities, a working fluid, often an organic chemical with a low boiling point, is boiled using the heat from the hot water.

The working fluid of a heat exchanger is transformed into vapor and is used to power a turbine. The water is then injected back into the ground to warm. There are very few or no air emissions since the working fluid and the water are kept apart throughout the entire process. This type of plant can be operated at comparatively low geothermal temperatures but these need higher investment and operating cost for working fluid.

Figure 4.5 summaries the possible applications of geothermal energy in the agricultural and agro-industrial sectors (Nguyen et al. 2015). Geothermal reservoirs of low to moderate temperatures (20–150 °C) are used directly for heating, cooling, space heating, greenhouses, aquaculture, and industrial activities, primarily in heat pumps. Geothermal resources with an operating temperature range of 70 to 149 °C are also used to produce electricity (i.e., binary electric generation). High-temperature geothermal reservoirs (between 150 and 300 °C) are used for indirect purposes including the generation of steam and power (i.e., conventional electric generation). Industrial operations utilize energy from conventional or binary power plants, and direct industrial uses can be made of binary power plants' hot water. In various agro-industrial processes that demand high temperatures, steam and superheated water are

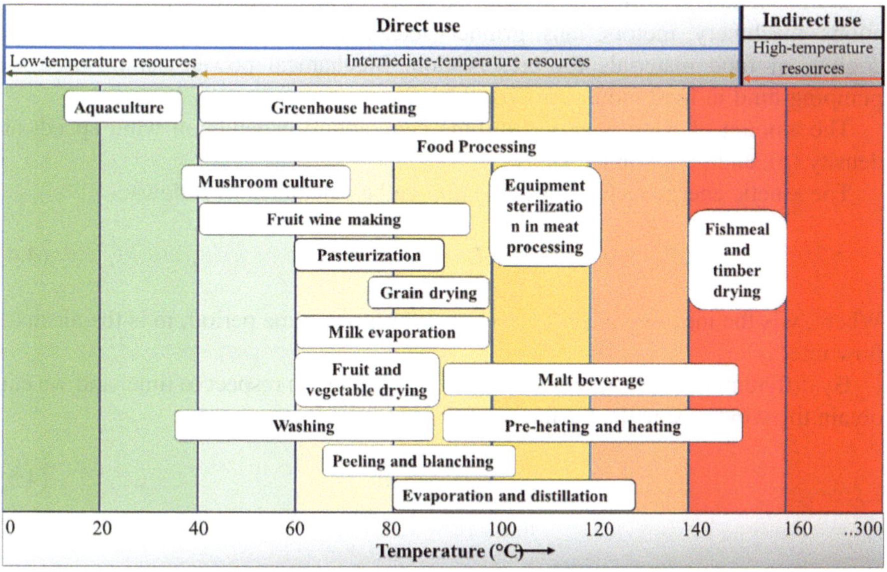

Fig. 4.5 Applications of geothermal reservoirs in the food processing sector

typically employed; however lower temperatures can occasionally be used, especially for preheating and drying agricultural products (Nguyen et al. 2015).

4.2.3 Wind Energy

The wind is the natural flow or relative movement of air above the earth surface due to pressure difference between two zones. Wind energy, often known as wind power, is the technique of harnessing the kinetic potential of the wind to produce mechanical or electrical energy. The factors contributing to wind are: inconsistent heating of the atmosphere by the sun, changes in the surface temperature and topography, and the rotation of the globe.

Wind turbines work by revolving blades like propellers that are mounted around a rotor to convert wind energy into power. The drive shaft of the electric generator is rotated by the rotor. Wind turbines transform wind's kinetic energy into mechanical power that can be applied to a variety of mechanical tasks, such as: grinding, conveying, stirring, or pumping water.

In a wind turbine, the blades are specifically made so that the air pressure on one side of the blade is lowered when the wind flows across it. The difference in air pressure between the blade's two sides causes both lift and drag forces. The lift force outweighs the drag force, which causes the rotor to spin. The rotor is connected to the generator directly if it is a direct drive turbine; otherwise, a gearbox that speeds up the rotation and enables a physically smaller generator is used. The conversion of aerodynamic force into generator rotation results in the generation of electricity.

In agro-industries, wind power can be used for operating milk processing operations, machinery, motors, fans, grinders, etc. It can be employed in heating and cooling of food materials for preservation. Mechanical power can be used for pumping fluid in food industries (Tong 2010).

The amount of wind energy a turbine can capture depends on wind speed, air density (ρ), and swept area (A).

The kinetic energy (KE) of moving air can be calculated as follows:

$$KE = ½ \times \dot{m} \tag{4.4}$$

Where, v is the mean wind speed over an appropriate time period; ṁ is the air mass flow rate.

By differentiating the kinetic energy of the wind with respect to time, and we can obtain the wind power (P_w) using the following equation:

$$P_w = ½ \times \dot{m} \tag{4.5}$$

$$\text{mass flow rate } (\dot{m} \tag{4.6}$$

$$\dot{m} \tag{4.7}$$

$$P_w = ½ \times \rho \times A \times v^3 \tag{4.8}$$

Turbine power output (P_T) is given using the following formula:

$$P_T = ½ \times \rho \times A \times v^3 \times C_p \tag{4.9}$$

Only a limited fraction of wind energy can, however, be turned into electricity. To convert wind energy into electrical energy, wind's kinetic energy must first be converted into mechanical energy to turn a wind generator's shaft. The crucial energy converters are wind blades, which require careful design with wind energy harvesting as the top priority. The power coefficient (C_p) is related to the converting efficiency, which is calculated as the proportion of the mechanical power that the blades actually catch to the available wind power.

$$C_p = P_T/P_W = 0.59 \tag{4.10}$$

The real power coefficient (C_p) usually ranges from 30 to 45%. which is much lower than its theoretical limit. The cube of the wind speed determines how much energy is in the wind. The quantity of power that is available in the wind is greatly affected by even small changes in wind speed. The turbine receives more energy when the air is denser. The size of the area through which the rotor sweeps will determine how much power the turbine can extract from the wind. The power that the turbine is capable of producing increases more significantly with a minor increase in blade length. At the same time, mechanical stability of the wind turbine should be assured when designing the blades of the wind turbine (Tong 2010).

4.2.4 Biomass Energy

Significant source of renewable energy biomass consists of biological elements from plants and animals (such as: wood from forests, waste products from forestry and agriculture), and organic wastes from both human and animal-based businesses.

In plants, atmospheric carbon dioxide is converted into various carbon-containing compounds (such as: sugars, starches, and cellulose) through photosynthesis. Biomass energy, also known as bioenergy, is the chemical energy that was initially stored in the sun and is found in both plants and animals (animals eat both plants and other animals) as well as in their waste. The widely accessible biomass resources in this area makes bioenergy an attractive option.

Variety of biomass conversion processes (such as: thermal, biochemical, or mechanical ones) can be used alone or in combination to convert biomass into bioenergy in the form of heat, electricity, or transportation fuels. From basic solid-wood combustor to boilers and biodigesters, which create fuel gas for steam production needed for heating or power production, biomass can be turned into energy using a variety of technologies. There are several ways to turn biomass into fuel gas, but the three processes that are most common are gasification, pyrolysis, and digesting.

4.2.4.1 Gasification

In a gasification process, solid biomass is burned with a limited quantity of oxygen through a thermochemical process at temperatures of around 800–1000 °C. Under these conditions, the majority of the fuel is converted into synthesis gas, also referred to as "syngas," which is a combination of methane, hydrogen, carbon monoxide, carbon dioxide, and nitrogen. In addition, char is produced as a secondary byproduct of the gasification process from a small fraction of biomass.

4.2.4.2 Pyrolysis

This is the heat breakdown of biomass when there is no air, little air, or not enough oxygen present. Depending on the process speed and temperature, the biomass is transformed into liquid and/or gaseous compounds with minor amounts of particles. Bioenergy can be made using chemicals and gases that are produced from liquids.

4.2.4.3 Anaerobic Digestion

This describes the biological breakdown of biomass in an oxygen-free environment by the action of microbiological agents (such as: bacteria and enzymes) under appropriate media conditions. It can be found in both natural settings, such as, oxygen-starved peat swamps, and man-made settings, such as, landfills, effluent lagoons, and specially designed biodigesters. Several types of anaerobic digesters have been designed and developed to maximize the conversion of the biomass to bioenergy.

4.3 Applications of Renewable Energy Sources in Food Drying

Drying can be expressed as a multidisciplinary unit operation having versatile applications in the processing of food, pharmaceuticals, paper, wood and production of chemical and biological materials. The primary objective of food drying is the reduction of water content of the food materials to achieve safe level so that deterioration pace due to microbial, physical and chemical parameters can be reduced; and simultaneously sensory, nutritional and quality aspects of the food materials can be maintained.

Drying process consumes high energy, which generally comes from fossil fuels, electricity, natural gas, etc. Thermal drying forms about 12–20% of the energy requirement of national industries in developed countries. Therefore, efforts of reducing the energy consumption have always been a challenging issue in the food and dairy industries (Akbulut 2012). Replacement of conventional sources of energy with renewable energy sources is a promising method to provide uninterrupted, reliable, and clean energy for sustainable and inexpensive drying of foods, which is highly important to balance the growing population and food supply.

4.3.1 Solar Dryers

Open sun drying is the most primitive traditional method utilizing renewable energy sources for preservation of agricultural products in the developing countries. This method deals with spreading the crops in thin layers on the ground. Sun drying is cheap and cost—effective but involves longer periods of food drying, food losses and produces products of low quality because of possible contamination (Rani and Tripathy 2021). The improved designs of solar dryers provide better quality and improved drying performances. The solar dryers are classified into main three categories, e.g., direct solar dryer, indirect solar dryer and mixed mode solar dryers. The drying of food materials in solar dryer is significantly affected by temperature, velocity and relative humidity of the working fluid and solar intensity, etc.

- **The direct type of solar dryer** consists of a box-type chamber consisting of a transparent plastic or glass covering at the top, which transmits solar radiation directly to reach the food materials that are kept on trays.
- In the case of **indirect solar dryer**, food materials are not directly exposed to the radiations coming from sun, rather solar energy is used to preheat the air using the solar collector. This preheated air is then employed for drying of food materials kept in the drying chamber, which remains integrated with solar collector.
- The design of **mixed mode solar dryer** is similar to indirect type solar dryer, but there is transparent cover at the top of drying chamber. Therefore, drying of food materials in mixed mode solar dryer involves the combination of direct solar radiation incidence and air preheating in the solar collector (Tiwari 2002).

Generally, flat plate types of solar air collectors are combined with drying chamber to construct indirect and mixed type of solar collector. Recently, many researchers have focused upon heat transfer improvement of solar collector by adding obstacles and baffles to the absorber plate (Rani and Tripathy 2022). Some collectors of these designs have been combined with drying chamber to enhance the dryer performance (Ssemwanga et al. 2020). The solar drying may sometimes be discontinuing because of discrepancy in the solar radiation resulting from the sunsets and cloudy weather, which has negative effect on the quality of food product.

The hybrid solar dryers are useful in solving this problem by utilizing an auxiliary heat source or thermal energy storage. The sensible heat (water, oil, organic solvents) and latent heat (phase-change materials) systems are utilized to store the thermal energy during the off-sunshine period for continuous drying. Photovoltaic integrated hybrid solar dryer consists of a thermal system to harness heat energy and photovoltaic panels to produce electricity (Jha and Tripathy 2017; Tiwari 2002).

4.3.2 Geothermal Dryer

Numerous countries including Kenya, Costa Rice, China, India and Thailand are using the geothermal energy for drying of the agro-products. The conventional method of geothermal dryer uses hot water and steam directly from the reservoir well. The energy harnessed through geothermal wells of hot water or steams and recovered waste heat from the geothermal plant provides the temperatures up to 150 °C sufficient for food drying.

The use of geothermal energy offers lower cost of hot water or steam than the oil and electricity. Air blown by the fan in a batch type of deep bed dryer for drying of grains is passed through a geothermal heat exchanger and it gains thermal energy. The heated air is distributed on the product uniformly using a perforated floor. Adjusting the flow rate of hot water in the heat exchanger helps in controlling the temperature of air exiting from heat exchanger and entering the drying chamber. The temperature of grains in some cases may reach up to 90 °C, while air temperature in range of 50–60 °C at RH of 40% is sufficient for drying products like coffee beans (Bundschuh et al. 2017; Nguyen et al. 2015).

The hot water from a geothermal well has been utilized for rice drying by directing the heating process (Fig. 4.6) (Eswara and Ramakrishnarao 2013). The dryer capacity and heating capacity were 10 t/h and 1360 kW. A water–air heat exchanger is used to heat the ambient air at 15 °C and 60% relative humidity to a temperature of 36 °C. The water enters the heat exchange at temperature of 75 °C and exit at 50 °C after releasing the heat energy to air.

The heated air from the heat exchanger is used for drying rice by circulated into the drying zone. As rice column moves downwards in the drier at fixed velocity resulting into a gravitation mixing and moisture of rice is reduced to 14% from the initial value of 20%. To avoid the rice cracking, drying air temperature is limited to 40 °C. The dried rice is air cooled.

A geothermal dryer at Kamojang geothermal field of West Java for dry beans and grains, utilizes the steam from a geothermal well. The steam at temperature of about 160 °C is used to heat the air with a geothermal tube-bank heat exchanger having heat exchange rate of 1000 W. The heated air is further blown into the drying chamber at air velocity ranging from 4 to 9 m/s and drying chamber varying from 45 °C to 60 °C (Nguyen et al. 2015). Rice, wheat, tomatoes, onions, cotton, chiles, and garlic have been dried using geothermal energy.

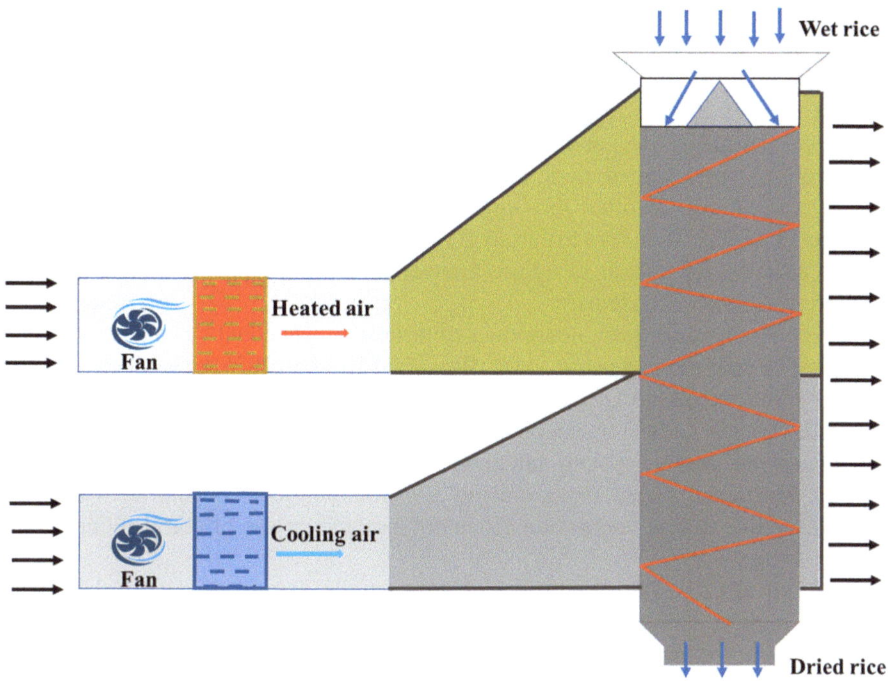

Fig. 4.6 Geothermal energy operated dryer for rice

The geothermal heat exchanger is one of the most crucial components of a dryer system that uses geothermal energy. Steel or copper pipes with copper or aluminum fins attached are used to increase the surface area for heat transfer. The pipelines are filled with geothermal hot water or steam, and a propeller fan is used to drive air through the heat exchanger. Geothermal hot water or steam heats the air before it is blown into the drying chamber to begin the drying process (Nguyen et al. 2015).

4.4 Role of Sustainable Energy in Food Cold Chain

Milk has a paramount role in providing the nutrition and economic development of farmers in remote regions of developing countries like India. However, improper handling and preservation strategies have led to spoilage of about 20–25% milk. The most effective method for preservation of milk is the heat treatment and refrigeration. Milk is rapidly cooled to 4 °C after milking (Edwin and Joseph Sekhar 2016). Cold-chain management in India has been accelerated in last decades to reduce postharvest losses of perishable crops. Most of the cold storages available are electric/diesel-driven, which impacts the environment adversely due to carbon emission.

For most of the cooling and storage applications, vapor compression refrigeration cycle using conventional energy source is used. As per International Institute of Refrigeration (IIR), the applications of air conditioning and refrigeration accounts 20% of total electricity consumption of world. The rapid depletion of fossil fuels (like coal, petroleum etc.) and the high energy demands in a cold storage presents challenging issues. Nonetheless, high running cost of available diesel/electricity-driven cold storage is often unaffordable for small scale farmers and industries (Gado et al. 2022). Therefore to curtail the increasing electricity demands, renewable energy sources are crucial to adopt. Hence, there is critical need of developing off-grid refrigeration systems.

Solar energy being clean, non-exhaustible, sustainable could be an alternative for operation of cold storage using VAR. Bharj and Kumar (2015) developed a hybrid solar system for cold storage using PV modules. Ganguly and De (2018) developed a parabolic trough collector supported multi-commodity cold storage based on LiBr-water absorption system integrated with the solar hydrogen system as a backup. Harnessing solar energy using PV modules and concentrators requires large space due to low conversion efficiency because of use of conventional fluids like air, water.

Rashid et al. (2013) studied the use of AgO nanofluids with weight 0.1–0.4% with water to improve the heat transfer ability in a solar absorption refrigeration system. However, implementation of nanofluid for system efficiency improvement will subsequently reduce the cost and increase the possibility of scale-up of system.

4.4.1 Solar Thermal Energy- Driven Adsorption Refrigeration System

The conventional vapor compression refrigeration system consists: mechanical compressor, condenser, expansion valve and evaporator. The consumption of fossil fuels and electricity for operation of vapor compression refrigeration system is a big problem as grid electricity is not continuously available in the rural areas and fossil fuels are going to be limited in the coming decades. The adsorption refrigeration system powered by solar thermal energy is a promising, eco-friendly, low-cost solution to replace the conventional refrigeration cycles. The compressor in conventional refrigeration uses the mechanical power driven by electrical energy; but in case of adsorption refrigeration system, the mechanical compressor is replaced by the thermal compressor operation with heat. The general adsorption cycle consists of four thermodynamics steps (Goyal et al. 2016):

- **Heating and pressurization phase**, also known as isosteric heating, involves the increase of temperature of adsorbent with the heat absorption. This creates elevation of evaporation pressure to reach the condensation pressure. The vapor of an adsorbent rises and moves toward condenser, where its temperature gets reduced.

- **Heating, desorption, and condensation phase**: Desorption of vapor occurred is followed by liquefaction of desorbed vapor. The temperature of adsorbent bed continuous to rise. This phase is indicated by the isobaric desorption process.
- **Cooling and depressurization phase** involves the entering of low-pressure vapors into adsorbed after coming from evaporator and process is also known as isosteric cooling. Absorber releases its heat leading to reduction of its temperature and pressure from condensation pressure to evaporation pressure.
- **Cooling, adsorption, and evaporation phase**: The adsorbent temperature is further reduced thus causing vapor adsorption. Due to heat supplied by heat source in the evaporator, the adsorbed vapors are vaporized. This adsorption cycle only needs heat energy, which can be provided by solar thermal systems.

Researchers have used flat plate solar collector, compound parabolic collector or evacuated tube collector for running heat-based refrigeration systems (Goyal et al. 2016).

4.4.2 Solar PV Integrated Vapor Absorption System

Several researchers have developed household refrigerator, compression air conditioning systems, which are driven by solar photovoltaic systems. These systems require PV panels, charge controller and battery for electricity storage. The working fluid is separated into refrigerant and absorbent by heating; and heat energy is supplied by an electric heater. The separated refrigerant is moved to the condenser, where heat is rejected to the external environment to lower the temperature. After condensation and expansion, the thermodynamic state of refrigerant is low temperature and low pressure facilitating the absorption of heat from the evaporator. The ammonia absorbs the water present in the absorber and leaves the evaporator. Here, heater is the only component run by electricity, which can be supplied by the PV panel. The energy achieved from the solar panel passes through the charge controller and output from the charger controller is used to supply power to the refrigeration unit as well as charge the battery. Therefore during no sun-shine hours, charge controller permits the power flow from the battery to the refrigeration unit (Selvaraj and Victor 2021).

The PV powered refrigeration system designed by Selvaraj and Victor (2021) was able to reduce temperature from 33 °C to 8.9 °C in 5 h in the conventional AC system, while temperature was reduced to 10 °C in the solar PV—powered DC system, which is enough for storing of fresh fruits and vegetables.

4.4.3 Cascaded Adsorption-Compression Refrigeration System Using Biomass-Solar-Wind Energies

To ensure a consistent power supply with solar intensity and wind speed, the PV panels and wind turbines are vital sources of electricity for vapor compression systems. The excess solar power is stored in the battery banks to be utilized to overcome the intermittency issue.

Hybrid RES refrigeration system for cooling of food materials proposed by Gado et al. (2022) has benefits of biomass energy, solar energy, and wind energy. The Photo voltaic thermal (PVT) collectors deliver electric and thermal energy, which drives the vapor compression cycle and adsorption system, respectively, in the cascaded adsorption-compression system. During the off-sunshine hours of night and winter season, the thermal energy to operate adsorption system is harnessed through the biomass energy.

In addition to PVT collectors, the wind energy may also supply the electricity to compression system. The basic adsorption cycle encompasses the desorber and adsorber reactor beds, expansion valve, condenser, and evaporator. The compression refrigeration cycle runs through the compressor, expansion valve, condenser, and evaporator. The refrigerant R134a is used in the compression cycle and pair of silica gel/water is used for the adsorption cycle, while brine (i.e., mixture of ethylene glycol/water) is selected as an anti-freezing fluid. The biomass-PVT battery system was more cost effective than the biomass-PVT-wind battery system.

4.5 Food Milling Using Renewable Energy

Milling is a one of the important post-harvest unit operations. From the ancient times, mills and presses are being driven using the water or wind power. Nowadays, the diesel engines and electric power have replaced the water and wind energy for running the modern mills. Any improvement of ancient machines with new techniques could result in better productivity and efficiency of wind mills.

4.5.1 Grain Milling

The investment and operation costs of diesel are high and fuel supply is also unreliable. The daily solar radiation available in tropical and subtropical areas can be a cost-effective alternative for grinding of grains. A mill operated using the PV-generator was analyzed by Beshada et al. (2006) for grinding of the grains. The mill consisted of a feeding mechanism, two millstones, which were arranged in vertical direction, gap adjusting mechanism, 12-VDC motor, spur gear transmission system and battery buffered PV-generator, which was used as a source of power to run the mill. One of the mill stones was kept fixed while other one was rotating at a speed of 400 rpm. The DC motor, having efficiency of 80%, was connected with a gearbox with speed transmission ratio of 13.5:1 and transmission efficiency of 90%. The motor was connected to a photovoltaic system that included 50 Wp PV panels, an 85 Ah battery, and a battery management system. The best rotational speed to enable a smooth operation and a low power need to run the mill was between 200 to 300 rpm.

In general, maize milling requires the most power, followed by wheat and barley. The mill operated more smoothly overall as a result of the reduction in the amplitude of the torque curve at the same time. The need for a maximum power was reduced by around 25 W during conditioning of the grain from its storage moisture content of 10–14%, while the specific energy consumption was decreased to 7–15 Wh/kg. This has a significant impact on the mill's milling capacity and enhances the mill's smooth functioning. However, treating the grain will result in more effort that is being done overall throughout the milling process.

The PV-powered grain mill works on the same principle as any other traditional mill. The designed system stands out since it may be utilized without any batteries to store energy. The PV panels directly convert solar radiation into electricity, and this electricity can then be fed directly to the mill's motor, minimizing any additional energy losses (such as: conversion losses, energy storage losses in batteries), and additional costs associated with battery maintenance and replacement. The designed mill can operate directly from a single-phase 230 V generator (Sharma et al. 2022).

The use of solar grain milling has the potential to improve farmer's income, lower operating costs and environmental effects. Due to the availability of solar milling technology, small-scale farmers in India can benefit from solar milling. General Mills has been successfully using the windmills to grind the grains using the water power. Recently, the company formed a cooperation to create a 200MW wind farm in Texas with Steelhead Americans and Renewable Energy Systems.

4.5.2 Oil Pressing

A consumption and production of about 180,354 and 39,400 MJ energy, respectively has been observed from one ton of sunflower oil (Nabavi-Pelesaraei et al. 2021). Therefore, there is a need of efficient and economical system for oil expression in rural areas. The solar PV and PV-T systems have also been used for sunflower oil production for both agricultural and industrial purposes.

A simple, batch type, solar powered oil expression unit capable of yielding high quality oil was developed by Mpagalile et al. (2006) and Mpagalile et al. (2007). The machine was powered using PV panels with 190 W, 12 VDC motor was designed, fabricated and tested for oil expression from coconut and groundnut that were grounded to particle size between 500 mm and 2 mm with moisture content of 12% (Mpagalile et al. 2007). The oil expression press provided overage oil extraction efficiency of 73% and 70% for coconut and groundnut, respectively. Increasing of pressing time increased the power consumption. The specific energy consumption showed increasing trend till 8 min of pressing and was correlated with compaction of cake. The produced oil was of excellent quality, fresh, and free of roots, with an average moisture content of 0.019% for groundnut oil and 0.015% for coconut oil. At 25 °C, the viscosities of groundnut and coconut oils were 42.1 MPa and 59.1 MPa, respectively.

4.6 Renewable Energy Utilization in Milk Processing

The dairy products are widely consumed foodstuffs. Figure 4.7 shows the water and energy requirements at different stages of milk processing (FAO 2017). To full fill the current dairy demands and maintaining self-sufficiency at farm level, use of renewable energy sources is very crucial. The renewable energy sources (such as: solar, geothermal and biogas energy) along with their hybrid utilization can provide effective solutions to meet dairy energy demands. This section presents an overview for potential use of renewable energy in milk processing operations.

4.6.1 Use of Renewable Energy Sources in Milk Pasteurization and Sterilization

Pasteurization of milk is generally carried at temperature a range of 60–80 °C with suitable time combinations. The solar energy has been efficiently utilized for pasteurization of milk in dairy industries. There are various solar based pasteurization

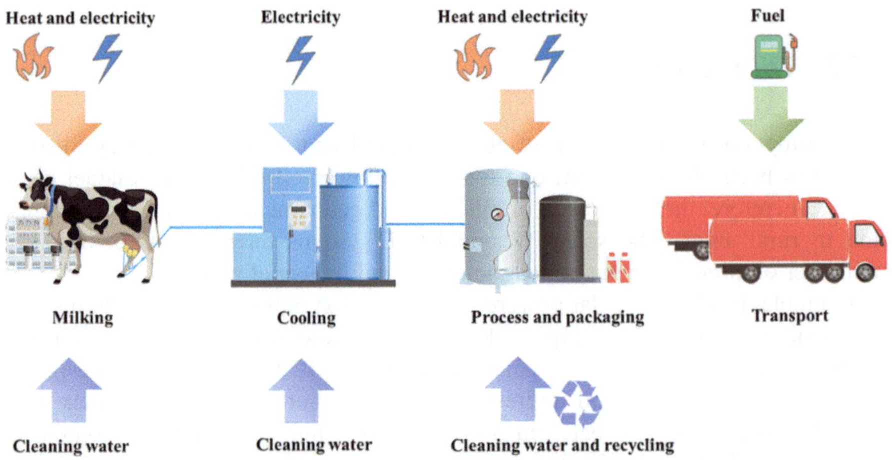

Fig. 4.7 Energy and water requirement at different stages of milk processing

methods, such as: solar panel milk pasteurization, solar concentrator pasteurization system, and evacuated tube solar collector system (Panchal et al. 2021).

Solar panel type milk pasteurization system makes use of photovoltaic solar panel to generate electricity for heating and pumping purpose. Solar panel-based system was able to provide temperatures up to 71 °C that can effectively pasteurize the milk (Meraj et al. 2021). Another method of solar pasteurization is through low-cost solar concentrators that can be flat plate collectors or parabolic trough collectors (Taj et al. 2022). These have an edge over other systems as they have lesser working parts, low operating and construction cost. An improved system consisting of Fresnel lens type solar concentration system was able to pasteurize 100 l of goat milk in 1 h (Franco et al. 2008). The solar photovoltaic panel and concentrator type of system can be used in hybrid mode as well providing more efficient pasteurization (Panchal et al. 2021).

Another system for a processing large quantity of milk (500 l), evacuated tubes type solar collectors, has been more suitable. These can attain higher temperature quicker than other systems; the only drawback was its greater head loss (Dobrowsky et al. 2015). The use of solar energy for milk sterilization is a perspective area of research, and only few studies have been reported and more detailed investigations are still required.

Another potential renewable source for milk pasteurization and sterilization is biogas energy. The electricity generation has efficiency of 8–54%, 1 m^3 of biogas can produce 4.69 kWh of electricity (Akbulut 2012). The generated electricity can be effectively be used for heating purpose in milk pasteurizing units.

Geothermal energy has also been exploited for milk pasteurization systems. By utilizing the geothermal energy for hot fluid in plate heat exchanger, the milk can be heated to a temperature of 78 °C for 15 s providing UHT treatment. The inlet temperature of geothermal hot water is about 87 °C and the outlet temperature of 77 °C. These systems can efficiently be used in high temperature short-time milk pasteurization units providing temperatures up to 72 °C to milk flowing at 3 l/s (Nguyen et al. 2015).

4.6.2 Use of Renewable Energy Sources in Milk Cooling Systems

The fresh milk must be cooled in few hours to maintain the quality and to avoid deterioration. To avoid the cooling related quality loss in milk, electrical power and refrigeration systems are required. The requirement of cooling facility is essential in the close vicinity, where the milk is obtained, use of renewable energy sources is very essential due to poor electrical infrastructure in rural areas. To utilize the solar potential for cooling, solar adsorption type chiller and photovoltaic chillers have been developed.

The solar adsorption milk chiller makes use of activated carbon and methanol as adsorption working pair, and it can effectively cool and store 150 l of milk for 10–12 h (Meraj et al. 2021). Solar photovoltaics can also be employed for cooling of the milk by using the photovoltaic electricity directly for compressor motors of refrigeration system. These systems have milk holding capacity of 150 l/day with coefficient of performance (COP) of 2.5 (Salilih and Birhane 2019). Apart from solar energy, biogas energy-based cooling systems are also being developed based on adsorption cycle making use of waste heat from generators. Utilization of geothermal energy for refrigeration could be more promising as compared to power generation. Geothermal energy-based system has also been developed; and a low-enthalpy geothermal fluid for space heating and cooling can be utilized for both cooling and heating having cool side COP of 2.1. To further enhance the performance of renewable energy cooling systems, hybrid utilization of multiple green energy sources must be used. A hybrid biomass, biogas and solar energy-based adsorption cooling system has been developed that can replace the dependency on fossil-fuel based electricity in Indian hilly areas (Edwin and Sekhar 2015; Panchal et al. 2021). A similar hybrid system with biomass and biogas combination is also being developed with maximum COP of 0.2 and payback period of 5.3 years (Edwin and Sekhar 2015). All these systems have potential to replace conventional energy sources employed in dairy industries. Although more promotion is still required for popularizing these systems.

Solar energy due to its wide availability and clean nature becomes a potential option to achieve green energy based cold storage system. However, it has been observed that use of any single renewable energy is not enough to provide continuous energy supply because of varying climate and weather conditions, therefore combination of two or more energy sources could be more effective for improvement of reliable power supply.

Hybrid-Res proposed by Edwin and Sekhar (2015) consisted of two major segments: (1) Energy conversion system; (2) Vapor absorption refrigeration system (VARS). The biomass gasifier, biogas plant, and Gober gas plants are being used by the energy conversion system depending on their availability and cooling load requirements. These can convert feedstock energy into thermal energy for heating the VARS generator. Wood chips, tapioca stems, and other energy crops are used in the biogas gasification process. Municipal solid waste and food waste from homes are used as energy sources in the biogas plant. Cow dung is used as an energy source in the Gober gas plant. The cooling system for the vapor absorption refrigeration system is either the LiBr-water or aqua ammonia system. The amount of cooling water per day to cool all the milk produced from 32 to 4 °C in 4 h is used to determine the cooling load. The suggested VARS system's payback period is between 4.8 to 6.4 years.

Development of Green cold storages will improve the socio-economic status of the farmers, utilization of agro-wastes for biogas production and reduce grid dependency.

4.7 RES Utilization for Heating

Water heating with forced and natural convection is one of the most common applications of RES in both domestic and industrial applications. Global energy consumption is expected to increase to 1000 EJ (1 EJ = 10^{18}) by 2050 (Moriarty and Honnery 2012) and world's population has been rapidly expanding, resulting in increased energy consumption for food production and supply. According to a preliminary assessment by the Food and Agricultural Organization (FAO), large increase in food production will be necessary to meet demand, which is predicted to expand by 70% by 2050 compared to year 2005. The UN has established 17 Sustainable Development Goals (SDGs) that must be achieved by 2030. Better natural resource management, the use of energy-efficient machinery, the creation of valuable products from waste and residues, the reduction of waste and losses, and the promotion of recycling at all supply chain stages are among the SDGs for the food industry (Malliaroudaki et al. 2022).

In the food business, renewable energy systems have a lot of potential because they can be utilized to reduce moisture in products and make them shelf stable (El-Mesery et al. 2022). Focusing on sustainable development approaches, photovoltaic (PV) modules have created a favorable economic climate. PV has wide scale application in mild winter region. The hybrid PVT technology in the form of thermosyphon helps in increasing the energy output from the same collector area that is used for water-heating purposes. In the long run, the payback and overall performance of a PVT system is more cost effective in water heating application (Chow et al. 2007).

In isolated places, CSP is extensively employed to create thermal energy for direct lighting needs (Owusu and Asumadu-Sarkodie 2016). Despite solar energy, alternative sustainable energy sources (such as: wind, hydro, geothermal, ocean, and biomass) have significant promise. Ocean thermal energy conversion could produce constant power output at least for small islands. Food wastes in the form of biomass energy conversion is very innovative and complementary to the other renewable energy sources (Moriarty and Honnery 2012). This section will focus on RES related to the food industry, notably food equipment washing, and related peeling, blanching, evaporation, and distillation operations.

4.7.1 *Washing OF Food Equipments*

The US Food and Drug Administration (FDA) issued the Preventive Controls for Human Food regulation in 2011 to require food manufacturers to implement suitable sanitization practices in their facilities to significantly reduce contamination and stop hazards, such as (Demirci et al. 2020): environmental pathogens, biological hazards associated with staff handling, and food allergen hazards. Renewable energy sources can be used to recycle used water, which can then be sterilized. In industries, major heat requirement is for drying, water heating, sterilization, pasteurization, distillation, blanching, cleaning, cooking, steam generation, washing, distillation, evaporation, and many food processing applications. Almost all of these applications require temperatures below 250 °C. Water heating or washing accounts for 36%, of total solar heating systems installed at a plant. In the food industry, washing vessels and equipments related with production lines is a standard procedure. Temperatures used for washing in the food and beverage industries typically range from 30 to 80 °C (Farjana et al. 2018).

4.7.2 Peeling and Blanching

Blanching is a food preservation technique, where food is briefly heated using hot water and steam. Depending on the nature of food, the time for blanching normally ranges from 30 to 60 s. It inhibits enzyme activity, which would otherwise result in flavor, color, and texture degradation. It is done before freezing, canning, or drying procedures for fruits or vegetables.

Integrated electrification technologies are more cost-effective and eco-friendlier than other conventional food processing techniques. Hot water blanching has several drawbacks, including the loss of nutrients that are soluble in water due to leaching and diffusion as well as the creation of an excessive amount of waste water. However, steam blanching inhibits the loss of water-soluble nutrients with little waste of water. It has some limitations like no uniform heating and it is not suitable for large size product. While vacuum-steam pulsed blanching is a cutting-edge steam blanching method. This method can efficiently inactivate the polyphenol oxidase and polyphenol peroxidase enzymes to impede browning (Zhang et al. 2021). Blanching food before canning has several benefits, such as: easier can-filling, reduced strain on the can during heating, and less corrosion.

4.7.3 Evaporation and Distillation

The electricity grid decarbonization can significantly reduce food processing greenhouse gas emissions. Majority of energy consumption in food manufacturing systems is for heating and one major pathway to its decarbonization is the electrification of this heat. Microwave, ultrasonic, radiofrequency, pulsed electric field, pulsed light, infrared and ohmic heating are examples of non-thermal technologies that are alternatives to traditional thermal treatments (Atuonwu and Tassou 2021). In contrast, open cycle mechanical vapor recompression (MVR) is used in evaporation and distillation systems, which uses mechanical compression to develop the pressure inside the system.

In the food and beverage industry, MVR heat pumps are frequently used (Gellings 2011). Electrification technologies can reduce heat transfer losses and microstructural degradation by improving the overall heat and mass transfer phenomenon. Muster-Slawitsch et al. (2011) demonstrated the concept of green Brewery concept for the potential reduction of thermal energy used in the brew house's evaporation and distillation.

4.8 Biogas Tri-generation System

Trigeneration systems combine cooling, heating, and power generation technologies to reduce global energy use. This system allows to save more energies and to reduce greenhouse emissions, compared to the traditional system. This system is used to reduce energy waste in a sustainable way, for example, it can reduce carbon dioxide emissions by 10–50%.

Leonzio (2018) analyzed the trigeneration system with biogas as renewable energy system, which consisted of absorption heat pump, mechanical compression heat pump, steam plant, and thermal recovery heat plant. ChemCad 6.3 was used to build the modelling and simulation portion of the setup. In the trigeneration plant system, the biogas combustion inside the boiler of the steam plant section produces the electrical energy with steam at low pressure, which is supplied to the absorption heat pump and heat recovery plant section. The low-temperature heat generated by the absorption chiller is transported to and valorized in a 2nd compression heat pump, which receives processed water at lower temperatures from the heat recovery plant and heats it to higher temperature.

Biogas combustion produces 3280 kW of energy, 925 kW of electrical energy, 2523 kW of thermal energy, and 473 kW of cooling effect. This system has an overall efficiency of 88.76%, with an electrical efficiency and a thermal efficiency of 27.68% and 89.84%, respectively. The carbon dioxide emission was nearly 0.27 kg/kWh. Trigeneration energy has greater economic and environmental advantages than other energy systems. These kinds of systems are eco-friendlier with low carbon foot prints.

4.9 Climate Change Mitigation Approaches in Food Industries

Renewable energy sources contribute to minimize global pollution by lowering overall greenhouse gas emissions, which help to prevent global warming, as well as the environmental and health risks connected with pollutants emitted by fossil fuels (Owusu and Asumadu-Sarkodie 2016). Fossils fuels are responsible for 74% of all carbon dioxide emissions (Moriarty and Honnery 2012) and thus have huge impact on the global temperature. Renewable wastes from agricultural sector, industries and residential areas have great potential in waste-to-energy routes. The most recent renewable energy technologies have enormous promise for waste conversion into biogas and biohydrogen generation (Kothari et al. 2010).

According to Global Trends in Renewable Energy Investment 2020, renewable energy projects in India has garnered \$64.2 billion (Rs. 4.7 lakh crore) from 2014 to 2019. Globally, India stands 4th in renewable energy capacity, 4th in wind, and 5th in solar power capacity. *Pradhan Mantri Kisan Urja Suraksha Evam Uttham Mahabhiyan* (PM KUSUM) is one of the world's greatest efforts to provide clean energy to over 3.5 million Indian farmers.

On September 8, 2020, the International Solar Alliance, the Government of India, and the World Bank signed a tripartite Memorandum of Understanding to implement the One Sun One Word One Grid (OSOWOG) initiative. Mondelez India Foods Pvt. Ltd. is utilizing LPG/High Speed Diesel in centralized heating system for processing needs. They are using Paraboloid dish with fully two axes tracking based Concentrated Solar Thermal system for process heating application. The system produces 0.515 million kcal/day, saving 51 kg of conventional fuel per day, or 3500 kcal/day.

An innovative Solar Dryer cum Space Heating Technology with thermal storage system was built and patented by the National Institute of Solar Energy. This latest system is ideally suited for drying agricultural items, such as: crops, fruits, vegetables, spices, chips, seafood, and tea, among other things (MNRE 2021).

4.10 Conclusion

This chapter presents a brief overview of applications of different renewable energy sources for food processing operations. The need of the switching from the fossil fuel to renewable energy for carrying out different unit operations of food and dairy industry has been recognized. It is possible to use the solar energy, geothermal energy, biomass energy for heating water and other liquids, pasteurization and sterilization of milk, drying of grains, fruits and vegetables, refrigeration and cold storages. Researchers have utilized wind mill and PV-powered solar mills for grain grinding and expressing the oil from sunflower, coconut and groundnut. The chapter highlighted that application of renewable energy sources (such as: solar, geothermal and biogas energy) along with their hybrid utilization can provide effective solutions to meet food and dairy energy demands and can improve the socio-economic status of the farmers, can reduce grid dependency in the country. Renewable energy systems have the potential to replace conventional energy sources employed in dairy industries, however, more promotion and extension is still required for popularizing this system.

References

Akbulut A (2012) Techno-economic analysis of electricity and heat generation from farm-scale biogas plant: Çiçekdağı case study. Energy 44(1):381–390

Atuonwu J, Tassou S (2021) Decarbonization of food manufacturing by the electrification of heat: a review of developments, technology options and future directions. Trends Food Sci Technol 107:168–182

Bakry AI, El-Samadouny YAF, El-Agouz SA, Alshrombably AM, Abdelfatah KS, Said MA (2018) Performance of the one-ended evacuated tubes as medium temperature solar air heaters at low flow rates. Sustainable Energy Technol Assess 30:174–182

Beshada E, Bux M, Waldenmaier T (2006) Design and optimization of a photovoltaic powered grain mill. Agric Eng Int CIGR E J 8:1–11

Bharj RS, Kumar S (2015) Energy efficient hybrid solar system for cold storage in remote areas. Int J Eng Res Technol 4(12):315–318

Bundschuh J, Chen G, Tomaszewska B, Ghaffour N, Mushtaq S, Hamawand I, Reardon-Smith K, Maraseni T, Banhazi T, Mahmoudi H, Goosen M, Antille DL (2017) Solar, wind and geothermal energy applications in agriculture: back to the future. In: Bundschuh J, Chen G, Chandrasekharam D, Piechock J (eds) Geothermal, wind and solar energy applications in agriculture and aquaculture, 1st edn. CRC Press, London, pp 1–32

Chow TT, He W, Ji J, Chan ALS (2007) Performance evaluation of photovoltaic-thermosyphon system for subtropical climate application. Sol Energy 81(1):123–130

Demirci A, Feng H, Krishnamurthy K (eds) (2020) Food safety engineering, 1st edn. Cham, Springer, p 760

Devadiga D, Selvakumar M, Shetty P, Santosh MS (2022) The integration of flexible dye-sensitized solar cells and storage devices towards wearable self-charging power systems: a review. Renew Sust Energ Rev 159:112252

Dobrowsky PH, Carstens M, De Villiers J, Cloete TE, Khan W (2015) Efficiency of a closed-coupled solar pasteurization system in treating roof harvested rainwater. Sci Total Environ 536: 206–214

Duffie JA, Beckman WA (eds) (2013) Solar engineering of thermal processes, 4th edn. New York, Wiley, p 928

Edwin M, Joseph Sekhar S (2016) Thermo-economic assessment of hybrid renewable energy-based cooling system for food preservation in hilly terrain. Renew Energy 87(Part 1):493–500

Edwin M, Sekhar SJ (2015) Thermal performance of milk chilling units in remote villages working with the combination of biomass, biogas and solar energies. Energy 91(C):842–851

El-Mesery HS, El-Seesy AI, Hu Z, Li Y (2022) Recent developments in solar drying technology of food and agricultural products: a review. Renew Sust Energ Rev 157:112070

Eswara AR, Ramakrishnarao M (2013) Solar energy in food processing - a critical appraisal. J Food Sci Technol 50(2):209–227

FAO (2017) The future of food and agriculture: trends and challenges, vol 4. Food & Organization of the United Nations, Rome, p 86

Farjana SH, Huda N, Mahmud MAP, Saidur R (2018) Solar process heat in industrial systems – a global review. Renew Sust Energ Rev 82:2270–2286

Franco J, Saravia L, Javi V, Caso R, Fernandez C (2008) Pasteurization of goat milk using a low-cost solar concentrator. Sol Energy 82(11):1088–1094

Gado MG, Nada S, Ookawara S, Hassan H (2022) Energy management of standalone cascaded adsorption-compression refrigeration system using hybrid biomass-solar-wind energies. Energy Convers Manag 258:115387

Ganguly A, De RK (2018) Conceptual design and performance analysis of a parabolic trough collector supported multi-commodity cold storage. IOP Conf Ser 402(1):1–11

Gellings C (2011) New electrification technology. Strat Plan Energy Environ 31(2):69–78

Goyal P, Baredar P, Mittal A, Siddiqui AR (2016) Adsorption refrigeration technology - an overview of theory and its solar energy applications. Renew Sust Energ Rev 53:1389–1410

Jha A, Tripathy PP (2017) Clean energy technologies for sustainable food security. In: Mujtaba IM, Srinivasan R, Elbashir NO (eds) The water-food-energy nexus: Processes, technologies, and challenges, 1st edn. CRC Press, Boca Raton, pp 197–219

Jha A, Tripathy PP (2019) Heat transfer modeling and performance evaluation of photovoltaic system in different seasonal and climatic conditions. Renew Energy 135:856–865

Kothari R, Tyagi VV, Pathak A (2010) Waste-to-energy: a way from renewable energy sources to sustainable development. Renew Sust Energ Rev 14(9):3164–3170

Leonzio G (2018) An innovative trigeneration system using biogas as renewable energy. Chin J Chem Eng 26(5):1179–1191

Machrafi H (2012) Philosophy for controlling auto-ignition in an HCCI engine. In: Machrafi H (ed) Green energy and technology, vol 1. Bentham Science Publisher, Sharjah, pp 323–366

Malliaroudaki MI, Watson NJ, Ferrari R, Nchari LN, Gomes RL (2022) Energy management for a net zero dairy supply chain under climate change. Trends Food Sci Technol 126:153–167

Meraj M, Mahmood SM, Khan ME, Azhar M, Tiwari GN (2021) Effect of N-photovoltaic thermal integrated parabolic concentrator on milk temperature for pasteurization: a simulation study. Renew Energy 163:2153–2164

MNRE (2021) Annual report. Government of India, New Delhi, pp 1689–1699

Moriarty P, Honnery D (2012) What is the global potential for renewable energy? Renew Sust Energ Rev 16(1):244–252

Mpagalile JJ, Hanna MA, Weber R (2006) Design and testing of a solar photovoltaic operated multi-seeds oil press. Renew Energy 31(12):1855–1866

Mpagalile JJ, Hanna MA, Weber R (2007) Seed oil extraction using a solar powered screw press. Ind Crop Prod 25(1):101–107

Muster-Slawitsch B, Weiss W, Schnitzer H, Brunner C (2011) The green brewery concept - energy efficiency and the use of renewable energy sources in breweries. Appl Therm Eng 31(13): 2123–2134

Nabavi-Pelesaraei A, Azadi H, Van Passel S, Saber Z, Hosseini-Fashami F, Mostashari-Rad F, Ghasemi-Mobtaker H (2021) Prospects of solar systems in production chain of sunflower oil using cold press method with concentrating energy and life cycle assessment. Energy 223: 120117

Nguyen MV, Arason S, Gissurarson M, Pálsson PG (2015) Uses of geothermal energy in food and agriculture. FAO, Rome, p 62

Owusu PA, Asumadu-Sarkodie S (2016) Review of renewable energy sources, sustainability issues and climate change mitigation. Cogent Eng 3(1):1167990

Panchal H, Patel R, Sathyamurthy R (2021) Investigation and performance analysis of solar milk pasteurization system. Int J Ambient Energy 42(5):522–529

Rani P, Tripathy PP (2020) Thermal characteristics of a flat plate solar collector: Influence of air mass flow rate and correlation analysis among process parameters. Sol Energy 211:464–477

Rani P, Tripathy PP (2021) Drying characteristics, and energetic investigation during mixed-mode solar drying of pineapple slices at varied air mass flow rates. Renew Energy 167:508–519

Rani P, Tripathy PP (2022) Experimental investigation on heat transfer performance of solar collector with baffles and semicircular loops fins under varied air mass flow rates. Int J Therm Sci 178:107597

Rashid FL, Eleiwi MA, Hoshi HA (2013) Design study of nanofluid solar absorption refrigeration system. Int J Innov Res Eng Sci 6(2):66–73

Salilih EM, Birhane YT (2019) Modelling and performance analysis of directly coupled vapor compression solar refrigeration system. Sol Energy 190:228–238

Selvaraj DA, Victor K (2021) Design and performance of solar PV integrated domestic vapor absorption refrigeration system. Int J Photoenergy 2021:6655113

Sharma K, Kothari S, Panwar NL, Rathore N (2022) Design and development of solar energy powered maize milling machine. Int J Ambient Energy 43(1):1671–1676

Ssemwanga M, Makule E, Kayondo SI (2020) Performance analysis of an improved solar dryer integrated with multiple metallic solar concentrators for drying fruits. Sol Energy 204:419–428

Taj F, Dar A, Dwivedi M, Naik HR, Farzana W, Khan SA (2022) Design strategies and recent advances in utilization of solar energy for pasteurization. MRS Energy Sustainability 9(1): 49–63

Tiwari GN (ed) (2002) Solar energy: fundamentals, design, modelling and applications. Alpha Science International Ltd., London, p 546

Tong W (2010) Fundamentals of wind energy. In: Tomg W (ed) Wind power generation and wind turbine design, vol 44. WIT Press, Southampton, pp 3–48

Zhang Y, Sun BH, Pei YP, Vidyarthi SK, Zhang WP, Zhang WK, Ju HY, Gao ZJ, Xiao HW (2021) Vacuum-steam pulsed blanching: an emerging blanching technology for beetroot. LWT 147: 111532

Part II
Recent Trends in Food Quality and Management

Chapter 5
Applications of Edible Coatings to Extend Shelf-life of Fresh Fruits

Amit Kotiyal and Pooja Singh

Contents

Abstract Bioagents, fast respiration and degradation contribute to short shelf-life of fresh fruits. Water quality degrades soon after harvest because of rapid evaporation. As a consequence, natural preservatives have been developed by scientists to keep fruits fresh. Eco-friendly and biodegradable edible coatings have been created to reduce waste. There are coatings that are antimicrobial, anti-bacterial, anti-viral and anti-browning all in one. By minimizing moisture loss, slowing ripening and stopping microbial growth, edible coatings can extend the shelf-life of fresh fruits. Polysaccharide, protein, lipid, plant and composite-based edible coatings are also available. As an edible coating, polysaccharides (such as: chitosan and gum) are often employed as an alternative to starch, cellulose and alginate, and pectin and chitosan. Soy protein, wheat gluten, and maize zein are all examples of protein-based coatings. Shellac resins, fatty acids and waxes are all components of lipid-based coatings. Aloe vera gel, herbal extracts and essential oils are examples of plant-based coatings. Protein, polysaccharides and lipids are all components of composite coatings. This chapter examines contemporary developments in edible fruit coverings, their qualities, categorization and their influence on fruit shelf-life. To meet the

A. Kotiyal (✉) · P. Singh
Department of Horticulture, Lovely Professional University (LPU), Phagwara, Punjab, India

J. A. Malik et al. (eds.), *Food Process Engineering and Technology*,
https://doi.org/10.1007/978-981-99-6831-2_5

growing demand for chemical-free fruits, this chapter on naturally edible coatings on fruits documents that no adverse effects were developed.

Keywords Composite coating · Food coating · Fruit shelf-life · Lipid · Plant based coating · Polysaccharide · Protein

5.1 Introduction

Fruits are an excellent source of minerals and nutrients, but they are also very perishable owing to their high-water content. When fruit is removed from the plant, the respiration process causes evaporation of water, which can reduce the fruit quality. These changes affect appearance and biochemical composition of fruits. Consumer preference and marketability are negatively impacted by loss of fruit shelf-life owing to shrivelling, wilting, flaccidity and decline (Mahajan et al. 2018). In modern food technology, the fruits are protected against oxidative browning, poor quality, loss of nutrients, etc. by variety of natural preservatives (Lin and Zhao 2007).

Biodegradable coatings have been developed to cut down on the amount of trash. A unique technique to maintain the shelf-life of food is to utilize edible coatings that provide antimicrobial, antibacterial, antiviral and anti-browning effects. Vegetable and animal polymers are utilized to create edible coatings that can be consumed (Jemilakshmi et al. 2020). Reduced moisture loss and decreased oxygen absorption due to fruit coatings are two benefits of these products (Mahajan et al. 2018). Aside from the three forms of edible coatings, hydrocolloids (e.g.: polysaccharides, starch and proteins) are most common than the lipids (such as: Fatty acids, waxes) and composites coatings (Jemilakshmi et al. 2020). Polysaccharides prevent oxygen and other gases from entering, whereas lipids inhibit water transport, and protein coatings give mechanical stability (Bourtoom 2008).

To preserve fresh fruits and vegetables, new edible coating and films comprising of variety of herbs and antibacterial compounds have recently been produced (Valencia-Chamorro et al. 2011). To enhance the food stability, performance, use and security, edible coatings may add functional substances, such as, antioxidants, antimicrobials and minerals (Lin and Zhao 2007). However, edible coatings do not always work as planned because the internal gas composition is critical to the effectiveness of edible coatings for fresh fruits (Dhall 2013). This chapter explores potential and adaptability of edible coatings for fresh fruits, which can intensify the effect on their shelf-life.

5.2 Edible Coatings

The food business has been using edible coatings for decades, and it is not a new preservation strategy. Conventional coating and waxing of fruits and vegetables has been a common practice (Skudlarek 2012). To build a protective layer on fruits and vegetables, an edible ingredient is applied to the surface in the form of a thin film of <0.3 mm thick (Fig. 5.1). Moisture, oxygen and solutes cannot get through it. As a preservative and barrier against postharvest and environmental damage, edible coating is a major advantage for both fresh and processed foods (Tharanathan 2003).

The horticultural produce (fresh whole or cut) can be coated with a edible coating (Dhall 2013). Because of its eco-friendliness, it can improve the aesthetic and protection of food. Proteins, lipids, polysaccharides or resins may all be utilized as edible coverings. In addition to reducing water and moisture loss, edible coatings also reduce ripening time and restrict the development of bacteria, all of which help to prolong the shelf-life of produce. Fresh and lightly processed items can be kept fresher for longer by applying edible coatings, which will reduce packaging waste (Vaishali et al. 2019).

To improve food quality, stability and functionality, edible coatings may include wide variety of functional substances. Dipping, washing, or spraying produce with edible stuff is a common method of coating in horticultural produce (Raghav et al. 2016). In recent years, edible coating has become particularly successful because of wide variety of available edible plants. Adding herbs to edible coatings is known as herbal edible coating. Among the plants typically utilized in herbal edible coatings are cinnamon, turmeric, lemongrass, tulsi, neem, aloe vera, and other herbs. Aloe vera extract edible coatings are mostly used in foods to improve the shelf-life of

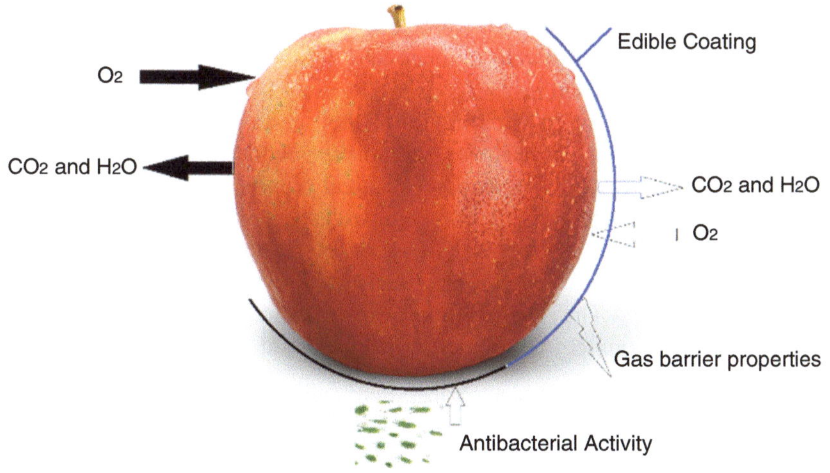

Fig. 5.1 Effect of coating on fresh fruit

fruits. Unlike other coatings (such as: wax and chemical coatings), the herbal edible coating has no negative health impacts on users (Jemilakshmi et al. 2020).

5.2.1 Properties of Edible Coating

- A high amount of adhesion is required.
- Antimicrobial and antibacterial properties should be present.
- Edible coatings should be stable in high relative humidity.
- Fruits and vegetable's nutritional and organoleptic properties should not be harmed.
- They need to be strong and durable.
- Toxic or allergic compounds should not be found in it, and it should be easy to digest.
- Water, moisture, O2, CO2 and ethylene should be well-repelled by edible coatings.

5.2.2 Classification of Edible Coatings

Water, alcohol, or their combination is required for the dispersion and dissolving of film products. Colors, flavors, and antibacterial agents may be added at this point (Bourtoom 2008). There are edible coatings based on lipids, proteins, polysaccharides, plants and combination of these.

5.2.2.1 Polysaccharide - Based Coatings

Chitosan, pectin, cellulose starch, alginate and gum are common polysaccharides that are used for edible coatings in foods. Oxygen and carbon dioxide cannot pass through polysaccharide-based films and this is their great characteristics for keeping out gases. As a result, these films have poor moisture and water permeability. Tensile strength and elongation are significant mechanical properties of this type of coating (Cazon et al. 2017). Water-soluble polysaccharides are made up of chains of polymers and are hence polymers themselves. A variety of edible coatings benefit from polysaccharide's crispiness, toughness, compactness, consistency thickening, adhesiveness, and viscosity. The respiration rate of fruits and vegetables is slowed down by the changed environment around these (Raghav et al. 2016).

- **Starch** content is high in wheat, maize, rice, and other cereal grains. Starch-rich foods include tubers like potatoes and lentils (Wittaya 2012). Starch is a carbohydrate found in abundance in nature. Starch contains anhydro-glucose units. The starch homopolymer consists of amylose and amylo-pectin (two types of

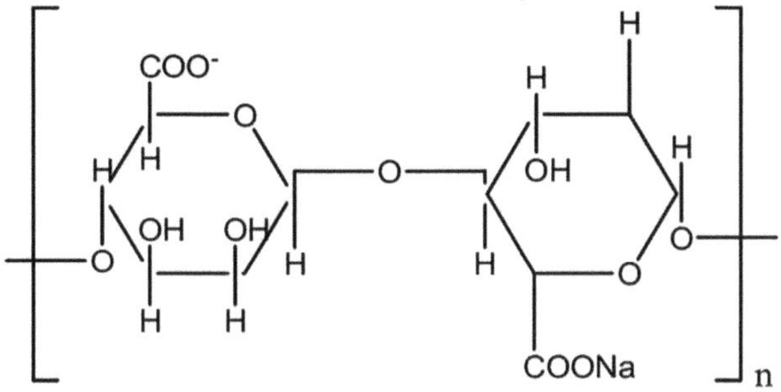

Fig. 5.2 Alginate structure

polysaccharides). Amylose is water soluble but amylo-pectin is not soluble in water. Linear or unbranched D-glucose chains (such as: Amylose and amylopectin), on the other hand, are D-glucose chains. Transparent or translucent, tasteless, less oxygen permeability and colorless at moderately less relative humidity (RH) are some properties of starch films (Lin and Zhao 2007). In water, the starch-based coating dissolves fast, and it has a low water-vapor barrier (Cazon et al. 2017). Ataulfo stenospermocarpy mango fruit (*Ataulfo*) has 15 days of shelf-life, when starch-based coating is applied. Fruit held at 10 °C for 10 days after starch coating exhibited no negative impacts on its firmness, color, or total soluble solids (Hernández-Guerrero et al. 2020).

- **Cellulose** is present in plant cell walls. Cellulose has a chemical structure similar to that of starch; but unlike starch, it is highly rigid. It is difficult to employ cellulose derivatives in large-scale commercial production because of their high cost. In cellulose, D-glucose chains are linked to each other by the bonds of β-1,4-glycosidic (Bourtoom 2008). Hydroxymethyl groups are found in the polymer backbone in both above and below plane orientations. Because of tight packaging, the polymer chains form a crystalline structure, which creates resistance in an aqueous medium (Jemilakshmi et al. 2020). The most popular and commercial cellulose derivatives are: MC (methyl cellulose), HPMC (hydroxypropyl methyl cellulose), HPC (Hydroxypropyl cellulose) and CMC (carboxymethyl cellulose). These derivatives have excellent film-forming characteristics. These derivatives are generally oil- and fat resistant, versatile, tasteless, moderately strong, odorless, clear water-soluble and have a low moisture and oxygen transmission (Al-Tayyar et al. 2020; Dhall 2013; Šuput et al. 2015). In an open box, methyl cellulose (MC) coated avocados had reduced respiration rates, green color and firmness up to 10 days at 20 °C (Liu et al. 2006). Guava cv. *Allahabad Safeda* coated with carboxymethyl cellulose (at 1.5 g/L) retained the fruit quality up to 12 days at optimum storage temperature; and its storage life was improved (Kumar et al. 2021).
- **Alginate** is a natural polysaccharide (Fig. 5.2), which is extracted from Phaseophytceae family member (known as brown seaweed). As salt of alginic acid, alginate is an alginate salt. There are two monomers in alginate, i.e., D-mannuronic acid and L-guluronic acid, which forms a linear chain copolymer. The powder of it comes in white, yellow and fibrous forms. Alginate has a good

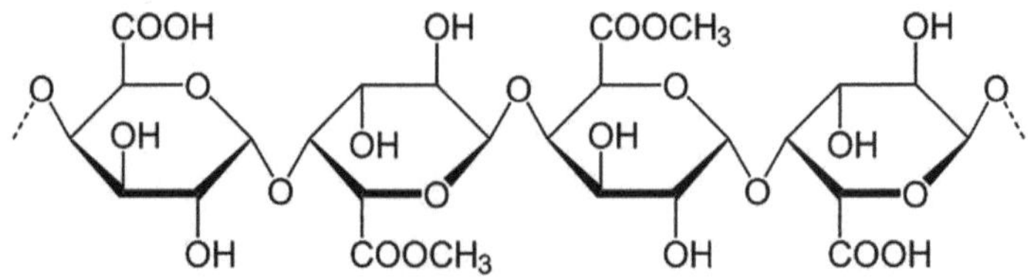

Fig. 5.3 Pectin structure

barrier property against moisture (Senturk-Parreidt et al. 2018). Alginate is a thickening and stabilizing agent because of its special colloidal characteristics (Acevedo et al. 2012). It reduces dehydration, can control respiration, improves mechanical properties and delays ripening process in horticultural produces. As a result of their hydrophilic nature, alginate is utilized to create robust, edible coatings and films (Borchard et al. 2005). A delay in postharvest ripening, softness and acidity loss, and a reduction in respiration rate in fruits of sweet cherry were observed, which had a coat of sodium alginate at concentrations of 1–5%. Edible coatings positively influence on retaining a greater antioxidant activity and phenolic concentration. The alginate-coating of cherries improved storability time from 8 days to 16 days at 2 °C but it was 2 days only at 20 °C (Díaz-Mula et al. 2012). Peaches covered with rhubarb-sodium alginate were able to keep their physiological and sensory qualities. It also has a good effect on the prevention of *Penicillium expansum* degradation and the extension of shelf-life (Li et al. 2019). Incorporating garlic oil into an alginate-based edible film acts as a natural antibacterial agent. Viable cell counts for Bacillus cereus, Staph aureus, Salmonella Typhimurium and E. coli reduces by 5.61, 4.31, 1.24 and 2.28 log cycles after 24 h of incubation in nutritive broth of above coating (Pranoto et al. 2005a).

- **Pectin** (Fig. 5.3) is used for thick gel ingredient as well as a medication carrier because of its anionic polysaccharide structure (Lopes da Silva and Rao 2006). It is typically found in citrus fruit peels and its by-products (Lin and Zhao 2007; Liu et al. 2006). It also has good mechanical qualities and acts as a gas- and oil barrier (Ciolacu et al. 2014). The 1,4-linked d-galactouronic acid residues result in the complicated structure of pectin (Dhanapal et al. 2012). The potential of pectin coatings has been widely studied to provide an efficient CO_2 and O_2 movement, as well as to prevent moisture loss and lipid migration while preserving the fruit's quality attributes (Valdés et al. 2015). The pectin-coated sapota fruits demonstrated a slower rate of change in terms of pH, ascorbic acid, firmness, weight loss, acidity, TSS, and color. The shelf-life of these fruits was 6–11 days at room temperature (Menezes and Athmaselvi 2016).
- **Chitosan** (Fig. 5.4) is an edible chitin-derived polymer. Crustacean shells are most common source of chitin. Natural and non-toxic coatings made from chitosan are most frequent (Raghav et al. 2016). The 2-Amino-2-deoxy-D-glucan linear polymer is known as chitosan. Deacetylated chitin is a source of chitosan.

Fig. 5.4 Chitosan structure

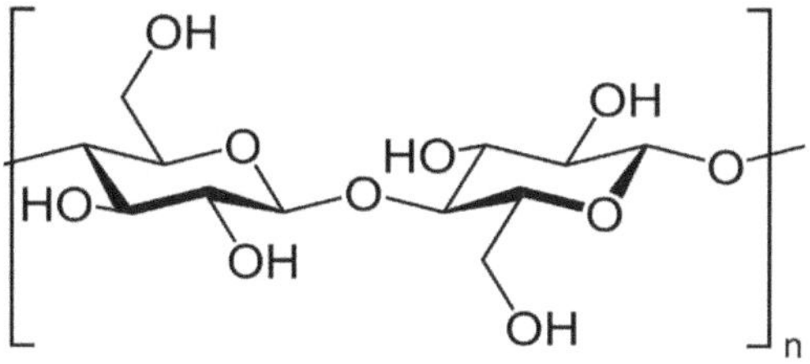

Chitin is a naturally occurring cationic biopolymer (Tharanathan and Kittur 2003). It is also effective against microorganisms and has an excellent CO_2 permeability and O_2 barrier. Chitosan is a biopolymer that has outstanding mechanical characteristics. Chitosan has a viscosity similar to that of natural gums (Skurtys et al. 2014). Fruit and vegetable coatings with chitosan have been demonstrated to delay ripening and reduce respiration rates (Vargas et al. 2006). When compared to the control, chitosan treatment reduced the deterioration and weight loss in 'El-Bayadi' grapes following storage. It also had more antioxidant capacity, hardness, phenols, and flavonoids than the control group (Al-Qurashi and Awad 2015). Chitosan (Fig. 5.4) coating delayed PPO activity and minimize the respiration rate, weight loss and alter color and eating quality. It helps to slow down the deterioration of longan fruit during storage (Al-Qurashi and Awad 2015). Coating the peeled Litchi fruit cv. Huaizhi with chitosan effectively preserved qualitative features and increased the shelf-life because it increases the levels of ascorbic acid, citric acid and total soluble solids, while suppressing PPO and POD activities (Dong et al. 2004). Genes involved in fruit ripening and the tricarboxylic acid cycle are downregulated by layer-by-layer coating, but genes involved in the glycolysis pathway are increased (Hira et al. 2022). In Banana, chitosan-montmorillonite coating improved the quality and color of fruits by reducing the respiration rate and preventing plasma membrane destruction (Wantat et al. 2022).

- **Gums** are polysaccharides. Because of their texture, gums are used to make edible coverings for horticultural produces. Gums are soluble in water. Adding polysaccharide gum coatings to food decreases respiration and weight loss while keeping its nutritious content (Tahir et al. 2019). Natural gum-based edible coatings may extend shelf-life, delay ripening, reduce respiration rate, biocompatibility, environmental safety, economic affordability, and serve as transporters for food additives (such as: antibacterial chemicals, antioxidants and vitamins (Salehi 2020). The three kinds of gums are exudate gums, extractive gums, and microbial fermentation gums (e.g., gum Arabic, guar gum and xanthan gum). A combination of xanthan, guar and Arabic gum may be utilized in creation of fair coatings and high adherence in batters during its wet conditions (Mei et al. 2002). In a study, gum Arabic (GA) was discovered to have the ability to reduce fresh food losses in persimmon fruits. Moreover, GA improved the storage life and sustained the excellence of persimmon fruits (Saleem et al. 2020).

5.2.2.2 Protein Based Coatings

Proteins from both animals and plants are used to make edible coatings. Other plant-based proteins, e.g., soy protein, zein (from maize), gluten (from wheat), casein and etc. are all known as edible coating materials (Baldwin et al. 2012). Great air barrier and mechanical properties and organoleptic qualities are found in such coatings and films (Krochta 2002). However, films prepared by protein have poor moisture blocking strength because of their hydrophilicity (Chen et al. 2019a). Protein-based edible coatings provide excellent O_2 barrier characteristics when the relative humidity is low. There is low opposition to the use of protein coatings in food items because of their additional value (Al-Tayyar et al. 2020). The rich cohesive characteristics of protein polymers make the films and coatings fragile and crack-prone (Lim et al. 2002). Edible films made from maize zein, collagen, wheat gluten, gelatine and soy protein have become more common.

- **Collagen** is a protein present in the bone, skin, tendons, connective tissue and ligaments. This protein is fibrous that provides tensile strength to animal tissue by self-assembling collagen molecules in the extracellular matrix. They make up about a third of the protein in animals. It is used to make edible protein films. It's simple to separate and purify in big quantities. Collagen fibers may be dissolved in acidic, alkaline, or neutral solutions (Wittaya 2012). Collagen films act as barrier by protecting food products from oxygen, moisture, and solute migration while also provide structural integrity and vapor permeability (Hashim et al. 2015), and better shelf-life of foods (Hernández-Guerrero et al. 2020). Collagen based composite edible coatings could minimize gas movement in mango and apple fruits, and hence have become useful for extending their shelf-life (Lima et al. 2010).
- **Gelatins:** Glutinous gelatin is made by hydrolyzing collagen (a fibrous insoluble protein). Gelatin is formed up of a specific sequence of amino acids. Proline, hydroxyproline and glycine are all found in high concentrations in gelatin, which distinguish it from other forms of gelatin. Physical, thermos-reversible gels occur, when gelatin aqueous solutions in the sol state are chilled to around 40 °C. Water vapor barrier is not present in gelatin films (Bourtoom 2008; Wittaya 2012). In mango fruits, coatings of 5% zein and 10% gelatin was employed to delay ripening, retain qualitative features, and to increase the shelf-life during mango storage (Gol and Rao 2014).
- **Corn Zein:** Maize gluten meal yields zein, a water-insoluble corn protein that is soluble in glycol esters, glycols and aqueous alcohol but not in water. Zein based protein films are more durable, and resistant to grease and water (Dangaran et al. 2009). Prolamines found in maize endosperm make up this substance. It exhibits film-forming, binding, and adhesive characteristics, as well as an effective oxygen barrier (Bourtoom 2008). Zein is a thermoplastic that is also relatively hydrophobic. Zein's hydrophobic characteristics are due to presence of sufficient amount of non-polar amino acids (Shukla and Cheryan 2001). Corn-Zein protein helps to keep fruits and vegetables fresh by avoiding color shift, hardness, and

weight loss, as well as increasing their shelf-life. It has good O_2 barrier characteristic (Raghav et al. 2016). Water vapor barrier properties of zein films are roughly 800 times higher than other edible coatings. It also helps fresh fruit to retain moisture and firmness while delaying color change (Wittaya 2012).

- **Wheat glutens** are globular polypeptide units that are present in wheat gluten, a group of water-insoluble proteins found in wheat flour. Aqueous solutions with low or high pH and low ionic pressure dissolve wheat gluten, even though it is insoluble in natural water (Bourtoom 2008). Water vapor can still pass-through wheat gluten films, although they are less efficient in blocking oxygen transport. Wheat gluten is hydrophilic. Hence the films have a low water vapor resistance. Flexible gluten films need the addition of plasticizers, such as, glycerine (Wittaya 2012). Palmitic acid, stearic acid and beeswax use in bilayer coating of wheat gluten to increase firmness, weight reduction, and physicochemical of strawberry fruits. The strawberry's flavor and appearance were also preserved by the gluten covering (Tanada-Palmu and Grosso 2005).
- **Soy protein**: Soybeans have a protein content of 38–44% compared with 8–15% for cereal grains. In most cases, water does not dissolve soybean protein. However, neutral salt solutions may be used to dissolve it (Wittaya 2012). Because of hydrophilic properties, soy protein-based edible coatings have poor moisture resistance. At low RH, soy protein-based coating has good O_2 barrier properties (Vaishali et al. 2019). Either heated soymilk or soy protein isolate (SPI) solutions may be used to create edible films. Soymilk is prepared by combining water and grinding soybeans, then extracting the milk from the crushed soybeans (Bourtoom 2008). Weight loss and hardness of fresh-cut apples were controlled by soy protein based edible coating (Alves et al. 2017). It has been observed that SPI (soy protein isolate)-coating forming solutions were successful in reducing water loss, maintaining color, and managing the liberation of active agents on Persian lime (*Citrus latifolia* Tanaka) depending on storage life (González-Estrada et al. 2017).

5.2.2.3 Lipid Based Coatings

The hydrophobic properties of lipid coatings, which make them good moisture barriers, are the primary reason for their utilization. Coating of lipid is used for its distinct quality, which can maintain the apparency of food by minimizing the respiration process in addition to preventing water loss (Dhall 2013). As the concentration of hydrophobicity process increases, water vapor permeability decreases. Mechanical strength is usually provided by a film prepared by lipid on a polymer structural medium, most commonly a polysaccharide (Bourtoom 2008). Paraffin wax, beeswax mineral or vegetable oil and carnauba wax are most prevalent lipid-based coating materials.

Aliphatic and alicyclic hydroxyl acid polymers are combined in the shellac resins. This resin is secreted by *Laccifer lacca*. Shellac resins can be dissolved in alcohol and alkaline solutions. This is compatible in coating preparation with most waxes, which give excellent moisture resistance and luster. It is only allowed in food coatings and adhesives as an indirect food additive (Bourtoom 2008). Shellac coating could enhance Valencia orange fruit's look and gloss. Besides this, it also prevents physiological loss in the fruit weight. The shellac coating has non-sticky, odorless, and quick to dry characteristics (Khorram et al. 2017). Mixing cinnamon essential oil into shellac is efficient to maintain the fruit attributes of Thomson navel oranges (Khorram and Ramezanian 2021).

- **Fatty acids** are polar lipids that function primarily as emulsifiers and dispersants. Fatty acids generated from vegetable oils (viz., soybean oil, olive oil, palm oil, coconut oil, sunflower oil and others) are commonly utilized to manufacture edible films and coatings. In water, the fatty acids are poorly soluble in their undissociated state, although they are moderately hydrophilic when separated (Rustan and Drevon 2005). Saturated fatty acids with the same chain length can melt at higher temperature than the unsaturated ones. Rate of saturation and fatty acid chain length both can influence fatty acid film's water vapor permeability. Lauric, myristic, oleic, capric, linoleic, linolenic, arachidonic, stearic and behenic fatty acids are major essential fatty acids for the creation of edible films.
- **Waxes** (viz., paraffin, beeswax, carnauba and others) are being used to preserve freshness of horticultural produces from moisture transmission and surface abrasion. The moisture-transfer protection by wax covering is higher compared with other lipid- or nonlipid coatings. Palm tree leaves are used to exude carnauba wax. There are commercial uses for this product that include protecting citrus and apple fruits from drying and shriveling, increasing their attractiveness and extending their shelf-life. This wax with a glycerol monolaurate coating on Ber (*Zizyphus mauritiana* Lamk.) fruit was able to reduce decay, increase in redness, firmness, shelf-life, and to maintain more vitamin C and chlorophyll in ber fruits (Chen et al. 2019b). Beeswax is a wax generated by honeybees that has been utilized in composite coatings with various materials (Maftoonazad and Ramaswamy 2005). For fresh fruits and vegetables, paraffin wax is created though distillation fraction of crude oil. The candelilla plant extract is used to make candelilla wax to make gas and moisture barrier coatings, as well as to improve the look of certain meals. They must be scraped-off before consumption if used in heavy layers, but they may be eaten if applied thinly (Bourtoom 2008). Using carnauba wax and orange peel essential oil coating on strawberries at room temperature helped in prevention from mold formation. Also these can minimize fresh food losses and extend shelf-life while retaining quality qualities of strawberry fruits (Saeed et al. 2021).

5.2.2.4 Composite Edible Coatings

These are made from a grouping of lipids, polysaccharides and proteins. Such coatings and films may benefit from this process to increase their mechanical and moisture barrier characteristics (Raghav et al. 2016). Layers of non-miscible emulsion, suspension, or dispersion, or a solution in a routinely used solvent are utilized to create these heterogeneous films (Bourtoom 2008). There are two types of composite films: bi-layer and emulsion, based on the process of preparation. Over a manufactured hydrophilic film, a bi-layer of a hydrophobic lipid layer is laminated to provide a separate hydrophobic layer that may be found within or on top of the film. The lipid content of an emulsion film, on the other hand, is equally distributed across the hydrophilic film. Bi-layer films have a high-water vapor barrier, whereas emulsions have a low water vapor barrier (Rhim and Shellhammer 2005). Aloe gel (20%) and banana-starch-chitosan based composite film could increase the storage life of strawberry fruits up to 15 days (Pinzon et al. 2020). Sapota cv. Kalipatti fruit may benefit from an edible covering of palm oil and methyl cellulose to help retain its quality. At 24 °C and 65% relative humidity, the composite coating increased the shelf-life (3 days) of sapota fruit and maintaining for seven days fruit quality (Vishwasrao and Ananthanarayan 2017). A composite edible covering made up of citric acid, ascorbic acid, olive oil and sodium alginate has the ability to prevent Ber cv. Gola fruit from deteriorating and extend its storage life and nutritional value (Ramana Rao et al. 2016).

5.2.2.5 Plant- Based Edible Coatings

Some major herbs used in edible coatings include rosemary, ginger, tulsi, turmeric, neem, garlic, lemon grass and aloe vera gel. Herbs, vitamins, enzymes and vital minerals all have antimicrobial characteristics (Douglas et al. 2005; Jiang and Li 2001).

- **Aloe gel** is commonly utilized for fruit coatings due to its antibacterial qualities. Moisture and water losses are also reduced. "Aloin" and "aloe-emodin" are the two primary substances present in Aloe gel (Jemilakshmi et al. 2020). Aloe vera contain yellow exudate color and an adhesive gel in its parenchymatous cell. Only adhesive gel is used for food coating, while yellow gel has some toxic effect (Ni et al. 2004). Water and gas barriers are provided by coatings made from Aloe vera gel, which may be eaten (Priya et al. 2020). It may reduce respiration, prevent microbial development and extend the shelf-life of horticultural produces. It can prevent softening and oxidative browning in fruits and vegetables by preserving volatile flavor components (Jemilakshmi et al. 2020). When stored at 12 °C and 96–98% relative humidity, Aloe-coated and uncoated oranges show no modification in TSS and total acidity concentrations over time but coated oranges had a greater ascorbic acid concentration than uncoated oranges (Arowora et al. 2013). Papaya fruits were preserved in aloe vera gel (100%) at room temperature and 84% relative humidity. After storage, weight loss was 22.5% in uncoated papayas compared with 7.9% for Aloe gel-covered papayas (Brishti et al. 2013). There seems to be no off-flavor or odor from the aloe vera coated table grapes (Tripathi and Dubey 2004). In the case of 'Crimson' table grapes and 'Starking' cherry, aloe vera gel coating was efficient even without the inclusion of extra antibacterial materials, viz., potassium sorbate, garlic oil and nisin (Pranoto et al. 2005b). Extension in shelf-life of blackberries up to three weeks was achieved by a starch, polymer-based Aloe vera and mandarin oil coating that preserved physicochemical qualities of these fruits (Arrubla Vélez et al. 2021).

- **Neem as herbal extract for edible coatings**: Neem is revered as non-toxic medicinal plant with antibacterial capabilities (Baswa et al. 2001). Azadirachtin, nimbidin, and nimoid are bioactive compounds in neem. The antibacterial properties of neem can affect *Salmonella enterica*, *Staphylococcus aureus*, *E. coli* and other organisms (Mahfuzul-Hoque et al. 2001). Neem extract, which has an antimicrobial property, may enhance the shelf-life of fruits like apples (Shweta et al. 2014). At 24 °C for 8 days and at 10 °C for 16 days, a pre-harvest coating of neem oil (0.5%) and chitosan (1%) was beneficial in maintaining quality and extending the shelf-life of 'Paluma' cultivar of guava (Silva et al. 2020). After harvest, adding Aloe vera gel or neem oil to apples can considerably enhance their shelf-life at room temperature (Attri et al. 2018).
- **Tulsi as herbal extract for edible coatings**: Tulsi (*Ocimum sanctum*) is classified into three varieties: Krishna Tulsi, Rama Tulsi, and Vana Tulsi (Douglas et al. 2005). Tulsi (referred to as Queen of Herbs) has a range of biochemicals including alkaloids, saponins, tannins, glycosides, and fragrant compounds that help in the treatment of a variety of illnesses. It also increases the shelf-life of fruits after harvest. Tulsi extract and oil can offer several health benefits, including antibacterial, antifungal, antiviral, antioxidative and antioxidant properties (Kayastha 2014). Tulsi herbal extract-based corn starch and beeswax coatings have the capacity to increase pear fruit life. Corn starch herbal edible coating could increase pear shelf-life by 60 days at 4 °C and 45 days at room temperature, respectively. Additionally, coated fruits had enhanced sensory properties (Saini and Raghav 2019).
- **Cinnamon as herbal extract for edible coatings**: It gives flavor and performs antibacterial, antioxidant, and inhibitory actions against pathogenic bacteria. Cinnamaldehyde and eugenol, which are found naturally in cinnamon, are antibacterial compounds; consequently, cinnamon is used in the manufacturing of food coating materials (Sharma et al. 2017). Cinnamon bark contains cinnamon aldehydes, which inhibited Gram negative and positive bacteria in sliced apples and prolonged shelf-life up to 12 days at 6°C (Muthuswamy et al. 2008).
- **Ginger as herbal extract for edible coatings**: Ginger essential oils have shown to increase the antioxidant capacity of edible films. Because of the presence of phenolic chemicals, it can serve as an antioxidant. Many studies have shown that ginger components have antifungal properties (Sharma et al. 2013, 2017). It has antimicrobial, antibacterial, anticoagulant, anti-inflammatory and antiviral activities. It slows down the pace of transpiration and prevents shrinking of fruits (Jemilakshmi et al. 2020).
- **Garlic as herbal extract for edible coatings**: Garlic inhibits the oxidation processes, giving fruits and vegetables a longer shelf-life. Garlic has antimicrobial properties because it contains phenolic and organosulfur chemicals. Many

microbial diseases can be treated by this (Karuppiah and Rajaram 2012). Strawberry quality was improved by applying a composite coating of CMC (carboxymethyl cellulose) and GEO (garlic essential oil, 2%). It also helped to sustain higher levels of total phenols and anthocyanins in the fruit (Dong and Wang 2017).

- **Turmeric (*Curcuma longa L.*) as herbal extract for edible coatings**: It has antifungal, antiviral and antioxidant characteristics (Jayaprakasha et al. 2005). It increases the shelf-life of vegetables and enhances their quality by reducing transpiration. It protects the fruit by coating, which helps in avoiding oxidative enzyme browning due to its antioxidant properties (Jemilakshmi et al. 2020). Chitosan and turmeric hydroalcoholic extract extended the shelf-life (33.33%) of processed pineapple, which resulted in superior consumer quality and no apparent bacterial and fungal rot after two weeks of storage (Rhim and Shellhammer 2005).
- **Essential oil as herbal extract for edible coatings**: This oil is obtained from various spices or other plant extracted chemicals; and is utilized for coating or augmented in packaging sheets to enhance the storage quality and adsorbing bad odors due to their antioxidant qualities. The application of essential oils can prevent postharvest infections while retaining fruit quality. When essential oil is incorporated into polymeric matrices, it gives them antibacterial and antioxidant characteristics (Maftoonazad and Ramaswamy 2005; Sánchez-González et al. 2011). Terpenolic and phenolic chemicals found in essential oils have antioxidant properties. Thymol, eugenol and carvacrol are the primary antioxidants in basil and thyme oils (Lee et al. 2005). For example, the oils of clove, oregano and rosemary contain phenolic groups that are most powerful. A higher proportion of Gram-positive bacteria succumb to these inhibitors than Gram-negative ones (Marino et al. 2001). However, fruit peels may also be used to remove ethanolic residue from other goods, such as, Coalho cheese, to increase their shelf-life (Leandro et al. 2021). Essential oils are used for control of postharvest infections while retaining fruit quality and reducing moisture loss. This technique eliminates the need for synthetic fungicides, allowing consumers' preferences, organic needs and pollution reduction (Du Plooy et al. 2009). A pectin film coated with cinnamon leaves essential oil has proven to accelerate the antioxidant content and limit the growth of bacterial in fresh-cut peaches (Ayala-Zavala et al. 2013). Blue mold rot in oranges was considerably inhibited by essential oils of mint, tulsi, and zinger (Tripathi 2001). Oil prevents mold and yeast development due to its high antioxidant properties. Essential oil of lemon, orange and mandarin maintained physiochemical status, consumer preference and shelf-life (18 days) of strawberry fruits (Shehata et al. 2020). Modified atmospheric packaging and essential oil combination can limit the postharvest deterioration and can preserve mango fruits' physicochemical quality, increasing their shelf-life (Perumal et al. 2021). Nano emulsions may also be used as an eco-friendly method. It is possible to create edible coatings with antibacterial qualities due to incorporation of naturally occurring chemicals into nano emulsions, as opposed to traditional emulsions (Mahajan et al. 2018).

5.3 Impact of Coatings on Shelf-Life of Fresh Fruits

Impact of glazing materials on shelf-life of different fruits is presented in Table 5.1. Coating organic fruits of mango and avocado with beeswax or carnauba wax can decrease the moisture loss from fruits, shrinkage of skin, pigment conversion, chilling injury and spoilage. BeeCoat (a beeswax-based organic wax) decreases the fruit softening, weight loss, color, and breakdown of acid resulting in a longer shelf-life. Signs of chilling damage (such as: distal end browning) are reduced by coating bio-organic avocado 'Ettinger' before cold storage at 5 °C for 3 weeks and later at 20 °C for eight days (Feygenberg et al. 2005). For Crispy Crimson table seedless grapes after harvest, treatment with aloe vera gel minimized quality losses and increased shelf-life (35 days) at 1 °C. This edible coating reduces early microbial growth for mesophilic aerobic bacteria, yeasts and molds. In addition, it protects the berry's beauty by preventing rachis browning and drying (Valverde et al. 2005).

The shelf-life of *Kagzi* lime is significantly increased at 25–30 °C and 60% relative humidity by emulsification of fruits in edible oils (such as: liquid paraffin wax, mustard oil, coco oil, mustard oil, sesame oil and castor oil). By emulsifying with the coconut oil, fruit acidity rose to 1.52%, and the juice content rose to 42.34%. Quality factors (such as: flavor and look, exterior color and no mold formation) can be maintained for 18 days using oil emulsion coating (Bisen et al. 2012).

Adding a little amount of cassava starch-based edible coating to strawberries had neutral effect on their physicochemical attributes. In addition to slowing weight and firmness loss, it also lowered respiration rates in strawberries during storage. Strawberries can have 12 days of shelf-life when coated with cassava starch but not with an antibacterial agent and 9 days with potassium sorbate coating at 5 °C (Garcia et al. 2012). It was observed that *A. vera* gel-coated papayas at 18–20 °C showed no discernible effect on fruit ripening, freshness, fungal activity, or quality preservation, after 15 days in storage. On the other hand, if fresh papayas are not treated before storage, they will deteriorate within 12 days (Mendy et al. 2019).

5.4 Conclusion

Water, minerals, and nutrients in fruits make them more susceptible to wilting and decay. Produce losses may be reduced by using edible coatings that are biodegradable and eco-friendly. It is a game-changing concept for foods to employ edible coatings to extend their shelf-life. Antimicrobial, antibacterial, antiviral and anti-browning properties are provided by these edible coatings. The ripening process might be delayed and microorganisms can be prevented from multiplying if edible covering is applied. There are several types of food-grade coatings, the most popular of which are those generated from polysaccharides and proteins, lipids, plants and composites. Composite coatings are made from combination of polysaccharides,

Table 5.1 Impact of glazing material on shelf-life of different fruits

Fruit	Glazing/coating material	Shelf-life	References
Avocado (cv. Hass)	Methyl cellulose	10 days at 20 °C	Maftoonazad and Ramaswamy (2005)
Avocado (ettinger) mango (tommy atkins)	Beeswax and carnauba wax	8 days at 20 °C	Feygenberg et al. (2005)
Black-berries	Aloe vera, starch and mandarin essential oil	19 days	Arrubla Vélez et al. (2021)
Guava cv. 'Allahabad Safeda'	Carboxymethyl cellulose	12 days	Kumar et al. (2021)
Guava cv. Gola	*A. vera* gel	12 days	Rehman et al. (2020)
Guava cv. Lalit	HPMC and palm oil	12 days	Vishwasrao and Ananthanarayan (2016)
Kagzi lime	Castor oil, liquid paraffin wax, $CaCl_2$ + $KMnO_4$, sesamum oil, mustard oil and coco oil	18 days	Bisen et al. (2012)
Mango	combinations of pectin, beeswax, sorbitol and monoglyceride	2 weeks	Moalemiyan et al. 2012)
Mango cv. Rang Kuai	Aloe gel wax, chitosan, carboxy methyl cellulose, Olive oil	12 days	Mandal et al. (2018)
Orange	*A. vera* gel	8 weeks	Arowora et al. (2013)
Papaya	*A. vera* gel	15 days	Mendy et al. (2019)
Plum cv. Blackamber	Alginate	3 weeks	Valero et al. (2013)
Sapota	Pectin	11 days	Menezes and Athmaselvi (2016)
Sapota cv. Kalipatti	Methyl cellulose and palm oil composite	7 days	Vishwasrao and Ananthanarayan (2017)
Strawberry	Cassava starch with potassium sorbate	12 days	Garcia et al. (2012)
	Aloe vera gel and banana starch-chitosan	15 days	Pinzon et al. (2020)
Sweet cherry	*A. vera* gel and nitric oxide	30 days	Asghari et al. (2013)
	Sodium alginate	16 days (2 °C) 2 days (20 °C)	Díaz-Mula et al. (2012)
Table grapes (crimson seedless)	*Aloe vera* gel	35 days at 1 °C	Valverde et al. (2005)

protein and lipids. Fresh or processed fruits can benefit from edible coatings that extend their shelf-life while also guarding against postharvest and environmental deterioration.

References

Acevedo CA, López DA, Tapia MJ, Enrione J, Skurtys O, Pedreschi F, Osorio F (2012) Using RGB image processing for designing an alginate edible film. Food Bioprocess Technol 5(5): 1511–1520

Al-Qurashi AD, Awad MA (2015) Postharvest chitosan treatment affects quality, antioxidant capacity, antioxidant compounds and enzymes activities of El-Bayadi table grapes after storage. Sci Hortic 197:392–398

Al-Tayyar NA, Youssef AM, Al-Hindi RR (2020) Edible coatings and antimicrobial nano emulsions for enhancing shelf-life and reducing foodborne pathogens of fruits and vegetables: a review. Sustain Mater Technol 2020:00215

Alves MM, Gonçalves MP, Rocha CM (2017) Effect of ferulic acid on the performance of soy protein isolate-based edible coatings applied to fresh-cut apples. LWT 80:409–415

Arowora KA, Williams JO, Adetunji CO, Fawole OB, Afolayan SS, Olaleye OO, Ogundele BA (2013) Effects of Aloe vera coatings on quality characteristics of oranges stored under cold storage. Greener J Agric Sci 3(1):39–47

Arrubla Vélez JP, Guerrero Álvarez GE, Vargas Soto MC, Cardona Hurtado N, Pinzón MI, Villa CC (2021) Aloe vera gel edible coating for shelf-life and antioxidant Proprieties preservation of Andean blackberries. Processes 9(6):999–1005

Asghari M, Khalili H, Rasmi Y, Mohammadzadeh A (2013) Influence of postharvest nitric oxide and aloe vera gel application on sweet cherry quality indices and storage life. Int J Agron Plant Prod 4(9):2393–2398

Attri BL, Mer MS, Kishor A, Kumar A (2018) Shelf-life and quality of apple at ambient conditions affected by different concentrations of Aloe vera gel and neem oil. Indian J Hortic 75(4): 667–677

Ayala-Zavala JF, Silva-Espinoza BA, Cruz-Valenzuela MR, Leyva JM, Ortega-Ramírez LA, Carrazco-Lugo DK, Miranda MRA (2013) Pectin–cinnamon leaf oil coatings add antioxidant and antibacterial properties to fresh-cut peach. Flavor Fragrance J 28(1):39–45

Baldwin EA, Hagenmaier R, Bai J (eds) (2012) Edible coatings and films to improve food quality, 2nd edn. CRC Press, Boca Raton, pp 13–78

Baswa M, Rath CC, Dash SK, Mishra RK (2001) Antibacterial activity of Karanj (Pongamia pinnata) and Neem (Azadirachta indica) seed oil: a preliminary report. Microbios 105(412): 183–189

Bisen A, Pandey SK, Patel N (2012) Effect of skin coatings on prolonging shelf-life of Kagzi lime (Citrus aurantifolia Swingle). J Food Sci Technol 49(6):753–759

Borchard W, Kenning A, Kapp A, Mayer C (2005) Phase diagram of the system sodium alginate/ water: a model for biofilms. Int J Biol Macromol 35(5):247–256

Bourtoom T (2008) Edible films and coatings: characteristics and properties. Int Food Res J 15(3): 237–248

Brishti FH, Misir J, Sarker A (2013) Effect of bio preservatives on storage life of papaya fruit (Carica Papaya L.). Int J Food Stud 2(1):126–136

Cazon P, Velazquez G, Ramírez JA, Vázquez M (2017) Polysaccharide-based films and coatings for food packaging: a review. Food Hydrocoll 68:136–148

Chen H, Sun Z, Yang H (2019a) Effect of carnauba wax-based coating containing glycerol monolaurate on the quality maintenance and shelf-life of Indian jujube (Zizyphus mauritiana) fruits during storage. Sci Hortic 244:157–164

Chen H, Wang J, Cheng Y, Wang C, Liu H, Bian H, Han W (2019b) Application of protein-based films and coatings for food packaging: a review. Polymers 11(12):2039

Ciolacu L, Nicolau AI, Hoorfar J (2014) Edible coatings for fresh and minimally processed fruits and vegetables. In: Hoorfar J (ed) Global safety of fresh produce. Woodhead Publishing Ltd, London, pp 233–244

Dangaran K, Tomasula PM, Qi P (2009) Structure and function of protein-based edible films and coatings. In: Embuscado ME, Huber KC (eds) Edible films and coatings for food applications. Springer, New York, pp 25–56

Dhall RK (2013) Advances in edible coatings for fresh fruits and vegetables: a review. Crit Rev Food Sci Nutr 53(5):435–450

Dhanapal A, Sasikala P, Rajamani L, Kavitha V, Yazhini G, Banu MS (2012) Edible films from polysaccharides. Food Sci Qual Manage 3:9–14

Díaz-Mula HM, Serrano M, Valero D (2012) Alginate coatings preserve fruit quality and bioactive compounds during storage of sweet cherry fruit. Food Bioprocess Technol 5(8):2990–2997

Dong F, Wang X (2017) Effects of carboxymethyl cellulose incorporated with garlic essential oil composite coatings for improving quality of strawberries. Int J Biol Macromol 104(Part A):821–826

Dong H, Cheng L, Tan J, Zheng K, Jiang Y (2004) Effects of chitosan coating on quality and shelf-life of peeled litchi fruit. J Food Eng 64(3):355–358

Douglas M, Heyes J, Smallfield B (2005) Herbs, spices and essential oils: post-harvest operations in developing countries. UNIDO FAO 61:1–8

Du Plooy W, Regnier T, Combrinck S (2009) Essential oil amended coatings as alternatives to synthetic fungicides in citrus postharvest management. Postharvest Biol Technol 53(3):117–122

Feygenberg O, Hershkovitz V, Ben-Arie R, Jacob S, Pesis E, Nikitenko T (2005) Postharvest use of organic coating for maintaining bio-organic avocado and mango quality. Acta Hortic 682(1): 507–512

Garcia LC, Pereira LM, de Luca Sarantopoulos CI, Hubinger MD (2012) Effect of antimicrobial starch edible coating on shelf-life of fresh strawberries. Packag Technol Sci 25(7):413–425

Gol NB, Rao TR (2014) Influence of zein and gelatin coatings on the postharvest quality and shelf-life extension of mango (Mangifera indica L.). Fruits 69(2):101–115

González-Estrada RR, Chalier P, Ragazzo-Sánchez JA, Konuk D, Calderón-Santoyo M (2017) Antimicrobial soy protein - based coatings: application to Persian lime (Citrus latifolia) for protection and preservation. Postharvest Biol Technol 132:138–144

Hashim P, Ridzwan MM, Bakar J, Hashim MD (2015) Collagen in food and beverage industries. Int Food Res J 22(1):1–5

Hernández-Guerrero SE, Balois-Morales R, Palomino-Hermosillo YA, López-Guzmán GG, Berumen-Varela G, Bautista-Rosales PU, Alejo-Santiago G (2020) Novel edible coating of starch-based stenospermo-carpic mango prolongs the shelf-life of mango fruit. J Food Qual 2020:132035

Hira N, Mitalo OW, Okada R, Sangawa M, Masuda K, Fujita N, Ushijima K, Akagi T, Kubo Y (2022) The effect of layer-by-layer edible coating on the shelf-life and transcriptome of Japanese pear fruit. Postharvest Biol Technol 185:111787

Jayaprakasha GK, Rao LJM, Sakariah KK (2005) Chemistry and biological activities of C. longa. Trends Food Sci Technol 16(12):533–548

Jemilakshmi TV, Rakshana L, Priya SK, Aishwarya B, Anithaee C (2020) Postharvest quality enhancement of fruits and vegetables using edible coatings: a review. J Crit Rev 7(7):786–790

Jiang Y, Li Y (2001) Effects of chitosan coating on postharvest life and quality of longan fruit. Food Chem 73(2):139–143

Karuppiah P, Rajaram S (2012) Antibacterial effect of Allium sativum cloves and Zingiber officinale rhizomes against multiple-drug resistant clinical pathogens. Asian Pac J Trop Biomed 2(8):597–601

Kayastha BL (2014) Queen of herbs, Tulsi (Ocimum sanctum), removes impurities from water and plays disinfectant role. J Med Plants Stud 2(2):1–8

Khorram F, Ramezanian A (2021) Cinnamon essential oil incorporated in shellac, a novel bio-product to maintain quality of navel orange fruit. J Food Sci Technol 58(8):2963–2972

Khorram F, Ramezanian A, Hosseini SMH (2017) Shellac, gelatin and Persian gum as alternative coating for orange fruit. Sci Hortic 225:22–28

Krochta JM (2002) Proteins as raw materials for films and coatings: definitions, current status, and opportunities. Protein-Based Films Coatings 1:1–40

Kumar S, Baswal AK, Ramezanian A, Gill KS, Mirza AA (2021) Impact of carboxymethyl cellulose based edible coating on storage life and quality of guava fruit under ambient storage conditions. J Food Measure Charact 15(5):4805–4812

Leandro GR, de Souza OF, de Medeiros TKF, de Oliveira JPF, de Medeiros RS, de Albuquerque PBS, de Souza MP (2021) Quality and safety of the Coalho cheese using a new edible coating based on Ziziphus joazeiro fruit pulp. Future Foods 4:100089

Lee SJ, Umano K, Shibamoto T, Lee KG (2005) Identification of volatile components in basil (Ocimum basilicum L.) and thyme (Thymus vulgaris L.) leaves and their antioxidant properties. Food Chem 91(1):131–137

Li XY, Du XL, Liu Y, Tong LJ, Wang Q, Li JL (2019) Rhubarb extract incorporated into an alginate-based edible coating for peach preservation. Sci Hortic 257:108685

Lim LT, Mine Y, Britt IJ, Tung MA (2002) Formation and properties of egg white protein films and coatings. In: Gennadios A (ed) Protein-based films and coatings, 1st edn. CRC Press, Boca Raton, pp 233–252

Lima ÁM, Cerqueira MA, Souza BW, Santos ECM, Teixeira JA, Moreira RA, Vicente AA (2010) New edible coatings composed of galactomannans and collagen blends to improve the postharvest quality of fruits – influence on fruits gas transfer rate. J Food Eng 97(1):101–109

Lin D, Zhao Y (2007) Innovations in the development and application of edible coatings for fresh and minimally processed fruits and vegetables. Compr Rev Food Sci Food Saf 6(3):60–75

Liu Y, Shi J, Langrish TAG (2006) Water-based extraction of pectin from flavedo and albedo of orange peels. Chem Eng J 120(3):203–209

Lopes da Silva JA, Rao MA (2006) Pectins: structure, functionality, and uses. In: Stephen AM, Phillips GO, Williams PA (eds) Food polysaccharides and their applications. CRC Press, Boca Raton, pp 353–412

Maftoonazad N, Ramaswamy HS (2005) Postharvest shelf-life extension of avocados using methyl cellulose-based coating. LWT-Food Sci Technol 38(6):617–624

Mahajan BC, Tandon R, Kapoor S, Sidhu MK (2018) Natural coatings for shelf-life enhancement and quality maintenance of fresh fruits and vegetables - a review. J Postharvest Technol 6(1): 12–26

Mahfuzul-Hoque MD, Bari ML, Inatsu Y, Juneja VK, Kawamoto S (2001) Antibacterial activity of guava (Psidium guajava L.) and neem (Azadirachta indica A. Juss.) extracts against foodborne pathogens and spoilage bacteria. Foodborne Pathog Dis 4(4):481–488

Mandal D, Sailo L, Hazarika TK, Shukla AC (2018) Effect of edible coating on shelf-life and quality of local mango of Mizoram. Res Crops 19(3):419–424

Marino M et al (2001) Modulation of anaerobic energy metabolism of Bacillus subtilis by arfM (ywiD). J Bacteriol 183(23):6815–6821

Mei Y, Zhao Y, Yang J, Furr HC (2002) Using edible coating to enhance nutritional and sensory qualities of baby carrots. J Food Sci 67(5):1964–1968

Mendy TK, Misran A, Mahmud TMM, Ismail SI (2019) Application of aloe vera coating delays ripening and extends the shelf-life of papaya fruit. Sci Hortic 246:769–776

Menezes J, Athmaselvi KA (2016) Study on the effect of pectin-based edible coating on the shelf-life of sapota fruits. Biosci Biotechnol Res Asia 13(2):1195–1199

Moalemiyan M, Ramaswamy HS, Maftoonazad N (2012) Pectin-based edible coating for shelf-life extension of mango. J Food Process Eng 35(4):572–600

Muthuswamy S, Rupasinghe HPV, Stratton GW (2008) Antimicrobial effect of cinnamon bark extract on Escherichia coli O157: H7, Listeria innocua, and fresh-cut apple slices. J Food Saf 28(4):534–549

Ni Y, Turner D, Yates KÁ, Tizard I (2004) Isolation and characterization of structural components of Aloe vera leaf pulp. Int Immunopharmacol 4(14):1745–1755
Perumal AB, Nambiar RB, Sellamuthu PS, Emmanuel RS (2021) Use of modified atmosphere packaging combined with essential oils for prolonging post-harvest shelf-life of mango. LWT 148:111662
Pinzon MI, Sanchez LT, Garcia OR, Gutierrez R, Luna JC, Villa CC (2020) Increasing shelf-life of strawberries (Fragaria ssp.) by using a banana starch-chitosan-aloe vera gel composite edible coating. Int J Food Sci Technol 55(1):92–98
Pranoto Y, Rakshit SK, Salokhe VM (2005a) Enhancing antimicrobial activity of chitosan films by incorporating garlic oil, potassium sorbate, and nisin. LWT-Food Sci Technol 38(8):859–865
Pranoto Y, Salokhe VM, Rakshit SK (2005b) Physical and antibacterial properties of alginate-based edible film incorporated with garlic oil. Food Res Int 38(3):267–272
Priya N, Mirza A, Suresh N (2020) Influence of different herbal edible coatings on fruit crops. Eur J Mol Clin Med 7(7):2367–2378
Raghav PK, Agarwal N, Saini M (2016) Edible coating of fruits and vegetables: a review. Int J Sci Res Mod Educ 1(1):188–204
Ramana Rao TV, Baraiya NS, Vyas PB, Patel DM (2016) Composite coating of alginate-olive oil enriched with antioxidants enhances postharvest quality and shelf-life of Ber (Ziziphus mauritiana) fruits. J Food Sci Technol 53(1):748–756
Rehman MA, Asi MR, Hameed A, Bourquin LD (2020) Effect of postharvest application of aloe vera gel on shelf-life, activities of antioxidative enzymes, and quality of guava fruit. Foods 9(10):1361–1368
Rhim JW, Shellhammer TH (2005) Lipid-based edible films and coatings. In: Han JH (ed) Innovations in food packaging. Academic Press, San Diego, pp 362–383
Rustan AC, Drevon CA (2005) Fatty acids: structures and properties. In: Encyclopedia of life sciences. Wiley, New York, pp 1–7
Saeed M, Azam M, Saeed F, Arshad U, Afzaal M, Bader Ul Ain H, Nasir Z (2021) Development of antifungal edible coating for strawberry using fruit waste. J Food Process Preserv 2021: 15956
Saini M, Raghav PK (2019) Shelf-life enhancement of fresh pears using tulsi (Ocimum sanctum) herbal edible coatings. J Appl Hortic 21(3):223–230
Saleem MS, Ejaz S, Anjum MA, Nawaz A, Naz S, Hussain S, Canan İ (2020) Postharvest application of gum Arabic edible coating delays ripening and maintains quality of persimmon fruits during storage. J Food Process Preserv 44(8):14583
Salehi F (2020) Edible coating of fruits and vegetables using natural gums: a review. Int J Fruit Sci 20(suppl. 2):S570–S589
Sánchez-González L, Vargas M, González-Martínez C, Chiralt A, Chafer M (2011) Use of essential oils in bioactive edible coatings: a review. Food Eng Rev 3(1):1–16
Senturk-Parreidt T, Müller K, Schmid M (2018) Alginate-based edible films and coatings for food packaging applications. Foods 7(10):170–177
Sharma N, Tiwari R, Srivastava MP (2013) Zingiber officinale Roscoe. Oil: a preservative of stored commodities against storage mycoflora. Int J Curr Microbiol App Sci 2(7):123–134
Sharma D, Dhanjal DS, Mittal B (2017) Development of edible biofilm containing cinnamon to control food-borne pathogens. J Appl Pharm Sci 7(1):160–164
Shehata SA, Abdeldaym EA, Ali MR, Mohamed RM, Bob RI, Abdelgawad KF (2020) Effect of some citrus essential oils on post-harvest shelf-life and physicochemical quality of strawberries during cold storage. Agronomy 10(10):1466–1470
Shukla R, Cheryan M (2001) Zein: the industrial protein from corn. Ind Crop Prod 13(3):171–192
Shweta C, Gupta KC, Mukesh A (2014) Efficacy of natural extracts on the storage quality of apple. Int J Curr Microbiol App Sci 3(3):706–711
Silva RS, Martins LP, Araujo RDC, Sousa SD, Santos AF (2020) The use of neem oil and chitosan during pre-harvest and in the post-harvest quality of guava. J Agron Sci 51:e2018649

Skudlarek JRG (2012) Antimicrobial efficacy of edible soy protein isolate films and coatings incorporated with hop ethanol extract and the influence on shelf-life and sensory attributes of bologna. M.Sc. Thesis, Department of Horticulture, University of Kentucky, USA, p 123

Skurtys O, Acevedo C, Pedreschi F, Enronoe J, Osorio F, Aguilera JM (eds) (2014) Food hydrocolloid edible films and coatings. Hauppauge, Nova Science Publishers, p 66

Šuput DZ, Lazić VL, Popović SZ, Hromiš NM (2015) Edible films and coatings: sources, properties, and application. Food Feed Res 42(1):11–22

Tahir HE, Xiaobo Z, Mahunu GK, Arslan M, Abdalhai M, Zhihua L (2019) Recent developments in gum edible coating applications for fruits and vegetables preservation: a review. Carbohydr Polym 224:115141

Tanada-Palmu PS, Grosso CR (2005) Effect of edible wheat gluten-based films and coatings on refrigerated strawberry (Fragaria ananassa) quality. Postharvest Biol Technol 36(2):199–208

Tharanathan RN (2003) Biodegradable films and composite coatings: past, present, and future. Trends Food Sci Technol 14(3):71–78

Tharanathan RN, Kittur FS (2003) Chitin: the undisputed biomolecule of great potential. Crit Rev Food Sci Nutr 43(1):61–87

Tripathi P (2001) Evaluation of some plant products against fungi causing postharvest diseases of some fruits. Doctoral dissertation, Department of Botany, Banaras Hindu University, Uttar Pradesh, India, p 173

Tripathi P, Dubey NK (2004) Exploitation of natural products as an alternative strategy to control postharvest fungal rotting of fruit and vegetables. Postharvest Biol Technol 32(3):235–245

Vaishali HP, Sharma S, Chaudhary V, Kumar M (2019) Importance of edible coating on fruits and vegetables: a review. J Pharmacogn Phytochem 8(3):4104–4110

Valdés A, Burgos N, Jiménez A, Garrigós MC (2015) Natural pectin polysaccharides as edible coatings. Coatings 5(4):865–886

Valencia-Chamorro SA, Palou L, Del Río MA, Perez-Gago MB (2011) Antimicrobial edible films and coatings for fresh and minimally processed fruits and vegetables: a review. Crit Rev Food Sci Nutr 51(9):872–900

Valero D, Díaz-Mula HM, Zapata PJ, Guillén F, Martínez-Romero D, Castillo S, Serrano M (2013) Effects of alginate edible coating on preserving fruit quality in four plum cultivars during postharvest storage. Postharvest Biol Technol 77:1–6

Valverde JM, Valero D, Martínez-Romero D, Guillén F, Castillo S, Serrano M (2005) Novel edible coating based on aloe vera gel to maintain table grape quality and safety. J Agric Food Chem 53(20):7807–7813

Vargas M, Albors A, Chiralt A, González-Martínez C (2006) Quality of cold-stored strawberries as affected by chitosan-oleic acid edible coatings. Postharvest Biol Technol 41(2):164–171

Vishwasrao C, Ananthanarayan L (2016) Postharvest shelf-life extension of pink guava (Psidium guajava L.) using HPMC-based edible surface coatings. J Food Sci Technol 53(4):1966–1974

Vishwasrao C, Ananthanarayan L (2017) Delayed postharvest ripening-associated changes in Manilkara zapota L. with composite edible coating. J Sci Food Agric 97(2):536–542

Wantat A, Seraypheap K, Rojsitthisak P (2022) Effect of chitosan coatings supplemented with chitosan-montmorillonite nanocomposites on postharvest quality of banana fruit. Food Chem 374:131731

Wittaya T (2012) Protein-based edible films: characteristics and improvement of properties. In: Eissa AA (ed) Structure and function of food engineering. IntechOpen, London, pp 43–70

Chapter 6
Food Processing and Management of Food Supply Chain: From Farm to Fork

S. Supreetha, Harshita Sonarthi, and Shweta Mall

Contents

S. Supreetha (✉) · H. Sonarthi
Dairy Technology Section, ICAR-National Dairy Research Institute, Southern Regional Station (SRS), Bengaluru, Karnataka, India

S. Mall
Animal Genetics and Breeding Section, ICAR-National Dairy Research Institute, Southern Regional Station (SRS), Bengaluru, Karnataka, India

J. A. Malik et al. (eds.), *Food Process Engineering and Technology*,
https://doi.org/10.1007/978-981-99-6831-2_6

Abstract Food supply chain management is a vast industrial process involving application of numerous technologies from farm to fork. With modernization and technology innovation, there is a wide scope for innovative, cost-effective and efficient technologies for its application in food industry throughout the food processing and management of food supply chain. The major concerns encountered by the food processors, consumers and regulatory bodies at various stages of food transit are food quality and safety. Due to lack of uniform and efficient technologies to monitor the processing, there is variation in the supply chain management. Adoption of precision technologies and artificial intelligence will reduce these variations in the food industry to enhance the efficiency of food processing and distribution. In the recent years, there are considerable changes in the food supply chain by application of artificial intelligence, robotics, sensor- based packaging, *etc.* to increase the marketing potential of food products by ensuring food quality and safety. Similarly, use of blockchain technology, Internet of Things (IoT), precision technology enhances the efficiency of the supply chain management, especially, the products with specific requirements of temperature, pressure, and gas concentration.

Keywords Artificial intelligence · Food safety · Food supply chain management · Industry 4.0 · Innovative technologies · Precision technology · Product packaging · Product traceability

6.1 Introduction

Food supply chain is the most essential and crucial part of a country's development. The needs and demands of individuals can be delivered with efficient management of supply chain. Hence, supply chain management is a vital part of the everyday life. Food is the essential requirement and basic need of all living beings. As the land is diverse with different geographical and climatic conditions, following a specific supply chain management system is difficult to achieve. However, measures are taken by adopting various transportation modes for efficient management of supply chain, such as: roadways, railways, waterways and airways. The infrastructure development to support food supply chain is also being carried out. Although several measures have already been taken, yet there are certain issues that have to be considered to enhance the efficiency of supply chain management.

With increasing technology innovations, application of suitable technology to achieve the process is significant. Food is a perishable commodity, and the factors affecting the shelf-life of foods are moisture content, water activity, nutrients present in it, temperature, pH, storage conditions, *etc.*, As food is perishable, it has to be efficiently processed to avoid food losses by enhancing the shelf-life of fresh produce. The processing parameters selected have to be properly controlled to obtain uniform quality of food product with required food safety, free from contamination that has to be measured continuously. Similarly, after the food reaches consumers, the consumer must be able to trace the history of food from farm to fork. The manufacturer should also be able to track the product to achieve food safety and to take the feedback from consumers to meet the requirements.

Implementation of innovative technologies in the entire food supply chain enables the producer, manufacturer and consumer to be aware of the product history, traceability, food safety and several other issues. Application of innovative technologies of soft computing and digital data base enables higher control and monitoring of the process, such as: real-time data collection, remote sensing and controlling, traceability, process parameters controlling, preservation, pilferation control, counterfeit control, inhibition of microbial proliferation, waste reduction, *etc.* To achieve these goals, innovative technologies of Industry 4.0 are adopted at industry level, such as: artificial intelligence, block chain technology, big data analysis, cloud computing, precision farming, machine learning, Internet of Things, etc.

This chapter focuses on food supply chain, problems faced by food industry in supply chain management, adoption of innovative technologies for food supply chain management, application of innovative technologies in food supply chain management at farm level, food processing level, consumer end and advantages of innovative technologies in management of food supply chain.

6.2 Challenges Faced by Food Industry for Food Supply Chain Management

India is a tropical country witnessing different seasons annually, with high temperature and humidity during summer, moderate to heavy rainfall during monsoon, cold temperature during winters and warm and humid climate during spring season (Krishnan et al. 2020). Due to its diverse geographical conditions ranging from mountains, hills covered with snow to desert, adoption of uniform logistic structure is complex and practically impossible. In the last decades, the agricultural productivity in India has increased enormously. However, the major concern is availability of safe food due to improper storage and packaging (Bhoi et al. 2019). High volatility in the demand and supply is the pivotal challenge among the frequently faced issues. Owing to massive food supply chain in the country, monitoring the food quality is a huge hurdle (Zhong et al. 2017).

The major concern for the supply chain logistics is no observable differentiation between the cost of production and cost involved in transportation. Lack of proper tool for assessing the price fluctuation, prediction of future patterns reduces the efficiency of proper functioning of food supply chain. As majority of food supply chain is impacted by absence of technology, it acts as a barrier for efficient utilization of resource (Negi and Anand 2015).

To overcome the problems faced by the food supply chain in India and in the world, various innovative technologies can be applied to help in the key element identification, real-time data collection, data analysis and interpretation, control and monitoring of various aspects of food production, processing, storage and distribution.

6.3 Innovative Technologies for Management of Food Supply Chain

To overcome the difficulties faced by the food supply chain, there is a need for adoption of several innovative technologies to provide simple solution for the problems of transportation, storage, distribution, consumer feedback including the technical issues like enhancing the shelf-life, waste reduction, product traceability and tracking, real-time data acquisition, *etc.* With the continuous evolution of technologies, recent developments in digital technologies and soft computing are being widely studied for application in food supply chain management, such as: Industry 4.0, cloud computing, block chain, Internet of Things (IoT), artificial intelligence, *etc.*

6.3.1 Industry 4.0

Industry 4.0 is the fourth industrial revolution, which is defined as, "*the new extent of organization and it comprises of different emerging concepts of innovative technologies, such as, radio frequency identification (RFID), additive manufacturing, artificial intelligence (AI), big data, cloud computing, machine learning (ML), smart sensing, robotics, augmented reality and Internet of Things (IoT)*" (Kumar et al. 2019).

Application of innovative technologies aids in flexibility, increased efficiency of production, processing with enhanced productivity, which results in tremendous competitiveness, growth and improves current industrial sustainability (Falcacer and Kumar 2018). Implementation of industry 4.0 is revolutionizing the method of manufacturing, improved storage and distribution of food products by smart technologies. The fourth industrial revolution mainly focuses on deepening the impact of digital technologies for efficient production strategies (Kumar et al. 2018).

The main goal of Industry 4.0 is to enable autonomous decision making, monitoring assets, real- time control and monitoring by connecting network through horizontal and vertical integration. It involves combination of physical, chemical, biological and digital world impacting across the industries. The aim of adoption of Industry 4.0 are automation, process monitoring and control, product optimization and service control (Alcácer and Cruz-Machado 2019; Hasnan and Yusoff 2018; Hassoun et al. 2022).

6.3.2 *Big Data*

Big data refers to integration of multiple data-sets and analyzing them, which are fundamentally complex, voluminous and are of insufficient quality, quantity and structure. The application of big data can offer new arena for food business to address emerging risks and operational challenges (Rupali and Amit 2017). Big data is defined as the, "*conglomeration of enormous volume of heterogenous data sets, which is huge and intricate that becomes difficult for processing using the existing database management tools*" (Suprem et al. 2013).

The source of various data-sets are online interactions, commercial transactions, data retrieved from instrumentation (such as: sensors, monitoring system, consumer response and food product tracking tools) (Li and Mahalik 2019; Li et al. 2021). Big data encompasses enormous amount of data in the form of digital information with respect to production, processing, storage, distribution and consumer activities that are acquired by several tools and instruments throughout the supply line using various technologies (Masudin et al. 2021). In the context of food supply chain, big data is a rapidly growing field that supports decision making, identification and differentiation during processing and distribution; and it is also used to study market demands and for achieving food quality and safety (Armbruster and McDonell 2014).

6.3.3 *Internet of Things (IoT)*

The Internet of Things (IoT) applied for commercial use in industry is a critical component of Industry 4.0. It encompasses a framework that includes numerous industrial sensors and RFID tags to form network and synchronize a huge number of equipments, machines and instruments for collection, processing and inference of real-time data regarding their operation, performance and other parameters like energy utilization, production efficiency, service and maintenance (Bigliardi et al. 2022). Every component in the manufacturing and processing line can be connected thus forming a network to generate the data for efficient operation (Samara and Natsheh 2019). Application of IoT in logistics and warehousing is present to greater extent; however, advent have to be carried out for deployment of warehouse facilities to effectively manage operational conditions and suitability with spatial zoning, maintenance of operational conditions of temperature, pressure, oxygen and carbon dioxide concentrations (Jones and Barnes 2014).

Adoption of IoT has proven to increase efficiency of operation and profits of supply chain management. Apart from financial benefits, IoT leads to gain in visibility, transparency, adaption, flexibility and virtualization across the food supply chain, which is crucial for the food processing sector (Tyagi 2017). IoT also helps in enhancing the output and productivity from postharvest, storage and distribution stages. The applications of IoT throughout the food supply chain includes efficient supply chain management, monitoring polyhouse conditions, intelligent farm machines and crop imaging based on drones, food quality assessment and monitoring by sensor and spectral methods, gene sequencing and block chain - based digital traceability (Sivakumar and Nanjundaswamy 2021).

IoT has widened the possibilities of achieving maximum production and has aided in achieving sustainability with environmental benefits. Tools of IoT (such as: cloud computing and sensors) help in managing water and soil parameters (such as: pH, mineral density, available nutrient profile for better water distribution and these access metrological information directly. Increased awareness among consumers for product origin, stability, storage and distribution is enhancing the requirement for adopting several technologies in agriculture at farm level (Nafrees et al. 2021).

IoT is applied in controlling irrigation and use of fertilizers in conventional agriculture and hydroponics, in continuous monitoring of the quality and quality of grains in storage bins and silos, application of RFID tags, sensor and detectors for monitoring the health, stage of maturation, fertilization in hens and eggs, use of multispectral images to analyze and control crop quality, application of neural network analysis, support vector machines and electronic sensor techniques for detection of fruit and vegetable quality during storage, use of insect surveillance system in open fields using vibro- acoustic sensors (Naresh et al. 2021; Ullo and Sinha 2021).

6.3.4 Cloud Computing

Cloud computing is the distributed architecture which enables centralization of server resources for scaling up the data to provide computing resources and servers as per specific requirements. It involves conventional service providers classified as infrastructure and service providers (Subudhi et al. 2019). Cloud computing is involved in hosting services over the internet. It is generally classified into platform as service (PaaS), software as service (SaaS) and infrastructure as service (IaaS) (Rahul et al. 2022).

The service provided is differentiated as public or private. The public cloud provides services to anybody connected over internet, whereas private cloud is exclusive data center or network to supply services to restricted number of individuals with restrained access and specific settings (Ramirez-Asis et al. 2022). However, cloud computing provides easy and scalable access to resources of computing and information technology services. The infrastructure for cloud computing includes hardware and software components required for implementation to work based on the demand for computing or as utility computing (Pandey et al. 2022).

The computing resources, applications, servers, data storage, developmental tools, networking and to host remote data center are provided by cloud computing, which is managed by cloud service providers (Rejeb et al. 2022). Cloud computing is considered as the method of influencing the internet network to apply software and other IT services with respect to the requirement. The users of cloud share bandwidth, processing power, memory, storage space and software. It is also referred to as the "assembly of computing software" and services are accessible from decentralized network of servers. It offers streamlined, simplified solution to complexity and expenditure of network installation (Srivastava and Khan 2018).

6.3.5 Artificial Intelligence

Artificial Intelligence (AI) refers to the imitation of human intelligence in machines, which are programmed to process thinking similar to humans and mimic their actions. The remarkable property of the artificial intelligence is the capability to rationalize and take necessary actions to accomplish a specific goal (Mavrnico et al. 2019).

Machine learning is a segment of artificial intelligence that deals with the concept that computer programs is able to learn and adapt to new data mechanically without being assessed by humans; and it is achieved by deep learning techniques to enable automatic learning by implementation of huge amount of unorganized data (Rylander and Olofsson 2021). Artificial intelligence is the capacity of a digital computer to perform specific tasks, which are allied with controllers. It is often applied to development of project structures endowed with intellectual process characteristic to human rational, such as, the ability to think, learns, find reasons, understand, summarize or learn from past experience (Mavani et al. 2021; Carisalo et al. 2019).

6.3.6 Block Chain Technology

A block chain is a disseminated database, which is distributed among nodes of a network of computer. The information is stored in the digital format to carryout operations in secured decentralized format. Blockchain assures reliability and safety of data records (Kitsantas et al. 2019). The major difference of blockchain technology from that of typical database rests in the structural format. Blockchain collects information in groups termed as blocks and stores it until maximum storage space is achieved; after filling up the specific storage space, the particular blocks are completed and are added to previously filled blocks forming a sequence of data known as blockchain, which continues the sequence of formation until the information is being supplied (Pierpaoli et al. 2013).

Data in the form of blocks makes it inherently an irreversible chronology of data when implemented in a decentralized form by tagging each block with specific time of addition and memory space consumed. It is basically a shared database in which information is channelized by cryptography. The purpose of blockchain technology is to allow record and distribute the digital information, which is not further allowed to edit. Since, blockchain is the basis of immutable records of transactions that cannot be altered, deleted or destructed, therefore it is termed as distributed ledger technology (DLT) (Manasi and Bhat 2020; Onyegbula et al. 2011).

Block chain technology achieves decentralized security as the information is stored linearly and chronologically. After addition of the block to the blockchain network, it is extremely difficult to change the information unless majority of the network has reached the consensus to access. Since the hash codes at the end of each block are created by mathematical function, therefore it converts digital information to string of alphanumeric characters (Sharma and Tiwari 2016).

6.3.7 Precision Technology

Recent developments in the field of soft computing with respect to support system of decision making and information technology have paved way for the progress of precision technology. By suitable application of innovative tools and technologies (such as: soft computing techniques, wireless sensor technology (WSN) and IoT), precision technology helps to achieve improved economic and environmental sustainability (Demirbaş 2018).

Precision technology includes planned, efficient, systematic and justified use and application of resources for increased efficiency of the process. To achieve efficiency of the process several technologies (such as: artificial neural network (ANN), decision trees (DT), fuzzy logic (FL), support vector machines (SVM) and geographic information system) are used (Jones and Barnes 2014). The key goal of the pression technology is to improve and to enhance the efficiency of production, processing and transportation phenomena by minimizing negative effects and the environmental degradation with sustainable use of required associates, such as: raw material, services like water, steam, refrigerant, processing aids, storage space, distribution method and manpower (Cisternas et al. 2020; Pierpaoli et al. 2013).

6.4 Applications of Innovative Technologies in Food Supply Chain

6.4.1 At Farm Level

Farm is the basic unit of production for all foods. Both plant-based and animal-based food products originate from the farm either by crop production, livestock rearing, poultry, fisheries, *etc*. Hence, monitoring of primary produce plays a key role in the supply chain, where the plant and animal products are temporarily stored and transported to processing facilities for conversion into shelf stable and value-added products (Naresh et al. 2021). Several technologies are used at various activities of farm to increase the efficiency of monitoring, ease of adoption, technology driven agriculture and animal husbandry, to achieve overall increased productivity (Shallo et al. 2021; Abbasi et al. 2022).

6.4.1.1 Livestock Monitoring

IoT is applied in a large farm that utilizes wireless applications to track the location, health and well- being of cattle. It aids in identification of sick animals in the herd, systematic treatment by tracking the history of medication, and helps to prevent spreading of diseases among the herd and also reduces the cost involved in labor management (Frost et al. 1997). Enzyme and hormone sensors aid in growth monitoring, animal metabolism, heat detection, fertilization and pregnancy monitoring (Qiao et al. 2021).

6.4.1.2 Smart Green Housing

Adoption of greenhouse farming is synonymous with increased harvest yield. Greenhouses control the environmental factors by manual intervention and proportional control mechanism. Application of digital technologies and soft computing aids in intelligent monitoring and climate control without manual intervention (Zhen and Li 2022). Sensors aids in measurement of several environmental factors, such as: temperature, humidity, pressure and sunlight. Network of sensors connected to remote access via cloud with the help of IoT eliminates the need for a constant manual monitoring. The cloud server can be used to process data and to control the actions inside the greenhouse (Hamdy et al. 2018).

6.4.1.3 Monitoring Climatic Conditions

Climatic conditions are crucial for crop production, harvesting, storage and processing. The requirement of specific climatic conditions varies with the specific agricultural produce (Sekaran et al. 2020; Sivakumar and Nanjundaswamy 2021). IoT based data solutions empower farmers to recognize the real-time weather conditions. Sensors positioned in the agricultural field collects data from the surrounding environment in real-time, which interprets on the factors of climatic conditions like temperature, humidity, rainfall that can be used for growing a particular crop (Samara and Natsheh 2019).

6.4.1.4 Remote Sensing

The sensors placed in the agricultural land collects data from weather stations, which are interpreted by analytical tools for analysis by remote sensing. The crops can be monitored based on the results obtained; and monitoring and controlling actions can be taken remotely based on severity of the conditions, which also aids in continuous crop assessment (Ullo and Sinha 2021).

6.4.1.5 Assessment of Soil Quality

Determination of soil quality helps to analyze the nutrient value of soil, drainage capacity and pH of soil to grow appropriate types of crops with suitable irrigation method and irrigation intensity, that can enhance the productivity of soil (Spandana and Pabboju 2019).

6.4.1.6 Computer Imaging

It involves placing of cameras at various locations of the farm to generate images, which undergo digital image processing. The image processing along with application of machine learning enables comparison of crops by real-time imaging to conclude the size, shape, color and maturity. It can also be used in sorting and grading of harvested crops based on color, shape and size (Sekaran et al. 2020; Sivakumar and Nanjundaswamy 2021).

6.4.2 Food Processing

Food industry plays a major role in converting raw materials from farm into processed products with enhanced shelf-life, value addition to meet the consumer demands. Uniform quality and food safety are main criteria for processed products. To achieve the goals targeted and to meet consumer demands, innovative technologies are used at various stages of food processing (Carisalo et al. 2019).

6.4.2.1 Logistics and Operational Efficiency

Food processors strive to streamline the logistics and to maximize the operational efficiency for efficient and competitive production. With the adoption of digital technologies and soft computing technologies, manufacturing can reduce cost, saves time and help to increase profits. With integrated work flow, the streamlining of system can be carried out with process automation (Lagorio et al. 2020; Katoch 2022). By adoption of innovative technologies, the food industries are able to collect data and optimize the operation continuously to increase efficiency.

6.4.2.2 Food Production and Processing

Maintenance of machinery and equipments are integral part of food processing industry. Application of IoT technologies aids in tracking and predicting machinery maintenance, which avoids unexpected interruptions (Castillo-Vergara et al. 2021). Preventive maintenance of equipments results in higher overall equipment efficiency. Monitoring of processing conditions (such as: temperature, flow rate, pressure with aid of sensors and actuators), the processing can be regulated to uniform product quality. In case of multiple product line, they are beneficial for seamless, resource—efficient changing (Fayomi et al. 2019). Application of calibrated smart valves helps to maintain food quality from raw material to final product, for which the quality has to be monitored constantly.

6.4.3 Warehouse and Inventory Management

Real-time tracking through IoT-based applications functions to monitor the ingredients and finished product during storage and distribution (Masudin et al. 2021). Product storage parameters (such as: humidity, temperature, gas concertation, moisture content, exposure to sunlight) can be detected, controlled and monitored (Hamdy et al. 2018). With the implementation of tracking devices for warehouse inventory, producers, manufactures and distributors of food products will be able to optimize the process of food supply chain management in real-time (Rylander and Olofsson 2021).

Consumer behavioural trend and real-time demand of consumers can be determined with the help of data retrieved from sensors and product tracking system to utilize and meet the demand (Zhen and Li 2022).

6.4.4 Packaging and Delivery of Food Products

Sensors implemented in packaging materials can be used as active packaging technique to determine the internal and external environment of food packaging. It can be used to track the food product, pilferation, damage allowing both manufacturers and consumers to be aware of product quality and flow (Kumperščak et al. 2019).

6.5 Advantages for Adoption Innovative Technologies in Food Supply Chain

Technology has huge impact in revolutionizing the existing industries for increased efficiency in production, to achieve the requirements, to meet consumer demands with less energy utilization, to increase production with enhanced shelf-life.

6.5.1 *Traceability and Transparency*

Traceability is the factor on demand for producers, manufactures and consumers to ensure quality of the product throughout the supply chain. It includes transit of food from production site to consumer through various stages of processing, packaging through several unit operations. The global food traceability market is expected to reach CAGR of 9.3% by 2025, which includes a share of USD 22 billion industry according to Global Newswire (Kumperščak et al. 2019).

End-to-end continuous traceability can be achieved by rapid and efficient collection, analysis and sharing of data through IoT and blockchain technology application. It enables manufacturers and consumers to track the food product along the whole food supply chain to ensure that the information is always accessible (Badia-Melis et al. 2015).

6.5.2 *Food Safety*

With the aid of digital technologies, real-time safety data (such as: gas concertation, presence of heavy metals, humidity, temperature and storage conditions) can be accessed; and it is frequently termed as active food supply chain management. With the data collected through sensors connected to cloud, presence of pathogens can be identified and measures can be taken to avoid the proliferation of pathogens. Innovative technologies can help in achieving process control with application of hazard analysis and critical control point (HACCP) to monitor the production process to ensure food quality and safety (Kourkoutas et al. 2016).

The digital identification tags (such as: radiofrequency identification (RFID), quick response (QR) codes, microchips, sensors, anti- pilferage tags) are very beneficial in tracking the food products, determination of counterfeit products, storage conditions and pilferation to build consumer's trust.

6.5.3 *Reduction of Food Waste*

During production and processing of food products, considerable amount of waste is generated that has to be either recycled or reduced to decrease the harmful effects on the environment. With the application of innovative technologies, the food products can be tracked efficiently on real-time continuously, which provides actual information regarding the product rather than an estimate (Manasi and Bhat 2020).

Further, the process can be automated for efficient control of parameters to reduce wastage during processing where, uniform quality can be achieved. Food traceability ensures real- time tracking of the product, which enables waste reduction during storage and distribution (Ion and Gheorghe 2014). By actual values of the market, sales and production can be monitored accordingly, services like water, refrigerant, steam can be utilized efficiently. Thus, food waste reduction and sustainability can be achieved in food supply chain.

6.6 Conclusion

Food supply chain is an essential prerequisite for a country to deliver the needs of citizens. Due to varying climatic conditions and geographical locations, adoption of uniform supply chain management is difficult in tropical countries. Although sufficient amount of food is produced in India, yet most of it gets wasted due to poor supply chain management. Hence, innovative technologies have been adopted throughout the food supply chain to monitor the food transit. Application of innovative technologies (such as: Industry 4.0, Internet of Things, blockchain technology, cloud computing, artificial intelligence, precision technology, etc.) aids in efficient food supply chain management by using sensors, actuators, process control devices that can be monitored on real-time with remote control tools. Application of these technologies have ensured precise data collection, processing, controlling and monitoring actions with reduced human intervention. Benefits of application of innovative technologies include product traceability and tracking, waste reduction, product quality and safety with reduced resource utilization.

References

Abbasi R, Martinez P, Ahmad R (2022) The digitization of agricultural industry–a systematic literature review on agriculture 4.0. Smart Agric Technol 2:100042

Alcácer V, Cruz-Machado V (2019) Scanning the industry 4.0: a literature review on technologies for manufacturing systems. Eng Sci Technol 22(3):899–919

Armbruster WJ, McDonell MM (2014) Informatics to support international food safety. In: Proc. 28th EnvironInfo 2014 Conference, Oldenburg, Germany, pp 127–134. http://enviroinfo.eu/sites/default/files/pdfs/vol85.14/0127.Pdf. Accessed 14 November 2022

Badia-Melis R, Mishra P, Ruiz-García L (2015) Food traceability: new trends and recent advances. A review. Food Control 57:393–401
Bhoi BB, Kundu S, Kishore V, Suganthi D (2019) Supply chain dynamics and food inflation in India. RBI Monthly Bulletin, New Delhi, pp 1–17
Bigliardi B, Bottani E, Filippelli S (2022) A study on IoT application in the food industry using keywords analysis. Proc Comput Sci 200:1826–1835
Carisalo F, Dasaklis TK, Patsakis C (2019) A systematic literature review of blockchain-based applications: current status, classification and open issues. Telematics Inform 36:55–81
Castillo-Vergara M, Quispe-Fuentes I, Poblete J (2021) Technological innovation in the food industry: bibliometric analysis. Eng Econ 32(3):197–209
Cisternas I, Velásquez I, Caro A, Rodríguez A (2020) Systematic literature review of implementations of precision agriculture. Comput Electron Agric 176:105626
Demirbaş N (2018) Precision agriculture in terms of food security: needs for the future. In: X. IBANESS congress series, Macedonia, Greece, pp 308–313
Falcacer V, Kumar S (2018) Review of literature on industry 4.0. In: National convention of IIIE and international conference. Rotre of Industrial Engineering in Industry 4.0 Paradigm, Bhubaneswar, pp 37–38
Fayomi OSI, Adelakun JO, Babaremu KO (2019) The impact of technological innovation on production. J Phys Conf Ser 1378(2):022014
Frost AR, Schofield CP, Beaulah SA, Mottram TT, Lines JA, Wathes CM (1997) Review of livestock monitoring and the need for integrated systems. Comput Electron Agric 17(2): 139–159
Hamdy W, Mostafa N, Elawady H (2018) Towards a smart warehouse management system. Proc Int Conf Ind Eng Oper Manage 2018:2555–2563
Hasnan NZN, Yusoff YM (2018) Short review: Application areas of industry 4.0 technologies in food processing sector. In: 2018 IEEE student conference on research and development (SCOReD). Piscataway, IEEE, pp 1–6
Hassoun A, Aït-Kaddour A, Abu-Mahfouz AM, Rathod NB, Bader F, Barba FJ, Regenstein J (2022) The fourth industrial revolution in the food industry, part I: industry 4.0 technologies. Crit Rev Food Sci Nutr 2022:1–17. https://doi.org/10.1080/10408398.2022.2034735
Ion I, Gheorghe FF (2014) The innovator role of technologies in waste management towards the sustainable development. Proc Econ Finance 8:420–428
Jones D, Barnes M (2014) Fuzzy composite programming to combine remote sensing and crop models for decision support in precision crop management. J Agric Syst 6(2):137–158
Katoch R (2022) IoT research in supply chain management and logistics: a bibliometric analysis using VosViewer software. Mater Today Proc 56(10):2505–2515
Kitsantas T, Vazakidis A, Chytis E (2019) A review of blockchain technology and its applications in the business environment. In: Unpublished paper at international conference on enterprise, systems, accounting, logistics & management, Chania, Crete, Greece, p 8
Kourkoutas Y, Chorianopoulos N, Nisiotou A, Valdramidis VP, Karatzas KA (2016) Application of innovative technologies for improved food quality and safety. Biomed Res Int 2016:9160375
Krishnan R, Sanjay J, Gnanaseelan C, Mujumdar M, Kulkarni A, Chakraborty S (2020) Assessment of climate change over the Indian region. In: Report of the Ministry of Earth Sciences (MOES), Government of India. Springer, Cham, p 226
Kumar S, Narkhede B, Jain K (2018) Industry 4.0: literature review and future research directions. Rotre of Industrial Engineering in Industry 4.0 Paradigm, Bhubaneswar, p 10
Kumar S, Prayag Tiwari P, Zymbler M (2019) Internet of things is a revolutionary approach for future technology enhancement: a review. J Big Data 6:111
Kumperščak S, Medved M, Terglav M, Wrzalik A, Obrecht M (2019) Traceability systems and technologies for better food supply chain management. Sciendo 1(1):567–574
Lagorio A, Zenezini G, Mangano G, Pinto R (2020) Systematic literature review of innovative technologies adopted in logistics management. Int J Log Res Appl 25(7):1043–1066

Li J, Mahalik NP (2019) A big data and cloud computing specification, standards and architecture: agricultural and food informatics. Int J Inf Commun Technol 14(2):159–174

Li J, Li X, Peng Y (2019) Application of big data in agricultural internet of things. Rev Fac Agron 36:1521–1529

Li K, Lee JY, Gharehgozli A (2021) Blockchain in food supply chains: literature review and synthesis analysis of platforms, benefits and challenges. Int J Prod Res 2021:1–20. https://doi.org/10.1080/00207543.2021.1970849

Manasi S, Bhat H (2020) Eco-innovations in waste management a review of high point cases. Institute for Social and Economic Change, Bangalore Working Papers, p 497. https://ideas.repec.org/p/sch/wpaper/497.html. Accessed 9 December 2022

Masudin I, Sumah B, Zulfikarijah F, Restuputri DP (2021) Effect of information technology on warehousing and inventory management for competitive advantage: theoretical framework. In: Almunawar MN, Ali MA, Lim SA (eds) Handbook of research on innovation and development of E-commerce and E-business in ASEAN. New Delhi, IGI Global, pp 570–593

Mavani NR, Ali JM, Othman S, Hussain MA, Hashim H, Rahman NA (2021) Application of artificial intelligence in food industry - a guideline. Food Eng Rev 14:134–175

Mavrnico T, Sami N, Sohail SS (2019) A review paper on Internet of Things (IOT). Indian J Appl Res 9(8):27–36

Nafrees ACM, Majeed UA, Kariapper R, Razith S, Pirapuraj P (2021) Internet of Things (IoT) enabled food technologies: a systematic review approach. Sri Lankan J Technol 2(2):6–13

Naresh RK, Kumar S, Chandra MS, Pathak SO, Gupta SK, Gawdiya S, Singh PK (2021) Application of digital technologies for the next level of agriculture growth and transformation under changing the Indian agriculture: a critical review. Int J Environ Clim Change 11(8):47–67

Negi S, Anand N (2015) Issues and challenges in the supply chain of fruits & vegetables sector in India: a review. Int J Manag Value Supply Chains 6(2):47–62

Onyegbula F, Dawson M, Stevens J (2011) Understanding the need and importance of the cloud computing environment within the National Institute of Food and Agriculture (United States Department of Agriculture). J Inf Syst Technol Plan 4(8):17–42

Pandey V, Pant M, Snasel V (2022) Blockchain technology in food supply chains: review and bibliometric analysis. Technol Soc 69:101954

Pierpaoli E, Carli G, Pignatti E, Canavari M (2013) Drivers of precision agriculture technologies adoption: literature review. Proc Technol 8:61–69

Qiao Y, Kong H, Clark C, Lomax S, Su D, Eiffert S, Sukkarieh S (2021) Intelligent perception for cattle monitoring: a review for cattle identification, body condition score evaluation, and weight estimation. Comput Electron Agric 185:106143

Rahul K, Banyal RK, Kumar V, Hooda D (2022) Cloud computing: technological innovations in the food industry. In: Mor RS, Kamble SS, Sangwan KS (eds) Operations and supply chain management in the food industry. Springer, Singapore, pp 127–141

Ramirez-Asis E, Vilchez-Carcamo J, Thakar CM, Phasinam K, Kassanuk T, Naved M (2022) A review on role of artificial intelligence in food processing and manufacturing industry. Mater Today Proc 51(8):2462–2465

Rejeb A, Keogh JG, Rejeb K (2022) Big data in the food supply chain: a literature review. J Data Inf Manage 4:33–47

Rupali M, Amit P (2017) Review paper on general concepts of artificial intelligence and machine learning. Int Adv Res J Sci Eng Technol 4(4):79–82

Rylander L, Olofsson F (2021) Implementation of technology in warehouse operations: a case study on internal logistics in warehouses from a sustainable perspective, p 56. https://www.diva-portal.org/smash/get/diva2:1572242/FULLTEXT01.pdf. Accessed 10 December 2022

Samara S, Natsheh E (2019) Intelligent real-time photovoltaic panel monitoring system using artificial neural networks. IEEE Access 7:50287–50299

Sekaran K, Meqdad MN, Kumar P, Rajan S, Kadry S (2020) Smart agriculture management system using internet of things. Telkomnika 18(3):1275–1284

Shalloo L, Byrne T, Leso L, Ruelle E, Starsmore K, Geoghegan A, O'Leary N (2021) A review of precision technologies in pasture-based dairying systems. Irish J Agric Food Res 59(2):279–291
Sharma V, Tiwari R (2016) A review paper on IOT & its smart applications. Int J Sci Eng Technol Res 5(2):472–476
Sivakumar B, Nanjundaswamy C (2021) Weather monitoring and forecasting system using IoT. Glob J Eng Technol Adv 8(2):8–16
Spandana K, Pabboju S (2019) Applications of IoT for soil quality. In: Gunjan VK, Diaz VG, Cardona M, Solanki VJ, Sunitha KVN (eds) ICICCT 2019 – system reliability, quality control, safety, maintenance and management: applications to electrical, electronics and computer science and engineering. Springer, Singapore, pp 277–286
Srivastava P, Khan R (2018) Review paper on cloud computing. Int J Adv Res Comput Sci Softw Eng 8(6):17–20
Subudhi BN, Rout DK, Ghosh A (2019) Big data analytics for video surveillance. Multimed Tools Appl 78:26129–26162
Suprem A, Mahalik N, Kim K (2013) A review on application of technology systems, standards and interfaces for agriculture and food sector. Comput Standards Interf 35(4):355–364
Tyagi A (2017) A review paper on cloud computing. Int J Eng Res Technol 23(5):1–2
Ullo SL, Sinha GR (2021) Advances in IoT and smart sensors for remote sensing and agriculture applications. Remote Sens 13(13):2585–2590
Zhen L, Li H (2022) Literature review of smart warehouse operations management. Front Eng Manage 9:31–55
Zhong R, Xu X, Wang L (2017) Food supply chain management: systems, implementations, and future research. Ind Manag Data Syst 117(9):2085–2114

Chapter 7
Microplastics in Foods: An Emerging Food Safety Threat

Shalini Sehgal, R. Kurup Krishna, and A. R. Yeswanth

Contents

Abstract The use of plastics has been increasing since World War One. Improper disposal and increased usage of plastics have led to the pileup of plastics in our surroundings. Microplastics (MPs) are small plastic particles of size below 5 mm. They are formed either due to the degradation of plastics or by deliberate addition to different products. In this paper, we have discussed how microplastics enter the environment and their food safety implications. Various studies have confirmed the presence of MPs in several foods including drinking water, tap water, seafood, salts, etc. The concentrations of MPs in Indian seafood, drinking water, salt and tap water has also been compiled. Chemicals and biological contaminants are adsorbed to MPs in different ways. Studies have shown that MPs can have biofilms called plastisphere. MPs were also found in samples of the placenta. Based on the studies conducted, the immediate risk to human health from MPs alone are minimal but the ability to transport and host different contaminants are concerning.

Keywords Microplastics · Pollution · Food chain · Environment · Food safety

S. Sehgal (✉) · R. K. Krishna · A. R. Yeswanth
Department of Food Technology, Bhaskaracharya College of Applied Sciences, University of Delhi, New Delhi, India

J. A. Malik et al. (eds.), *Food Process Engineering and Technology*,
https://doi.org/10.1007/978-981-99-6831-2_7

7.1 Introduction

Plastics have been extensively used for products in almost everything in our day to day life because of their user-friendliness, versatility and cost-effectiveness. These properties of plastics that make it omnipresent around us (Rao 2019; De-la-Torre 2019). Due to the environmental repercussions of plastics, it has become a widespread issue, which affects not only living organisms but also the natural cycles of the planet (Frias et al. 2021).

Microplastics (MPs) are microscopic plastic particles of size 5 mm and below (Lechthaler et al. 2021; Rao 2019; Munier and Bendell 2018; De-la-Torre 2019; Acquah et al. 2021). MPs are omnipresent and can be extracted from almost every living organism. MP concentrations in rivers such as Brahmaputra, Indus, Kosasthalaiyar and the Adyar have been quantified (Lechthaler et al. 2021; Shopova et al. 2020). These studies showed that the MP concentrations were much higher than earlier studies. This can be a result of the differences in the quantification techniques. MPs are essentially inert but they generally have an inclination to soak up poisonous pollutants and dangerous microorganisms (Rao 2019). Data like the response of immune cells to microplastics, the degree of uptake and therefore the response of various organs or the after effects of microplastics and also the toxicity of different compound varieties are all little or not better-known (Rao 2019).

Numerical simulations were utilised by a research group for assessing the inflow of plastics to the ocean by different watercourses. This explicit study concluded and showed us findings that include, Asian rivers are the major pathway for discharging of bulk of the plastics and Ganges River in India, releases regarding 1–3 billion microplastics to the ocean on a usual basis is the second most contaminated river within the world (Lechthaler et al. 2021; Lebreton et al. 2017). Another study emphasised the lacking information on baseline stages of microplastic pollution of southern indian rivers alongside the south west coast of the Indian peninsula (Lechthaler et al. 2021; Tympa et al. 2021). The results when compared suggested that in the Kosasthalaiyar River concentration of microplastics was double as much as the Adyar River maybe because of Kosasthalaiyar River's high catchment area and due to the industrialization around its catchment area (Lechthaler et al. 2021).

The pandemic has already and will also aggravate the MP contamination in the coming years. This is due to the usage of single-use plastics in masks, PPE kits etc. The MP fibers that are released from masks alone in South Korea and were found to be more than 1381 million particles per day (Dissanayake et al. 2021).

7.2 Sources of Microplastics in the Environment

All organisms interact with their environment and any contaminants if present. The versatility of MPs allows them to enter these interactions as shown in Fig. 7.1. MPs have two sources, primary and secondary which is how it is introduced to the

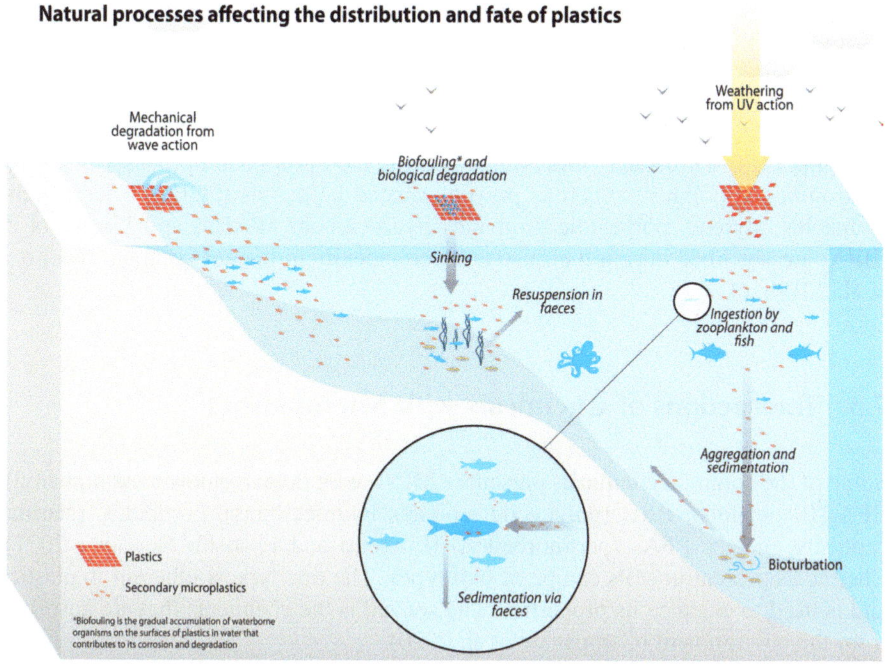

Fig. 7.1 Distribution of plastics in the environment. (Source: Maphoto/Riccardo Pravettoni 2018, www.grida.no/resources/6911)

environment. Primary sources mainly are made up of the microbeads present in cosmetic products. These are intentionally added to the products to make them abrasive and their size is 1 mm and below (Rao 2019). Primary microplastics can enter the surroundings in various ways like accidentally or may be lost throughout the transportation or cargo or conjointly discharged from manufacturing facilities. These can be carried on to oceans or freshwater ecosystems if the waste products treatment systems are too little to sustain (Acquah et al. 2021). Secondary sources are through the degradation of plastics (Rao 2019; Acquah et al. 2021). There are multiple ways for the entry of secondary microplastics into the environment such as fabric substances that enter sewage through the air after cleaning or drying, etc. They also come from bigger plastic objects when they undergo fragmentation (Acquah et al. 2021).

Entry of microplastics into the marine environment is mainly from land to water bodies by accidental loss or mishandling of plastics, littering, mismanaged landfills etc. Wastewater management plants are also taken into consideration as a critical pathway of microplastics to the marine environment (Kazour et al. 2019). MPs were found even in the snow in the Italian Alps, this can be due to human activity or movement of MPs through the air and then coming down of these airborne MPs due to precipitation or settling due to gravity (Oliveri Conti et al. 2020; Evangeliou et al. 2020).

As Microplastics get accumulated in different types of ecosystems at an increasing rate these days, they are an issue of great concern. Real environmental risks posed by MPs remain uncertain as to the two main factors, Precariousness and mutability in the data, hinder a realistic assessment (Barboza et al. 2018a; Kontrick 2018). In the case of marine ecosystems, microplastics may act as chemical vehicles, including those deliberately incorporated during their production process, as well as environmental pollutants such as polychlorinated biphenyls (PCBs), toxic metals, phthalates, styrene, polycyclic aromatics hydrocarbons (PAHs) and bisphenol A (BPA) are absorbed during their use and permanently in the environment (Barboza et al. 2018a, c).

7.3 Interactions of Chemicals with Microplastics

Some of the common chemicals present in MPs are PCB (polychlorinated biphenyl), BPA (Bisphenols), PAH (polycyclic aromatic hydrocarbons), pesticides, pharmaceuticals, metals, PFAs (perfluoroalkyl) (Lionetto and Esposito Corcione 2021). Chemicals present in MPs can be of two types. The first type is inherent in plastics and is used to enhance its properties. The second is the chemicals that are adsorbed from the environment (Campanale et al. 2020).

The chemicals that adsorb to MPs fall into three categories due to hydrophobic interaction, biofilm interaction and interaction between inherent chemicals and pollutants in the environment (Veerasingam et al. 2020). Organic contaminants are adsorbed to MPs by following interactions pore filling, hydrogen bonding, hydrophobic partitioning, electron donor-acceptor, Electrostatic, van der Waals force (Verla et al. 2019).

Studies conducted in Burrard Inlet in Canada provide an insight into which trace metals are adsorbed in MPs, which suggests that microplastics will be a sink for trace metals and an inherent source of chemicals. They also stated that PVC in particular will be a sink for copper and lead. It will also contain inherent chemicals like cadmium and zinc (Mason et al. 2018). MPs act as a sink for chemicals due to their lipophilicity and higher surface area of small particles (Al Naggar et al. 2021).

Dicentrarchus labrax juveniles when subjected to mercury along with MPs showed that MPs increased the uptake of mercury significantly (Barboza et al. 2018b). Mercury in the gills and liver were examined and this accumulation of more mercury can be attributed to the adsorption properties of MPs. Phthalate esters can get adsorbed to MPs and this adsorption was in the order DEHP (Bis (2-Ethylhexyl)phthalate) > DBP (DibutylPhthalate) > DEP (Diethyl Phthalate) > DMP (Dimethyl phthalate) (Deng et al. 2020).

7.4 Food and Microplastics

MPs were found in all aquatic organisms, salts, drinking water, honey, beverages, packaged foods, vegetables, organic fertilizers and even dust (Jin et al. 2021; Kontrick 2018). The entire food web is contaminated with the microplastics as reflected in Fig. 7.2. A study assessed the MP concentrations in vegetables and fruits in Italy. Apple (*M. domestica*), pear (*P. communis*), broccoli (*B. oleraceaitalica*), lettuce (*L. sativa*) and carrots (*D. carota*) were found to be contaminated with MPs (Munier and Bendell 2018). MPs can enter root vegetables through the 'crack entry' method and can localise within root tissues (Tsering et al. 2021).

In recent studies microplastic concentrations of 0.44 and 0.11 per g were found in sugar and salt respectively. Whereas in alcohol the concentration was 0.03 per g and 0.09 per g in case bottled water (Cox et al. 2019). Human intake of MPs through plant diet is approximately 80 g each day. The plants take up them from polluted soil and water bodies (Ebere et al. 2019). MPs in seafoods like the filter feeders are well documented. Some of the examples from European countries are the *Mytilus galloprovincialis* and *Mytilus edulis*, the concentration of MPs is in the range of 3–5 fibres/10 g of mussels. So human consumption of microplastics is inevitable and MPs smaller than 150 μm can move through the gastrointestinal epithelium (De-la-Torre 2019).

Superior-quality protein, omega-3 fatty acids, minerals and vitamins (A, B, D) are some of the major nutrients that are present in fish meat that is beneficial for human health. The nutrients in fish especially omega-3 fatty acids reduce the risk of triglyceride levels, arrhythmias, atherosclerotic plaques and lower blood pressure, intake of fish can reduce the instances of heart problems in the population. Even

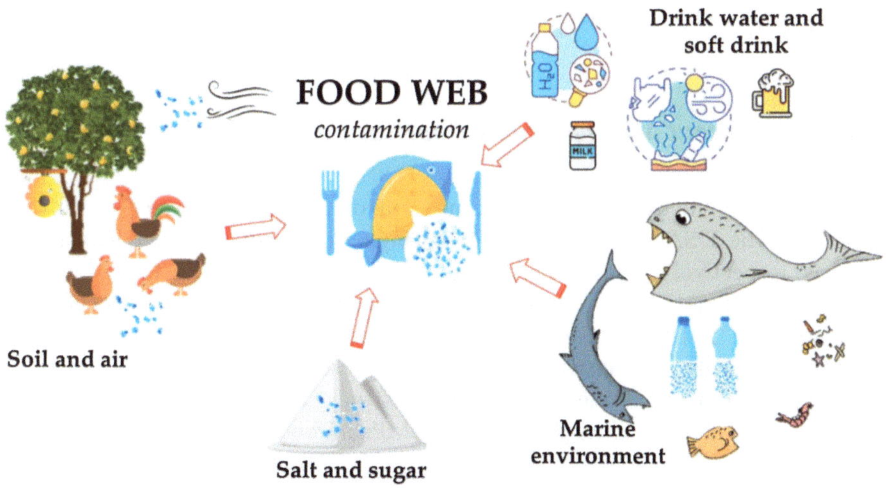

Fig. 7.2 Microplastics as contaminants in the food ecosystem. (Source: Pironti et al. 2021)

Though fish is good for health, post-harvest quality parameters such as microbial numbers, protein destruction, Oxidisation of fat, pathogens and toxicants, etc. determine its consumption. Small plastic particles have been reported in more than 690 marine species which was ingested by them and this is in different trophic levels. The colorful, buoyant small sized particles is perceived as food by the fish and consequently gets ingested (Rao 2019). Tissues of different organs like gills, stomach, intestine, liver etc., of the organisms is where microplastics were detected mostly (Jin et al. 2021). Microplastic Consumption by the marine fauna is accidental in food chain. Biological functions are affected when the accumulation takes place in tissues and cells (Abihssira-García et al. 2020).

The degradation of MPs by organisms in the plant's rhizosphere is not well researched (Wang et al. 2020). However, the structural makeup of plants mainly their biomass and tissues, their symbiosis could be significantly altered by the microplastics in soil. The MPs are also found in terrestrial edible snails such as *Helix aspersa Helix pomatia and Helix aperta* which can be used as a gauge for the food safety risks from food consumption. Microplastics were also found in honey and sugar tap water, drinking mineral water, beer, table salt and canned foods. Several reports have pointed out that microplastics that have become widespread in human consumption. Even though the intake of MPs is low, chronic exposure and consumption of microplastics by humans can pose a potential food safety hazard to human health (De-la-Torre 2019). MPs have also been isolated from various processed liquid foods such as milk, honey, and beer. The concentrations of microplastics ranged from being not detectable to few 100 particles/l (Kwon et al. 2020).

The impact of MPs in the species of the *Araceae* family was quantified and the MPs caused stress by decreasing the root growth in the plant because of mechanical blockage. The Mps also blocked the pores of the seed coat of the *Lepidium sativum* (garden cress) thereby restricting the water uptake and delaying the germination (Carrasco Silva et al. 2021).

7.4.1 Different Microplastic Concentrations of Indian Foods

A recent study focused on the concentration of MPs of tap water in different countries. The USA and UK were having higher concentrations per liter at 7.73 and 9.24 respectively. But Indian concentrations were lower at 6.24, Other European countries such as France, Germany and Switzerland had lower concentrations (Kosuth et al. 2018). The differences can be attributed to many things such as population density or industrialization but the presence in tap water points to the widespread nature of MPs and one of the easiest ways through which it can enter the food system. Table 7.1 gives the amount of MPs in food commodities in India.

Table 7.1 shows the concentration of MPs in food commodities and water in India

Food commodity	Source	Concentration (particles)	References
Salt	Eight different brands	103 ± 39 to 56 ± 49 per kg	Sathish et al. (2020)
Tap water	New Delhi	6.24 per liter	Kosuth et al. (2018)
Pelagic fish species	Kerala	0.005 ± 0.02 per g (edible tissue)	Daniel et al. (2020a)
		0.054 ± 0.098 per g (inedible tissue)	
Shellfish	Kerala	2.7 ± 12 per kg (edible tissue)	Daniel et al. (2020a)
Squid (*U.* (*P*) *duvaucelii*)	Kerala	7.7 ± 20 per kg (edible tissue)	Daniel et al. (2021)
Flower crab (*P. pelagicus*)	Kerala	3.2 ± 10 per kg (edible tissue)	Daniel et al. (2021)
Indian white shrimps (*Fenneropenaeus indicus*)	Kerala	0.04 ± 0.07 per g	Daniel et al. (2020b)
Indian white shrimps (*Fenneropenaeus indicus*)	Indian ocean	5570 ± 100 per g w.w	Curren et al. (2020)
Oysters (*Crassostrea sp.*)	Goa (sal estuary)	4 ± 2 per g	Ragusa et al. (2021)
Asian green mussel (*Perna viridis*)	Goa (sal estuary)	3.2 ± 1.8 per g	Ragusa et al. (2021)
Short neck clam (*Paphia malbarica*)	Goa (sal estuary)	0.7 ± 0.3 per g	Parolini et al. (2021)
Flathead grey mullet (*Mugil cephalus*)	Goa (sal estuary)	7.8 ± 4 per g	Ragusa et al. (2021)
Whipfin silver-biddy (*Gerres filamentosus*)	Goa (sal estuary)	5.3 ± 4.9 per g	Ragusa et al. (2021)
Blackfin Sea catfish (*Arius jella*)	Goa (sal estuary)	4.6 ± 2.6 per g	Ragusa et al. (2021)
Green chromide (*Etroplus suratensis*)	Goa (sal estuary)	1.4 ± 0.3 per g	Ragusa et al. 2021)
Bottled drinking water (Bisleri)	Chennai	826 per liter	
Bottled drinking water (Bisleri)	Mumbai	203 per liter	
Bottled drinking water (Bisleri)	New Delhi	3.72 per liter	
Bottled drinking water (Aquafina)	Chennai	174 per liter	
Sea salt	Tuticorin, Tamil Nadu	35 ± 15 to 72 ± 40/kg	Saha et al. (2021)
Borewell salt	Tuticorin, Tamil Nadu	2 ± 1 to 29 ± 11/kg	Saha et al. (2021)

7.5 Sources of Microplastics in Human Body

MPs get introduced into the body through the air by inspiration (Amato-Lourenço et al. 2020), through consumption and through skin contact (Rao 2019; Jin et al. 2021; Seth and Shriwastav 2018; Munier and Bendell 2018; Kosuth et al. 2018). Though human skin inhibits microplastics to directly pass through, attainable pathways are hair follicles, open skin injuries and sweat glands. Although these are all pathways of microplastics to the human body, yet those that enter through the food cause a larger warning due to residual monomers, longtime interactions with other toxic pollutants, leaching of plastics additives, weathering and also due to pathogenic microorganisms from the environment. The estimation of microplastic ingestion by humans depends on various factors such as content, consumption habits, etc. For e.g., Ingestion of microplastics through contaminated mussels are less than through airborne fibers that fall into our meal (De-la-Torre 2019; Campanale et al. 2020; Brennecke et al. 2016; Yang et al. 2019).

In Order to facilitate the uptake of microplastics to the human body, there are multiple particle characteristics which include size, surface functionalization, hydrophobicity, etc. (Weis and Palmquist 2021).

7.6 Food Safety Concerns and Human Health

FAO defines food security as "Food security (is) a situation that exists when all people, at all times, have physical, social and economic access to sufficient, safe and nutritious food that meets their dietary needs and food preferences for an active and healthy life" (Reforms 2003). According to FAO 2009, "the four main pillars of food security are food availability, access, utilization and stability" (FAO 2009). Microplastics in food suggests that the food accessibility and approach pillars of food safety have been compromised (De-la-Torre 2019).

Human consumption of microplastics is unavoidable and is a threat to both food safety and health of humans (De-la-Torre 2019). Studies have shown that MPs can have biofilms called "plastisphere". This plastisphere can host pathogens and serve as a source for the transfer of Metal and Antibiotic resistant genes (MRGs, ARGs) (Bowley et al. 2021; Wu et al. 2020). Filter feeders such as oysters can reject microplastics when ingested. When oysters were given virgin microbeads and *E. coli* coated microbeads. *E. coli* coated microbeads were ingested more than the virgin microbeads, this shows that the ability of microplastics to form biofilms can trick the organisms such as filter feeders to ingest more microplastics (Fabra et al. 2021). The organisms in turn get contaminated with these pathogens and can pose food safety issues for humans. But these methods of using primary MPs cannot give accurate results and their effects are underestimated (Amato-Lourenço et al. 2020; Wang et al. 2019). Secondary MPs have different physical and chemical properties due to weathering by different physical, biological and chemical agents in the environment (Amato-Lourenço et al. 2020).

The placental samples of four women using Raman spectroscopy were taken to study the presence of MPs. The samples were taken from the fetal side, maternal side, chorioamniotic membranes and found 5,4,3 MPs respectively. These MPs consisted of polypropylene particles, stains used for plastics and MPs from cosmetic use and had sizes between 5 and 10 μm (Prata et al. 2020). When bisphenol concentrations from PVC microplastics were analysed and were set at 0.1 mg per liter bisphenol desorption was 1.8–15.3% in water and 43.7–54.0% in intestinal-like conditions. When it was set at 0.2 mg per liter the desorption in water was 24.9–43.1% and in intestinal conditions was 58.4–63.1% (Wright and Kelly 2017). This shows that chemicals can be transferred from microplastics to the human body. MPs can also carry and emit phthalate esters into the body and result in intestinal metabolic disarrays and swelling in the mouse gut (Deng et al. 2020).

In marine organisms, after ingestion the synthetic polymers are not digested via enzymatic activity and hence the MPs are retained. These retained MPs transfer through the food chain and accumulate in more complex organisms like human beings. MPs of size lower than 150 μm can move through the gastrointestinal epithelium. The main sites for the uptake of microparticles to enter the circulatory system are the microfiber present in Peyer's patches and it happens through endocytosis (Campanale et al. 2020; De-la-Torre 2019; Yang et al. 2019). The smallest of microplastics may access all organs and also transport through all cellular membranes (De-la-Torre 2019). This presence of microplastics at the level of gastro-intestine was confirmed by finding microplastics in human stool (Shopova et al. 2020). Microplastics can also move into other tissues and cause persistent swelling and oxidative strain which can lead to chronic inflammation and increased risk of neoplasia (Parolini et al. 2021).

7.7 Conclusion

Though we have a robust food industry and different foods are consumed in different parts of the world, only limited studies are available which analyse the presence of microplastics in food. The majority of all these studies emphasise more on the physical characteristics of the particles and the concentration of MPs in food, especially aquatic organisms. The overall human exposure to MP through food consumption and its effects is more difficult to quantify due to lack of studies related to MPs. The decontamination of MPs during the processing of food plays a vital role as the primary intake of MPs is from the consumption of finished food products.

The microplastic concentrations are low but the ability to adsorb pollutants is a serious concern. They also have the ability to transfer both biological and chemical contaminants to humans through food. Taking into account the increasing usage of plastics and their improper disposal, microplastics could be a big concern for food safety and human health in the coming years.

References

Abihssira-García IS, Park Y, Kiron V, Olsvik PA (2020) Fluorescent microplastic uptake by immune cells of Atlantic Salmon (Salmo salar L.). Front Environ Sci 8:1–11. https://doi.org/10.3389/fenvs.2020.560206

Acquah J, Liu H, Hao S, Ling Y, Ji J (2021) Microplastics in freshwater environments and implications for aquatic ecosystems: a mini review and future directions in Ghana. J Geosci Environ Prot 09(3):58–74. https://doi.org/10.4236/gep.2021.93005

Al Naggar Y, Brinkmann M, Sayes CM, Al-Kahtani SN, Dar SA, El-Seedi HR, Grünewald B, Giesy JP (2021) Are honey bees at risk from microplastics? Toxics 9(5):2–5. https://doi.org/10.3390/toxics9050109

Amato-Lourenço LF, dos Santos GL, de Weger LA, Hiemstra PS, Vijver MG, Mauad T (2020) An emerging class of air pollutants: potential effects of microplastics to respiratory human health? Sci Total Environ 749:141676. https://doi.org/10.1016/j.scitotenv.2020.141676

Barboza LGA, Dick Vethaak A, Lavorante BR, Lundebye AK, Guilhermino L (2018a) Marine microplastic debris: an emerging issue for food security, food safety, and human health. Mar Pollut Bull 133:336–348. https://doi.org/10.1016/j.marpolbul.2018.05.047

Barboza LGA, Vieira LR, Branco V, Carvalho C, Guilhermino L (2018b) Microplastics increase mercury bioconcentration in gills and bioaccumulation in the liver and cause oxidative stress and damage in Dicentrarchus labrax juveniles. Sci Rep 8(1):1–4. https://doi.org/10.1038/s41598-018-34125-z

Barboza LGA, Vieira LR, Branco V, Figueiredo N, Carvalho F, Carvalho C, Guilhermino L (2018c) Microplastics cause neurotoxicity, oxidative damage, and energy-related changes and interact with the bioaccumulation of mercury in the European seabass Dicentrarchus labrax (Linnaeus 1758). Aquat Toxicol 195:49–57. https://doi.org/10.1016/j.aquatox.2017.12.008

Bowley J, Baker-Austin C, Porter A, Hartnell R, Lewis C (2021) Oceanic Hitchhikers – assessing pathogen risks from marine microplastic. Trends Microbiol 29(2):107–116. https://doi.org/10.1016/j.tim.2020.06.011

Brennecke D, Duarte B, Paiva F, Caçador I, Canning-Clode J (2016) Microplastics as a vector for heavy metal contamination from the marine environment. Estuar Coast Shelf Sci 178:189–195. https://doi.org/10.1016/j.ecss.2015.12.003

Campanale C, Massarelli C, Savino I, Locaputo V, Uricchio VD (2020) A detailed review study on potential effects of microplastics and additives of concern on human health. Int J Environ Res Public Health 17(4):3–18. https://doi.org/10.3390/ijerph17041212

Carrasco Silva G, Galleguillos Madrid F, Hernández DM, Pincheira G, Peralta AK, Urrestarazu Gavilán M, Vergara-Carmona V, Fuentes-Peñailillo F (2021) Microplastics and their effect in horticultural crops: food safety and plant stress. Agronomy 11(8):1528. https://doi.org/10.3390/agronomy11081528

Cox KD, Covernton GA, Davies HL, Dower JF, Juanes F, Dudas SE (2019) Human consumption of microplastics. Environ Sci Technol 53(12):7068–7074. https://doi.org/10.1021/acs.est.9b01517

Curren E, Leaw CP, Lim PT, Leong SCY (2020) Evidence of marine microplastics in commercially harvested seafood. Front Bioeng Biotechnol 8:3–6. https://doi.org/10.3389/fbioe.2020.562760

Daniel DB, Ashraf PM, Thomas SN (2020a) Microplastics in the edible and inedible tissues of pelagic fishes sold for human consumption in Kerala, India. Environ Pollut 266:115365. https://doi.org/10.1016/j.envpol.2020.115365

Daniel DB, Ashraf PM, Thomas SN (2020b) Abundance, characteristics, and seasonal variation of microplastics in Indian white shrimps (Fenneropenaeus indicus) from coastal waters off Cochin, Kerala, India. Sci Total Environ 737:139839. https://doi.org/10.1016/j.scitotenv.2020.139839

Daniel DB, Ashraf PM, Thomas SN, Thomson K (2021) Microplastics in the edible tissues of shellfishes sold for human consumption. Chemosphere 264:128554. https://doi.org/10.1016/j.chemosphere.2020.128554

De-la-Torre GE (2019) Microplastics: an emerging threat to food security and human health. J Food Sci Technol 57(5):1601–1608. https://doi.org/10.1007/s13197-019-04138-1

Deng Y, Yan Z, Shen R, Wang M, Huang Y, Ren H, Zhang Y, Lemos B (2020) Microplastics release phthalate esters and cause aggravated adverse effects in the mouse gut. Environ Int 143: 105916. https://doi.org/10.1016/j.envint.2020.105916

Dissanayake J, Torres-Quiroz C, Mahato J, Park J (2021) Facemasks: a looming microplastic crisis. Int J Environ Res Public Health 18(13):2–8. https://doi.org/10.3390/ijerph18137068

Ebere EC, Wirnkor VA, Ngozi VE (2019) Uptake of microplastics by plants: a reason to worry or to be happy? World Sci News 131:256–267

Evangeliou N, Grythe H, Klimont Z, Heyes C, Eckhardt S, Lopez-Aparicio S, Stohl A (2020) Atmospheric transport is a major pathway of microplastics to remote regions. Nat Commun 11(1):3381. https://doi.org/10.1038/s41467-020-17201-9

Fabra M, Williams L, Watts JE, Hale MS, Couceiro F, Preston J (2021) The plastic Trojan horse: biofilms increase microplastic uptake in marine filter feeders impacting microbial transfer and organism health. Sci Total Environ 797:4–7. https://doi.org/10.1016/j.scitotenv.2021.149217

FAO (2009) Declaration of the World Summit on Food Security. WSFS 2009/2, Rome

Frias JP, Ivar do Sul JA, Panti C, ARA L (2021) Editorial: microplastics in the marine environment: sources, distribution, biological effects, and socio-economic impacts. Front Environ Sci 9:1–4. https://doi.org/10.3389/fenvs.2021.676011

Jin M, Wang X, Ren T, Wang J, Shan J (2021) Microplastics contamination in food and beverages: direct exposure to humans. J Food Sci 86:2816–2837. https://doi.org/10.1111/1750-3841.15802

Kazour M, Terki S, Rabhi K, Jemaa S, Khalaf G, Amara R (2019) Sources of microplastics pollution in the marine environment: importance of wastewater treatment plant and coastal landfill. Mar Pollut Bull 146:608–618. https://doi.org/10.1016/j.marpolbul.2019.06.066

Kontrick AV (2018) Microplastics and human health: our great future to think about now. J Med Toxicol 14(2):117–119. https://doi.org/10.1007/s13181-018-0661-9

Kosuth M, Mason SA, Wattenberg EV (2018) Anthropogenic contamination of tap water, beer, and sea salt. PloS One 13(4):e0194970. https://doi.org/10.1371/journal.pone.0194970

Kwon JH, Kim JW, Pham TD, Tarafdar A, Hong S, Chun SH, Lee SH, Kang D, Kim JY, Kim SB, Jung J (2020) Microplastics in food: a review on analytical methods and challenges. Int J Environ Res Public Health 17(18):6710. https://doi.org/10.3390/ijerph17186710

Lebreton LCM, van der Zwet J, Damsteeg J-W, Slat B, Andrady A, Reisser J (2017) River plastic emissions to the world's oceans. Nat Commun 8(1):15611. https://doi.org/10.1038/ncomms15611

Lechthaler S, Waldschläger K, Sandhani CG, Sannasiraj SA, Sundar V, Schwarzbauer J, Schüttrumpf H (2021) Baseline study on microplastics in Indian Rivers under different anthropogenic influences. Water 13(12):1–13. https://doi.org/10.3390/w13121648

Lionetto F, Esposito Corcione C (2021) An overview of the sorption studies of contaminants on poly(ethylene terephthalate) microplastics in the marine environment. J Mar Sci Eng 9(4):445. https://doi.org/10.3390/jmse9040445

Maphoto/Riccardo Pravettoni (2018). www.grida.no/resources/6911

Mason SA, Welch VG, Neratko J (2018) Synthetic polymer contamination in bottled water. Front Chem 6:2–10. https://doi.org/10.3389/fchem.2018.00407

Munier B, Bendell LI (2018) Macro and micro plastics sorb and desorb metals and act as a point source of trace metals to coastal ecosystems. PLoS One 13(2):e0191759. https://doi.org/10.1371/journal.pone.0191759

Oliveri Conti G, Ferrante M, Banni M, Favara C, Nicolosi I, Cristaldi A, Fiore M, Zuccarello P (2020) Micro- and nano-plastics in edible fruit and vegetables. The first diet risks assessment for the general population. Environ Res 187:109677. https://doi.org/10.1016/j.envres.2020.109677

Parolini M, Antonioli D, Borgogno F, Gibellino MC, Fresta J, Albonico C, de Felice B, Canuto S, Concedi D, Romani A, Rosio E, Gianotti V, Laus M, Ambrosini R, Cavallo R (2021) Microplastic contamination in snow from Western Italian Alps. Int J Environ Res Public Health 18(2):768. https://doi.org/10.3390/ijerph18020768

Pironti C, Ricciardi M, Motta O, Miele Y, Proto A, Montano L (2021) Microplastics in the environment: intake through the food web, human exposure and toxicological effects. Toxics 9(9):224. https://doi.org/10.3390/toxics9090224

Prata JC, da Costa JP, Lopes I, Duarte AC, Rocha-Santos T (2020) Environmental exposure to microplastics: an overview on possible human health effects. Sci Total Environ 702:134455. https://doi.org/10.1016/j.scitotenv.2019.134455

Ragusa A, Svelato A, Santacroce C, Catalano P, Notarstefano V, Carnevali O, Papa F, Rongioletti MCA, Baiocco F, Draghi S, D'Amore E, Rinaldo D, Matta M, Giorgini E (2021) Plasticenta: first evidence of microplastics in human placenta. Environ Int 146:106274. https://doi.org/10.1016/j.envint.2020.106274

Rao MB (2019) Microplastics in the aquatic environment: implications for post-harvest fish quality. Indian J Fish 66(1):143–147. https://doi.org/10.21077/ijf.2019.66.1.83125-20

Trade Reforms (2003) Food security: conceptualizing the linkages. Rome, FAO

Saha M, Naik A, Desai A, Nanajkar M, Rathore C, Kumar M, Gupta P (2021) Microplastics in seafood as an emerging threat to marine environment: a case study in Goa, west coast of India. Chemosphere 270:129359. https://doi.org/10.1016/j.chemosphere.2020.129359

Sathish MN, Jeyasanta I, Patterson J (2020) Microplastics in salt of Tuticorin, Southeast Coast of India. Arch Environ Contam Toxicol 79(1):111–121. https://doi.org/10.1007/s00244-020-00731-0

Seth CK, Shriwastav A (2018) Contamination of Indian sea salts with microplastics and a potential prevention strategy. Environ Sci Pollut Res 25(30):30534–30540. https://doi.org/10.1007/s11356-018-3028-5

Shopova S, Sieg H, Braeuning A (2020) Risk assessment and toxicological research on micro- and nanoplastics after oral exposure via food products. EFSA J 18:e181102. https://doi.org/10.2903/j.efsa.2020.e181102

Tsering T, Sillanpää M, Sillanpää M, Viitala M, Reinikainen SP (2021) Microplastics pollution in the Brahmaputra River and the Indus River of the Indian Himalaya. Sci Total Environ 789:147968. https://doi.org/10.1016/j.scitotenv.2021.147968

Tympa LE, Katsara K, Moschou PN, Kenanakis G, Papadakis VM (2021) Do microplastics enter our food chain via root vegetables? A Raman based spectroscopic study on Raphanus sativus. Materials 14(9):2329. https://doi.org/10.3390/ma14092329

Veerasingam S, Ranjani M, Venkatachalapathy R, Bagaev A, Mukhanov V, Litvinyuk D, Verzhevskaia L, Guganathan L, Vethamony P (2020) Microplastics in different environmental compartments in India: analytical methods, distribution, associated contaminants, and research needs. TrAC Trends Anal Chem 133:116071. https://doi.org/10.1016/j.trac.2020.116071

Verla AW, Enyoh CE, Verla EN, Nwarnorh KO (2019) Microplastic-toxic chemical interaction: a review study on quantified levels, mechanism and implication. SN Appl Sci 1(11):1400. https://doi.org/10.1007/s42452-019-1352-0

Wang J, Liu X, Li Y, Powell T, Wang X, Wang G, Zhang P (2019) Microplastics as contaminants in the soil environment: a mini-review. Sci Total Environ 691:848–857. https://doi.org/10.1016/j.scitotenv.2019.07.209

Wang F, Zhang M, Sha W, Wang Y, Hao H, Dou Y, Li Y (2020) Sorption behavior and mechanisms of organic contaminants to nano and microplastics. Molecules 25(8):1827. https://doi.org/10.3390/molecules25081827

Weis JS, Palmquist KH (2021) Reality check: experimental studies on microplastics lack realism. Appl Sci 11(18):8529. https://doi.org/10.3390/app11188529

Wright SL, Kelly FJ (2017) Plastic and human health: a micro issue? Environ Sci Technol 51(12):6634–6647. https://doi.org/10.1021/acs.est.7b00423

Wu P, Tang Y, Jin H, Song Y, Liu Y, Cai Z (2020) Consequential fate of bisphenol-attached PVC microplastics in water and simulated intestinal fluids. Environ Sci Ecotechnol 2:100027. https://doi.org/10.1016/j.ese.2020.100027

Yang Y, Liu G, Song W, Ye C, Lin H, Li Z, Liu W (2019) Plastics in the marine environment are reservoirs for antibiotic and metal resistance genes. Environ Int 123:79–86. https://doi.org/10.1016/j.envint.2018.11.061

Chapter 8
Scope of Genetically Engineered Organisms in Food Science and Pest Management Strategy

Kaushik Pramanik, Mainak Barman, Rimpa Kundu, and Satish Kumar Singh

Contents

K. Pramanik
Department of Agricultural Entomology, Bidhan Chandra Krishi Viswavidyalaya, Nadia, West Bengal, India

M. Barman (✉) · R. Kundu
Department of Genetics and Plant Breeding, Bidhan Chandra Krishi Viswavidyalaya, Nadia, West Bengal, India

S. K. Singh
Department Plant Breeding and Genetics, Dr. Rajendra Prasad Central Agricultural University, Samastipur, Bihar, India

J. A. Malik et al. (eds.), *Food Process Engineering and Technology*,
https://doi.org/10.1007/978-981-99-6831-2_8

Abstract The expansion of successful bioagents for the management of insect pests and mites has received increasing attention to prevent ecological pollution from chemical pesticides. Genetic engineering tools can help researchers to learn more about an organism's biology and pathogenicity. The process of changing an organism's genome by using current molecular technology is known as genetic engineering (also known as genetic alteration). It requires introducing synthetic genes or foreign DNA into the organism under study. The empathy of advantageous characters, cloning of genes impacting the corresponding characters and the advancement of procedures for presenting those genes in natural enemy species to increase their efficiency as biocontrol agents are all highlighted in this chapter. Artificial selective breeding and increasing genetic diversity are two methods for enhancing the genetic diversity of beneficial organisms; and various researchers have selectively bred beneficial species, with variable degrees of success. This chapter deals with these issues in an elaborative and simplified manner to make a clear understanding of the methodologies and current progress in this aspect of agriculture.

Keywords Artificial selection · Biological control · Crop protection · Genetic modification · Hybridization

8.1 Introduction

The habitat of any living creature is divided into two distinct areas: exterior and interior. The bulk of biological or other approaches for controlling insects and pests focus on how the outer environment can be handled or altered to reduce the population of pests. Internal biological environments, on the other hand, lend themselves to changes that can function against pests. In this regard, the genetic makeup of the pest and its potential for helpful alteration has received special attention. Insect pests used in agricultural aspects have changed dramatically in the first decade of the twenty-first century due to environmental and technological modifications.

Since the green revolution, India's crop losses have fallen from 17.5 to 23.3% (Dhaliwal et al. 2010). Biological control is a commercially feasible, efficient, technically appropriate, non-degrading, and socially acceptable pest management strategy. Its objective is to use naturally occurring organisms (parasites, predators and pathogens) to combat insects and pests that damage crops or other undesirable species. Nowadays, the improvement of successful bioagents for controlling insect and mite infestations has received renewed care. Natural enemy utilization—

whether by establishing novel species or enhancing the potency of those already present in the environment—is an intentional attempt at biocontrol (Sankaran 1986).

From Australia, the coccinellid beetle (*Cryptolaemus montrouzieri* (Muls.)) was introduced to India in 1898 and was the first and most effective biological control against an insect issue (Rao et al. 1971). The major advantages of using bioagents include: non-target organism and environmental safety, lack of mammalian toxicity, and lack of residual toxicity. In contrast, some drawbacks include: slow action, small host range, and low efficacy. In addition to improving an insecticidal product, genetic engineering can help researchers in better understanding of a specific organism's biology and manner of toxicity. Aside from pest control, this has found to hold widespread use in many aspects of modern agriculture.

To make an organism genetically modified, several characteristics like, brief life cycles, elevated reproductive potential, ease of mass rearing etc., are considered. This is especially true for studying insects, e.g., the honeybee (*Apis mellifera* L.), vinegar fly (*Drosophila melanogaster* Meigen) and silkworm (*Bombyx mori* L.); and these form the basis of most of the modern genetics to lessen environmental pest infestation. Imposing a genetic variation in insects is essential, whether caused by natural races or different mutagenic substances to which they are susceptible, as well as their genetic plasticity (as demonstrated by the widespread evolution of pesticide resistance).

Whitten (1970) noted that when a large number of bioagents are available for release, genetic alteration for insect pest destruction rarely involves far-less fundamental biological and environmental evidence on the organism of target than other tools of biocontrol. The most essential component is the ability of random mating between the released insects and the wild population. On the other hand, a substantial basic ecology study could prevent many of the issues associated with low-budget genetic manipulation programs (Whitten and Foster 1975).

Genetic modification involves utilizing a live organism to reduce the community of an insect pest that is believed to be harmful. Compared to the other pest control approaches, this technique has exceptional specificity and environmental safety. Modern DNA technology is used in genetic engineering (sometimes referred to as genetic alteration), which is the intentional direct manipulation of an organism's genome. It entails introducing synthetic genes or foreign DNA into the target organism.

In genetic engineering methods, two terms are used, namely: (a) genetically engineered organisms; and (b) genetically modified organisms. These two terms refer to creatures that have had their genetic makeup transformed through gene manipulation. At the same time, recombinant DNA technology refers to the processes that combine DNA molecules from several sources into a single molecule to produce a new one. The introduction of altered or new genes results from the subsequent implantation of DNA into an organism. Recombinant DNA (rDNA) molecules are DNA sequences created by combining genetic material from numerous sources using laboratory methods (molecular cloning), thus resulting in sequences that would not usually be present in those biological organisms.

In recent research, the parasites resistant to pesticides are picked in the laboratory and can be established in the field to improve integrated pest management (IPM) programs. Natural enemies, whether laboratory-bred or genetically modified, may play a significant role in IPM programs and can reduce pesticide consumption in near future. It is possible to genetically alter natural enemies of insects and pests to increase their effectiveness in agricultural cropping systems. Insect's natural enemies have been used in genetic enhancement programs for:

- Changes in host preference
- Enhanced host finding ability
- Improved climatic tolerances
- Improved synchronization with the host
- Inducing the lytokous reproduction
- Insecticide resistance
- Non-diapause

Monogenic characteristics are prone to switching into polygenic ones. Manipulating biological control agents in this manner appears to be a potential technique to control pests in the field. This chapter focuses on scope of genetically engineering organisms in food science and pest management strategy.

8.2 Scope of Genetic Engineering in Food Science and Food Process Engineering

Genetically modified organisms (GMOs) have been proved successful for improving the productivity of cereals and other crops as well. Using DNA-coated beads as carriers, the targeted genes are substantially battered into the plants using the biolistic transformation technique. Minimized costs for drug or food making, augmented crop yields, enhanced food quality and mineral composition, lessening of pesticide usages, resistance to pathogenic attacks, better food safety, and medicinal benefits for escalating population of the world are among few advantages of GMOs in agriculture. Furthermore, advancement has been put up in designing crops that mature early and can resist environmental hazards including flood, frost, drought, and salt. Invention of ornamental plant or bioplastic products are examples of further applications. Numerous creatures have also undergone genetic modifications in an effort to increase efficiency and decrease illness vulnerability. For example, farm animals have been developed to exhibit resistance to mad cow syndrome, whereas salmon has been designed to grow larger and mature faster.

While it is asserted that food biotechnology offers a promising solution to malnutrition and food scarcities by growing plant productivity and generating primary foods fortified with minerals, yet the accumulating evidence over the past 20 years since the introduction of genetically modified food to the marketplace does not entirely agree on these allegations. The longstanding effects of biotech plants on humans (such as: abnormal nutritional variations, allergic disorder, antibiotic resistance, and toxicity) are of specific interest to consumers. Furthermore, GM food stuffs are advertised as an ecologically safe crop by biotech companies and the scientists that work for them. Several scientific findings show that GM crops consuming animals have suffered with severe injury or even there is occurrence of death. When foraging on GM cottonseed, *Bt*-maize, and some GM-corn: pigs, cows, buffalo, goats, and other farm animals suffered difficulties, such as: infertility, miscarriages, early pregnancies, and many also dead. Rats fed on transgenic soya or potatoes, had aberrant immature sperm. However, this is a contentious topic because studies carried out by the biotech crop manufacturer did not demonstrate any adverse impacts of GM crops on mice. There are certain examples provided by the opponents, despite the fact that biotech corporations do not agree to the causal correlation between the intake of Genetically Modified Foods (GMFs) and problems with human health. For instance, foodborne illnesses like soya allergies have been more prevalent during the past 10 years in the UK and the US and there is a Morgellons infection epidemic in the US. Hundreds of cotton cultivators in India are also said to have had skin allergies, according to reports. According to recent research, *Bt*-corn expresses an allergenic protein that modifies the body's overall immune responses.

The rDNA technology is thought to be an effective tool for boosting food production, and the quality of food or drug factories for the creation of vaccinations and/or medical drugs. Genetically Modified Organisms are crucial for advancing organic agriculture, because they may be capable to decrease agriculture's ecological footprint, decreasing the use of chemicals, decreasing emissions of CO_2, conserving fossil fuels, and preserving soil and moisture. GM crops are crucial in tackling the serious global nutrition and food safety issues in growing countries. Although GM crops are not described as absolute solution, yet they could definitely make a substantial impact to a variety of dimensions and enticements to this constantly expanding problem. Enough knowledge of the product's safety tests is needed to allow the general public to make an informed choice regarding the consumption of GMF. Unfortunately, there are not many of these data available at present. Since the structure of crops cultivated in distinct regions may differ due to the growth and agronomic circumstances, therefore it is difficult to assess the dietary contents of GM crops with those of their traditional equivalents.

Obviously, this debate has no definitive answer, and it will probably go on for years. Plant biotechnology has the ability to address numerous issues in food production and offers number of advantages. However, it is also essential that the importation of GM crops into the ecosystem does not provide any fresh dangers, irreversible effects, or hazards to public health. To combat the enduring proliferation of agricultural productivity, it is preferable to achieve worldwide accession of regulations and legislations of GM crops within this framework.

8.3 Budding Scientific Doors to Boost Natural Enemies

8.3.1 *Artificial Selection*

Roush and Hoy (1981) have proposed the artificial choice of natural enemies for pesticide resistance as a process of enhancing natural enemies' utility in IPMs.

8.3.1.1 Development of Endosulfan-Tolerant *Trichogramma* Strain

In India, consumption of chemical pesticides accounts for 50% use in cotton (FICCI 2013). Endosulfan was one of India's most extensively applied pesticides, mostly applied on cotton, vegetable crops, and pulses to combat pests, such as: whiteflies, thrips, leafhoppers, aphids, and lepidopteran larvae (Jayashree and Vasudevan 2007). It had a tremendous harmful exposure impact on egg parasitoids (Trichogrammatids) (Jalali and Singh 1993; Stinner et al. 1974). Pesticides can have a long-term harmful effects on parasitoids, resulting in delayed development, longevity, parasitism capacity, lower emergence rates, fecundity, and mating probability.

Trichogramma chilonis has been developed to be endosulfan-tolerant by the National Bureau of Agricultural Insect Resources (NBAIR, Bangalore) (Jalali et al. 2006). Besides this *Trichogramma* strain, various tolerant strains like the Bio Scl and Bio H3 for graminaceous tissue borers, the Bio C1 and Bio C6 for cotton bollworms, and the Bio H3 for *Helicoverpa armigera* have been developed by NBAIR. Among these, few are 60–100% more productive compared to prior strains (Singh 2001). A strain of *T. chilonis* that is endosulfan—resistant has been produced, licensed and marketed as Endogramma. At NBAIR, this strain was again modified for multiple insecticidal leniences, viz. fenvalerate and monocrotophos.

8.3.1.2 Development of *Trichogramma* Strain Tolerant to High Temperature

The capability of *T. chilonis* to parasitize and survive declines as the temperature rises beyond 32 °C (Jayashree and Vasudevan 2007). The *T. chilonis* strains TcUP2 and TcUP1 were recently shown to exhibit greater biotic potential and resistant to higher temperatures (32–40 °C). These strains might be employed in warmer climates to combat sugarcane borers (NAIP 2012). *T. chilonis,* a genetically enriched Ludhiana strain tolerant to higher temperature, had host finding ranges as far away from the release site as 10 and 8 m (Singh and Shenhmar 2008).

8.3.1.3 Development of *Trichogramma* Strain Tolerant to Multiple Pesticides

T. chilonis strains tolerant to monocrotophos, endosulfan, and fenvalerate (MITS-TC) were produced and field-tested against *Helicoverpa armigera.* For crossings with tolerant male and female parents, endosulfan tolerance appeared to be recessive, Spinosad tolerance was seen to be semi-dominant, and lambda-cyhalothrin tolerance was discovered to be entirely dominant because selection leads to minor incremental improvement in pesticide resistance over time. It has been proposed that a polygenic form of inheritance will most likely result from laboratory selection for pesticide resistance (Roush and McKenzie 1987). For 34 generations, M1TS-TC was exposed to Spinosad, Indoxacarb, and Tebufenozide, which improved tolerance levels by 372.5, 761.4, and 137.0% over vulnerable populations. Endosulfan ($LC_{50} = 76.46$), Spinosad (2×10^5), Lambda-cyhalothrin (2×10^4), and Fenvalerate tolerance were found in *T. chilonis* strains TcTl, TcCbl, TcT4 and TcT5 (1×10^3), respectively (NAIP 2012).

According to Charles et al. (2011), Spinosad, Endosulfan, and Lambda-cyhalothrin tolerant strains of *T. chilonis* had LC_{50} values of 9.84, 278.03, and 6.23 ppm, respectively, compared to susceptible strains with LC_{50} values of 4.78, 106.03, and 3.45 ppm. However, the assumption that a single gene controlling the resistance does not exclude the use of a large population greater than 8 ppm for resistance selection (Jalali et al. 2006).

8.3.1.4 Genetically Engineered *Metaseialus occidentalis*

Metaseialus occidentalis is a predator of mite, which is remarkably adaptable that thrives in hot weather (80–110 °F) and survives in low humidity in inland California's lowlands. However, it does not succeed in cool coastal locations, where it enters a state of dormancy. This mite has been genetically engineered for carbaryl and permethrin resistance in California and scientific proper use of these could save $60–$110 per ha. Selection for improved fertility, temperature tolerance, non-diapause, and pesticide resistance has been made with various phytoseiid species (Hoy 1984). The phytoseiids' peculiar genetic system, which is para-haploid, could empower for rapid resistance evolution. However, ecological characteristics of these mites (low vagility and quick breeding period) predispose them to create pesticide resistance quickly (Hoy 1985).

8.3.1.5 Modification of *Aphytis melinus* (Parasitoid) Resistant to Carbaryl

Among various armored scale species parasitized by *Aphytis melinus,* the important ones are: the Oleander scale, San Jose scale, Latania scale, and California red scale.

In citrus orchards, *A. melinus*, readily convenient for commercial use, is frequently used to minimize California red scale. Before releasing the parasites, one of the most crucial steps is to monitor pesticide residues, because many chemicals are harmful to *Aphytis spp*. In 1989, a laboratory-out strain (R) of *A. melinus* was found in *Citrus sinensis* trees in California; and this strain was resistant to carbaryl residues (Spollen and Hoy 1992). The strong carbaryl resistance of the native populations of *A. melinus* prevented the formation and recuperation of R-strain in citrus orchard.

8.3.1.6 Development of Azinphosmethyl Resistant Population of Trioxys pallidus (Parasitoids)

In 1988, the California Agriculture Magazine reported an experimental assortment of a parasite strain of *Trioxys pallidus* Haliday (walnut aphid) with azinphosmethyl (Guthion) resistance. The ability of these aphids (resistant to Guthion) was investigated to flourish on walnut leaves, which were treated at field rates of Guthion. According to the findings, a Guthion-resistant strain of *T. pallidus* was selected, which was more advantageous than the previous. Furthermore, the strain is resistant to Supracide, Thiodan (methidathion), and Lorsban (chlorpyrifos); and these three herbicides were employed in the walnut IPM program.

Temporal response tests were utilized by Spollen and Hoy (1992) to investigate how Guthion resistance is passed down. The resistant and susceptible colonies were mass multiplied to produce male and female F_1 progeny. By allowing F_1 individuals to mate with one another, F_2 offsprings were created. To rule out any maternal effects, similar temporal responses were seen in the reciprocal F_1 female progeny. The resistance of Guthion is connoted as a dominant factor and is regulated by several key genes, according to the LT_{50} levels of the F_1 females.

8.3.2 *Hybridization*

8.3.2.1 Development of Several Hybrid Strains of *Trichogammatid*

The *T. chilonis* strain (hybrid) was created to tolerate the resistance to high temperatures and repeated pesticide treatment. The numerous *T. chilonis* strains tolerant to insecticides were given a temperature tolerance (32–38 °C) to increase its proficiency to parasitize and survive *C. cephalonica* eggs. Prior to being subjected to insecticides at a high temperature, both strains were united and given 24 h to mate. Monocrotophos (1.5 ml/l), Fenvalerate (0.4 ml/l) and Endosulfan (2.0 ml/l) insecticides were used. At 6 and 24 h, the parasitoids' survival and the percentage of parasitism after 5 days were evaluated. In addition, a *T. chilonis* strain developed resistance to modern insecticides (Avaunt, Tracer, and Mimic), as well as to temperature range of 32–40 °C and 50% relative humidity.

On the other side, the kind of genes responsible for these traits will control the tolerance to temperature and pesticides. The traits may be polygenic or monogenic. Only when the corresponding genes are prevalent in a F_1 hybrid, one can anticipate both traits to be displayed. Only homozygotes can be used to study polygenic or recessive traits because hybrids are unlikely to convey the features.

8.3.2.2 Development of Hybrid *Chrysoperla carnea* (Stephens) Strain Resistant to Insecticides

Chrysoperla carnea (Stephens) is an efficient prodigious predator. Numerous insects (such as: aphids, Lepidoptera eggs, whiteflies, mites, jassids, mealybugs, thrips) are successfully controlled by using it. This predator is a cosmopolitan species because of its excellent efficiency, which makes it economically available. Including pyrethroids, it has showed an incredible resistance to various different groups of insecticides.

As compared to the susceptible strain, after 15 generations, the methoxy-SEL strain of *C. carnea* secured 3531.67-fold resistance with methoxyfenozide, which has a very selective effect and extremely low acute toxicity for mammals. This demonstrated enormous potential for this natural enemy to acquire pesticide resistance through natural selection. Numerous findings on *C. carnea*'s pesticide resistance (Mansoor et al. 2013, Abbas et al. 2014, Mansoor et al. 2017, Mansoor and Shad 2019, 2020) show its potential to thrive under variety of field environments.

8.3.3 *Biotechnological Techniques*

While rDNA (recombinant DNA) technology and mutagenesis are optimistic new tools for genetic enhancement, yet heterosis as a strategy to boost efficiency remains largely untapped. This procedure calls for discovery, cloning, and introduction of the desired gene(s) into the natural enemies' genomes. These genes must be fixed, extracted in the appropriate tissues at the proper times, and passed onto the offspring. Therefore, the modified strains of natural enemy must be able to thrive in agricultural landscapes. Using genetic engineering techniques would enable quick success without the requirement for years of conventional upbringing and the utilization of genes from any species rather than simply the gene pool of the arthropods' natural enemies.

8.3.3.1 Genetic Engineered Strains of *Bacillus thuringiensis*

The field-level restriction of *Bacillus thuringiensis* (Bt) involves low persistence under field conditions, which calls for repeated spraying, and the dispersion of enormous quantities of spores. These two issues have been approached simultaneously in two different methods; and both are involved in creating a bacterium that is not typically infective to insects. First, *Pseudomonas fluorescence* was used to clone Intrahepatic cholestasis of pregnancy (ICP) genes into a non-pathogenic strain (Carlton 1996). Encapsulated ICPs with superior field residual characteristics but no Bt spores remained after the bacteria were eliminated. The first recombinant Bt product to undergo field testing was this one, which was given legal permission by the Environmental Protection Agency (EPA) in 1985 for small-scale field trials.

In the next case, spoOA (CryIIIA-active *B. subtilis* mutant strain) was produced. It works well against Coleopterans (Lereclus et al. 1995). The disruption of the spoOA gene prevents sporulation since it is essential for beginning of sporulation. This approach allows the mutant strain to be cultivated indefinitely, prevents the formation of spores, and partially preserves the insecticidal toxin by enclosing it inside a cell. These two cutting-edge formulations offer Bt-based biopesticides that are stable and safe for the environment. The δ-endotoxins of Bt have been introduced into the *Pseudomonas fluorescens* chromosomes, a common nonpathogenic soil bacteria that also colonize maize roots (Obukowicz et al. 1986). Expanding the range of environments, where Bt-based insect pest management can be applied may be made possible by engineering of this bacterium.

8.3.3.2 Development of Modified Predatory Mite, *M. occidentalis*, Through DNA Sequencing

According to Hoy (2009), this predator contains variety of genomes, including its own mitochondrial and nuclear genomes as well as those of microbial symbionts. The 25 kb size of the mitochondrial genome is mainly mitey because of gene duplication and triplication. On the other hand, the nuclear genome is 88 Mb, which is little compared to mitey. In the past, this predator's genetic makeup was changed for agricultural purposes to produce strains that can be genetically adjusted by the use of rDNA technology and lack the capability to overwinter in diapause or be pesticide resistant. The nuclear genome sequencing would yield valuable knowledge for projects involving genetic augmentation.

8.3.3.3 Genetic Modification of Entomopathogenic Fungus (EPF)

The potency of control depends on relative humidity after the spores are applied. It is one major drawback of using EPF (like *Metarhizium anisopliae, Beuveria bassiana* etc.) to control insects. After infection, the recombinant strains of the fungus led to considerable melanization in the larvae and partial hydrolysis of hemolymph proteins.

Genetically Modified Strain of *Metarhizium anisopliae*

Genes pertaining to appressorium formation, pathogenicity, and nutritive stress have been duplicated using *M. anisopliae*. To increase its insecticidal effectiveness, extra copies of the Prolactin (*prl*) gene were inserted into the genome, which sets a subtilin-like protease implicated in host cuticle penetration.

It was intended to promote the expression of Glycerol-3-Phosphate Dehydrogenase (gpd), an *Aspergillus nidulans* constitutive promotor. Three to 6 *prl* genes were found in each of the chosen transformants. A recombinant fungus produced *Prl* in the larval hemocoel of lepidoptera contaminated with a procedure concerned in the start of the melanization activity and general metabolism of tyrosine is however typically only exhibited as the cuticle grows. Recombinant strain-infected larvae died 25% quicker than the wild-type *M. anisopliae*-infected larvae, and feeding injury was 40% lower. Figure 8.1 shows how *M. anisopliae* FM-03 attacks *P. lilacinus*. Figure 8.2 delineates the process of evolution of genetically modified natural enemies.

8.3.3.4 Genetic Engineered Strain of Entomopathogenic Virus (EPV)

The rDNA techniques and advancements in genome mapping have boosted the insecticidal capabilities of baculoviruses (BV). Because of their host specificity, BVs are standard elements of IPM programs.

Recombinant viruses of *Helicoverpa zea* nucleo-poly-hedro-virus (HzSNPV) and *Autographa californica* NPV (AcMNPV) have been created using selective insect

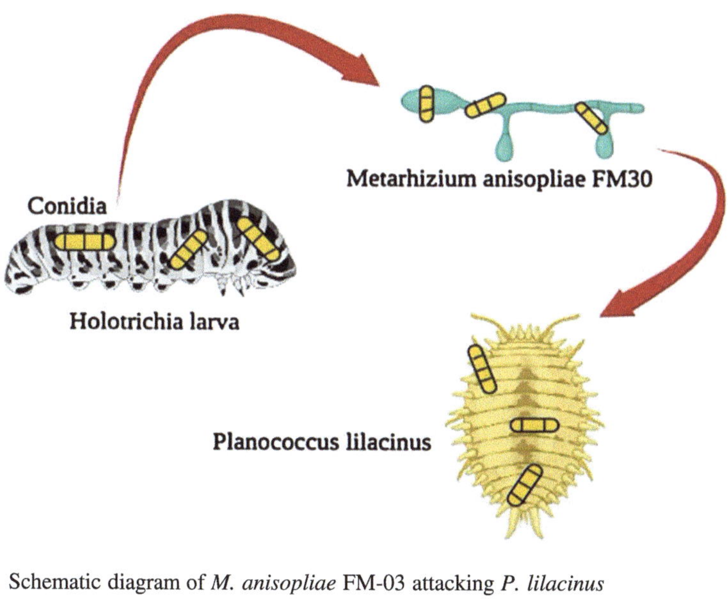

Fig. 8.1 Schematic diagram of *M. anisopliae* FM-03 attacking *P. lilacinus*

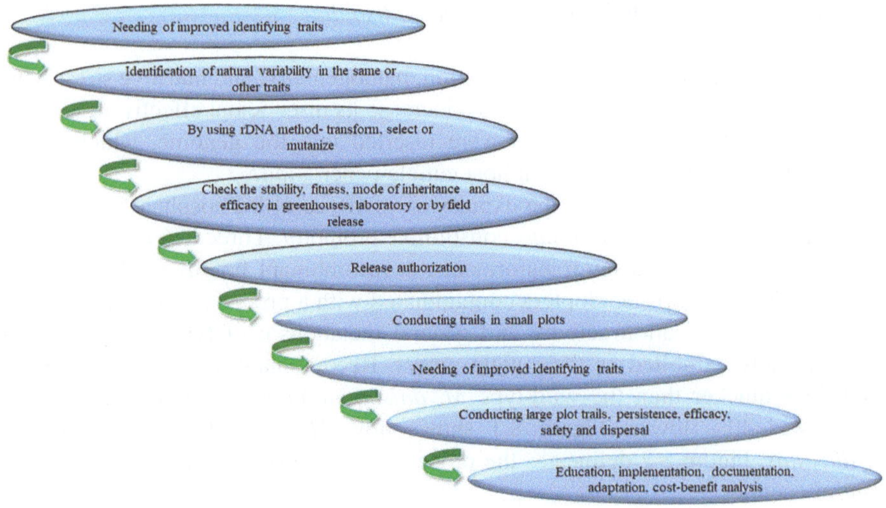

Fig. 8.2 Flowchart showing how genetically modified natural enemies evolved

toxin genes of the *Leiurus hebraeus* (Birula). Two significant pests of cotton (*Gossypium hirsutum*), *Heliothis virescens* (F.) and *Helicoverpa zea* (Boddie) are naturally infected by these viruses. Parasitoids and Predators of the heliothine pest complex are non-target organisms that cannot support virus replication (Huang et al. 1997).

8.3.3.5 Development of Modified Entomopathogenic Nematodes (EPN) Through Genetic Engineering

There is a symbiotic interaction between *Xenorhabdus* bacteria and entomopathogenic nematodes. The interaction is distinct because each species of nematode brings its own strain of bacteria. Since the host finding ability, desiccation tolerance, virulence, and low-temperature activity of nematode strains vary spontaneously, selection method can be employed to enrich these characteristics (Gaugler and Hashmi 1996).

The model organism for molecular genetics, *Caenorhabditis elegans*, is a member of the superfamilies Steinermatidae and Heterorhabditidae. Genes that are conserved in *C. elegans* are probably similar to those in the Heterorhabditidae and Steinernamatidae. The knowledge and resources for manipulating the *C. elegans* genome have been heavily utilized in engineering entomopathogenic worms. This study aimed to increase the heat tolerance and environmental stability of the HP88 strain of nematode, *H. bacteriophora*. The easiness of mass production, rapid reproduction, and potential to create pure lines from hermaphrodite populations make the genetic investigations of Heterohabditidae easier. *C. elegans* roll over due to the rol-6 allele of a collagen-producing gene. The hsp16-lacZ fusion construct was used to build and evaluate a unique nematode transformation device.

8.4 Developmental Constraints of Genetically Modified Organisms

A variety of issues and difficulties with conducting laboratory selections were noted by Simmonds (1963) and Messenger and van den Bosch (1971). Due to the inherent complexity of the natural environment, it is frequently impossible to identify the precise causes of an advantageous organism's ex-situ maladaptation and consequent inability to be remedied in the lab. Due to lack of understanding of the genetic basis for acquiring advantageous traits in beneficial species, sensible selective breeding is challenging and time-consuming. It is crucial to achieve a suitably wide genetic variation to guarantee the accessibility of convenient features on which to base selections. Last but not the least, once a suitable strain can be created and released, and it will continue to behave as intended in the wild, revert to the wild type, or will natural selection produce the strain that is most suitable regardless of how much work was done in the lab prior to colonization?

8.5 Conclusion

Although predators and parasitoids do not harm the ecosystem, yet their use is restricted due to how well they work in different environments. The enhancement of their potency may increase their use as an alternative to chemical pest control. Classic biological control has a very commend B:C ratio (250:1), while augmentative biological control has a cost-benefit ratio equivalent to insecticides (1:2 to 1:5) but substantially reduced development expenses. The most tenable, economical, and environmentally benign pest management technique is biological control, which will play a large part in future IPM program (Bale et al. 2008). Biological traits that could enhance natural enemy's efficacy include quickening the pace of development, boosting modifying the sex ratio, progeny production, expanding temperature and RH tolerances, and changing host or habitat inclinations.

References

Abbas N, Mansoor MM, Shad SA, Pathan AK, Waheed A, Ejaz M, Zulfiqar MA (2014) Fitness cost and realized heritability of resistance to spinosad in Chrysoperla carnea (Neuroptera: Chrysopidae). Bull Entomol Res 104(6):707–715

Bale JS, Van Lenteren JC, Bigler F (2008) Biological control and sustainable food production. Philos Trans R Soc B Biol Sci 363(1492):761–776

Carlton BC (1996) Development and commercialization of new and improved biopesticides. Ann N Y Acad Sci 792(1):154–163

Charles V, Puneeth P, Vijayan VA, Jalali SK (2011) Genetic aspects of insecticide tolerance in Trichogramma chilonis Ishii strains. J Biol Control 25(3):207–212

Dhaliwal GS, Jindal V, Dhawan AK (2010) Insect pest problems and crop losses: changing trends. Indian J Ecol 37(1):1–7

FICCI (2013) Knowledge and strategy paper released at 3rd national agrochemicals conclave July, 2013. Indian Agrochemicals Industry, New Delhi, p 17

Gaugler R, Hashmi S (1996) Genetic engineering of an insect parasite. In: Setlow JK (ed) Genetic engineering, vol 18. Springer, Boston, pp 135–155

Hoy MA (1984) Genetic improvement of a biological control agent: multiple pesticide resistance and non-diapause in Metaseialus occidentalis (Nesbitt) (Phytoseiidae). In: Griffiths DA, Bowman CE (eds) Proceedings VI international congress of acarology, Edinburgh, vol 2. Halsted Press, New York, pp 673–679

Hoy MA (1985) Recent advances in genetics and genetic improvement of the Phytoseiidae. Annu Rev Entomol 30(1):345–370

Hoy MA (2009) The predatory mite Metaseiulus occidentalis: Mitey small and Mitey large genomes. Bioessays 31(5):581–590

Huang XP, Davis TR, Hughes P, Wood A (1997) Potential replication of recombinant baculoviruses in nontarget insect species: reporter gene products as indicators of infection. J Invertebr Pathol 69(3):234–245

Jalali SK, Singh SP (1993) Susceptibility of various stages of Trichogrammatoidea armigera to some pesticides and effect of residues on survival and parasitizing ability. Biocontrol Sci Technol 3(1):21–27

Jalali SK, Singh SP, Venkatesan T, Murthy KS, Lalitha Y (2006) Development of endosulfan tolerant strain of an egg parasitoid Trichogramma chilonis Ishii. Indian J Exp Biol 44(2): 584–590

Jayashree R, Vasudevan N (2007) Persistence and distribution of endosulfan under field condition. Environ Monit Assess 131(1):475–487

Lereclus D, Agaisse H, Gominet M, Chaufaux J (1995) Overproduction of encapsulated insecticidal crystal proteins in a Bacillus thuringiensis spoOA mutant. Biotechnology 13(1):67–71

Mansoor MM, Shad SA (2019) Resistance, its stability and reversion rate of resistance to imidacloprid, indoxacarb and chlorfenapyr in a field population of green lacewing Chrysoperla carnea (Stephens). Arch Phytopathol Plant Prot 52(9–10):884–894

Mansoor MM, Shad SA (2020) Inheritance of polygenic but stable pyriproxyfen resistance in a bio-control agent Chrysoperla carnea: cross-resistance and realized heritability. Pest Manag Sci 76(12):4009–4017

Mansoor MM, Abbas N, Shad SA, Pathan AK, Razaq M (2013) Increased fitness and realized heritability in emamectin benzoate-resistant Chrysoperla carnea. Ecotoxicology 22(8): 1232–1240

Mansoor MM, Raza ABM, Abbas N, Aqueel MA, Afzal M (2017) Resistance of green lacewing, Chrysoperla carnea Stephens to nitenpyram: cross-resistance patterns, mechanism, stability, and realized heritability. Pest Biochem Physiol 135(2):59–63

Messenger PS, van den Bosch R (1971) The adaptability of introduced biological control agents. In: Huffaker CB (ed) Biological control. Springer, Boston, pp 68–92

NAIP (2012) Effect of abiotic stresses on the natural enemies of crop pests: Trichogramma, Chrysoperla, Trichoderma and Pseudomonas and mechanism of tolerance to these stresses. Annual report 2011-12, New Delhi, pp 1–6. www.nbair.res.in/pdbc-naip/home.html. Accessed 15 Aug 2022

Obukowicz MG, Perlak FJ, Kusano-Kretzmer K, Mayer EJ, Watrud LS (1986) Integration of the delta-endotoxin gene of Bacillus thuringiensis into the chromosome of root-colonizing strains of Pseudomonads using Tn5. Gene 45(3):327–331

Rao VP, Ghani MA, Sankaran T, Mathur KC (1971) A review of the biological control of insects and other pests in South-East Asia and the Pacific region. Trinidad 6:59–65

Roush RT, Hoy MA (1981) Genetic improvement of *Metaseiulus occidentalis*: selection with methomyl, dimethoate, and carbaryl and genetic analysis of carbaryl resistance. J Econ Entomol 74(2):138–141

Roush RT, McKenzie JA (1987) Ecological genetics of insecticide and acaricide resistance. Annu Rev Entomol 32(1):361–380

Sankaran T (1986) Current status and future projections for biological control of insect pests in India. Proc Indian Natl Sci Acad Biol Sci 52(1):108–116

Simmonds FJ (1963) Genetics and biological control. Can Entomol 95:561–567

Singh SP (2001) Biological control of pest and diseases. In: Paper presented at the Eighth General Body Meeting of NAAS. National Academy of Agricultural Sciences, New Delhi, p 23

Singh S, Shenhmar M (2008) Host searching ability of genetically improved high temperature tolerant strain of *Trichogramma chilonis* Ishii in sugarcane. Ann Plant Prot Sci 16(1):107–110

Spollen KM, Hoy MA (1992) Genetic improvement of an arthropod natural enemy: relative fitness of a carbaryl-resistant strain of the California red scale parasite *Aphytis melinus* DeBach. Biol Control 2(2):87–94

Stinner RE, Ridgway RL, Coppedge JR, Morrison RK, Dickerson WA Jr (1974) Parasitism of Heliothis eggs after field releases of *Trichogramma pretiosum* in cotton. Environ Entomol 3(3): 497–500

Whitten MJ (1970) Genetics of pests in their management. In: Rabb RL, Guthrie FE (eds) Concepts of pest management. North Carolina State University, Raleigh, pp 119–137

Whitten MJ, Foster GG (1975) Genetical methods of pest control. Annu Rev Entomol 20(1): 461–476

Part III
Innovations in Food Packaging

Chapter 9
New-Age Packaging for Foods and Food Products

Deeptimayee Mahapatra, Soumitra Goswami, and Mamoni Das

Contents

Abstract The food packaging industry has undergone a revolution, leading to the event of recent packaging materials to reflect advancements in food technology, material and environmental sciences, and consumer preferences. Enormous number of packaging techniques has been developed so that we can enjoy food throughout the year. The new-age packaging can not only provide safety or works as a barrier to the food, rather it is capable of preserving essential nutrients in them and also efficient in incorporating healthy bioactive compounds to food through diffusion. It can also detect any unwanted deterioration of food without opening the package and without using any lab equipment. Of these, novel benefits are contributed by the

D. Mahapatra (✉) · M. Das
Department of Food Science and Nutrition, College of Community Science, Assam Agriculture University, Jorhat, Assam, India

S. Goswami
Department of Food Science and Technology Programme, College of Agriculture, Assam Agriculture University, Jorhat, Assam, India

J. A. Malik et al. (eds.), *Food Process Engineering and Technology*,
https://doi.org/10.1007/978-981-99-6831-2_9

newly developed packaging technologies like intelligent/smart packaging, active packaging, antimicrobial packaging, use of nanotechnology in packaging, sustainable packaging, and bioactive packaging. This chapter has compiled and interpreted research studies for last 5-years citing principles and applications of these recent innovations in food packaging technologies.

Keywords Active packaging · Antimicrobial packaging · Bioactive packaging · Intelligent/smart packaging · Sustainable packaging

9.1 Introduction

The fundamental role of packaging in food industry has always been understood as a cover that provides safety from environment and confirms the integrity of food product during handling, transportation, storage, and retailing. However, it has developed well beyond this understanding overtime. Now it is not limited to the function of covering or holding the product as a container, rather it provides protection from environmental, chemical, and physical challenges (Abdullahi 2018). Apart from providing passive protection the new-age packaging also contributes to the quality of the products by modifying the atmosphere around it (Pawar and Aachal 2013; Risch 2009). It also functions as an integral part of a brand's messaging, its image as well as consumer experience (Yano et al. 1995).This continuous evolution of the packaging industry is contributed by several pivotal reasons, such as:

- firstly change in lifestyle, dietary habits and consumer preferences (Dobrucka et al. 2015);
- secondly the exploitation of convenient, ready to eat, less processed, nutritionally rich and high-quality packaged food products (Dobrucka and Cierpiszewski 2014);
- thirdly the flooded demand for prepared meals like microwave meals, ready to reconstitute meals and smaller packages of food etc. (Restuccia et al. 2010), and
- fourthly the escalating need for packaging with anti-microbial properties due to the increasing outbreaks of food borne diseases.

Food packaging is a dynamic field and continuous developments are its characteristics. Revolution in food packaging industry has revolutionized the packaging materials. The conventional packaging materials like leaves, woods, bamboo, metallic cans, paper board, glass, and aluminum foils etc. are now replaced by polyester, polypropylene, polyethylene and ethylene vinyl alcohol polymers etc. (Majid et al. 2016). Moreover, the twentieth century has witnessed some major development of packaging technologies like intelligent or smart packaging, antimicrobial packaging, sustainable packaging and active packaging etc. Over these years, these novel developments have improved the food safety, enhanced the product quality and minimized product loss, which has ultimately strengthened the economy as well. This chapter highlights developments in conventional packaging materials and modern packaging technologies for foods and food products.

9.2 Active Packaging

An active packaging system is one that involves the intentional interaction of food with its packaging. Active packaging materials significantly raise shelf-life, enhance safety and sensory attributes, and maintain product quality by removing unwanted components from the package or the food itself (Dainelli et al. 2008). The functions of this type of packaging are focused mostly on the preservation of moisture and volatile components of the product, retardation of microbial growth, oxidation of the product, and controlling respiration rate of the product. To perform these functions, active packaging includes the application of diverse compounds (Ozdemir and Floros 2004) such as:

- anti-microbial agents (sorbates, Ag-zeolite, peroxide, benzoates, SO_2, propionates, C_2H_5OH, ozone, antibiotics, quaternary ammonium salts), etc.;
- antioxidant emitters (butylated hydroxyanisole, butylated hydroxytoluene, tertiary butylhydroquinone, ascorbic acid, tocopherol);
- aroma absorbers (baking soda, active charcoal);
- ethylene binding agents (activated charcoal, silica gel, $KMnO_4$, bentonite, Fuller's earth, SiO_2, and zeolite);
- insect repelling agents (pyrethrins, permethrin);
- oxygen scavengers (Fe_2O_3, catechole, $FeCO_3$, iron-sulfur, sulfite salt, $CuSO_4$); and
- respiration controllers (packaging with surface-treated, perforated or microporous films).

Literature has shown that antioxidant and antimicrobial functions have emerged as major developments in the field of active packaging materials.

9.2.1 Oxygen Scavenging Active Packaging

Lee et al. (2018b) created sulfite-based oxygen scavenger with sodium metabisulfite and incorporated it into kimchi packaging material. The sulfite-based oxygen scavenger outperformed commercial oxygen scavengers in the oxygen-scavenging capability for kimchi. In comparison to commercial oxygen scavengers, the SOS performed superior in the oxygen-scavenging capability test. For a practical application test, a 12-week storage study for the kimchi was performed at 0 and 10 °C with and without the sulfite-based oxygen scavenger. The results showed that sulfite-based oxygen scavenger prevented kimchi from deteriorating due to oxygen, preserved its color, and had negligible effect on kimchi fermentation.

Similarly, a nonferrous oxygen scavenger was developed especially for preserving raw meatloaves, with activated carbon and ascorbic acid (sodium-l-ascorbate). The oxygen-scavenging capacity of the developed sachet was tested on raw meatloaves at regular intervals. Furthermore, after 4-days of storage (at 4 °C) study alterations in the microbiological content and lipid oxidation were investigated. The storage data proved that the nonferrous oxygen scavenger sachet has improved storage properties (low thiobarbituric acid and microbial contents) of meat which advocates the practical applications of these sachets in the food packaging business (Lee et al. 2018a).

Singh and his co-researchers attempted an enrichment of low-density polyethylene with Na_2CO_3 and pyrogallic acid coatings (10 and 20%). The developed film was utilized to preserve peeled garlic for 30 days at 25 and 5 °C. The changes in quality and packaging atmosphere were monitored periodically. The oxygen scavenger package stabilized oxygen and CO_2 concentrations were faster at 5 °C than at 25 °C. On the otherhand due to respiration, and decreased O_2 concentrations (approached zero by day 15 of storage), the control sample did not settle down. The produced oxygen scavenger package exhibited good results in minimizing fungal decay, weight loss, and aerobic count, as well as keeping firmness, color loss, and storage stability (Singh et al. 2019).

Another study on similar concept was performed by Gaikwad et al. (2020a). A UV-activated oxygen scavenger film was developed with 5, 10, 20% of polyisoprene incorporated into LDPE in which UV light acted as the activating factor for the oxygen-scavenging reaction to start in the film. The results reported improved barrier properties, and thermal stability as compared to pure LDPE films. Also, when polyisoprene LDPE films were studied for packing beef jerky, the outcomes showed an increased trend in stability, color, and pH with reduced lipid oxidation during storage for 90 days (Gaikwad et al. 2020a).

An oxygen barrier film made of chitosan, gallic acid and sodium carbonate was created using solution casting technology. The changes in physical, mechanical, structural, and oxygen-scavenging efficacies of the developed film due to gallic acid were investigated. Though gallic acid and sodium carbonate together reduced tensile strength and break elongation, yet the developed oxygen barrier films were thicker and more water-soluble than straight chitosan films. X-ray diffraction of the films revealed excellent compatibility between gallic acid, sodium carbonate, and chitosan, resulting in increased water and oxygen permeability in the chitosan matrix/chitosan-gallic acid-sodium carbonate films (Singh et al. 2021).

Gaikwad et al. (2020b) produced a trans-polyisoprene-based rubber/packaging that functioned as an oxygen-scavenging film, which needed light for activation. It was studied for roasted peanuts packaging to study its preventive properties against oxidative deteriorations. The O_2 scavenging analysis revealed that trans-polyisoprene packaging had fair rates of oxygen scavenging efficacies. The packaging was also efficient in scavenging oxygen than the pure low density polyethylene packaging when utilized for preserving peanuts. Experimental data also showed that generated oxygen scavenging packaging prevented oxidation of roasted peanuts even after 90 days of storage, suggesting potential use of this packaging in food packaging industry.

9.2.2 Antioxidant Films/Active Packaging

Tanwar et al. (2021) developed an antioxidant film by combining polyvinyl alcohol and corn starch with different concentrations of coconut shell extract (3, 5, 10, and 20% v/v) and sepiolite clay. The evaluation of microstructure and physical property revealed that the addition of sepiolite clay to polyvinyl alcohol enhanced the elongation break; and inclusion of coconut shell extract increased the water vapor permeability. Due to the addition of coconut shell extract to the films, the antioxidant activity of the films was increased by 80%. Additionally, increasing the amount of coconut shell extract had a minor impact on the color of the active films but improved their thermal characteristics. The practical utilization of the developed polyvinyl alcohol-corn starch film-based sachets showed remarkable antioxidation properties in soyabean oil packaging.

Kadam et al. (2021) used pine needles (Cedrus deodara) extract that was comprised of curcuminoid pigments for production of a chitosan-based antioxidant film. The researchers used scanning electron microscopy to look at the effects of adding 5, 10, and 20% pine needle extract (PNE) to the antioxidant assay, physical properties, oxygen permeability, water vapor permeability, and microstructures of the films. The films containing pine needle extract (both 10 and 20%) showed higher antioxidant activities and lowered oxygen and water vapor permeability. By introducing pine needle extract, the chitosan films developed a strong antioxidant property and distinctive features that proved them suitable for food packaging.

Bi et al. (2020) developed an antioxidant and antibacterial packaging by incorporating 1% (wt.) of flavones like chrysin, apigenin and luteolin into chitosan and synthetic vitamin-E (D-α-tocopheryl polyethylene glycol 1000 succinate/TPGS) composite films. The incorporation of flavones affected the homogeneity of the films and the crystallinity of the films by forming hydrogen bonds with other film components. All the films containing flavones showed lower tensile strength but better UV–light barrier properties and antibacterial properties compared to the chitosan-TPGS film. The film containing luteolin outperformed the other flavone-containing films in terms of its highest water vapor, oxygen and UV light barrier properties, tensile strength and antioxidant activity. However, highest antibacterial activity against Escherichia coli, Salmonella typhimurium, Staphylococcus aureus and Listeria monocytogenes was exhibited by the film containing chrysin.

García et al. (2020) developed an antioxidant film incorporating almond skin extract in a polycaprolactone base for fried almonds packaging. The addition of almond skin extract had no effect on the structural properties (oxygen permeability, structural, tensile, thermal, or antioxidant properties) of the developed film. According to the DPPH radical scavenging activity, almond skin extract gave poly-caprolactone films antioxidant characteristics. Furthermore, the created film's packing performance was tested for 56 days at 40 °C. Active coatings increased stability of fried almonds as measured by peroxide values, p-anisidine values, hexanal concentration, fatty acid profile, and distinctive spectroscopic bands. The findings suggested that poly-caprolactone/almond skin extract films could be used as a sustainable food packaging material with antioxidant properties to preserve foods from lipid pero-oxidation.

9.3 Intelligent/Smart Packaging

The intelligent/smart packaging is developed to perform intelligent functions like detection, communication, recording, tracing, and sensing etc. either by incorporating smart devices/systems to the packaging material either to the inside or to the outside of the package (Kour et al. 2013). This kind of packaging is confused with active packaging because of their simultaneous development.

The basic functional difference between active and intelligent packaging is that active packaging involves use of certain chemical compounds for preserving the quality and extending the shelf-life of the product on the other hand intelligent packaging involves systems/devices that gives signals about the quality and undesirable changes in the product during storage and transport. Examples of intelligent packaging systems/devices are: Time-temperature indicator (TTI), gas indicator, seal and leak indicators, freshness and ripening indicators, microbial growth indicators, and physical shock indicators etc. (Dobrucka et al. 2015; Prasad and Kochhar 2014; Soltani et al. 2021).

9.3.1 Time-Temperature Indicator (TTI)

Temperature is considered as a foundation factor in cold chain management systems because of its vast influence on product quality (Tsironi et al. 2017). A product has to travel through different temperatures from processing, production and packaging to storage and consumption. To monitor the effects of these temperature differences, time-temperature indicators have emerged as useful aids. These visual indicators change color due to any temperature abuse. The most common type of TTI in food industries is diffusion-based indicator or microbial indicator (Soltani et al. 2021).

Diffusion based TTI are porous wick materials with dyed esters of fatty acids, which discharges dye when temperature increases above a specific level causing an irreversible color change of the indicator. The dyes used for these indicators are specific to each food and storage condition (Taoukis and Labuza 1989; Yildirim et al. 2018). Generally, diffusion based TTIs are either placed on the outer surface of the heat-sensitive product packages or on the export boxes.

The microbial TTI are basically labels containing bacteria especially Lactic acid bacteria (ex. Weissell akoreensis, L. sakei, Janthino bacterium sp.) placed inside the food packages (Hsiao and Chang 2017; Lim et al. 2014); and with increase in temperature above a specific value (Mataragas et al. 2019), the bacteria are absorbed by the product. Under such storage conditions, the bacteria start growing, which causes pH change followed by color change of the indicators (Soltani et al. 2021). Microbial TTI are used for perishable foods to monitor their shelf-life in a cold chain.

For example, Jang et al. (2018) developed a TTI with vision based detection system for frozen fish products. The indicator was created to determine thawing of fish and fish products during frozen storage. A 3 mm internal diameter polypropylene tube was manufactured and trimmed to a length of 14–20 mm. The tube was then frozen at 20 °C after one end was thermally sealed and the other end was injected with 0.1% acetic acid. The open side of the frozen tube was then plugged by immersing it in a 10% gelatin solution. Along with blue litmus paper, the tube was affixed to a polyvinyl packing bag, which was then frozen at (−20) °C. After freezing, the bag was brought to room temperature (20 °C) to observe the time dependence of the litmus paper color change. With the length, the hue shifted from blue to red. Hence, the degree of thawing was determined by the length of the red zone of the litmus paper. As a result, the developed TTI is a useful instrument for controlling the temperature of frozen fish and fisheries goods.

9.4 Gas Indicators

Gas indicators are useful to indicate mere change in gas composition inside the package due to the activity of the food products, leakages, nature of packaging, and environmental conditions. This is why these indicators need to be in contact with the product or placed on the inner side of the package. Usually, it is utilized to detect the presence of gases like O_2, CO_2 or C_2H_4 (De Jong et al. 2005). However, indicators that can monitor alteration of gases (like CO_2, NH_3, H_2S, dimethylamine ($(CH_3)_2NH$) and trimethylamine (C_3H_9N)) are now being developed (Lee et al. 2019). Gases like O_2 and CO_2 are also utilized to monitor changes in food quality that are attributed by leakage or faulty packaging and sometimes termed as seal and leak indicators (Prasad and Kochhar 2014).

Generally, gas indicators are not recommended for a wide range of foods; hence, configuration of gas indicators and selection of proper materials are important for developing the best indicator and enhancing the accuracy of information regarding food quality (Soltani et al. 2021).

In this digital era, every possible field of science is adopting digital options or artificial intelligent for solving many problems. Food packaging industry is not different from these industries and a constant trial has been seen among researchers to include smart phones, digital codes, chips or devices for disseminating information on product safety, quality, storage condition and many more.

Araque et al. (2018) developed a method for determining the gaseous O_2 concentration inside the packed food using a luminous membrane sensitive to O_2 that was optically activated and could read by a smart phone. The light source for the optical stimulation of the membrane was a smart phone flash combined with an optical filter. The back camera of the same system records the created light, which is dampened by the surrounding gaseous oxygen. The response parameter is calculated by combining the registered intensities at two different wavelength ranges that correspond to the sensitive membrane's emission and absorption peaks. This research proved that an oxygen sensor with a luminous substance as an indication could be a useful tool in the field of food packaging.

9.5 Freshness and/or Ripening Indicators

Preserving the freshness of food products especially fruits and vegetables, is a burning issue for the packaging industry. Several indicators are being developed as aids for this problem and collectively categorized as freshness or ripening indicators that detects the level of freshness or most accurately the level of deterioration of the product by detecting the activities of several metabolites (such as: diacetyl, amines, CO_2, NH_3 and H_2S) produced during the ageing of food products (Nopwinyuwong et al. 2010; Smolander and Biotechnology 2003). Metabolites like hydrogen sulfide, n-butyrate, L-lactate, D-Lactate and acetic acid are also proved as viable indicators for fruits, vegetables and meat products.

Certain biogenic amines like putrescine, cadaverine, and histamine, which are produced during degradation of protein to amino byproducts act as freshness indicating metabolites specifically in meat and fish (Pacquit et al. 2006, 2007; Rokka et al. 2004).

Change in pH is another potential indicator of microbial metabolites and ripeness which can be efficiently detected by color change of the indicator on the package (Hong and Park 2000; Rokka et al. 2004). For example, a food packaging film containing Amaranthus leaf extract (ALE) was developed with polyvinyl alcohol and gelatin. Betalain present in ALE changes the color under basic pH and it was the novelty/technique (color change) of his research. The impact of adding ALE to the film, its physical, mechanical, and functional properties was assessed. Films containing ALE had improved UV protection, lower water solubility, and lower water vapor permeability, as well as better mechanical characteristics. The addition of ALE imparted to the film antioxidant, antibacterial, and intelligent characteristics. The films were effective in delaying microbial development and reducing oxidative rancidity when used to preserve cold seafood and chicken. The shelf-life of samples packaged in neat films was 3 days, while those in active films degraded after 12 days. On deterioration, the film's color changed from red to yellow, which was supported by increases in pH, total volatile basic nitrogen, and microbiological counts in the spoilt samples (Kanatt 2020).

Lee et al. (2019) constructed a freshness indicator based on the principle of pH change for chicken breast. The indicator had Tyvek sheets (a gas and vapor impermeable sheet) at the innermost layer, bromocresol green (a pH sensitive dye) embedded sheet as middle layer and low-density ethylene sheet as last layer. With change in carbon dioxide content or TVBN content (total volatile basic nitrogen) due to bacterial actions inside the chicken breast package, the indicator changed its color from green to yellow, which indicated spoilage of the product. The level of spoilage or quantification of spoilage was facilitated via a smart phone through red-green-blue digital image analysis system. The results implied that this indicator was a very simple, accurate and consumer friendly device.

9.6 Bacterial Growth Indicator

Bacterial growth indicators are vital for preserved meat and fish products as microbial growth in them are difficult to identify by the consumers and can directly affect their health. Organic acids like D-lactate and acetate are proved promising indicators in fresh meat and fish respectively (Kakouri et al. 1997; Shu et al. 1993). In case of fish some volatile nitrogen compounds (ammonia, dimethylamine and trimethylamine) can be used as indicator of microbial growth/spoilage (Ohlenschlager 1997).

Carbon dioxide is commonly produced during microbial growth and can act as a spoilage indicator. But it has limited use because of the development of modified atmosphere packaging where both CO_2 and inert gases are typically used as bacteriostatic agents (Smolander and Biotechnology 2003). However, it is effective in certain food products like pea and tomato soup (Mattila et al. 1990).

K-value is the ratio of the sum of hypoxanthine and inosine to total ATP-related compounds (Henehan et al. 1997). K-value is another indicator of microbial growth in fish and some type of meat (Watanabe et al. 1989; Yano et al. 1995). For fresh meat, this value is low, which gradually increases with increase in microbial growth (Smolander and Biotechnology 2003).

9.7 Role of Nanotechnology in Food Packaging

Different concepts for use of nanotechnology in the field of food packaging have been presented in the scientific literature. The major contributions of nanotechnology are basically aiding functional benefits to the packaging material and acting as indicator for gases, chemical contaminants, biological contaminants and environmental changes (Cushen et al. 2012; Joyner and Kumar 2015). Functional properties of packaging material are proved to be enhanced by incorporating nano-particles in them. Nanoparticles are generally defined as particles of diameter 1–100 nm.

The well-established nanostructures/nanoparticles are silver (Ag) based nanoparticles, zinc oxide nanoparticles, grapheme materials, and titanium oxide nanoparticles (Mei and Wang 2020). Ag-based particles are widely studied for their protecting effects against bacteria, fungi and viruses (Siddiqui et al. 2018). Zinc Oxide (ZnO) nanoparticles provided UV protection, anti-oxidation protection, better mechanical and barrier properties etc. when added to packaging materials (Noshirvani et al. 2017). ZnO nanoparticles are also effective in protection against bacteria(Gram-positive and Gram-negative) (Agarwal et al. 2018), fungi (Sun et al. 2018), and some spores (Li et al. 2016).

However, some recent studies have also highlighted the improved mechanical properties like water-holding capacity (Oun and Rhim 2017), oxygen and moisture permeability (Jafarzadeh et al. 2016), tensile strength (Krepker et al. 2017; Vahedikia et al. 2019; Yang et al. 2016), and anti-microbial activity (Jafarzadeh et al. 2016; Krepker et al. 2017; Vahedikia et al. 2019) due to the combined use of several nanoparticles together in packaging materials. Study also proved that nanoparticles contribute to extended freshness of food by elevating the thermal stability of packaging material up to 350% (Abdullahi 2018).

Hasani-Javanmardi et al. (2021) used nanoemulsion technique to develop an oxygen absorber (OA) packaging using safflower oil nanoemulsion (SNE) and cumin essential oil (CEO). During storage, the produced films restricted mesophilic and psychro-trophic bacterial proliferations (especially enterobacteria, and lactic acid bacteria). These films had negatively affected the rate of total volatile nitrogen content, lipid oxidation, protein oxidation, met-myoglobin production, and color or sensory deterioration. To postpone microbiological development, to extend shelf-life, and to retard physicochemical and sensory changes in lamb loin samples, a combination of safflower oil nano-emulsion and cumin essential oil followed by oxygen absorber packaging was most appropriate and effective treatment in this study.

An active packaging film incorporated with zein (principal protein of maize) and pomegranate peel extract encapsulated in chitosan nanoparticles was prepared by Cui et al. (2020). The chitosan nanoparticles/pomegranate peel extract nanocomposites were prepared using the ionic gelation process and were successfully integrated into the zein film. FTIR and X-Ray diffusion tests validated the efficacy of pomegranate peel extract encapsulation and the molecular interactions between zein, chitosan nanoparticles, and pomegranate peel extract E. Sustainable release of pomegranate polyphenols from the nanocomposite film was achieved by altering the film surface using cold nitrogen plasma. X-ray photoelectron spectroscopy (XPS) was used to reveal the surface alteration of the nanocomposite film. The produced nanocomposite film outperformed the neat zein film in terms of thermal stability. The antibacterial ability of nanocomposite film was established for a pork meat sample inoculated with L. monocytogenes. Plasma-treated nanocomposite films had a slower rate of PE release and effectively limited the development of L. monocytogenes under refrigerated storage.

A titanium oxide nano-particle based bio-nanocomposite packaging film incorporated with sesame protein isolate was developed by Fathi et al. (2019) by using casting method. The addition of 3% of TiO_2 reduced water vapor permeability and water solubility while increased tensile strength, water contact angle, and opacity of the films. However, when TiO_2 level reached 5%, unfavorable impacts were recorded for both mechanical and water barrier properties. The photocatalytic activity of bio-nanocomposites was tested, and it was discovered that sesame protein isolate—TiO_2 5% films can scavenge 10.96% of atmospheric oxygen in 6 h. Hence the study concluded that the developed sesame protein isolate—TiO_2 bionanocomposite films could be a promising option as an active packaging material for food.

TiO_2 was used in another study in the form of a nanotube by Wen et al. (2019) to develop a visible colorimetric oxygen indicator. The indicator was composed of silver-loaded titanium dioxide nanotube, glycerol, methylene blue, and hydroxyethyl cellulose. The efficient photocatalytic performance of silver-loaded TiO_2 was observed through X-ray diffraction, X-ray photoelectron spectroscopy, transmission electron microscopy, etc. These findings established the developed packaging as an efficient oxygen indicator. It also displayed two noteworthy properties: inhibiting the indicator's reaction to natural light (wavelength > 380 nm) and reducing the time required for bleaching after activation. This type of oxygen indicator can greatly improve the accuracy of package non-destructive testings.

Use of nano particles in antibacterial and antioxidant packaging is not new one such research study was conducted by Eskandarabadi et al. (2019). In their study, the food packaging films were made of ethylene-vinyl acetate copolymer, nanocomposites containing extracted anthocyanin, rosemary extract and ZnO/Fe-MMT nanoparticles. Antioxidant capacity test with DPPH, minimum bactericidal concentration (MBC), minimum inhibition concentration (MIC), antibacterial test, and capacity of oxygen absorption were used to study the performance of the produced films. When compared to pure ethylene vinyl acetate film, films containing rosemary extract, zinc oxide, and modified montmorillonite displayed exceptional antibacterial and antioxidant activity.

SiO_2 nanoparticles are often used nanoparticle in food packaging. Hou et al. (2019) developed an agar/sodium alginate nanocomposite film with SiO_2 incorporated at different concentrations (2.5, 5, 7.5, and 10% by wt). Using Fourier Transform Infrared, the nanocomposite film was examined and creation of hydrogen bonding between nano-SiO_2 and polysaccharide was observed. Tensile strength and elongation at break all increased to maximum levels as nano-SiO_2 concentration climbed from 0 to 10% by weight. The film containing 10% nano-SiO_2 outperformed the film manufactured without nano-SiO_2, in terms of the tensile strength and elongation at break properties. The addition of nano-SiO_2 to the film increased its UV light resistance and thermal stability. Overall, adding nano-SiO_2 to the polysaccharide layer improved its mechanical characteristics, water resistance, and thermal stability.

Silver (Ag) nanoparticles/gelatin-modified montmorillonite nanofiller were used to develop nano-biocomposite films based on plasticized cellulose acetate/triethyl citrate. Along with thymol *Curcuma longa* (*C. longa*) tuber extract was used to biosynthesize Ag nanoparticles in situ in the clay. The clay and the formed films were characterized for morphological, physical, and functional qualities. The tensile strength and oxygen barrier properties of the clay were improved, but the triethyl had an antagonistic effect. Furthermore, the DPPH assays revealed that triethyl citrate included films have antioxidant properties. The films demonstrated antibacterial activity against bacteria and fungus, with Escherichia coli being the most sensitive, and growth suppression in the vapor-phase was proven the most effective method (Dairi et al. 2019).

Echeverría et al. (2018) assessed the potential application of soy protein isolate (SPI), montmorillonite (MMT) and clove essential oil (CEO)-based active nanocomposite films for the preservation of blue fin tuna (*Thunnus thynnus*) muscle fillets during the refrigerated storage. Tuna fillets were covered with the prepared films for 17 days of storage at 2 °C. As a control, polyethylene films were utilized. Microbial growth and lipid auto-oxidation of tuna fillets were reduced by protein films containing 10 g MMT/100 g SPI and activated with CEO during storage. The inclusion of clay appeared to encourage the release of clove oil's active components by extending its antibacterial (particularly effective against *Pseudomonas* spp.) and antioxidant activity over time without monitoring the clay's diffusion.

Other than enhancement of mechanical properties of packaging, nanotechnology plays vital role as indicators for chemical, physical and biological contaminants. Nanotechnology includes nano sensors that are being developed to detect or respond to physicochemical (sensors) and biological signals (biosensors) and to convert that response to human identifiable outputs (Abdullahi 2018). Various nano sensors are now developed to detect contaminants, pollutants, or microbes in food packages (Joyner and Kumar 2015; Omanović-Mikličanin and Maksimović 2016). The general indicator of freshness of food is expiry date, which is provided by the producers. Expiry date of any product depends on some required storage and processing conditions which can be easily abused causing deterioration of food without any signs on the packages. Hence nano sensors are helpful to tackle these problems in packaging industry.

9.8 Antimicrobial Packaging Systems

Antimicrobial packaging systems have been developed as a solution to the outbreak of worldwide food borne diseases. This packaging system involves elimination of essentials growth requirements of microbes which can either directly act on them and reduce the microbial counts or indirectly extend their lag phages or reduce their growth rates. The basic principle behind antimicrobial packaging is hurdle technology that involves use of hurdles (pH, heat treatments, water activity, redox potential etc.) in combinations to increase their lethality on the microbes (Singh and Shalini 2016). Antimicrobial packages are comprised of two components the packaging material and the antimicrobial agents. The packaging materials can be biodegradable (edible coating and films from proteins, lipids and starch, chitosan, polylactic acid, polyhydroxy butyrate and polyhydroxy alkanoate) or non-biodegradable (nylon, low density polyethylene (LDPE), high density polyethylene (HDPE) and polystyrene, etc.).Apart from these conventional packaging materials, some novel materials like poly lactic acid (PLA), poly vinyl alcohol (PVA), and chitosan-poly vinyl alcohol blends are also used as antimicrobial packaging materials (Sofi et al. 2018).

Antimicrobial agents attack any microbe that try to penetrate the packaging of product and can be categorized as (Abreu et al. 2015; Coma 2008; Cooksey 2000; Cutter 2002; Halek and Garg 1989; Huang et al. 1997; Jabeen et al. 2015; Padgett et al. 1998; Pires et al. 2009; Plackett et al. 2007; Winther and Nielsen 2006; Yi et al. 1998):

- chemical antimicrobials (BHT, benzoic acid, sorbate, EDTA etc.),
- natural agents (lysozyme, citrate, nisin, clove, grape seeds, chitosan),
- biotechnology products (silver zeolite, lysozyme) and
- gases (Chlorine dioxide, sulfur dioxide, carbon dioxide).

Literature has suggested that certain essential oils and plant extracts are also being studied for potential antimicrobial properties in packaging systems. Examples of such widely used essential oils and plant extracts are: linalool, thymol, carvacrol, clove oil, cinnamaldehyde, basil essential oils, oregano essential oil, pimento oil, and grape seed extract, etc. (Ahmed et al. 2016; Emiroglu et al. 2010; Nadarajah et al. 2005; Oussalah et al. 2004; Sivarooban et al. 2008; Suppakul et al. 2006).These antimicrobial agents are mixed with the packaging materials and antimicrobial packaging films or coatings are then prepared. The antimicrobial packaging systems function through three different modes, namely, release, absorption and immobilization (Ahvenainen 2003).

Release type includes the diffusion of antimicrobial agents into food or headspace of the package whereas absorption mode involves removal of moisture, oxygen, CO_2 and pH, etc. that are essential for microbial growth. Immobilization system works differently from the other two modes, the antimicrobial agent like immobilized lysozyme and glucose oxidase enzyme fights with the microbe on the contact surface and restrains their entry inside the packages (Sofi et al. 2018).

Traditional herbal/medicinal plants possess several properties out of which antimicrobial and anti-viral properties is a topic of discussion in this chapter. These antibacterial compounds have made their ways to packaging industries now. The study by Mahcene et al. (2020) is a classic example of it. In this study, sodium alginate based edible antimicrobial films incorporated with essential oils (EO) of *R. officinalis* L, *A. herba alba* Asso, *O. basilicum* L and *M. pulegium* L. the films were developed. The results showed that the active films had a strong antibacterial effect against almost six pathogenic bacteria including species like E. coli, P. aeruginosa and S. aureus. Furthermore, the antioxidant capacity of the films ranged from 4.57 to 23.09% and improved barrier and thermal properties and biodegradability were recorded as well for the developed essential oil enriched edible films.

Antosik et al. (2021) prepared an antimicrobial packaging film with polypropylene, oregano oil and cedar oil. The enriched polypropylene films were made by regranulation method and cast extrusion method. Tests for mechanical properties and biocidal properties were conducted for the films and it was recorded that these films were absolutely effective against wide range of microbes (*B. subtilis*, *E. coli*, *S. aureus*, *P. putida*, *C. albicans*, *A. alternata*, *F. oxysporum*).

Xu et al. (2020) developed a packaging film with antifungal coatings in an effort to add antimicrobial compounds in the packaging material. For berry preservation, they used a sodium alginate covering supplemented with cyclolipo-peptides. During the storage of blueberries, the coated films demonstrated powerful antifungal and freshness-preserving characteristics. At the end of storage period, the 3% cyclolipopeptides added film promisingly reduced the overall fungal count. The sodium alginate-cyclolipopeptides films displayed low tensile strength and elongation at break, as well as low water vapor permeability and high-water solubility. The addition of cyclolipopeptides (3%) reduced the developed film's tensile strength, the water vapor permeability and elongation at break values, while it increased its water solubility. However, it was recorded that the developed coating film performed satisfactorily.

An antimicrobial packaging against *Staphylococcus aureus* (Gram-positive), *Escherichia coli* (Gram-negative), and *S. aureus* was developed by Gaikwad et al. (2019). Antimicrobial low-density polyethylene (LDPE) packaging films were prepared with varying quantities of pyrogallol (PGL) and polyurethane coatings. The films' antibacterial activity, barrier and color qualities, and coating role were all tested. The PGL content has an effect on the thermal stability of the LDPE/PGL films. The barrier characteristics of pyrogallol coatings for water and oxygen increased. For numerous microorganisms, the produced films showed satisfactory antibacterial activity and proved that it could be a boon for food packaging industry.

An antimicrobial packaging with combinations of zinc oxide (ZnO) and clove essential oil (CEO) at different concentrations was developed by Ejaz et al. (2018) for shrimps. The antimicrobial film was constituted with bovine skin gelatin (BSG), ZnO nanorods (2%), and CEO (25%, 50%). The ZnO generated a film that was both flexible and mechanically resistant. Incorporating ZnO-NR into BSG/CEO films improved their oxygen and UV barrier properties. The chemical organization of the BSG composite films was changed dramatically, according to FTIR spectra. The microstructural analysis revealed that adding ZnO to BSG/CEO films avoided porosity. During refrigerated storage, the antibacterial activity of composite films containing 50% CEO, especially when combined with ZnO, was highest against Listeria monocytogenes and Salmonella Typhimurium inoculated in shrimps. The findings suggest that these BSG/CEO/ZnO films could be used as potent antimicrobial packaging for peeled shrimp.

9.9 Sustainable Packaging

Packaging industry is considered as the biggest user of plastics (Cui et al. 2020). More than 350 million tons of plastics are produced every year from which a large portion is utilized by food packaging industry solely (Ritchie and Roser 2018).This exploitation of plastic has raised several threats for environment as well as mankind. Accumulation of plastic in soil, endangering sea lives, depletion of fossil fuel reserves, and elevated petroleum costs are some of the major threats contributed by the use of petroleum-based plastics. Hence, the new age research studies are directed towards development of eco-friendly packaging/sustainable packaging. Some well-known examples of sustainable materials are bioplastics and biopolymers.

Bioplastics can be defined as earth friendly hybrid plastics constructed from renewable resources which made them compostable and biodegradable (Kumar and Thakur 2017).These are either biobased (sugar cane pulp, fiber composites, and starch-based films etc.) or biodegradable (Polylactic acid plastics) in nature (Nitaigour 2014). Biopolymers on the other side show required physiochemical, thermal and mechanical properties that made them suitable for the production of bioplastics. Biopolymers such as starch, cellulose, pectin, keratin, chitin, gelatin and polyhydroxyalkanoates can be obtained from natural biomass sources. These natural sources can be either biological systems (microbes, plants, animals) or biological materials (sugars, starch, protein, lipid etc.) (Pawar and Aachal 2013).

Among all biological materials, starch is the most promising material because of its availability, low cost and biodegradable nature (Tang et al. 2011). Starch is utilized for producing biodegradable and edible films as well as for filling reinforced plastics (Bagheri 1999; Eskandarabadi et al. 2019). Most commonly used starches for packaging material production are cellulose (Peterson et al. 1999) chitin (Tang et al. 2011), pectin (Fishman et al. 2006).

Protein based packaging film production includes the use of gluten, soy protein, gelatin, corn zein, casein and whey protein etc. These films are more efficient in water vapor permeability than plastics because of their hydrophilic nature (Tang et al. 2011). Lipids such as waxes (paraffin, candelilla wax and beeswax), fatty acids and fats are used as constituent in packaging film production.

Mostly fatty acids with carbon length (Cushen et al. 2012; De Jong et al. 2005) are preferred for films and coating production. These lipid-based films possess excellent protection from moisture due to their hydrophobic nature and are proved to be excellent barrier for oxygen penetration as well. Usually, wax films are substantially more resistant to moisture migration than other lipid or non-lipid edible films. Despite of such vast progress in sustainable packaging materials, some materials are still unsuitable for direct use as packaging as a result of their poor or insufficient barrier and mechanical properties. Thus, multilayer packaging and composite material packaging are the new focal point of research in the field of food packaging materials.

Costa et al. (2021) studied the effect of chitosan and cellulose nanocrystal films in chicken meat packaging. The addition of cellulose nanocrystal increased the thermal stability and oxygen barrier while maintaining water vapor permeability. The rise in tensile strength and Young's modulus in chitosan/cellulose nanocrystal films also indicated an improvement in mechanical characteristics. These films showed bactericidal and fungicidal efficacy against Gram-positive and Gram-negative bacteria, as well as Candida albicans. Finally, when compared to commercial membranes, chitosan-based films reduced the growth of Pseudomonas and Enterobacteria bacteria in meat during the first days of storage, while chitosan/cellulose nanocrystal films reduced total volatile basic nitrogen (TVBN), indicating their effectiveness in preventing meat spoilage under refrigeration conditions. This study demonstrates how natural-based films can be used as green food preservation options.

A composite biodegradable film using chitosan, bacterial cellulose, and curcumin was constructed by Xu et al. (2021) with the objective of developing a biodegradable packaging. The findings of their study revealed that higher molecular weight chitosan had higher contact angles and mechanical qualities, as well as reduced moisture content, water vapor transfer rate, and oxygen transmission rate. The developed chitosan-cellulose-curcumin composite film had the best antioxidant activity in the fatty food system, indicating that it can be a good sustainable active packaging material for meat and other fat rich foods.

Swaroop and Shukla (2019) used industrial melt-processing to develop blown polylactic acid (PLA) and MgO nanocomposite food packaging sheets. The integration of 2wt percent MgO nanocomposites in PLA film increased tensile strength and plasticity, whereas 1wt percent incorporation improved antibacterial characteristics for E. coli. Overall, their study demonstrated the feasibility of producing large-scale, sustainable blown PLA/MgO nanocomposite food packaging films.

An eco-sustainable packaging based on poly lactic acid, diatomite and coffee extract was created by Cacciotti et al. (2018). For the first time, multifunctional eco-sustainable systems based on polylactic acid (PLA) as a biopolymeric matrix, diatomaceous earth as a reinforcing filler, and spent coffee grounds extract as an oxygen scavenger were created in order to improve mechanical and gas barrier properties simultaneously. The outcomes showed that the inclusion of diatomite and coffee grounds extract improved both mechanical and oxygen barrier qualities, implying a probable synergic effect of the two additions.

The concept of biodegradable rice starch/carboxymethyl chitosan film was developed by Suriyatem et al. (2018). Extracts of propolis (ppl) a resin-like material made by bees was incorporated in different proportion ranging from 0 to 10% w/w into the packaging films. The results showed that as ppl content increased, so did free radical scavenging and color values, while luminosity and swellability of the films decreased. Antimicrobial activity against Gram-positive bacteria was demonstrated by active films containing 5–10% ppl especially, against Staphylococcus aureus and Bacillus cereus. The developed film with 10% ppl demonstrated increased flexibility and thermal stability without affecting oxygen permeability. The results suggested that incorporating ppl into rice starch/chitosan films could improve their antioxidant and antimicrobial properties.

9.10 Bioactive Packaging

Bioactive packaging is a recent concept in the field of food packaging with a similar principle as active packaging. The principal difference between active packaging and bioactive packaging is that bioactive packaging deals directly with the health of consumer by diffusing bioactive compounds to food from packaging rather than only providing safety and maintaining quality of food as active packaging (Lopez-Rubio et al. 2006).

The current food technology industries demand retention of bioactive compounds in food for a longer time because of the ever-growing craze for healthy food. The bioactive packaging concept works in three ways—use of biodegradable packaging materials containing bioactive compounds as packaging or edible coating, micro and nano-encapsulation of bioactive compounds to food or packaging materials and incorporation of enzymes to packaging that can convert some components of food for better health properties. This technology utilizes bioactive compounds like probiotics, prebiotics, vitamins, phytochemicals, bioavailable flavonoids, and mineral oils etc. for packaging production (Lagaroan et al. 2007).

Chitosan, gelatin, poly lactic acid, polyglycolic acid, carrageenan and alginate are some of the examples of studied biomaterials for the production of bioactive packagings. Despite of the immeasurable benefits of this technology, more research is needed for development of method and materials to ensure proper encapsulation and to obtain desired release rates of bioactive compounds from packaging. Such an experiment was reported by Sanchez et al., where they constructed an edible active film with banana starch and curcumin loaded orange oil nanoemulsion. The developed films showed increased hydrophobicity, and elongation at break values than the control (without nanoemulsion). Tests for release of bioactive agents (curcumin) in both aqueous and non-aqueous food revealed that curcumin release was highest in aqueous food than non-aqueous food due to the presence of a diffusion gradient. Moreover the study results gave a new perspective/role of packaging material to the field of food packaging that requires more such research experiments (Sanchez et al. 2022).

9.11 Conclusion

Food packaging is an irreplaceable aspect of food industries which is always changing with time and demand. Numbers of technologies were developed in the past and are still evolving till today. Packaging technologies like intelligent packaging, active packaging and bioactive packaging needs to be explored more because of their higher potential benefits in packaging industries. It is also necessary to assess the limitations and risks of each developed packaging technique. However, with increasing concerns of environment more research and efforts are required for developing more biodegradable or eco-friendly packaging materials. It is quite obvious that the advancement in nanotechnology, biotechnology and electronics has changed the packaging industry to an extent that the conventional packaging materials are almost at the stage of extinction. People now-a-days donot need to read the labels of the food just a scan of the barcodes can provide them all the information of constituents, quality tests done, storage conditions and more. Hence it can be expected that in near future the packaging will be used not only as information carriers but also as miniaturized analytical tools.

References

Abdullahi N (2018) Advances in food packaging technology: a review. J Postharvest Technol 6(4): 55–64

Abreu AS, Oliveira M, Rodrigues RM, Cerqueira MA, Vicente AA, Machado AV (2015) Antimicrobial nanostructured starch-based films for packaging. Carbohydr Polym 129:127–134

Agarwal H, Menon S, Kumar SV, Rajeshkumar S (2018) Mechanistic study on antibacterial action of zinc oxide nanoparticles synthesized using green route. Chem Biol Interact 286:60–70

Ahmed J, Hiremath N, Jacob H (2016) Efficacy of antimicrobial properties of polylactide/cinnamon oil film with and without high-pressure treatment against Listeria monocytogenes and Salmonella typhimurium inoculated in chicken sample. Food Packag Shelf Life 10:72–78

Ahvenainen R (2003) Active and intelligent packaging: an introduction. In: Ahvenainen R (ed) Novel food packaging techniques, 1st edn. Woodhead Publishing, Cambridge, pp 5–21

Antosik AK, Kowalska U, Stobińska M, Dzięcioł P, Pieczykolan M, Kozłowska K, Bartkowiak A (2021) Development and characterization of bioactive polypropylene films for food packaging applications. Polymers 13(20):e3478

Araque PE, De Vargas-Sansalvador IM, Ruiz NL, Rodriguez MME, Rodriguez MAC, Olmos AM (2018) Non-invasive oxygen determination in intelligent packaging using a smartphone. IEEE Sens J 18(11):4351–4357

Bagheri R (1999) Effect of processing on the melt degradation of starch-filled polypropylene. Polym Int 48(12):1257–1263

Bi F, Yong H, Liu J, Zhang X, Shu Y, Liu J (2020) Development and characterization of chitosan and D-α-tocopheryl polyethylene glycol 1000 succinate composite films containing different flavones. Food Packag Shelf Life 25:e100531

Cacciotti I, Mori S, Cherubini V, Nanni F (2018) Eco-sustainable systems based on poly lactic acid, diatomite and coffee grounds extract for food packaging. Int J Biol Macromol 112:567–575

Coma V (2008) Bioactive packaging technologies for extended shelf-life of meat-based products. Meat Sci 78:90–103

Cooksey K (2000) Utilization of antimicrobial packaging films for inhibition of selected microorganisms. In: Risch S (ed) Food packaging: testing methods and applications. ACS Publications, Washington DC, pp 17–25

Costa SM, Ferreira DP, Teixeira P, Ballesteros LF, Teixeira JA, Fangueiro R (2021) Active natural-based films for food packaging applications: the combined effect of chitosan and nanocellulose. Int J Biol Macromol 177:241–251

Cui H, Surendhiran D, Li C, Lin L (2020) Biodegradable zein active film containing chitosan nanoparticle encapsulated with pomegranate peel extract for food packaging. Food Packag Shelf Life 24:e100511

Cushen M, Kerry J, Morris M, Cruz-Romero M, Cummins E (2012) Nanotechnologies in the food industry-recent developments, risks and regulation. Trends Food Sci Technol 24(1):30–46

Cutter CN (2002) Microbial control by packaging: a review. Crit Rev Food Sci Nutr 42(2):151–161

Dainelli D, Nathalie G, Dimitrios S, Esther ZB, Paul T (2008) Active and intelligent food packaging: legal aspects and safety concerns. Trends Food Sci Technol 19(suppl 1):S103–S112

Dairi N, Ferfera-Harrar H, Ramos M, Garrigós MC (2019) Cellulose acetate/AgNPs-organoclay and/or thymol nanobiocomposite films with combined antimicrobial/antioxidant properties for active food packaging use. Int J Biol Macromol 121:508–523

De Jong AR, Boumans H, Slaghek T, Van Veen J, Rijk R, Van Zandvoort M (2005) Active and intelligent packaging for food: is it the future? Food Addit Contam 22(10):975–979

Dobrucka R, Cierpiszewski R (2014) Active and intelligent packaging food – research and development – a review. Polish J Food Nutr Sci 64(1):7–15

Dobrucka R, Cierpiszewski R, Korzeniowski A (2015) Intelligent food packaging - research and development. Log Forum 11(1):7–14

Echeverría I, López-Caballero ME, Gómez-Guillén MC, Mauri AN, Montero MP (2018) Active nanocomposite films based on soy proteins-montmorillonite-clove essential oil for the preservation of refrigerated bluefin tuna (*Thunnus thynnus*) fillets. Int J Food Microbiol 266:142–149

Ejaz M, Arfat YA, Mulla M, Ahmed J (2018) Zinc oxide nanorods/clove essential oil incorporated type B gelatin composite films and its applicability for shrimp packaging. Food Packag Shelf Life 15:113–121

Emiroglu ZK, Yemis GP, Coşkun BK, Candogan K (2010) Antimicrobial activity of soy edible films incorporated with thyme and oregano essential oils on fresh ground beef patties. Meat Sci 86(2):283–288

Eskandarabadi SM, Mahmoudian M, Farah KR, Abdali A, Nozad E, Enayati M (2019) Active intelligent packaging film based on ethylene vinyl acetate nanocomposite containing extracted anthocyanin, rosemary extract and ZnO/Fe-MMT nanoparticles. Food Packag Shelf Life 22(1): e100389

Fathi N, Almasi H, Pirouzifard MK (2019) Sesame protein isolate based bionanocomposite films incorporated with TiO2 nanoparticles: study on morphological, physical and photocatalytic properties. Polym Test 77:e105919

Fishman ML, Coffin DR, Onwulata CI, Willett JL (2006) Two-stage extrusion of plasticized pectin/ poly(vinyl alcohol) blends. Carbohydr Polym 65(4):421–429

Gaikwad KK, Singh S, Lee YS (2019) Antimicrobial and improved barrier properties of natural phenolic compound-coated polymeric films for active packaging applications. J Coat Technol Res 16(1):147–157

Gaikwad KK, Singh S, Shin J, Lee YS (2020a) Novel polyisoprene based UV-activated oxygen scavenging films and their applications in packaging of beef jerky. LWT Food Sci Technol 117: e108643. https://doi.org/10.1016/j.lwt.2019.108643

Gaikwad KK, Singh S, Negi YS, Lee YS (2020b) The effect of trans-polyisoprene/LDPE based active films on oxidative stability in roasted peanuts. J Food Meas Charact 14(4):1857–1864

García AV, Serrano NJ, Sanahuja AB, Garrigós MC (2020) Novel antioxidant packaging films based on poly(ε-caprolactone) and almond skin extract: development and effect on the oxidative stability of fried almonds. Antioxidants 9(7):629–638

Halek GW, Garg A (1989) Fungal inhibition by a fungicide coupled to an ionomeric film. J Food Saf 9(3):215–222

Hasani-Javanmardi M, Fallah AA, Abbasvali M (2021) Effect of safflower oil nanoemulsion and cumin essential oil combined with oxygen absorber packaging on the quality and shelf-life of refrigerated lamb loins. LWT Food Sci Technol 147:e111557

Henehan G, Proctor M, Abu-Ghannam N, Wills C, McLoughlin JV (1997) Adenine nucleotides and their metabolites as determinants of fish freshness. In: Olafsdottir G, Luten J, Dalgaard P, Careche M, Verrez-Bagnis V, Martinsdóttir E, Heia K (eds) Methods to determine the freshness of fish in research and industry, EU-FAIR project AIR3CT94 2283. International Institute of Refrigeration, Paris, pp 266–272

Hong S, Park W (2000) Use of color indicators as an active packaging system for evaluating kimchi fermentation. J Food Eng 46(1):67–72

Hou X, Xue Z, Xia Y, Qin Y, Zhang G, Liu H, Li K (2019) Effect of SiO2 nanoparticle on the physical and chemical properties of eco-friendly agar/sodium alginate nanocomposite film. Int J Biol Macromol 125:1289–1298

Hsiao HI, Chang JN (2017) Developing a microbial time-temperature indicator to monitor total volatile basic nitrogen change in chilled vacuum-packed grouper fillets. J Food Process Preserv 41(5):113–158

Huang LJ, Huang CH, Weng YM (1997) Using antimicrobial polyethylene films and minimal microwave heating to control the microbial growth of tilapia fillets during cold storage. J Food Sci 24:263–268

Jabeen N, Majid I, Nayik GA (2015) Bioplastics and food packaging: a review. Cogent Food Agric 1(1):e1117749

Jafarzadeh S, Alias AK, Ariffin F, Mahmud S, Najafi A (2016) Preparation and characterization of bionanocomposite films reinforced with nano kaolin. J Food Sci Technol 53(2):11–19

Jang M, Hong C, Choi J, Kim K, Choi J (2018) Time-temperature indicator for a vision-based detection system for managing the storage temperature of frozen fish products. Korean J Fish Aquat Sci 51(1):91–94

Joyner JR, Kumar DV (2015) Nanosensors and their applications in food analysis: a review. Int J Sci Technol 3(4):80–90

Kadam AA, Singh S, Gaikwad KK (2021) Chitosan based antioxidant films incorporated with pine needles (Cedrus deodara) extract for active food packaging applications. Food Control 124: e107877

Kakouri A, Drosinos EH, Nychas G-JE (1997) Storage of Mediterranean fresh fish (Boops boops, and Sparus aurata) under modified atmospheres or vacuum at 3 and 10°C. In: Luten JB, Borresen T, Oehlenschlager J (eds) Seafood from producer to consumer: integrated approach to quality, vol 38. Elsevier Science, Amsterdam, pp 17–19

Kanatt SR (2020) Development of active/intelligent food packaging film containing Amaranthus leaf extract for shelf-life extension of chicken/fish during chilled storage. Food Packag Shelf Life 24:e100506

Kour H, Wani NAT, Malik A, Kaul R, Chauhan H, Gupta P, Singh J (2013) Advances in food packaging - a review. Stewart Postharvest Rev 9(4):1–7

Krepker M, Shemesh R, Danin Poleg Y, Kashi Y, Vaxman A, Segal E (2017) Active food packaging films with synergistic antimicrobial activity. Food Control 76:117–126

Kumar S, Thakur K (2017) Bioplastics - classification, production and their potential food applications. J Hill Agric 8(2):118–122

Lagaroan JM, Fernandez-Saiz P, Ocio MJ (2007) Using ATR±FTIR spectroscopy to design active antimicrobial food packaging structures based on high molecular weight chitosan polysaccharide. J Agric Food Chem 55(7):2554–2562

Lee JS, Chang Y, Lee ES, Song HG, Chang PS, Han J (2018a) Ascorbic acid-based oxygen scavenger in active food packaging system for raw meatloaf. J Food Sci 83(3):682–688

Lee JS, Jeong S, Lee HG, Cho CH, Yoo SR (2018b) Development of a sulfite-based oxygen scavenger and its application in kimchi packaging to prevent oxygen-mediated deterioration of kimchi quality. J Food Sci 83(12):3009–3018

Lee K, Baek S, Kim D, Seo J (2019) A freshness indicator for monitoring chicken breast spoilage using a Tyvek® sheet and RGB color analysis. Food Packag Shelf Life 19:40–46

Li J, Zhang Q, Xu M, Wu C, Li P (2016) Antimicrobial efficacy and cell adhesion inhibition of in situ synthesized ZnO nanoparticles/polyvinyl alcohol nanofibrous membranes. Adv Condens Matter Phys 2016:e6394124

Lim SH, Choe WY, Son BH, Hong KW (2014) Development of a microbial time-temperature integrator system using lactic acid bacteria. Food Sci Biotechnol 23(2):483–487

Lopez-Rubio A, Gavara R, Lagaroan JM (2006) Bioactive packaging: turning foods into healthier foods through biomaterials. Trends Food Sci Technol 17(10):567–575

Mahcene Z, Khelil A, Hasni S, Akman PK, Bozkurt F, Birech K, Goudjil MB, Tornuk F (2020) Development and characterization of sodium alginate-based active edible films incorporated with essential oils of some medicinal plants. Int J Biol Macromol 145:124–132

Majid I, Nayik GA, Dar SM, Nanda V (2016) Novel food packaging technologies: innovations and future prospective. J Saudi Soc Agric Sci 17(4):454–462

Mataragas M, Bikouli VC, Korre M, Sterioti A, Skandamis PN (2019) Development of a microbial time temperature indicator for monitoring the shelf-life of meat. Innov Food Sci Emerg Technol 52:89–99

Mattila T, Tawast J, Ahvenainen R (1990) New possibilities for quality control of aseptic packages: microbiological spoilage and seal defect detection using head-space indicators. Lebensmittel-Wissenschaft und -Technologie (LWT) 23(3):246–251

Mei L, Wang Q (2020) Advances in using nanotechnology structuring approaches for improving food packaging. Annu Rev Food Sci Technol 11:339–364

Nadarajah D, Han JH, Holley RA (2005) Inactivation of Escherichia coli O157:H7 in packaged ground beef by allyl isothiocyanate. Int J Food Microbiol 99(3):269–279

Nitaigour PM (2014) Advances in packaging methods, processes and systems. Challenges 5(3): 374–389

Nopwinyuwong A, Trevanich S, Suppakul P (2010) Development of a novel colorimetric indicator label for monitoring freshness of intermediate-moisture dessert spoilage. Talanta 81(3): 1126–1132

Noshirvani N, Bhanbarzadeh B, Mokarram RR, Hashemi M (2017) Novel active packaging based on carboxymethyl cellulose-chitosan-ZnO NPs nanocomposites for increasing the shelf-life of bread. Food Packag Shelf Life 11:106–114

Ohlenschlager J (1997) Suitability of ammonia-n, dimethylamine-n, the use of freshness indicators in packaging 141 trimethylamine-n, trimethylamine oxide-n, and total volatile basic nitrogen as freshness indicators in seafoods. In: Olafsdottir G, Luten J, Dalgaard P, Careche M, Verrez-Bagnis V, Martinsdóttir E, Heia K (eds) Methods to determine the freshness of fish in research and industry, EU-FAIR project AIR3CT94 2283. International Institute of Refrigeration, Paris, pp 266–272

Omanović-Mikličanin E, Maksimović M (2016) Nanosensors applications in agriculture and food industry. Bull Chem Technol Bosnia Herzegovina 47:59–70

Oun AA, Rhim JW (2017) Carrageenan-based hydrogels and films: effect of ZnO and CuO nanoparticles on the physical, mechanical, and antimicrobial properties. Food Hydrocoll 67: 45–53

Oussalah M, Caillet S, Salmieri S, Saucier L, Lacroix M (2004) Antimicrobial and antioxidant effects of milk protein-based film containing essential oils for the preservation of whole beef muscle. J Agric Food Chem 52(18):5598–5605

Ozdemir M, Floros JD (2004) Active food packaging technologies. Crit Rev Food Sci Nutr 44(3): 185–193

Pacquit A, Lau KT, McLaughlin H, Frisby J, Quilty B, Diamond D (2006) Development of a volatile amine sensor for the monitoring of fish spoilage. Talanta 69(2):515–520

Pacquit A, Frisby J, Diamond D, Lau KT, Farrell AP, Quilty B, Diamond D (2007) Development of a smart packaging for the monitoring of fish spoilage. Food Chem 102(2):466–470

Padgett T, Han IY, Dawson PL (1998) Effect of lauric acid addition on the antimicrobial efficacy and water permeability of corn zein films containing nisin. J Food Process Preserv 24(5): 423–432

Pawar PA, Aachal HP (2013) Biodegradable polymers in food packaging. Am J Eng Res 2(5): 151–164

Peterson K, Nielsen PV, Bertelsen G, Lawther M, Olsen MB, Nilsson NH, Mortensen G (1999) Potential of biobased materials for food packaging. Trends Food Sci Technol 10(2):52–68

Pires AD, Soares NF, de Andrade NJ, da Silva LM, Camilloto GP, Bernardes PC (2009) Increased preservation of sliced mozzarella cheese by antimicrobial sachet incorporated with allyl isothiocyanate. Braz J Microbiol 40(4):1002–1008

Plackett D, Ghanbari-Siahkali A, Szente L (2007) Behavior of alpha- and beta-cyclodextrin encapsulated allyl isothiocyanate as slow-release additives in polylactide-co-polycaprolactone films. J Appl Polym Sci 105(5):2850–2857

Prasad P, Kochhar A (2014) Active packaging in food industry: a review. IOSR J Environ Sci Toxicol Food Technol 8(5):1–7

Restuccia D, Spizzirri UG, Parisi OI, Cirillo G, Curcio M, Lemma F, Puoci F, Vinci G, Picci N (2010) New EU regulation aspects and global market of active and intelligent packaging for food industry applications. Food Control 21(11):1425–1435

Risch SJ (2009) Food packaging history and innovations. J Agric Food Chem 57(18):8089–8092
Ritchie H, Roser M (2018) Plastic pollution. ourworldindata.org/plastic-pollution. Accessed 31 Aug 2022
Rokka M, Eerola S, Smolander M, Alakomi H, Ahvenainen R (2004) Monitoring of the quality of modified atmosphere packaged broiler chicken cuts stored in different temperature conditions B. Biogenic amines as quality-indicating metabolites. Food Control 15(8):601–607
Sanchez LT, Pinzon MI, Villa CC (2022) Development of active edible films made from banana starch and curcumin-loaded nanoemulsions. Food Chem 371:e131121
Shu HC, Hakanson H, Mattiasson B (1993) D-lactic acid in pork as a freshness indicator monitored by immobilized D-lactate dehydrogenase using sequential injection analysis. Anal Chim Acta 283(2):727–737
Siddiqui N, Bhardwaj A, Hada R, Yadav VS, Goyal D (2018) Synthesis, characterization and antimicrobial study of poly(methyl methacrylate)/Ag nanocomposites. Vacuum 153:6–11
Singh S, Shalini R (2016) Effect of hurdle technology in food preservation: a review. Crit Rev Food Sci Nutr 56(4):641–649
Singh S, Gaikwad KK, Lee YS (2019) Development and application of a pyrogallic acid-based oxygen scavenging packaging system for shelf-life extension of peeled garlic. Sci Hortic 256: e108548
Singh G, Singh S, Kumar B, Gaikwad KK (2021) Active barrier chitosan films containing gallic acid - based oxygen scavenger. J Food Meas Character 15(1):585–593
Sivarooban T, Hettiarachchy NS, Johnson MG (2008) Physical and antimicrobial properties of grape seed extract, nisin, and EDTA incorporated soy protein edible films. Food Res Int 41(8): 781–785
Smolander M, Biotechnology VTT (2003) The use of freshness indicators. In: Ahvenainen R (ed) Novel food packaging techniques. Woodhead Publishing, London, pp 127–143
Sofi SA, Singh J, Rafiq S, Ashraf U, Dar BN, Nayik GA (2018) A comprehensive review on antimicrobial packaging and its use in food packaging. Curr Nutr Food Sci 14(4):305–312
Soltani M, Mohi-Alden K, Omid M (2021) A critical review on intelligent and active packaging in the food industry: research and development. Food Res Int 141:e110113
Sun Q, Li J, Le T (2018) Zinc oxide nanoparticle as a novel class of antifungal agents: current advances and future perspectives. J Agric Food Chem 66(43):11209–11220
Suppakul P, Miltz J, Sonneveld K, Bigger SW (2006) Characterization of antimicrobial films containing basil extracts. Packag Technol Sci 19(5):259–268
Suriyatem R, Auras RA, Rachtanapun C, Rachtanapun P (2018) Biodegradable rice starch/ carboxymethyl chitosan films with added propolis extract for potential use as active food packaging. Polymers 10(9):e954
Swaroop C, Shukla M (2019) Development of blown polylactic acid-MgO nanocomposite films for food packaging. Compos A: Appl Sci Manuf 124:e105482
Tang XZ, Kumar P, Alavi S, Sandeep KP (2011) Recent advances in biopolymers and biopolymer-based nanocomposites for food packaging materials. Crit Rev Food Sci Nutr 52(5):426–442
Tanwar R, Gupta V, Kumar P, Kumar A, Singh S, Gaikwad KK (2021) Development and characterization of PVA-starch incorporated with coconut shell extract and sepiolite clay as an antioxidant film for active food packaging applications. Int J Biol Macromol 185:451–461
Taoukis PS, Labuza TP (1989) Applicability of time-temperature indicators as shelf-life monitors of food products. J Food Sci 54(4):783–788
Tsironi T, Ronnow P, Giannoglou M, Taoukis P (2017) Developing suitable smart TTI labels to match specific monitoring requirements: the case of Vibrio spp. growth during transportation of oysters. Food Control 73:51–56
Vahedikia N, Garavand F, Tajeddin B, Cacciotti I, Jafari SM, Omidi T, Zahedi Z (2019) Biodegradable zein film composites reinforced with chitosan nanoparticles and cinnamon essential oil: physical, mechanical, structural and antimicrobial attributes. Colloids Surf B Biointerfaces 177:25–32

Watanabe A, Tsuneishi A, Takimoto Y (1989) Analysis of ATP and its breakdown products in beef by reversed-phase HPLC. J Food Sci 54(5):1169–1172

Wen J, Huang S, Jia L, Ding F, Li H, Chen L, Liu X (2019) Visible colorimetric oxygen indicator based on Ag-loaded TiO2 nanotubes for quick response and real-time monitoring of the integrity of modified atmosphere packaging. Adv Mater Technol 4(9):e1900121

Winther M, Nielsen PV (2006) Active packaging of cheese with allyl isothiocyanate, an alternative to modified atmosphere packaging. J Food Prot 69(10):2430–2435

Xu L, Zhang B, Qin Y, Li F, Yang S, Lu P, Wang L, Fan J (2020) Preparation and characterization of antifungal coating films composed of sodium alginate and cyclo-lipopeptides produced by Bacillus subtilis. Int J Biol Macromol 143:602–609

Xu Y, Liu X, Jiang Q, Yu D, Xu Y, Wang B, Xia W (2021) Development and properties of bacterial cellulose, curcumin, and chitosan composite biodegradable films for active packaging materials. Carbohydr Polym 260:e117778

Yang W, Fortunati E, Dominici F, Giovanale G, Mazzaglia A (2016) Synergic effect of cellulose and lignin nanostructures in PLA-based systems for food antibacterial packaging. Eur Polym J 79:1–12

Yano Y, Kataho N, Watanabe M, Nakamura T, Asano Y (1995) Evaluation of beef ageing by determination of hypoxanthine and xanthine contents: application of a xanthine sensor. Food Chem 52(4):439–445

Yi JH, Kim IH, Choe CH, Seo YB, Song KB (1998) Chitosan-coated packaging papers for storage of agricultural products. Appl Biol Chem 41(6):442–446

Yildirim S, Röcker B, Pettersen MK, Nilsen-Nygaard J, Ayhan Z, Rutkaite R, Radusin T, Suminska P, Marcos B, Coma V (2018) Active packaging applications for food. Compr Rev Food Sci Food Saf 17(1):165–199

Chapter 10
Biodegradable Packaging: Recent Advances and Applications in Food Industry

Ritika B. Yadav

Contents

R. B. Yadav (✉)
Department of Food Technology, Maharshi Dayanand University, Rohtak, Haryana, India

J. A. Malik et al. (eds.), *Food Process Engineering and Technology*,
https://doi.org/10.1007/978-981-99-6831-2_10

Abstract The development of biodegradable packaging materials can decrease the environment pollution generated by synthetic polymers like plastics. The utilization of biodegradable polymers over non-biodegradable polymers is a good choice for reducing environmental damage and preserving natural resources. The biodegradable polymers are categorized into natural, synthetic, and microbial biopolymers. These are non-toxic, compostable, and biodegradable. The aim of this chapter is to provide an overview of biodegradable packaging and recent developments regarding their applications for food products. The natural and synthetic biodegradable packaging materials, their sources, and film properties have been discussed in detail.

Keywords Biodegradable packaging materials · Natural biopolymers · Synthetic biopolymers · Food packaging · Biodegradation of biopolymers

10.1 Introduction

Packaging is a vital segment of the food processing industries. Packaging maintains and protects the food product quality, thus increasing shelf-life and reducing the wastage of food. The thermoplastic polymers i.e., polyethylene, polyethylene terephthalate (PET), polystyrene, and polyvinylidene chloride have been commonly used in the food sector due to their numerous benefits, like lightweight, low cost, good barrier properties, long-lasting, and good mechanical strength. These synthetic polymers are generally non-degradable, and only 10% of the discarded plastics are recycled (Zhao et al. 2020), thus contributing to a massive accumulation of plastic waste material in the environment.

A huge quantity of plastic waste is dropped into the rivers and ocean that disturbs the marine ecosystem. The incineration of plastic waste causes the release of gases such as carbon monoxide, carbon dioxide, amines, furans, dioxin etc. This degrades the quality of air and leads to various health disorders (Smith 2005). The increasing environmental concerns impose to alternative packaging material that is eco-friendly. Biodegradable packaging materials evolved as an innovative approach for food industries to control the threat caused by synthetic and non-biodegradable polymers.

Biopolymers are non-toxic, compostable and biodegradable polymers prepared from renewable sources. Biopolymers can be categorized into three main categories: polymers chemically prepared from biobased monomers, polymers prepared by microorganisms, and polymers extracted from natural resources (Cha and Chinnan 2004). Biodegradation is a process of degradation of polymer by microorganisms into gases (carbon dioxide and methane) and biomass (Meereboer et al. 2020). The polymer properties (crystallinity, composition, hydrophilicity) and environmental factors (temperature, anaerobic or aerobic conditions, humidity) affect the decomposition rate (Meereboer et al. 2020).

The biodegradable coatings and films have suitable mechanical and physicochemical properties to replace plastic polymers (Guerrero et al. 2020). The mechanical properties such as tensile strength (TS), bursting strength etc., of biodegradable polymers depends on their composition (Cho et al. 2001; Willett 1994), method of preparation, and storage (Lawton 1996). The modification of biodegradable polymeric materials caused enhancement of thermal, gas barrier, and mechanical characteristics while keeping their stability and also increasing the shelf-life of food like cheese, fish, and fruits (Freitas et al. 2020; Homayounpour et al. 2020; Zhou et al. 2021). Biodegradable packaging is the best attempt to solve the issues of non-biodegradable packaging materials as the later have a tendency to accumulate in the environment for many years (Jambeck et al. 2018).

This chapter provides an overview of biodegradable packaging and recent developments regarding their applications for food products. The natural, synthetic, and microbial biopolymers, their sources, and properties have been discussed in detail. Further, the advantages, limitations, and modification methods for improving the functional characteristics of the biopolymers are discussed.

10.2 Biodegradable Polymeric Materials

The utilization of biodegradable polymers over non-biodegradable polymers is a good choice for reducing environmental damage and preserving natural resources. The biodegradable characteristic is the ability of a polymeric material to be decomposed through the action of living organisms (fungi, bacteria etc.) and enzymes (Ndahebwa et al. 2018). Biodegradation can be accomplished by different techniques like anaerobic degradation and aerobic degradation.

During biodegradation, the polymers are converted into smaller fragments such as monomers, oligomers, and dimers via the action of enzymes produced by microorganisms which are finally converted into biomass, methane, carbon dioxide (CO_2), and water (Ndahebwa et al. 2018; Ragaert et al. 2017). Other processes such as oxidation, hydrolysis, and photo-degradation may also have some effect on the degradation of biopolymers. It is needed that biodegradable polymers should be durable, stable, and non-toxic when in use and should be degraded quickly after disposal. The degradation of such polymers is often initiated from end-groups due to the strong and firm backbones of carbon (Kumar et al. 2011).

Depending upon the source, biopolymers are divided into three different types such as natural, synthetic, and microbial biopolymers. The polysaccharides i.e., starch, cellulose, gums, alginate, and chitosan are used. The protein sources include whey protein, soy protein, zein, and gelatin. A polymer with both hydrophilic and hydrophobic fractions showed greater biodegradability than those with either hydrophilic or hydrophobic groups (Patwary et al. 2020). The addition of fillers could increase the degradation rate of polymeric materials by facilitating the microorganism action on the surface of polymers (Alshabanat 2019). To improve the functional characteristics of biodegradable films, the ingredients such as antioxidants, antimicrobials, flavonoids etc. are added to the polymers during processing (Lu et al. 2019; Salgado et al. 2015).

10.3 Classification of Biopolymers

Biodegradable materials are categorized into three groups on the basis of source (Fig. 10.1).

1. Natural biopolymers are obtained from natural sources, e.g., polysaccharides (starch, cellulose, alginates, pectin, and chitin) and proteins (corn zein, gelatin, soy protein, wheat gluten, and whey protein).
2. Synthetic biopolymers are synthesized through polymerization from biobased and oil-based materials (polylactic acid, polylactide, polybutylene succinate, etc.)
3. Microbial biopolymers are produced by microorganisms, such as, polyhydroxybutyrate and polyhydroxylalkanoate.

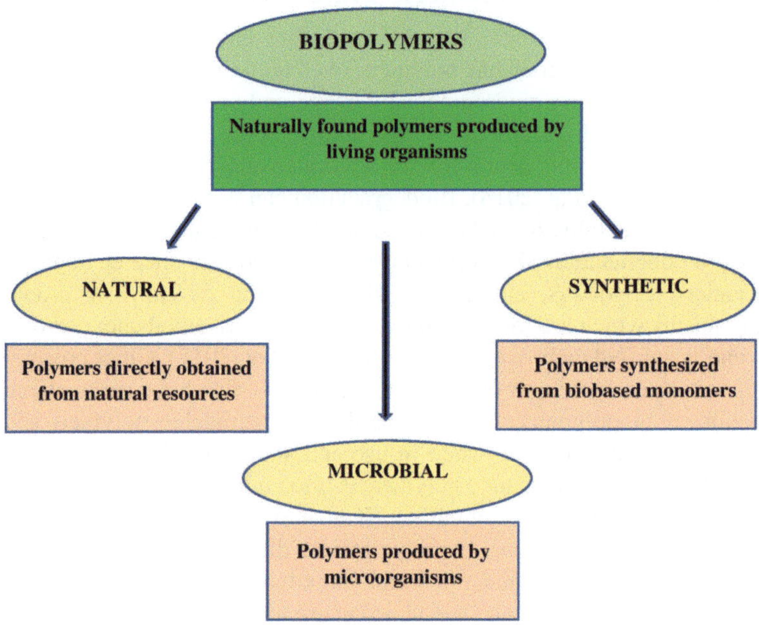

Fig. 10.1 Types of biopolymers

10.4 Natural Biopolymers

Natural polymers are generally organic compounds like polysaccharides, proteins, and their complexes. They are obtained from several natural sources i.e., fungi, bacteria, plants, and animals. The properties of a few ubiquitous natural biopolymers are discussed here.

10.4.1 Starch

Starch is one among the most utilized natural polymers due to its good functional characteristics like low-cost, eco-friendly, film-forming characteristics, and biodegradable nature (Fonseca-García et al. 2021; Jiang et al. 2020; Jimenez et al. 2012). The starch consists of amylose and amylopectin (linear and branched structure, respectively) of D-glucose monomeric units connected through α 1→4 and α 1→6 glycosidic bonds (Jimenez et al. 2012). Among the two polymers, amylose has crystalline structure and good film-forming ability (Acosta et al. 2015). The starch-based films have synthetic polymer-like properties such as good transparency, tasteless and odorless (Hazrati et al. 2021).

The main challenge with starch films is that they are hydrophilic and brittle. However, these characteristics of starch-films can be altered and improved by adding plasticizers. The different plasticizers (sorbitol, glycol, glycerol) are used to modify the starch into TPS (thermoplastic starch). TPS can be produced by heating the starch with a plasticizer at a temperature of 140–160 °C with high shear in an extruder (Abdul-Khalil et al. 2018). Cross-linking of native starch using plasticizers at an appropriate concentration can decrease the water loss (Reddy and Yang 2010). The addition of polyvinyl alcohol decreases the brittleness of the starch films (Gómez-Aldapa et al. 2020; Wu et al. 2017). It was also observed that the inclusion of lipids reduces the WVTR of films.

Oliveira et al. (2018) reported a decrease in WVTR by adding the beeswax to corn starch, gelatin, and cassava starch blend. The inclusion of essential oils (EOs) i.e., thyme oil, eugenol, cinnamon, and carvacrol, improve the antimicrobial behavior and thus, increase the shelf-life of foods (Marin et al. 2017).

10.4.2 Cellulose

Cellulose is a linear polysaccharide consisting of β 1→4 glucopyranosyl units. The cellulose derivatives like hydroxy-propyl cellulose (HPC), hydroxy-propyl methylcellulose, and carboxymethyl cellulose (CMC) possess film-forming characteristics. Cellulose has many benefits, such as non-toxic, biodegradable, low-cost, and provide films with the good barrier, thermal, and mechanical characteristics

(Achachlouei and Zahedi 2018; Noshirvani et al. 2017; Singh et al. 2019). HPC and CMC films have good oil resistance and gas barrier properties (Villalobos et al. 2005). Cellulose acetate (CA) is an important cellulose derivative having TS similar to polypropylene. CA is currently used for the packaging of fresh products and dry foods, although its application is limited because of moisture sensitivity (Cazón et al. 2017).

10.4.3 Alginates

Alginate is a biodegradable and nontoxic linear polysaccharide obtained from seaweeds. It consists of α-D manuronate (M) and β-L guluronate (G) linked by 1→4 linkages (Liakos et al. 2013; Tabassum and Khan 2020). The proportion and sequence of these two units (M and G) in alginates are responsible for their properties. The physicochemical characteristics of alginate films are related to the three-dimensional network developed in presence of the divalent ions (Nešić et al. 2017). The alginate-based films have good mechanical strength, TS, oxygen barrier property, and flexibility than other natural polymer-based films (Shahabi-Ghahfarrokhi et al. 2020). Several antioxidant and antimicrobial compounds have been added to alginate-based coatings and films to form bioactive packaging materials. Bazargani-Gilani (2018) developed an alginate-based coating with resveratrol and found increased shelf-life of trout fillet. The other applications of alginate-based coatings and films include peach (Li et al. 2019), mushrooms (Zhu et al. 2019a), lamb meat and fresh-cut mango (Guerrero et al. 2020).

10.4.4 Pectin

Pectin is a linear-chain polysaccharide comprising α 1→4 galacturonic acid having different degree of esterification (Šešlija et al. 2018). It is known for its biodegradability, compatibility, gel-forming ability, and non-toxicity (Chaichi et al. 2017; Espitia et al. 2014; Nisar et al. 2018). Pectin films have low mechanical strength and poor physico-chemical characteristics (Rai et al. 2019; Šešlija et al. 2018). To improve the mechanical and adhesion properties of the pectin films/coatings, plasticizers (glycerol, glycol, and sucrose) and flexible polymers (polyvinyl alcohol, CMC, and starch) are generally added. The application of pectin-based coatings on the lime increased shelf-life, decreased refrigeration cost, and enhanced fruit safety (Maftoonazad and Ramaswamy 2019).

10.4.5 Chitin and Chitosan

Chitin is generally present in the shells of crabs, shrimps, insects, etc. Chitin is a linear polymer consisting of N-glucosamine and N-acetyl-glucosamine bonded through β1→4 linkages. Chitosan is derived by deacetylation of chitin comprising of β (1–4)-2 amino-2 deoxy-D glucose monomer units (Ramos-García et al. 2012; Wu 2014). Chitin and chitosan are biodegradable polymers having good film formation and antimicrobial properties. Both are used for the preparation of film and coatings to improve the shelf-life of vegetables and fruits (Zhao and McDaniel 2005).

The Chitosan-coated strawberries showed a germicidal influence against fungi and bacteria and also maintained the appearance, flavor, and texture (Pavinatto et al. 2020). Gomes et al. (2018) developed a chitosan coating incorporated with nanoparticles of chitosan. The coating showed an antibacterial influence against gram −ve and gram +ve bacteria. The incorporation of oily-antimicrobial materials to chitosan could improve the adhesive characteristics of the coatings on the lipophilic surface like peels of fruits. Sanchez-Gonzalez et al. (2011) noted an increased antimicrobial effect of chitosan-based coatings after the addition of bergamot oil.

10.4.6 Zein

Zein is a protein (prolamin) present in the seeds of corn. It is hydrophobic in nature due to nonpolar amino acids (Sahraee et al. 2019). Zein is thermoplastic in nature, insoluble, and possesses good film-forming ability (Gao et al. 2017; Zhu et al. 2019b). Studies reported that zein films have good thermal stability and selective permeability (Gao et al. 2017; Zhang et al. 2019; Zhu et al. 2019b). These films also have a good barrier and antibacterial characteristics (Zhang et al. 2019) but carry poor mechanical strength (Sun et al. 2018). Ghasemi et al. (2015) prepared the antimicrobial zein based film by adding Zataria boiss essential oil and found the effectiveness of films against S. enteritidis, S. aureus, and E. coli growth, resulting in the increase of shelf-life of cheese.

Pereira et al. (2019) studied the influence of the inclusion of garlic and thyme EOs on the mechanical and antimicrobial characteristics of zein-based films. The results of the study indicated that the addition of EOs to the films showed an antimicrobial effect as well as performed as a plasticizer, as confirmed by the decrease in glass transition temperature (Tg), decrease in mechanical cracks, and increase of puncture strength.

10.4.7 Gelatin

Gelatin is a high molecular weight (HMW) protein that is derived by partial hydrolysis of collagen. Gelatin is tasteless, odorless, and soluble in the aqueous solution of alcohol. It is hydrophilic in nature and has a good film-forming ability. The thermo-reversible nature of gel formed by gelatin plays a vital role in widening the utilization of gelatin for encapsulation of drugs and oil. The encapsulation gives protection against light and oxygen. Ramos et al. (2016) applied gelatin film as coatings on the meats to decrease the water and oxygen permeability. Several studies reported the influence of the inclusion of strengthening additives, crosslinkers, antimicrobial substances, antioxidants, and plasticizers in gelatin-based coatings/films to enhance the barrier and mechanical characteristics (Ortiz-Zarama et al. 2016; Ramos et al. 2016).

10.4.8 Soy Protein

Soy protein is a natural biodegradable source used in the formation of coatings and edible films. Soy protein concentrate and isolate are the by-products of the oil industry. Soy protein consists of globulins i.e., β conglycinin and glycinin (Song et al. 2011). The soy protein-based films are hydrophilic and possess low barrier and mechanical properties compared to synthetic films. The addition of different polymers and lipids to soy protein increased the mechanical strength (MS) of these films (Tulamandi et al. 2016). The incorporation of stearic acid (25%) improves the tensile and thermal properties and decreases the loss of moisture (Lodha and Netravali 2005). The coating of poly lactic acid on soy protein-based films improves the barrier and mechanical properties of such films (Rhim et al. 2007).

10.4.9 Wheat Gluten

Wheat gluten is a readily obtainable by-product derived during wheat starch isolation. It has good film-forming properties but is slightly brittle in nature. Wheat gluten-based films have good oxygen-barrier properties but low moisture-barrier properties (Irissin-Mangata et al. 2001). The inclusion of plasticizer to wheat gluten reduces the brittleness and enhances the processability. The antimicrobial compounds have also been added to improve the antimicrobial properties of gluten-based films. Türe et al. (2012) prepared the antimicrobial gluten-based films by adding potassium sorbate (PS) and determined the antimicrobial effect against Fusarium incarnatum and A. niger. They found that the antimicrobial gluten-based films with >10% PS were effective against A. niger, however a level of 2.5% or more of PS was effective against growth of *F. incarnatum*.

10.4.10 *Whey Protein*

Whey protein is the by-product obtained during processing of casein or cheese and comprises of α lactalbumin, proteose peptones, β lactoglobulin, immunoglobulins, and bovine albumin. On the basis of protein content, whey proteins are either WPC with 25–80% protein or WPI with >90% protein (Moeini et al. 2021). WPI is a pure form of whey protein with good film forming properties and produces transparent and odorless films (Ozer et al. 2016). The whey protein films have good gas barrier and mechanical characteristics but low moisture barrier characteristics. The films of whey protein are also slightly brittle. The addition of plasticizers (glycerol) decreases the brittleness of film. The incorporation of natamycin, malic acid, and nisin in whey protein showed an inhibitory effect against P. aeruginosa, P. commune, and L. monocytogenes (Pintado et al. 2010).

10.5 Synthetic Biopolymers

The synthetic biopolymers are synthesized from monomers by polycondensation or ROP (such as: polylactic acid (PLA)), which has good potential for utilization as biodegradable packaging material (Weber et al. 2002). The properties of a few ubiquitous synthetic biopolymers are discussed here.

10.5.1 *Polylactic Acid*

PLA is an aliphatic polyester obtained from lactic acid by condensation polymerization or from lactide through ROP. The monomer lactic acid is generally obtained by the microbial fermentation of sugar, corn, starch, or biomass. It is biodegradable, compostable, and non-toxic with good mechanical strength (Yu et al. 2006). The commercial PLA which is a copolymer, consists of L-lactic acid and D-lactic acid. The properties of PLA can differ from amorphous to semicrystalline to according to the ratio of two enantiomers of a monomer (L- and D-lactic acid). The mechanical properties of PLA, such as TS can vary generally based on whether it is oriented or annealed, or its level of crystallinity (Garlotta 2001). The properties of PLA such as tensile stress, stiffness, transparency, gas permeability, and stability, are comparable to the properties of PET, PE (polyethylene), PS, and PP (Turco et al. 2019b). The PLA applications in food packaging sector include cups, plates, cutlery, bags, straws, lids, containers, etc. The drawbacks like high cost, less flexibility, less degradation rate, and brittleness reduce the use of PLA in the industries. However, the addition of TPS to PLA enhances biodegradability and mechanical properties (Turco et al. 2019a).

10.5.2 Polycaprolactone (PCL)

PCL is a biodegradable, semicrystalline polymer with good resistance to oil, water, and chlorine. It is obtained through the ROP of ε-caprolactone (Mochizuki and Hirami 1997). It has a low Tg of about −60 °C and a melting temperature (Tm) of 60–65 °C. The application of PCL for the packaging is limited due to its poor mechanical and barrier properties. The addition of chitosan to PCL increases the hydrophobicity and decreases the WVTR (Sarasam et al. 2006). The PCL is generally mixed with some other biopolymers like PLA, cellulose propionate, etc. to improve the dyeability, adhesion, and crack resistance (Navarro-Baena et al. 2016).

10.5.3 Polybutylene Succinate (PBS)

PBS is a biodegradable, semicrystalline, thermoplastic polymer with good thermal properties, chemical resistance, melt processability, mechanical properties, and biodegradability, comparable to those of the PP and PE (Gumede et al. 2018). It is obtained from poly-condensation of 1,4-butanediol and succinic acid. It has Tg of −45 to −10 °C and Tm of 90–120 °C. The different techniques can be used for processing of PBS, such as injection and blow molding, blown film, and fibers spinning. The applications of PBS include containers, cutlery, bags, foam trays, blown bottles, and packaging film.

10.5.4 Polyvinyl Alcohol (PVOH)

PVOH is a biodegradable semicrystalline synthetic polymer. It is synthesized by complete or partial hydrolysis of PVA (polyvinyl acetate). It has a Tm of 180–190 °C, a degree of hydrolysis in the range of 86.5–89%, with a molecular weight in the range of 26,300–30,000 (Ramaraj 2007). PVOH has good emulsifying, adhesive, biodegradability, and film-forming properties. PVOH film has high oxygen barrier properties, mechanical properties, flexibility, and TS. Conte et al. (2006) evaluated the antibacterial effect of lysozyme added PVOH films and found the effectiveness of films against Micrococcus lysodeikticus growth.

10.5.5 Polyglycolide (PGA)

PGA is prepared by condensation of glycolic acid or by ROP of glycolide. Injection molding, compression molding, extrusion, and solvent casting techniques are used for the processing of PGA. PGA has high crystallinity of 45–55%, and due to this, it is insoluble in many organic solvents. It has a Tm of 220–225 °C and Tg of 35–40 °C (Nair and Laurencin 2007). It has good mechanical properties and high biodegradability.

10.6 Microbial Biopolymers

These polymers are produced by microorganisms through a biosynthetic process and biodegraded by microorganisms. Examples are polyhydroxylalkanoate and polyhydroxybutyrate.

10.6.1 Polyhydroxylalkanoate (PHA)

PHAs are biodegradable natural biopolymers produced via microbial fermentation of glucose, sugar, or oil feedstock. Bacteria accumulate the PHA intracellularly as a reserve substance at a concentration of 30–80% in presence of abundant carbon and limited nitrogen (Mercan et al. 2002). Depending upon the monomers, the melting temperature of the PHA varies from 40 to 180 °C. The mechanical strength of PHA depends on the size of alkyl present in the monomeric unit. PHA is completely biodegradable, and degradation occurs through breakage of linkages at chain ends by esterases.

10.6.2 Polyhydroxybutyrate (PHB)

Polyhydroxybutyrate (PHB) is the most common PHA synthesized through radical polymerization of 3, hydroxybutyrate. PHB has good optical properties and resistivity to UV with 55 °C Tg and 180 °C Tm. It has crystallinity greater than 50%. PHB has good mechanical properties, moisture barrier properties, and biocompatibility (Cherpinski et al. 2018; Mlalila et al. 2018). PHB has wide applications as a packaging material due to its lamellar-structure, which provides good aroma-barrier properties. PHB has low thermal stability near its melting temperature (180 °C), which limits its applications (Grassie et al. 1984). To overcome this, plasticizers such as citrate ester can be added. PHB is biodegraded by microorganisms like bacteria, and fungi in different environments (Kim et al. 2000). The hydrolysis during the degradation process yields 3, hydroxybutyric acid.

Many attempts have been made to enhance the thermal, mechanical, and other properties for packaging purposes. Garrido-Miranda et al. (2018) analyzed the antifungal and antioxidant effects of eugenol incorporation in nanocomposites of PHB-TPS and found an increase in the shelf-life of refrigerated foods. Manikandan et al. (2020) prepared and characterized the PHB based GNCs (graphene nanocomposites) film and found a reduction of WVP (water vapor permeability). Ma et al. (2018) studied the influence of plasticizers (GMC and GML) in PLA-PHB composite film. The incorporation of plasticizers increased the TS of film. The shelf-life of salmon packed in PLA-PHB-plasticizers composite film was increased by 15–17 days.

The copolymer PHBV (polyhydroxy-butyrate-valerate) is formed by incorporating propionic acid (PA) to the feedstock provided to the bacteria (Sheu et al. 2009). It is highly crystalline with a Tm of 108 °C and Tg between −5 and 20 °C (Avella et al. 1996). It is less brittle than PHB. The Tm and mechanical strength of PHBV depend on the amount of hydroxy-valerate (HV) unit. The Tm, Tg, crystallinity, and TS decrease with the increase in the percentage of HV (Hsieh et al. 2009; Mohanty et al. 2000). PHBV has a high rate of degradation in comparison to PHB.

10.7 Advantages and Limitations of Biodegradable Polymers

Biopolymers are compostable and biodegradable polymers prepared from renewable sources (Fig. 10.2). These are non-toxic and thus can be utilized as packaging material for raw and processed foods. It is very easy to recycle them. The recycling of polymers helps to reduce the landfill problems. The recycled waste material can also be used as a renewable source for biogas or compost. The biopolymers require less energy during manufacturing in comparison to synthetic petroleum-based

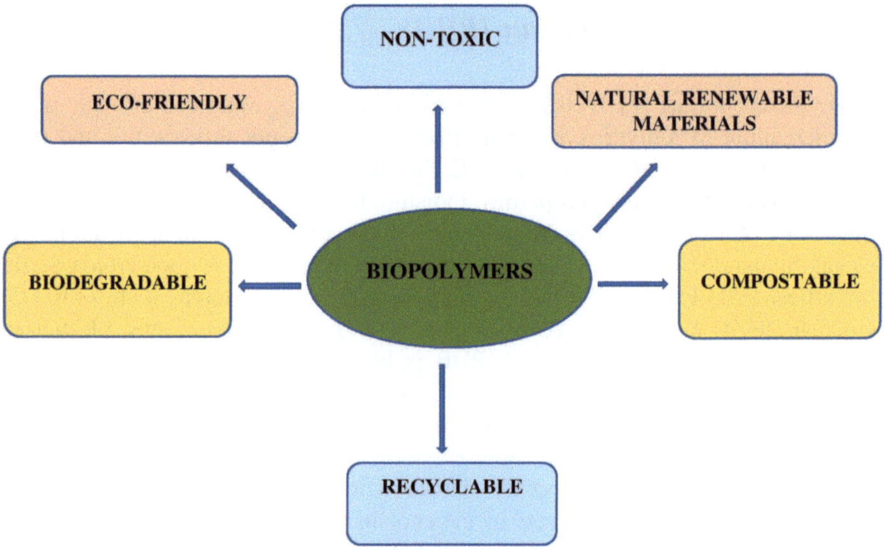

Fig. 10.2 Properties of biopolymers

polymers. The production of biopolymers requires about 65% less energy than synthetic plastics, thus contributing to reduced formation of greenhouse gases (Ahvenainen 2003; Halley 2002). Due to biodegradation, biopolymers are easily broken down and converted to carbon dioxide, water, inorganic substances, or biomass. There is no deposition of waste and thus acts as an environment friendly.

The films and coatings of natural biopolymers usually have good barrier characteristics, but overall, their properties such as low mechanical strength, brittleness, and higher WVP make them inferior to those of other synthetic plastics (Cazón et al. 2017). Due to biodegradable nature, these films are more susceptible to microbial spoilage and lipid oxidation resulting in short shelf-life than petroleum-based synthetic plastics. The hygroscopic nature of biopolymer films makes them unstable in a humid environment. Due to high hygroscopicity, the films are not suitable for high-moisture foods. However, the mechanical strength, moisture, and gas barrier characteristics of biopolymer films can be improved by different modification techniques.

10.8 Modification Methods for Enhancing the Functionality of Biofilms

To overcome the limitations related to the applications of biopolymers for the packaging of food products, the biopolymeric materials should be modified to enhance their mechanical, functional, and structural characteristics. Currently, the most commonly used techniques for modification include chemical and physical modifications.

10.8.1 Addition of Lipids

The hygroscopic nature of most of the biopolymers limited their uses as a packaging material. The addition of lipids has been noticed to decrease the WVP of the polymeric materials. The inclusion of 15% palm wax to fish gelatine films significantly decreased the WVP (Syahida et al. 2020). Several other studies also reported similar effects, such as PLA-beeswax (Lim et al. 2015) and gluten-beeswax (Rocca-Smith et al. 2016). The EOs are also used to provide additional effect to reduce the WVP of films.

10.8.2 Biopolymer Composites

The simple technique of modification is the blending of different types of polymers to obtain the film with enhanced functional properties. The structure of composite film depends upon the interaction among the polymers and the conditions of blending (pH, ionic strength, and temperature). The polysaccharides and proteins are blended to obtain composites, in which polysaccharides (anionic) and proteins (positively-charged) are interacted to make highly structured materials, and the interfacial properties, water-solubility, barrier properties, and mechanical strength of the blends are better than that of individual material (Turgeon et al. 2003; Ye 2008). The composites of chitosan-PVA (Haghighi et al. 2020), starch-PVA (Domene-López et al. 2018), and chitosan-starch films (Luchese et al. 2018) are some polymer blends studied for their food packaging uses.

10.8.3 Crosslinking

Crosslinking (CL) is the process when different polymeric chains interact either inter- or intramolecularly in the presence of cross-linking agents and develop chemical bonds between them to provide stronger 3-dimensional networks with enhanced stability, mechanical, and barrier characteristics. During CL, the mechanical and thermal characteristics of polymeric materials can be enhanced using calcium ions or carboxylic acid as cross-linking agents. Wu et al. (2019) studied the influence of citric acid induced CL on the functional properties of potato starch-chitosan film and noted that the inclusion of citric acid (15%) decreased the WVP and increased the TS of the composite film. Picchio et al. (2018) evaluated the influence of CL of casein with tannic acid and found an increase in TS of films with improved water resistance.

10.8.4 Addition of Antimicrobial Agents

Antimicrobial films can be formed by adding the bioactive compounds into the biopolymers during their processing or by coating them on the surface. The antimicrobial effect of the film depends on the quantity of antimicrobial compounds added to the polymer and the amount released during the storage time. Zahan et al. (2020) developed the antimicrobial bacterial cellulose (BC) film by incorporation of lauric acid (LA). The results described that the addition of LA to BC film provides an inhibitory effect against the growth of B. subtilis. The natural antimicrobial compounds such as vanillin and eugenol were added in PHB films at different levels to develop antimicrobial films. PHB films having eugenol and vanillin were found to control the growth of different bacteria (Xavier et al. 2015). The incorporation of

EOs also showed antimicrobial behavior with no adverse effect on the quality of food (Ali et al. 2019).

10.8.5 Bionanocomposites

The barrier and mechanical properties of biopolymers may be enhanced by incorporating the reinforcing compounds i.e., fillers, and thus, forming the composites. Polymer nanocomposites (PNC) are formed by adding a nano-filler into a polymeric matrix. The fillers include silica, nanoclay, nanocellulose, carbon nanotubes (Talegaonkar et al. 2017), graphene (Borriello et al. 2009), and starch nanoparticles (Chen et al. 2008). The uniform dispersal of nano-particles results in a polymer matrix-filler complex, that modifies the mobility of molecules, thermal, mechanical, and barrier properties of the composite. Arun et al. (2022) prepared the cellulose nanofibers (CNF) from coconut shell and used these nanofibers for reinforcing PVA based film. After the incorporation of CNF, elongation, TS, and Tm significantly increased in comparison to only PVA film. Swaroop and Shukla (2018) prepared nano MgO-reinforced-PLA film and found an increase in oxygen barrier property and TS.

10.8.6 Biodegradable Intelligent Packaging

The pigments and natural colors are added to biopolymers to monitor the quality attributes of packaged food materials. Several studies reported that the incorporation of natural pigments such as curcumin, anthocyanins, and betalains in films caused a color change due to chemical and physical changes in the headspace of package. Oxygen (Deshwal et al. 2018), carbon dioxide (Saliu and Della Pergola 2018), hydrogen sulfide (Zhai et al. 2019), ammonia (Liang et al. 2019), time-temperature (Maciel and Yoshida 2014), and pH (Badar et al. 2019) sensing films have been formed and studied for their effectiveness. Several pH indicating films have been used in monitoring the quality of milk with anthocyanins (Goodarzi et al. 2020; Ma and Wang 2016).

10.9 Mechanism of Biodegradation of Biopolymers

Biodegradation is the process of conversion of polymeric substances into CO_2, methane, water, and biomass via the action of enzymes produced by microbes (Ndahebwa et al. 2018; Ragaert et al. 2017). The process of biodegradation is affected by the chemical and physical characteristics of polymers. The characteristics such as surface area, hydrophobic and hydrophilic property, molecular weight, glass transition temperature, and chemical structure of polymers play a vital role in biodegradation. The crystallinity of polymer also affects the biodegradability because enzymes generally strike the amorphous regions of the polymer. The crystalline region is more resistant in comparison to the amorphous zone due to the intact and close arrangement of molecules (Patwary et al. 2020). The biodegradation process involves different steps (Lucas et al. 2008).

- The combined effort of microorganisms and abiotic factors (temperature, pH, moisture) convert the biodegradable material into small fractions. This process is known as biodeterioration.
- The microorganisms release catalysts i.e., enzymes that break the molecule into monomers, oligomers, and dimers, and the process is known as depolymerization.
- Some fragments are identified by the receptors present on microbial cells and thus pass through the plasma membrane, whereas other unidentified fragments remain in the extracellular zone.
- The fragments enter into the cytoplasm and assimilate to form primary metabolites, and secondary metabolites with energy and biomass. This process is known as assimilation.
- Few metabolites (aldehydes and organic acids) discharged by the microbial cells, CH_4, CO_2, H_2O, and some other salts are released in the surrounding.

The processes such as oxidation, hydrolysis, and photo-degradation may also have some effect on the degradation of biopolymers. The nature and type of enzymes, type of polymer, and environmental factors like pH, soil, light, moisture, oxygen, temperature, etc. also affect the rate of biodegradation. The enzymatic test and soil burial test are mainly used to assess the biodegradability of polymeric molecules. Zahan et al. (2020) reported that the complete degradation of BC film (100 × 150 mm) was achieved at the seventh day in the soil. The responsible microorganisms for the degradation were identified as *Rhizopus* sp. and *Bacillus* sp. Crosslinking of the polymers resulted in reduced hydrophilicity and decreased degradation rate of film because of the formation of complex structure (Mittal et al. 2020).

10.10 Applications and Significance of Biodegradable Polymers

Biodegradable materials are seeking attention due to their biodegradability and sustainability. Several scientists have studied the applications and effectiveness of these biopolymers as a suitable packaging material for fresh and processed foods (Table 10.1). The antimicrobial agents, bioactive compounds, plasticizers, and blends of different polymers are used to improve the functional characteristics of biopolymers. Several researchers have developed biodegradable coatings and films

Table 10.1 Applications of biodegradable polymers

Food product	Biodegradable polymer	Applications/benefits	References
Blueberry	PLA (container)	Decreased fungal growth and increased shelf-life when stored for 18 days at 10 °C	Almenar et al. (2008)
Cheese	Zein based film + essential oil (*Zataria multiflora boiss*)	Inhibited the growth of *L. monocytogenes*, *S. enteritidis*, & *E. coli* with improved the shelf-life of cheese during storage (70 days)	Ghasemi et al. (2015)
Cheese	Fish gelatin + orange peel pectin composite film	Restricted microbial attack, reduced weight loss, and retention of texture, color, and pH of cheese during storage	Jridi et al. (2020)
Cheese	WPI + natamycin + lactic acid coating	Decreased hardness, water loss, & color change	Ramos et al. (2012)
		No growth of pathogenic microorganisms	
Fresh berries	BC + guar gum based PVP-CMC films	Maintained the texture and color for 15 days, restricted WVP & OTR of the film keep the berries fresh	Bandyopadhyay et al. (2019)
Fresh cut apples	Bacterial cellulose nanocrystals (BCNCs) coating	Improved firmness, titrable acidity, & soluble solids; reduced weight loss	Zhai et al. (2020)
Fresh fruits	Alginate coating	Reduced the WVP and maintained the freshness of fruits	Campos et al. (2011)
Mango	Chitosan film	Increased shelf-life with no fungal growth, maintained the color and flavor of mangoes when stored at 27 °C (65% RH)	Srinivasa et al. (2002)
Melon	PLA (container)	The quality of freshly cut and packaged melons in PLA was maintained better in comparison to PET during storage for 10 days at 10 °C	Zhou et al. (2016)
Pears	CMC + candelilla wax coating	Prevented the pears from fungal attack, retained their firmness, and improved the shelf-life of pears	Kowalczyk et al. (2017)
Prato cheese	Gelatin + chitosan composite film with boldo extract	Protected the cheese against the microbial growth & oxidation	Bonilla and Sobral (2019)
Processed & fresh meat products	BC + nisin film	Good antimicrobial properties, inhibited aerobic bacteria & *L. monocytogenes* growth in vacuum-packed frankfurters	Nguyen et al. (2008)

(continued)

Table 10.1 (continued)

Food product	Biodegradable polymer	Applications/benefits	References
Strawberries	Chitosan + beeswax coating	Decreased the rate of respiration and metabolism, maintained the physicochemical properties of strawberries during storage	Velickova et al. (2013)
Strawberry (refrigerated)	Wheat gluten + lipid coating	Results showed a decrease of weight loss and retention of firmness	Tanada-Palmu and Grosso (2005)
Trout fillet	Sodium alginate-resveratrol coating	Increased the shelf-life during storage (4 °C for 15 days), completely inhibited the growth of *Enterobacteriaceae,* LAB, *Pseudomonas*, & yeasts	Bazargani-Gilani (2018)

to maintain the freshness, firmness, color, and texture of fresh fruits and vegetables. Biodegradable packaging films are also used for packaging of dairy and meat products.

10.11 Conclusion

Biodegradable polymers are useful in reducing the environmental concerns related to plastics. These are the good alternative to lower the dependence on conventional synthetic materials along with the evolution of eco-friendly packaging material with regard to their biocompatible, non-toxic, renewable and biodegradable nature. The use of biodegradable packaging material is still less in comparison to conventional synthetic materials because of various factors i.e., mechanical and barrier characteristics, safety, legislation, and cost. The various modification techniques such as the addition of plasticizers and lipids, crosslinking, composites, and use of antimicrobials enhance the functional properties of these biodegradable polymers. There is a need to develop new and innovative techniques for value addition to biodegradable polymers. The biodegradable packaging materials for fresh fruits and vegetables, milk products, and meat products have been developed. However, further research is needed for food products that involve processing at high temperatures and humidity. Other studies required to be focus on the use of sensors, nanotechnology, and active packaging systems that can give details in regards to the quality of products inside the package.

References

Abdul-Khalil HPS, Banerjee A, Saurabh CK, Tye YY, Suriani AB, Mohamed A, Karim AA, Rizal S, Paridah MT (2018) Biodegradable films for fruits and vegetables packaging application: preparation and properties. Food Eng Rev 10(3):139–153

Achachlouei BF, Zahedi Y (2018) Fabrication and characterization of CMC-based nanocomposites reinforced with sodium montmorillonite and TiO2 nanomaterials. Carbohydr Polym 199(33): 415–425

Acosta S, Jiménez A, Cháfer M, González-Martínez C, Chiralt A (2015) Physical properties and stability of starch-gelatin based films as affected by the addition of esters of fatty acids. Food Hydrocoll 49:135–143

Ahvenainen R (ed) (2003) Novel food packaging techniques, 1st edn. Woodhead Publishing, Cambridge, p 400

Ali A, Chen Y, Liu H, Yu L, Baloch Z, Khalid S, Chen L (2019) Starch-based antimicrobial films functionalized by pomegranate peel. Int J Biol Macromol 129:1120–1126

Almenar E, Samsudin H, Auras R, Harte B, Rubino M (2008) Postharvest shelf-life extension of blueberries using a biodegradable package. Food Chem 110(1):120–127

Alshabanat M (2019) Morphological, thermal, and biodegradation properties of LLDPE/treated date palm waste composite buried in a soil environment. J Saudi Chem Soc 23(3):355–364

Arun R, Shruthy R, Preetha R, Sreejit V (2022) Biodegradable nano composite reinforced with cellulose nano fibers from coconut industry waste for replacing synthetic plastic food packaging. Chemosphere 291:132786

Avella M, Immirzi B, Malinconico M, Martuscelli E, Volpe MG (1996) Reactive blending methodologies for Biopol. Polym Int 39(3):191–204

Badar H, Ariyawardana A, Collins R (2019) Dynamics of mango value chains in Pakistan. Pak J Agric Sci 56(2):523–530

Bandyopadhyay S, Saha N, Brodnjak UV, Sáha P (2019) Bacterial cellulose and guar gum based modified PVP-CMC hydrogel films: characterized for packaging fresh berries. Food Packag Shelf-life 22:100402

Bazargani-Gilani B (2018) Activating sodium alginate-based edible coating using a dietary supplement for increasing the shelf-life of rainbow trout fillet during refrigerated storage (4°C). J Food Saf 38(1):12395

Bonilla J, Sobral PJ (2019) Gelatin-chitosan edible film activated with Boldo extract for improving microbiological and antioxidant stability of sliced Prato cheese. Int J Food Sci Technol 54(5): 1617–1624

Borriello C, De Maria A, Jovic N, Montone A, Schwarz M, Antisari MV (2009) Mechanochemical exfoliation of graphite and its polyvinyl alcohol nanocomposites with enhanced barrier properties. Mater Manuf Process 24(10–11):1053–1057

Campos CA, Gerschenson LN, Flores SK (2011) Development of edible films and coatings with antimicrobial activity. Food Bioproc Tech 4(6):849–875

Cazón P, Velazquez G, Ramírez JA, Vázquez M (2017) Polysaccharide-based films and coatings for food packaging: a review. Food Hydrocoll 68:136–148

Cha DS, Chinnan MS (2004) Biopolymer-based antimicrobial packaging: a review. Crit Rev Food Sci Nutr 44(4):223–237

Chaichi M, Hashemi M, Badii F, Mohammadi A (2017) Preparation and characterization of a novel bionanocomposite edible film based on pectin and crystalline nanocellulose. Carbohydr Polym 157:167–175

Chen Y, Cao X, Chang PR, Huneault MA (2008) Comparative study on the films of poly (vinyl alcohol)/pea starch nanocrystals and poly(vinyl alcohol)/native pea starch. Carbohydr Polym 73(1):8–17

Cherpinski A, Gozutok M, Sasmazel HT, Torres S, Lagaron JM (2018) Electrospun oxygen scavenging films of poly(3-hydroxybutyrate) containing palladium nanoparticles for active packaging applications. Nanomaterials 8(7):469–475

Cho JW, Woo KS, Chun BC, Park JS (2001) Ultraviolet reflective and mechanical properties of polyethylene mulching films. Eur Polym J 37(6):1227–1232

Conte A, Buonocore GG, Bevilacqua A, Sinigaglia M, Del Nobile MA (2006) Immobilization of lysozyme on poly(vinylalcohol) films for active packaging applications. J Food Prot 69(4): 866–870

Deshwal GK, Panjagari NR, Badola R, Singh AK, Minz PS, Ganguly S, Alam T (2018) Characterization of biopolymer-based UV-activated intelligent oxygen indicator for food-packaging applications. J Packag Technol Res 2(1):29–43

Domene-López D, Guillén MM, Martin-Gullon I, García-Quesada JC, Montalbán MG (2018) Study of the behavior of biodegradable starch/polyvinyl alcohol/rosin blends. Carbohydr Polym 202:299–305

Espitia PJP, Du WX, de Jesús Avena-Bustillos R, Soares NDFF, McHugh TH (2014) Edible films from pectin: physical-mechanical and antimicrobial properties - a review. Food Hydrocoll 35: 287–296

Fonseca-García A, Jiménez-Regalado EJ, Aguirre-Loredo RY (2021) Preparation of a novel biodegradable packaging film based on corn starch-chitosan and poloxamers. Carbohydr Polym 251:117009

Freitas PA, Silva RR, de Oliveira TV, Soares RR, Soares NF (2020) Biodegradable film development by nisin Z addition into hydroxypropyl methyl cellulose matrix for mozzarella cheese preservation. Int J Food Stud 9(2):360–372

Gao P, Wang F, Gu F, Liang J, Li N, Ludescher RD (2017) Preparation and characterization of zein thermo-modified starch films. Carbohydr Polym 157:1254–1260

Garlotta D (2001) Literature review of poly(lactic acid). J Polym Environ 9(2):63–84

Garrido-Miranda KA, Rivas BL, Pérez-Rivera MA, Sanfuentes EA, Peña-Farfal C (2018) Antioxidant and antifungal effects of eugenol incorporated in bionanocomposites of poly (3-hydroxybutyrate) - thermoplastic starch. LWT Food Sci Technol 98:260–267

Ghasemi S, Javadi NHS, Moradi M, Khosravi-Darani K (2015) Application of zein antimicrobial edible film incorporating Zataria multiflora boiss essential oil for preservation of Iranian ultrafiltered Feta cheese. Afr J Biotechnol 14(24):2014–2021

Gomes LP, Souza HK, Campiña JM, Andrade CT, Silva AF, Gonçalves MP, Paschoalin VMF (2018) Edible chitosan films and their nanosized counterparts exhibit antimicrobial activity and enhanced mechanical and barrier properties. Molecules 24(1):127–134

Gómez-Aldapa CA, Velazquez G, Gutierrez MC, Rangel-Vargas E, Castro-Rosas J, Aguirre-Loredo RY (2020) Effect of polyvinyl alcohol on the physicochemical properties of biodegradable starch films. Mater Chem Phys 239:122027

Goodarzi MM, Moradi M, Tajik H, Forough M, Ezati P, Kuswandi B (2020) Development of an easy-to-use colorimetric pH label with starch and carrot anthocyanins for milk shelf-life assessment. Int J Biol Macromol 153:240–247

Grassie N, Murray EJ, Holmes PA (1984) The thermal degradation of poly (-(D)-β-hydroxybutyric acid), Part II: changes in molecular weight. Polym Degrad Stab 6(2):95–103

Guerrero A, Ferrero S, Barahona M, Boito B, Lisbinski E, Maggi F, Sañudo C (2020) Effects of active edible coating based on thyme and garlic essential oils on lamb meat shelf-life after long-term frozen storage. J Sci Food Agric 100(2):656–664

Gumede TP, Luyt AS, Muller AJ (2018) Review on PCL, PBS, and PCL/PBS blends containing carbon nanotubes. Express Polym Lett 12(6):505–529

Haghighi H, Leugoue SK, Pfeifer F, Siesler HW, Licciardello F, Fava P, Pulvirenti A (2020) Development of antimicrobial films based on chitosan-polyvinyl alcohol blend enriched with lauroyl arginate ethyl (LAE) for food packaging applications. Food Hydrocoll 100:105419

Halley P (2002) Biodegradable packaging for the food industry. Packag Bottling Int 4(4):56–57

Hazrati KZ, Sapuan SM, Zuhri MYM, Jumaidin R (2021) Effect of plasticizers on physical, thermal, and tensile properties of thermoplastic films based on Dioscorea hispida starch. Int J Biol Macromol 185:219–228

Homayounpour P, Jalali H, Shariatifar N, Amanlou M, Khanjari A (2020) Protective effect of nanochitosan incorporated with free/nanoliposome Cumin (Cuminum cyminum L.) aqueous extract on sardine fish. J Aquat Food Prod Technol 29(9):949–961
Hsieh WC, Wada Y, Mitobe T, Mitomo H, Seko N, Tamada M (2009) Effect of hydrophilic and hydrophobic monomers grafting on microbial poly (3-hydroxy butyrate). J Taiwan Inst Chem Eng 40(4):413–417
Irissin-Mangata J, Bauduin G, Boutevin B, Gontard N (2001) New plasticizers for wheat gluten films. Eur Polym J 37(8):1533–1541
Jambeck J, Hardesty BD, Brooks AL, Friend T, Teleki K, Fabres J, Beaudoin Y, Bamba A, Francis J, Ribbink AJ, Baleta T, Bouwman H, Knox J, Wilcox C (2018) Challenges and emerging solutions to the land-based plastic waste issue in Africa. Mar Policy 96:256–263
Jiang T, Duan Q, Zhu J, Liu H, Yu L (2020) Starch-based biodegradable materials: challenges and opportunities. Adv Ind Eng Polym Res 3(1):8–18
Jimenez A, Fabra MJ, Talens P, Chiralt A (2012) Edible and biodegradable starch films: a review. Food Bioproc Tech 5(6):2058–2076
Jridi M, Abdelhedi O, Salem A, Kechaou H, Nasri M, Menchari Y (2020) Physicochemical, antioxidant and antibacterial properties of fish gelatin-based edible films enriched with orange peel pectin: Wrapping application. Food Hydrocoll 103:105688
Kim MN, Lee AR, Yoon JS, Chin IJ (2000) Biodegradation of poly(3-hydroxy butyrate), Sky-Green®, and Mater-Bi® by fungi isolated from soils. Eur Polym J 36(8):1677–1685
Kowalczyk D, Kordowska-Wiater M, Zięba E, Baraniak B (2017) Effect of carboxymethylcellulose/candelilla wax coating containing potassium sorbate on microbiological and physicochemical attributes of pears. Sci Hortic 218:326–333
Kumar AA, Karthick K, Arumugam KP (2011) Properties of biodegradable polymers and degradation for sustainable development. Int J Chem Eng Appl 2(3):164–167
Lawton JW (1996) Effect of starch type on the properties of starch containing films. Carbohydr Polym 29(3):203–208
Li X, Du X, Liu Y, Tong L, Wang Q, Li J (2019) Rhubarb extract incorporated into an alginate-based edible coating for peach preservation. Sci Hortic 257:108685
Liakos I, Rizzello L, Bayer IS, Pompa PP, Cingolani R, Athanassiou A (2013) Controlled antiseptic release by alginate polymer films and beads. Carbohydr Polym 92(1):176–183
Liang T, Sun G, Cao L, Li J, Wang L (2019) A pH and NH3 sensing intelligent film based on Artemisia sphaerocephala gum and red cabbage anthocyanins anchored by carboxymethyl cellulose sodium added as a host complex. Food Hydrocoll 87:858–868
Lim JH, Kim JA, Ko JA, Park HJ (2015) Preparation and characterization of composites based on polylactic acid and beeswax with improved water vapor barrier properties. J Food Sci 80(11): E2471–E2477
Lodha P, Netravali AN (2005) Thermal and mechanical properties of environment-friendly 'green' plastics from stearic acid modified-soy protein isolate. Ind Crop Prod 21(1):49–64
Lu X, Chen J, Guo Z, Zheng Y, Rea MC, Su H, Zheng X, Zheng B, Miao S (2019) Using polysaccharides for the enhancement of functionality of foods: a review. Trends Food Sci Technol 86:311–327
Lucas N, Bienaime C, Belloy C, Queneudec M, Silvestre F, Nava-Saucedo JE (2008) Polymer biodegradation: mechanisms and estimation techniques–a review. Chemosphere 73(4):429–442
Luchese CL, Pavoni JMF, Dos Santos NZ, Quines LK, Pollo LD, Spada JC, Tessaro IC (2018) Effect of chitosan addition on the properties of films prepared with corn and cassava starches. J Food Sci Technol 55(8):2963–2973
Ma Q, Wang L (2016) Preparation of a visual pH-sensing film based on tara gum incorporating cellulose and extracts from grape skins. Sens Actuators B 235:401–407
Ma Y, Li L, Wang Y (2018) Development of PLA-PHB-based biodegradable active packaging and its application to salmon. Packag Technol Sci 31(11):739–746
Maciel VB, Yoshida CM (2014) Development of temperature indicator prototype: card paper coated with chitosan intelligent films. J Agric Chem Environ 3(1):5–10

Maftoonazad N, Ramaswamy HS (2019) Application and evaluation of a pectin-based edible coating process for quality change kinetics and shelf-life extension of lime fruit (Citrus aurantifolium). Coatings 9(5):285

Manikandan NA, Pakshirajan K, Pugazhenthi G (2020) Preparation and characterization of environmentally safe and highly biodegradable microbial polyhydroxy butyrate (PHB) based graphene nanocomposites for potential food packaging applications. Int J Biol Macromol 154: 866–877

Marin A, Atarés L, Chiralt A (2017) Improving function of biocontrol agents incorporated in antifungal fruit coatings: a review. Biocontrol Sci Technol 27(10):1220–1241

Meereboer KW, Misra M, Mohanty AK (2020) Review of recent advances in the biodegradability of polyhydroxy alkanoate (PHA) bioplastics and their composites. Green Chem 22(17): 5519–5558

Mercan N, Aslim B, Yüksekdağ ZN, Beyatli Y (2002) Production of poly-b-hydroxy butyrate (PHB) by some Rhizobium bacteria. Turk J Biol 26(4):215–219

Mittal A, Garg S, Bajpai S (2020) Fabrication and characteristics of poly (vinyl alcohol)-starch-cellulosic material based biodegradable composite film for packaging application. Mater Today Proc 21:1577–1582

Mlalila N, Hilonga A, Swai H, Devlieghere F, Ragaert P (2018) Antimicrobial packaging based on starch, poly (3-hydroxybutyrate) and poly (lactic-co-glycolide) materials and application challenges. Trends Food Sci Technol 74:1–11

Mochizuki M, Hirami M (1997) Structural effects on the biodegradation of aliphatic polyesters. Polym Adv Technol 8(4):203–209

Moeini A, Germann N, Malinconico M, Santagata G (2021) Formulation of secondary compounds as additives of biopolymer-based food packaging: a review. Trends Food Sci Technol 114:342–354

Mohanty AK, Misra MA, Hinrichsen GI (2000) Biofibers, biodegradable polymers and biocomposites: an overview. Macromol Mater Eng 276(1):1–24

Nair LS, Laurencin CT (2007) Biodegradable polymers as biomaterials. Prog Polym Sci 32(8): 762–798

Navarro-Baena I, Sessini V, Dominici F, Torre L, Kenny JM, Peponi L (2016) Design of biodegradable blends based on PLA and PCL: from morphological, thermal and mechanical studies to shape memory behavior. Polym Degrad Stab 132:97–108

Ndahebwa MC, Magoma G, Imbuga M, Makonde HM (2018) Molecular characterization of low-density polyethene (LDPE) degrading bacteria and fungi from Dandora dumpsite, Nairobi, Kenya. Int J Microbiol 2018:4167845

Nešić A, Onjia A, Davidović S, Dimitrijević S, Errico ME, Santagata G, Malinconico M (2017) Design of pectin-sodium alginate-based films for potential healthcare application: study of physicochemical interactions between the components of films and assessment of their antimicrobial activity. Carbohydr Polym 157:981–990

Nguyen VT, Gidley MJ, Dykes GA (2008) Potential of a nisin-containing bacterial cellulose film to inhibit Listeria monocytogenes on processed meats. Food Microbiol 25(3):471–478

Nisar T, Wang ZC, Yang X, Tian Y, Iqbal M, Guo Y (2018) Characterization of citrus pectin films integrated with clove bud essential oil: physical, thermal, barrier, antioxidant and antibacterial properties. Int J Biol Macromol 106:670–680

Noshirvani N, Ghanbarzadeh B, Gardrat C, Rezaei MR, Hashemi M, Le Coz C, Coma V (2017) Cinnamon and ginger essential oils to improve antifungal, physical and mechanical properties of chitosan-carboxymethyl cellulose films. Food Hydrocoll 70:36–45

Oliveira VRL, Santos FKG, Leite RHL, Aroucha EMM, Silva KNO (2018) Use of biopolymeric coating hydrophobized with beeswax in post-harvest conservation of guavas. Food Chem 259: 55–64

Ortiz-Zarama MA, Jiménez-Aparicio AR, Solorza-Feria J (2016) Obtainment and partial characterization of biodegradable gelatin films with tannic acid, bentonite and glycerol. J Sci Food Agric 96(10):3424–3431

Ozer BBP, Uz M, Oymaci P, Altinkaya SA (2016) Development of a novel strategy for controlled release of lysozyme from whey protein isolate based active food packaging films. Food Hydrocoll 61:877–886

Patwary MS, Surid SM, Gafur MA (2020) Properties and applications of biodegradable polymers. J Res Updates Polym Sci 9:32–41

Pavinatto A, de Almeida Mattos AV, Malpass ACG, Okura MH, Balogh DT, Sanfelice RC (2020) Coating with chitosan-based edible films for mechanical/biological protection of strawberries. Int J Biol Macromol 151:1004–1011

Pereira LAS, Silva PDC, Pagnossa JP, Miranda KWE, Medeiros ES, Piccoli RH, Oliveira JED (2019) Antimicrobial zein coatings plasticized with garlic and thyme essential oils. Braz J Food Technol 22:e2018135

Picchio ML, Linck YG, Monti GA, Gugliotta LM, Minari RJ, Igarzabal CIA (2018) Casein films crosslinked by tannic acid for food packaging applications. Food Hydrocoll 84:424–434

Pintado CM, Ferreira MA, Sousa I (2010) Control of pathogenic and spoilage microorganisms from cheese surface by whey protein films containing malic acid, nisin and natamycin. Food Control 21(3):240–246

Ragaert K, Delva L, Van-Geem K (2017) Mechanical and chemical recycling of solid plastic waste. Waste Manag 69:24–58

Rai SK, Chaturvedi K, Yadav SK (2019) Evaluation of structural integrity and functionality of commercial pectin based edible films incorporated with corn flour, beetroot, orange peel, muesli and rice flour. Food Hydrocoll 91:127–135

Ramaraj B (2007) Crosslinked poly (vinyl alcohol) and starch composite films, II: physicomechanical, thermal properties and swelling studies. J Appl Polym Sci 103(2):909–916

Ramos ÓL, Pereira JO, Silva SI, Fernandes JC, Franco MI, Lopes-da-Silva JA, Malcata FX (2012) Evaluation of antimicrobial edible coatings from a whey protein isolate base to improve the shelf-life of cheese. J Dairy Sci 95(11):6282–6292

Ramos M, Valdés A, Beltran A, Garrigós MC (2016) Gelatin-based films and coatings for food packaging applications. Coatings 6(4):41–48

Ramos-García M, Bosquez-Molina E, Hernández-Romano J, Zavala-Padilla G, Terrés-Rojas E, Alia-Tejacal I, Barrera-Necha L, Harnandez-Lopez M, Bautista-Baños S (2012) Use of chitosan-based edible coatings in combination with other natural compounds, to control Rhizopus stolonifer and Escherichia coli DH5α in fresh tomatoes. Crop Prot 38:1–6

Reddy N, Yang Y (2010) Citric acid cross-linking of starch films. Food Chem 118(3):702–711

Rhim JW, Lee JH, Ng PK (2007) Mechanical and barrier properties of biodegradable soy protein isolate-based films coated with polylactic acid. LWT Food Sci Technol 40(2):232–238

Rocca-Smith JR, Marcuzzo E, Karbowiak T, Centa J, Giacometti M, Scapin F, Debeaufort F (2016) Effect of lipid incorporation on functional properties of wheat gluten based edible films. J Cereal Sci 69:275–282

Sahraee S, Milani JM, Regenstein JM, Kafil HS (2019) Protection of foods against oxidative deterioration using edible films and coatings: a review. Food Biosci 32:e100451

Salgado PR, Ortiz CM, Musso YS, Di Giorgio L, Mauri AN (2015) Edible films and coatings containing bioactives. Curr Opin Food Sci 5:86–92

Saliu F, Della Pergola R (2018) Carbon dioxide colorimetric indicators for food packaging application: applicability of anthocyanin and poly-lysine mixtures. Sens Actuators B Chem 258:1117–1124

Sanchez-Gonzalez L, Cháfer M, Hernández M, Chiralt A, González-Martínez C (2011) Antimicrobial activity of polysaccharide films containing essential oils. Food Control 22(8):1302–1310

Sarasam AR, Krishnaswamy RK, Madihally SV (2006) Blending chitosan with polycaprolactone: effects on physicochemical and antibacterial properties. Biomacromolecules 7(4):1131–1138

Šešlija S, Nešić A, Ružić J, Krušić MK, Veličković S, Avolio R, Malinconico M (2018) Edible blend films of pectin and poly(ethylene glycol): preparation and physico-chemical evaluation. Food Hydrocoll 77:494–501

Shahabi-Ghahfarrokhi I, Almasi H, Babaei-Ghazvini A (2020) Characteristics of biopolymers from natural resources. In: Processing and development of polysaccharide-based biopolymers for packaging applications. Elsevier, New York, pp 49–95

Sheu DS, Chen WM, Yang JY, Chang RC (2009) Thermophilic bacterium Caldimonas taiwanensis produces poly(3-hydroxy butyrate-co-3-hydroxy valerate) from starch and valerate as carbon sources. Enzyme Microb Technol 44(5):289–294

Singh P, Magalhães S, Alves L, Antunes F, Miguel M, Lindman B, Medronho B (2019) Cellulose-based edible films for probiotic entrapment. Food Hydrocoll 88:68–74

Smith R (ed) (2005) Biodegradable Polymers for Industrial Applications. Woodhead Publishing, Cambridge, pp 223–226

Song F, Tang DL, Wang XL, Wang YZ (2011) Biodegradable soy protein isolate-based materials: a review. Biomacromolecules 12(10):3369–3380

Srinivasa P, Baskaran R, Ramesh M, Harish Prashanth K, Tharanathan RN (2002) Storage studies of mango packed using biodegradable chitosan film. Eur Food Res Technol 215(6):504–508

Sun H, Shao X, Jiang R, Shen Z, Ma Z (2018) Mechanical and barrier properties of corn starch phosphate-zein bilayer films by thermocompression. Int J Biol Macromol 118:2076–2081

Swaroop C, Shukla M (2018) Nano-magnesium oxide reinforced polylactic acid biofilms for food packaging applications. Int J Biol Macromol 113:729–736

Syahida SN, Ismail-Fitry MR, Ainun ZMAA, Hanani ZAN (2020) Effects of palm wax on the physical, mechanical and water barrier properties of fish gelatin films for food packaging application. Food Packag Shelf Life 23:e100437

Tabassum N, Khan MA (2020) Modified atmosphere packaging of fresh-cut papaya using alginate based edible coating: quality evaluation and shelf-life study. Sci Hortic 259:e108853

Talegaonkar S, Sharma H, Pandey S, Mishra PK, Wimmer R (2017) Bionanocomposites: smart biodegradable packaging material for food preservation. In: Grumezescu AM (ed) Food packaging, chap 3. Academic Press, New York, pp 79–110

Tanada-Palmu PS, Grosso CR (2005) Effect of edible wheat gluten-based films and coatings on refrigerated strawberry (Fragaria ananassa) quality. Postharvest Biol Technol 36(2):199–208

Tulamandi S, Rangarajan V, Rizvi SS, Singhal RS, Chattopadhyay SK, Saha NC (2016) A biodegradable and edible packaging film based on papaya puree, gelatin, and defatted soy protein. Food Packag Shelf Life 10:60–71

Turco R, Ortega-Toro R, Tesser R, Mallardo S, Collazo-Bigliardi S, Boix AC, Malinconico M, Rippa M, Di Serio M, Santagata G (2019a) Poly(lactic acid)/ thermoplastic starch films: effect of cardoon seed epoxidized oil on their physicochemical, mechanical, and barrier properties. Coatings 9(9):574–580

Turco R, Tesser R, Cucciolito ME, Fagnano M, Ottaiano L, Mallardo S, Malinconico M, Santagata G, Di Serio M (2019b) Cynara cardunculus biomass recovery: eco-sustainable, nonedible resource of vegetable oil for the production of poly (lactic acid) bioplasticizers. ACS Sustain Chem Eng 7(4):4069–4077

Türe H, Gällstedt M, Hedenqvist MS (2012) Antimicrobial compression-molded wheat gluten films containing potassium sorbate. Food Res Int 45(1):109–115

Turgeon SL, Beaulieu M, Schmitt C, Sanchez C (2003) Protein–polysaccharide interactions: phase-ordering kinetics, thermodynamic and structural aspects. Curr Opin Colloid Interface Sci 8(4-5): 401–414

Velickova E, Winkelhausen E, Kuzmanova S, Alves VD, Moldão-Martins M (2013) Impact of chitosan-beeswax edible coatings on the quality of fresh strawberries (Fragaria ananassa) under commercial storage conditions. LWT Food Sci Technol 52(2):80–92

Villalobos R, Chanona J, Hernández P, Gutiérrez G, Chiralt A (2005) Gloss and transparency of hydroxypropyl methylcellulose films containing surfactants as affected by their microstructure. Food Hydrocoll 19(1):53–61

Weber CJ, Haugaard V, Festersen R, Bertelsen G (2002) Production and applications of biobased packaging materials for the food industry. Food Addit Contam 19(S1):172–177

Willett JL (1994) Mechanical properties of LDPE/granular starch composites. J Appl Polym Sci 54(11):1685–1695

Wu S (2014) Effect of chitosan-based edible coating on preservation of white shrimp during partially frozen storage. Int J Biol Macromol 65:325–328

Wu Z, Wu J, Peng T, Li Y, Lin D, Xing B, Han G (2017) Preparation and application of starch/ polyvinyl alcohol/citric acid ternary blend antimicrobial functional food packaging films. Polymers 9(3):102

Wu H, Lei Y, Lu J, Zhu R, Xiao D, Jiao C, Xia R, Zhang Z, Shen G, Liu Y, Li M (2019) Effect of citric acid induced crosslinking on the structure and properties of potato starch/chitosan composite films. Food Hydrocoll 97:e105208

Xavier JR, Babusha ST, George J, Ramana KV (2015) Material properties and antimicrobial activity of polyhydroxy butyrate (PHB) films incorporated with vanillin. Appl Biochem Biotechnol 176(5):1498–1510

Ye A (2008) Complexation between milk proteins and polysaccharides via electrostatic interaction: principles and applications–a review. Int J Food Sci Technol 43(3):406–415

Yu L, Dean K, Li L (2006) Polymer blends and composites from renewable resources. Prog Polym Sci 31(6):576–602

Zahan KA, Azizul NM, Mustapha M, Tong WY, Rahman MSA (2020) Application of bacterial cellulose film as a biodegradable and antimicrobial packaging material. Mater Today Proc 31: 83–88

Zhai X, Li Z, Shi J, Huang X, Sun Z, Zhang D, Wang S (2019) A colorimetric hydrogen sulfide sensor based on gellan gum - silver nanoparticles bionanocomposite for monitoring of meat spoilage in intelligent packaging. Food Chem 290:135–143

Zhai X, Lin D, Li W, Yang X (2020) Improved characterization of nanofibers from bacterial cellulose and its potential application in fresh-cut apples. Int J Biol Macromol 149:178–186

Zhang L, Liu Z, Wang X, Dong S, Sun Y, Zhao Z (2019) The properties of chitosan/zein blend film and effect of film on quality of mushroom (Agaricus bisporus). Postharvest Biol Technol 155: 47–56

Zhao Y, McDaniel M (2005) Sensory quality of foods associated with edible film and coating systems and shelf-life extension. In: Innovations in food packaging. Academic Press, New York, pp 434–453

Zhao X, Cornish K, Vodovotz Y (2020) Narrowing the gap for bioplastic use in food packaging: an update. Environ Sci Technol 54(8):4712–4732

Zhou H, Kawamura S, Koseki S, Kimura T (2016) Comparative quality changes of fresh-cut melon in bio-based and petroleum-based plastic containers during storage. Environ Control Biol 54(2): 93–99

Zhou X, Cheng R, Wang B, Zeng J, Xu J, Li J, Kang L, Cheng Z, Gao W, Chen K (2021) Biodegradable sandwich-architectured films derived from pea starch and polylactic acid with enhanced shelf-life for fruit preservation. Carbohydr Polym 251:117117–117213

Zhu D, Guo R, Li W, Song J, Cheng F (2019a) Improved postharvest preservation effects of Pholiota nameko mushroom by sodium alginate–based edible composite coating. Food Bioproc Tech 12(4):587–598

Zhu Q, Lu H, Zhu J, Zhang M, Yin L (2019b) Development and characterization of pickering emulsion stabilized by zein/corn fiber gum (CFG) complex colloidal particles. Food Hydrocoll 91:204–213

Chapter 11
Recent Trends in Biodegradable Packaging of Foods and Food Products

Lakshita Rao

Contents

L. Rao (✉)
Department of Pharmaceutical Sciences, Gurugram University, Gurugram, Haryana, India

J. A. Malik et al. (eds.), *Food Process Engineering and Technology*,
https://doi.org/10.1007/978-981-99-6831-2_11

Abstract Plastics have elevated the food sector to new heights, because traditional petroleum-based plastics are not biodegradable, and these pose serious environmental issues, such as: endangering marine life and lowering air quality. For the food industry, biodegradable polymers or biopolymers can reduce the risks brought on by non-biodegradable plastic. They have been divided into three groups: polymers made from microbes, monomers, and biomass. Physical, mechanical, thermal, and barrier qualities are all important for determining how good a biopolymer is. Recently, biodegradable edible packaging materials have also been employed to carry functional components, thus demonstrating their promise for medicine delivery to address the issues associated with nutritional shortage. The potential of bio-based materials (such as: proteins, polysaccharides, lipids, etc.) to create biodegradable packaging materials is highlighted in this chapter. Additionally, the chapter looks at the additives that are utilized to enhance the mechanical and physicochemical characteristics of biodegradable packaging materials. The chapter also emphasizes the traits of various biopolymers and their mixtures, a comparison of the properties of biopolymers and non-biodegradable materials, the market potential for food packaging applications, and cutting-edge developments in biodegradable packaging from the perspectives of food safety and quality. Chapter also highlights several commercial forms, such as: films, trays, bags, coatings, and foamed goods for use as active packaging, modified atmosphere packaging, and edible packaging.

Keywords Biodegradable packaging · Biodegradable polymers · Degradation · Environmental impact · Food processing · Packaging forms

11.1 Introduction

Plastics are heavily used in the food sector for packaging of foods to preserve the food quality and safety. Historically, food manufacturers have packaged their goods in glass and metal containers, but thanks to plastic's many benefits, this practice has spread throughout (Alvarado 2009). It is inexpensive, low density, moldable, light, has flexible optical and mechanical qualities, is resistant to corrosive chemicals, impermeable to water and gases, and allows for the printing of important customer information.

There is growing concern about the impact on human health in addition to the physical effects on terrestrial and marine ecosystems due to the toxic materials (flame retardants, pigments, plasticizers, compatibilizers, etc.) used in plastic fabrication that can migrate to water or other contact media, including food. Similarly, biggest issues are the microplastics, which, contrary to bigger plastics, are difficult to notice with the unaided eye. Once these particles enter the ocean, it is impossible to recover them. Toxic, bioaccumulative, and transporting persistent organic pollutants (POPs) that can get into the food chain by consuming marine organisms is easier and bioavailable when there are microplastics present (Chonhenchob et al. 2007).

Owing to their variety of excellent characteristics and ease of production, polymers play a crucial role in modern life. Around 348 million tons of plastics (thermosets, thermoplastics, adhesives, elastomers, sealants, coatings and PP-Fibers) were produced globally in 2017, while 359 million tons were produced globally in 2018 (Netravali and Chabba 2003).

About 95–99% of plastic is produced by petrochemical industries using synthetic plastics, which come from non-renewable sources. Products made of synthetic plastic are utilized extensively in a variety of industries, including packaging, building materials, and medical equipment. The packaging industry uses 43% of the annual synthetic polymers produced in India (Ficci 2014).

Synthetic plastic, on the other hand, cannot degrade physically, chemically, or biologically, which ultimately increases garbage. Waste causes various serious issues with the environment and public health. They build up on the roads and streets, clogging the drain and causing it to overflow. Aquatic life is harmed by the vast amounts of plastic waste that are dumped in rivers and ocean. The release of toxic gases during incineration (such as: carbon monoxide, carbon dioxide, furans, chlorine, 1,3-butadiene, dioxin and amines) lowers air quality, heightens the risk of global warming, and poses a number of health risks (Smith 2005).

The shelf life of packed foods can be extended by adding antibacterial and/or antioxidant components to biodegradable food packaging materials. The contained antimicrobial agents inhibit the growth of bacteria on the surface of the food more efficiently than their direct inclusion into the meal because of their controlled release from the packaging to the product. Lower concentrations of active chemicals are needed since the antibacterial action can be prolonged throughout time in a more consistent way with controlled release. Antimicrobial compounds present in the material, however, have the potential to drastically change the biodegradation pattern of the transported bioplastic by interfering with the natural microbial population that is in charge of the biodegradation process (Amin et al. 2021).

This chapter examines several research creating biodegradable polymer-based antimicrobial materials. Additionally, the effect of adding these chemicals on the active components' biodegradation behavior was considered.

In a number of industries, including consumer electronics, packaging, horticulture, automotive, rigid and flexible packaging, building and construction, coatings, and others, bioplastics are employed. More than 53% of the entire amount of bioplastics generated in 2019 will be used in the packaging sector, which is the largest use. The first bioplastic product that was successfully produced and was recognised as being compostable in an industrial setting was biodegradable food packaging. Since then, the need for bioplastics for food packaging has significantly increased. Rigid packaging contributes primarily to the non-biodegradable packaging, whereas flexible packaging primarily uses biodegradable polymers.

Additionally, improved atmospheric storage for many fruits and vegetables uses biodegradable polymers (Mangaraj et al. 2019).

11.2 Classification Bio-Based Polymers

11.2.1 Polymers Isolated Directly from Natural Material or Biomass

11.2.1.1 Chitosan and Chitin

The chitin is linear copolymer with a N-acetylglucosamine and N-glucosamine -1,4 linkage. Depending on the processing technique, the monomers are distributed at random through the polymer. It is widely accessible and is known as the amino cellulose. It is mostly found in shells of crabs, insects, shrimp, and other animals. Growing fungi, which has a protein concentration of 10–15%, is another source of chitin. Chitin has a very low solubility, hence it is typically mixed for use in packaging (Teng et al. 2001).

Chitin undergoes partial N-deacetylation, resulting in the formation of chitosan, which has a tight crystalline structure, strong hydrogen bonds, and is insoluble in water and only a small number of acidic solutions. Chitosanase or lysozymes can break down chitosan. Uses of chitin and chitosan are constrained by their insolubility issues in the majority of solvents and cosolvants. The N-carboxymethyl chitosan or N-carboxyethyl chitosan is a modified form of the natural chitin that has been developed for use in a variety of companies. Both are used to create different biodegradable packaging films, with the biggest one being used as an edible covering of fresh products to extend the shelf life. The fungus, yeasts, and bacteria that can be found in food are all susceptible to the antibacterial capabilities of chitosan and chitin.

11.2.1.2 Cellulose and Starch

The amylopectin (poly—1,4-D-glucopyranoside and -1,6-D-glucopyranoside) and Amylose (poly—1,4-D-glucopyranoside) make up the starch. It may be derived from grains like wheat, rice, potatoes, and corn and is widely available. The amount of amylose and amylopectin varies with the source. The elongation and strength increase along with the amylose content. Because the connection breaks and the granules dissolve at temperatures between 150 and 250 °C, starch can either be used as a filler to create blends with different resins. Thermoplastic starch is the most common kind of starch (TPS). TPS is quite sensitive to humidity, and water content affects its thermal characteristics. TPS, also known as plasticized starch, serves as a substitute for synthetic polymers. Recent research has led to the development of fully biodegradable "bio-composites," or "green" composites in which natural fibres are combined with biodegradable polymers. The acetal bond of TPS is hydrolyzed during biodegradation. Glucosidases cleave the 1,6 bond, while amylases break the 1,4 linkage. BioplastTM from Biotec, SolanylTM from Rodenburg Biopolymers, EcoframTM from National Starch, BiocoolTM from Novamont, and PlanticTM from Plantic Technologies are few examples of commercially available starches and their blends. Cellobiose repeating units combine to produce the linear polymer known as cellulose. Because it is crystalline, organic solvents cannot dissolve it. Due to its low fluency and insolubility, it is changed into various forms for use (Netravali and Chabba 2003).

11.2.1.3 Collagen and Gelatin

A protein found in connective tissue called collagen is made up of several polypeptides, including glycine, hydroxyproline, lysine and proline. Collagen is very flexible because of the glycine it contains. They are included in PVA and cellulose films. Blended cellulose film is weak and fragile. Gelatine is of greater molecular weight polypeptide created by the chemical breakdown of collagen. It contains 19 amino acids and has great film-forming ability. The barrier and mechanical properties of the film are significantly influenced by the amino acid composition, molecular weight distribution and kind of plasticizer utilized. Applications are restricted by its limited heat stability during processing, which is a significant constraint. Various additives are added to the film to enhance or modify its mechanical and barrier qualities, resulting in excellent films for food packing. Sausag is packaged in a 4% gelatine film with 2.5% corn oil and 5% olive oil. The enzyme protease causes gelatine to degrade (Ramos et al. 2016).

11.2.1.4 Wheat Gluten and Soy Protein

Wheat gluten is a by-product in production of starch. Compared to other polymers, their rate of breakdown is the fastest and they produce no hazardous byproducts. Although it makes wonderful films, it becomes brittle without a plasticizer. Water-soluble carbohydrates are not present in soy protein concentrate. A 70% protein concentration can be found in it.. Because the protein is hydrophilic, TSP films have poor barrier or mechanical qualities. Isolated soy protein films are moisture-sensitive. Stearic acid, added at a rate of 25%, enhances the thermal and tensile qualities while lowering moisture sensitivity. Glycerol, gellan gum, or K-carrageenan are added to soy protein film to create biodegradable soybean-based packing containers (or trays) (Mohareb and Mittal 2007).

11.2.1.5 Polylactic Acid (PLA)

By polymerizing lactide monomer through ring opening, aliphatic polyester known as PLA is created. The fermentation commodities which are renewable like sugar, corn and other feedstocks, etc., yields the lactic acid monomers most frequently. It has a high molecular weight, is compostable, recyclable, and dissolves quickly due to its great transparency. The crystalline to amorphous nature of PLA's characteristics can be altered by altering the monomeric ratio. Glass transition temperature of PLA, which is marketed, ranges from 63 to 63.8 °C. Rate of deterioration of the entire polymer depends on its initial crystallinity and monomer concentration, with the highest monomer content exhibiting the lowest rate of degradation because of its highly crystalline nature (Kale et al. 2006).

11.2.2 Polycaprolactone (PCL)

PCL is a semi-crystalline, low-cost, fully biodegradable, simple polymer made from fossil fuels. Its solubility in a variety of organic and inorganic solvents and glass transition temperature (Tg) of under 60 °C promote its use as a compatibilizer in polyurethane formulations. The addition of PCL to the hydrophilic chitosan polymer increases the blend's overall hydrophobicity. Low water vapour transmission rate (WVTR) values are seen in the combination when compared to pure films. As a result of this property, food kept in these films has a longer shelf life. Under the brand names Tone® from Union Carbide, Celgreen® from Daicel, and CAPA® from Solvay, it is available commercially (Sarasam et al. 2006).

11.2.3 Polybutylene Succinate (PBS)

In polyalkenedicarboxylate family, polybutylene succinates are made up of polycondensing aliphatic dicarboxylic acids like succinic and adipic acid with glycols like 1,4-butanediol and ethylene glycol. It is a crystalline polymer white, with good processability, a Tg of 45 to 10 °C, a melting point of 90–120 °C, and a break elongation of 330%. PBS resembles polypropylene (PP) and polyethylene (PE) in terms of its mechanical qualities. They were created for the first time in Japan in 1990 under the brand name Bionolle (Wang et al. 2007). The characteristics and pace of decomposition of these polymers are influenced by the type of diols and diacids utilized for condensation.

11.2.4 Polylactide Aliphatic Copolymer (CPLA)

Dicarboxylic acid, an aliphatic polyester, and lactide, a renewable resource, are combined to create CPLA. With respect to the percentage of polyester contained in the mixture, it possesses qualities similar to PP and PS. Up to a temperature of 200 °C, it is stable. In comparison to the CO_2 produced by the burning of PP and PE, very little CO_2 is produced during combustion. The burning of CPLA does not result in the production of any hazardous substances. If combined with food, it starts to break down after 2 weeks. Degradation occurs naturally over a period of 5–6 months, and total decomposition takes 12 months. One film that has been commercialised and is being made in Japan by Dainippon Ink Chemicals under the brand name CPLATM (Siracusa et al. 2008).

11.2.5 Polyglycolide (PGA)

Glycolic acid polycondensation produces polyglycolide or polyglycolic acid. With a glass transition temperature (Tg) of 35–40 °C and a melting point (Tm) of roughly 222–250 °C, it is one of the most basic aliphatic polyesters. Due to its high crystallinity (40–55%), it is water insoluble but soluble in the most fluorinated solvents that can be used to produce films with high molecular weights of polymers. Within 5–6 months, the organism completely reabsorbed the polymer (Tiberiu 2011). Commercialized PGA includes KurudexTM. The Chemours Company also creates a low molecular weight film (Tiberiu 2011).

11.2.6 Polypropylene Carbonate (PPC)

CO_2 and propylene carbonate are copolymerized to create PPC, the most widely used aliphatic polycarbonate. PPC films have advantages over PBAT, LDPE, and PE/starch mix films, such as better tensile and barrier (O_2 and H_2O vapor) characteristics. Although the PPC films' tear resistance is less than that of the PBAT, it is still superior to that of LDPE and PE/starch mixes. The amorphous PPC, however, has a number of drawbacks, such as significant shrinkage, poor thermal stability, a low glass transition temperature (25–45 °C), Insufficient mechanical characteristics, and variable performance of the polymer depending on the type of catalyst employed to generate the PPC (Muthuraj et al. 2018).

11.3 Properties of Biopolymers

11.3.1 Tensile Strength

The tensile strength of a material is highest amount of stress it can withstand before failing. It is one of the most popular mechanical techniques for figuring out a material's strength. Protection of food packaging depends on the mechanical qualities. The kind of polymer, additives, production conditions, chemical modifications all affects tensile strength. Processing and storage affect a material's tensile strength. The mechanical characteristics of materials like PLA that have NPs (nanoparticles) added increase when bio-nano composites are formed. The tensile strength is compared as follows:

$$\text{PEN} > \text{PET} > \text{PVDC} > \text{PC} > \text{EVOH} > \text{PVC} > \text{PP} > \text{OPP} > \text{HDPE} > \text{HIPS} > \text{OPS} > \text{LLDPE} > \text{LDPE} > \text{EVA}$$

Among these, PET and EVA have the highest and lowest strengths, respectively.

In case of biodegradable polymers (CPLA > PLA > PBS > chitin > PHB > collagen > PCL > PGA > cellulose > starch > wheat gluten), PLA have the maximum tensile strength while starch has the lowest.

The sort of additives utilized to produce the film has an impact on the tensile strength range. Many biodegradable polymers (such as: CPLA, PLA, and PHB) have comparable tensile strengths to thermoplastics. This indicates that these biodegradable polymers can be an excellent substitute for thermoplastics.

11.3.2 Water Vapor Transmission Rate

The amount of water vapor that passes through packing material per unit area and time is known as the water vapor transmission rate (WVTR: $\text{kg-mm}^2\text{-s}^1$). Food products are susceptible to moisture, and when the moisture level rises, the food product's shelf life shortens. In some circumstances, the WVTR is a crucial consideration when choosing a packing material because various food products, such as dairy products, meat, and seafood, need a specific range of moisture levels inside their packages. Water vapor transmission rate is most measured at (38 °C), 90% RH. Thermoplastic follows the order as follows:

$$\text{PVC} > \text{OPS} > \text{HIPS} > \text{EVA} > \text{EVOH} > \text{PET} > \text{LDPE} > \text{PC} > \text{PEN} > \text{PP} > \text{HDPE} > \text{LLDPE} > \text{OPP} > \text{PVDC}$$

Biodegradable polymers, follows the order as below:

$$\text{PBS} > \text{Collagen} > \text{PLA} > \text{PCL} > \text{PGA} > \text{starch} > \text{cellulose} > \text{Wheat gluten} > \text{PHB} > \text{chitin}$$

Biodegradable plastics can be used to store dry goods as they are less permeable to water than thermoplastic polymers. Due of PBS's weak water vapour retention, its application alone is minimal. When PBS and PLA are combined in a 20:80 ratio, the resulting mixture has a strong water vapor retention property (Muthuraj et al. 2018).

11.3.3 Oxygen Transmission Rate

The amount of oxygen that can pass through a material per unit area and time under pressure is determined by the oxygen permeability coefficients (OPC) [mol/(m^2.s. Pa)]. Low OPC delays oxidation, thereby extending the product's shelf life. OTR of several polymers are typically measured at 25 °C and 0% RH. The OTR in thermoplastic polymers follows the order as below:

PC > EVA > LDPE > OPS > LLDPE > PP > HDPE > OPP > HIPS > PET > PVC > PVDC > EVOH > PEN

OTR for biodegradable polymers follows the order as below:

PCL > cellulose > PHB > chitin > starch > wheat gluten > PBS > PGA > PLA

Biodegradable polymers have a very low OTR, which means that only a small amount of oxygen can pass through them. Biodegradable polymers are sometimes combined to create blends in order to improve their barrier qualities. When compared to native chitosan films, chitosan: starch films demonstrated a reduction in OTR at 20:80. Chitosan was added to the film, which also boosted its hydrophobicity (Akter et al. 2014).

11.3.4 Elongation at Break

Elongation at break is the measure of modified length to initial length. It determines how far a material can stretch or lengthen before breaking. These numbers show how malleable a polymer is so that various shapes can be created. The higher the value, the stronger the polymer is measured as a percentage. Most thermoplastics, including LLDPE, LDPE, EVA, and HIPS, have a high elongation percentage. OPS, PEN, and PVC are some additional polymers with relatively low value. Except for PCL, which has an elongation of 250–300% and can be compared to HDPE because their values are quite similar, biodegradable polymers typically have low percent elongation. In composite films made of starch and chitosan, Elongation at break is increased when the chitosan to starch ratio is changed (Sun et al. 2019).

11.3.5 Melting Point

Melting temperature is the temperature at which a material begins to change its structure or the point at which a phase transition is noticed (Tm). Knowing the highest temperature, a polymer can withstand without deforming is a thermodynamic feature of the material. Temperatures that are mostly melting can be found in various ranges. The greatest Tm materials for bottles are PC and PET, with respective Tm of 240–280 °C and 245–270 °C.

Chitin has a Tm of 290–300 °C, which is equivalent to PET or PC when it comes to biodegradable polymers. But because chitin cannot be shaped into diverse shapes on its own, it is typically combined with other substances. The Tm in biodegradable polymers follows the order of:

$$\text{chitin} > \text{cellulose} > \text{PGA} > \text{PHB} > \text{PLA} > \text{CPLA} > \text{starch} > \text{Collagen} > \text{PBS} > \text{PCL}$$

Tm of thermoplastic polymers has the following order:

$$\text{PC} > \text{PET} > \text{PEN} > \text{EVOH} > \text{OPP} > \text{PP} > \text{PVDC} > \text{LLDPE} > \text{HDPE} > \text{LDPE} > \text{HIPS} > \text{PVC} > \text{OPS} > \text{EVA}$$

11.3.6 Thermal Stability

Regarding their thermal properties, polymeric materials could be utilized in a variety of consumer-focused applications. Designing materials with better characteristics requires a thorough understanding of polymer thermal breakdown. Thermal stability is the capacity of a polymeric material to survive the effects of heat and to maintain properties like strength, toughness, or elasticity at a particular temperature.

Thermogravimetric analysis is typically used to determine the heat stability of polymers. The chemical composition, degree of crystallinity, and molecular weight of a polymer all affect how stable it is at high temperatures. The cross-linking mechanism and aromatic structures in the polymer backbone increase the heat resilience of polymers. On the other hand, polymers are less resilient when the main chain contains double bonds or oxygen-containing groups (Krol-Morkisz and Pielichowska 2019).

11.4 Common Methods for Packaging in Food Industry

11.4.1 Modified Atmosphere Packaging (MAP)

Thermal characteristics are important. A perishable product is packaged in an environment that has been altered so that it has a different composition than air, according to the definition of MAP. One of the most popular methods for packaging and preserving agricultural items, primarily fruits and vegetables, involves altering the gaseous composition in the package's headspace. The final gaseous composition is influenced by the product weight when packed, storage temperature, commodity respiration rate, cultivar, and stage of ripening. By changing the O_2 and CO_2 content, MAP reduces respiration rates and delays fruit ripening. Keeping a constant temperature, it can help stop water loss and fruit shriveling. Biodegradable packaging is being used to package fresh agricultural food in place of polyethylene terephthalate, high-density polyethylene, and low-density polyethylene (Peelman et al. 2013).

11.4.2 Edible Packaging

Edible packaging is a fantastic option for food applications because of its ability to protect foods with its barrier and mechanical properties, control the release of active ingredients, and increase sensory aspects. They are consumed along with the food product and are a crucial component of the cuisine. The most common types of edible packaging are films, sheets, coatings, and pouches. The same formulation is used to create both films and coatings, but while coatings are applied as a liquid product, films are applied as solid sheets (Galus and Kadzinska 2015).

The main components of edible packaging are lipids, proteins, and polysaccharides. The two most often utilized ingredients are Chitosan and gelatin/collagen. The gelatine coatings cut down on oxygen, moisture, and oil migration. The most popular commercially available edible protein film is constructed of collagen sausage casing. Films placed over beef steak that has been thawed and without altering color or causing lipid oxidation, cooled and reduced exudation. Collagen-based films are used for a range of cooked meat products to encourage juiciness, lessen shrinkage, and absorb fluid exudates. Due to their great adhesion to the surfaces of cut fruits and vegetables and good gas barrier qualities, polysaccharide-based films have been used to increase the longevity of fruits and vegetables in storage. However, due to their hydrophilic nature, they do not have strong moisture resistance (Falguera et al. 2011).

11.4.3 Active Packaging

Active packaging refers to the use of elements that might leak or absorb substances into or from the packaged food or its surroundings. It is a cutting-edge technique for maintaining the shelf life and safety of the food product. The absorbers (scavenging system) in the active packaging system remove CO_2, O_2, moisture, smell, or ethylene while the emitters (releasing system) introduce diverse chemicals including anti-microbial, antioxidant, flavors, etc. into the headspace. When added directly to food, some active compounds may have their activity blocked or reduced due to interactions between the active substances and food components in the bulk food system. The regulated release of active chemicals in Active packaging is consequently more effective for the bulk food system. Antimicrobial packaging usually makes use of active packaging. Trays were made by baking cassava bagasse with polyvinyl alcohol and adding clove or oregano essential oils in order to investigate the antibacterial potential. Out of the two techniques tested direct inclusion and surface coating surface coating exhibited the best antibacterial performance against gram-positive and gram-negative bacteria, yeast, and molds. The inclusion of zinc oxide and oregano essential oil in a very thin "bio-paper" created by PHBV is one of the current active packaging solutions. *Escherichia coli* and *Staphylococcus aureus* are both resistant to the paper's antibacterial properties (Barrett 2020).

11.5 Types of Biodegradable Packaging

11.5.1 Films

In every industry, films are the most popular type of bio-packaging. Films made of biodegradable materials were first intended to take the place of PE film. They are more durable than non-biodegradable plastics. The following are crucial qualities of a quality packing film:

- Permitting controlled breathing
- Strong barrier qualities
- To keep the building's structural integrity intact
- To stop or lessen microbiological deterioration

The findings of a study on the carbon dioxide and oxygen permeability of biodegradable film used to package tomatoes demonstrated that the fruit could breathe well through the films with the best permeability, preventing microbial contamination and maintaining the fruit's quality (Muratore et al. 2005).

The usage of blown films in bags and other packaging applications is widespread. For blown film grading, PLA was employed as a basis because of its outstanding mechanical and transparency qualities. The seal ability property alters as the crystallinity level does. A single biodegradable polymer can't be employed in creation of blown films due to its sluggish crystallization and poor melting strength. Polyesters are laminated using the co-extrusion technique. For instance, thermoplastic starch (TPS) is coated with PHA and PHB while being film blown during the co extrusion process. In the packaging of cheese, Avebe's Paragon TM technology is used.

11.5.2 Containers or Trays

Vegetables, salads, and fruits can be packaged in thermoformed trays or containers because they need a regulated environment to retain their quality. First, sheets are produced from the polymer via melt extrusion, and then certain shapes are produced by heating the sheets above Tg and Tm. The vast majority of trays constructed from biodegradable polymers are fragile and moisture-resistant. The tray's structural characteristics don't change while it freezes. Mangoes, melon, and other tropical fruits were stored on trays made of orientated PLA. The fruits' shelf-life was identical to that of fruits packed in PET trays (Chonhenchob et al. 2007).

11.5.3 Foamed Product

Foams made of starch are used to provide loose fill. The creation of foamed items can be produced using a variety of methods, such as expandable bead molding, loose-fill molding, extrusion transfer moulding and foam extrusion. There are several foamed items made of starch, such as trays, clamshells, etc., that can be used for food packing, but they need to have direct food contact coatings. Paraffin and other polymers are favored over starch and other coatings for PLA. The foamed product's and coating's adhesion is crucial. Starch-based foam called Novamont, created in the USA, is utilized in a variety of packaging applications. The Landaal Packaging system created Green Cell foam TM as a sustainable replacement for PP foams. In 4 weeks, it fully decomposed in a damp soil environment (www.landaal.com).

11.5.4 Bags

Biodegradable bags are most frequently used in the food industry because to their composition of raw materials, which makes them flexible, strong, and resistant to breakage, moisture, and temperature change. Food items can be packaged and stored in the biodegradable bags. Additives must be added for these bags to be used in various sectors. The bags are all made with sustainable materials. After serving as packaging for a few weeks, they break down into carbon dioxide, water, and other substances. The excellent replacements for plastic bags are the biodegradable bags (Nampoothiri et al. 2010).

11.5.5 Gels

Hydrogels are a type of biodegradable gel that are frequently used to stop microbial infection. Complex hydrogels can be created as an alternative to bio-based polymer synthesis. When applied to lettuce, hydrogel had no beneficial benefits on retaining the pectic substance and quality content, but when applied to the fruits of *Solanum muricatum*, the gel had a beneficial impact on preserving beta-carotene. Certain fruits' shelf-life is decreased when hydrogels of different polymeric materials are combined, primarily due to water evaporation from the environment (Barrett 2020).

11.6 Biodegradation Method

A polymer undergoes biodegradation when it is broken down by microorganisms into biomass, carbon dioxide, water, or methane. There are several processes involved in the biodegradation of polymeric materials, such as:

- Assimilation: By interacting with the metabolism, molecules that enter the cytoplasm produce a wide range of primary and secondary metabolites that are high in energy and biomass.
- Biodeterioration: When other abiotic components and soil microorganisms cooperate, biodegradable material is broken down into minute parts.
- Depolymerization—Different catalytic agents, primarily enzymes, are released by microorganisms, cleaving the polymer to create oligomers, dimers, and monomers.
- Mineralization: Organic acids and aldehydes are just a couple of the metabolites that microbial cells discharge into the environment.
- Recognizability Some fragmented oligomers, dimers, and monomers pass through the microbial cell's plasma membrane and are recognised by microbe receptors. The extracellular environment still contains the unidentified piece.

Additionally, compounds such CO_2, CH_4, H_2O, and others are discharged into the atmosphere. The following reaction is involved in the biodegradable polymer degradation.

$$CO_2 + H_2O + \text{biodegradable polymers}$$

Depending on the environment, such as the soil, sea, or compost, microorganisms that are responsible for polymer breakdown may or may not be present. The primary mechanism involved in degradation is the colonisation of the exposed surface following the microbe adhesion on the polymer surfacethe type of substrate, the nature of the enzymes, their location (extracellular or intracellular), and environmental factors like moisture, pH, soil, light, oxygen, moisture, temperature, etc. are just a few of the variables that affect how quickly something degrades. The two primary techniques for assessing the biodegradability of a polymer are the soil burial test and the enzymatic test.

11.7 Composting

The accelerated breakdown of organic materials by a diverse microbial population under controlled circumstances in a aerobic environment moist, warm, aerobic environment is known as composting. It is a time-tested technique to turn organic material into fertile humus. Compost will be the main byproduct of the biodegradation of this organic material, along with carbon dioxide and water. Considering that the carbon dioxide produced is already a part of the biological carbon cycle, it has no effect on greenhouse gases. Compositing is advantageous for the soil because it improves soil aeration, enriches the soil with nutrients, boosts microbiological activity and helps to maintain soil moisture. Under controlled composting circumstances, PLA breaks down to CO_2, water, and biomass in less than 90 days. In huge plants, when the temperature can approach 140 °C, breakdown occurs. The enzyme PHA depolymerase, which is released by 55% of penicillium yeast, can break down PHA in a normal biological environment. Within 45 days, degradation within a controlled environment can be finished. Composting and biodegrading products provide a secure alternative for protecting the environment, preserving fossil resources, and reducing CO_2 emissions. The International Standards Organization (ISO) has developed ISO 17088, "Specification for Compostable Plastics," globally in response to these US and European standards. The following are the specifications of these standards for full biodegradation in composting conditions (Song et al. 2009):

- A biodegradation rate that is comparable to the decay of organic materials including leaves, paper, grass, and food wastes.
- If using radiolabelled polymers, the duration of the composting cycle or degrading process should not exceed 365 days.
- Microbial activity that transforms polymer material in the form of granules, films, or powder into CO_2, water, and biomass.
- The polymer's carbon converts to CO_2 at a rate of about 90%.

11.8 Toxicity

In order to enhance the properties of biodegradable plastics or bioplastics, various compounds, such as additives, antioxidants, stabilizers, chain initiators, cross-linking agents, nucleating agents, catalysts, etc., are added during manufacture. Many of these materials are not covalently attached to the matrix of the polymer, therefore chemical migration, a process that exposes people to these substances, is a possibility. In order to improve environmental performance, the toxicity of bioplastics is typically evaluated during the manufacture stage or the degradability stage. It is common practise to ignore the release of chemicals by the substance during use. Chemicals will be added in greater quantities as the use of bioplastic expands.

To test the in-vitro toxicity, various testing techniques are performed, including endocrine activity, oxidative stress response, and baseline toxicity. Some of them are starch- and cellulose-based. Aliivibrio fischeri's bioluminescence, a measure of baseline toxicity, was suppressed by bio-PE, bio-PET, PBAT, and PHA samples, mostly with high potency (low EC20). Tests on cell cultures demonstrated that the polybutylene succinate-based Bionolle manufactured by Showa Highpolymer has no harmful effects on cells. The bioassay method was used to test the degradation of PLA and nanocomposites (organoclays Cloisite 20A and Cloisite 30B) for the assessment of ecotoxicity. The results showed a decreasing mitotic index and increasing chromosomal abnormalities, which is likely because the products formed after degradation have an aneogenic effect.

Bioplastics' phytotoxic effects were demonstrated to be negligible when applied to agricultural and horticultural food products. It was discovered that Mater-bi, a commercially available bag manufactured from vegetable starch, could constitute a threat akin to that posed by other conventional plastics if left in the natural environment by interfering with the development of plant radicles. Standard phytotoxicity studies employing *Lepidium sativum* L. seedlings verified this. It significantly altered the characteristics of water as well. Studies on the toxicity of PHB and PBAT leachates revealed that Daphnia magna survival was reduced after only 48 h of exposure. A combination of PLA and Bio-PE was used to create films and pellets, which contained 37 non-volatile compounds.

The largest concentrations of cyclic oligomers, including adipic acid, phthalic acid, and butanediol, were discovered. Only a small number of research have been published on the nanotoxicology and nanoecotoxicology of various biodegradable polymers, despite the fact that nanoparticles (NPs) are employed to enhance the characteristics of polymer for food packaging applications. These particles can be inhaled, inhaled through contact, and exposed through contact. The chemical migration of NPs from the polymer to the food product is what causes oral ingestion. Smaller nanoparticles have greater absorption rates and are more easily disseminated throughout the body, harming cells. Studies on mice showed that injecting carbon nanotubes into the animal's peritoneal cavity resulted in asbestos-like, length-dependent, hazardous behaviour.

The sustainability of bioplastics must take into account a variety of aspects, including societal implications, land use, greenhouse gas emissions, and toxicity. Research on using secure chemical alternatives must be carried out based on scientific concepts such as the Tiered Protocol for Endocrine Disruption (TiPED). Despite all of these toxicological effects, the majority of biodegradable polymers and bioplastics have found use in the food industry thanks to advancements in their manufacturing process. Before being widely employed for commercial purposes, risks were found and described.

11.9 Consumer Perception of Biodegradable Packaging

Through their purchases of the product, customers play a very important part in the success of environmentally friendly food packaging. The choice is based on a number of trade-off factors, including shape, color, and design. For instance, certain products' flavors are influenced by color. There are many reasons why consumers can decide against buying a sustainable product, including the longer production process, perceived lower quality, and limited supply. Lack of recognition is one of the key drivers of non-development market growth. These ecological packaging can be recognized by consumers thanks to logos, labels, etc. Many individuals are unaware of biobased packaging and believe it to be a marketing ploy. Some research found that factors like pricing and product quality were more significant than the use of green packaging.

Consumers are only willing to pay a small premium price. To expand the market for ecologically friendly packaging, a number of ideas might be considered, such as the requirement that the packaging have a third-party label certification to improve consumer confidence. Businesses who package food in an environmentally friendly manner must clearly state any benefits and have strong labels on their product containers. Without proper communication, packaging may not be advantageous to firms in the food industry and the environment (Krol-Morkisz and Pielichowska 2019).

11.10 Conclusion

The use of bio-based polymers is increasing for the packaging of food and other applications at a great speed. Biodegradable polymers are made from renewable feedstocks, agricultural waste and there is a great opportunity for research work in harnessing this economic opportunity.

References

Akter N, Khan RA, Tuhin MO, Haque ME, Nurnabi M, Parvin F, Islam R (2014) Thermomechanical, barrier, and morphological properties of chitosan-reinforced starch-based biodegradable composite films. J Thermoplast Compos Mater 27(7):933–948

Alvarado MDL (2009) Envasado en la Industria de Alimentos y sus Nuevas Tendencias (Packaging in the food industries: new tendencies). Bachelor's thesis, Universidad Autónoma Agraria, Saltillo, Mexico DF, p 89

Amin U, Khan MU, Majeed Y, Rebezov M, Khayrullin M, Bobkova E, Shariati MA, Chung IM, Thiruvengadam M (2021) Potentials of polysaccharides, lipids and proteins in biodegradable food packaging applications. Int J Biol Macromol 31(183):2184–2198

Barrett A (2020) Biodegradable food packaging extending food shelf life. https://bioplasticsnews.com/2020/06/12/ypack-biodegradable-food-packaging-shelf-life. Accessed 14 Oct 2022

Chonhenchob V, Chantarasomboon Y, Singh SP (2007) Quality changes of treated fresh-cut tropical fruits in rigid modified atmosphere packaging containers. Packag Technol Sci 20(1): 27–37

Falguera V, Quintero JP, Jimenez A, Munoz JA, Ibarz A (2011) Edible films and coatings: structures, active functions and trends in their use. Trends Food Sci Technol 22(6):292–303

Ficci (2014) Potential of plastics industry in Northern India with special focus on plastic culture and food processing - 2014: a report on plastic industry. https://ficci.in/publication.asp?spid=20396. Accessed 21 Dec 2022

Galus S, Kadzinska J (2015) Food applications of emulsion-based edible films and coatings. Trends Food Sci Technol 45(2):273–283

Kale G, Auras R, Singh SP (2006) Degradation of commercial biodegradable packages under real composting and ambient exposure conditions. J Polym Environ 14(3):317–334

Krol-Morkisz K, Pielichowska K (2019) Thermal decomposition of polymer nanocomposites with functionalized nanoparticles. In: Pielichowski K, Majka TM (eds) Polymer composites with functionalized nanoparticles: synthesis, properties, and applications. Elsevier, Amsterdam, pp 405–435

Mangaraj S, Yadav A, Bal LM, Dash SK, Mahanti NK (2019) Application of biodegradable polymers in food packaging industry: a comprehensive review. J Packag Technol Res 3(1): 77–96

Mohareb E, Mittal GS (2007) Formulation and process conditions for biodegradable/edible soy-based packaging trays. J Packag Technol Sci 20(1):1–15

Muratore G, Nobile D, Buonocore GG, Lanza CM, Asmundo N (2005) The influence of using biodegradable packaging films on the quality decay kinetic of plum tomato. J Food Eng 67(4): 393–399

Muthuraj R, Misra M, Mohanty AK (2018) Biodegradable compatibilized polymer blends for packaging applications: review. J Appl Polym Sci 135(24):e45726

Nampoothiri KM, Nair NR, John RP (2010) An overview of the recent developments in polylactide (PLA) research. Bioresour Technol 101(22):8493–8501

Netravali AN, Chabba S (2003) Composites get greener. Mater Today 6(4):22–29

Peelman N, Ragaert P, De Meulenaer B, Adons D, Peeters R, Cardon L, Impe FV, Devlieghere F (2013) Application of bioplastics for food packaging. Trends Food Sci Technol 32(2):128–114

Ramos M, Valdes A, Beltran A, Garrigos MC (2016) Gelatin-based films and coatings for food packaging applications. Coatings 6(4):41

Sarasam AR, Krishnaswamy RK, Madihally SV (2006) Blending chitosan with polycaprolactone: effects on physicochemical and antibacterial properties. Biomacromolecules 7(4):1131–1138

Siracusa V, Rocculi P, Romani S, Rosa MD (2008) Biodegradable polymers for food packaging: a review. Trends Food Sci Technol 19(12):634–643

Smith R (ed) (2005) Biodegradable polymers for industrial applications. CRC Press, Boca Raton, p 531

Song JH, Murphy RJ, Narayan R, Davies GBH (2009) Biodegradable and compostable alternatives to conventional plastics. Philos Trans R Soc B 364(1526):2127–2139

Sun K, Li F, Li J, Li J, Zhang C, Chen S, Sun X, Cui J (2019) Optimization of compatibility for improving elongation at break of chitosan/starch films. RSC Adv 9(42):24451–24459

Teng WL, Khor E, Tan TK, Lim LY, Tan SC (2001) Concurrent production of chitin from shrimp shells and fungi. Carbohydr Res 332(3):305–316

Tiberiu N (2011) Concepts in biological analysis of resorbable materials in oromaxillo facial surgery. Oro Maxillofac Implantol (Romania) 2(1):33–38

Wang X, Zhou J, Li L (2007) Multiple melting behavior of poly (butylene succinate). Eur Polym J 43(8):3163–3170

Chapter 12
Scope of Herbal Extracts and Essential Oils for Extension of Shelf-Life of Packaged Foods

Pinar Gumus

Contents

Abstract In recent years, plant materials and their extracts have become prominent. Essential oils are volatile or semi-liquid liquids obtained from aromatic plants. Essential oils show many bioactivities, such as antimicrobial, antioxidant, antibacterial, anti-inflammatory, antifungal, anticancer, antidiabetic, and antianxiety activities, among others. Essential oils have become more popular in various industries, including food and beverages, household, agricultural, soap, perfumery, cosmetics, and pharmaceuticals. This chapter presents an overview of essential oils and herbal extracts and their potential applications in extending the shelf-life of food packaging. Additionally, innovative food packaging concepts are gaining popularity.

Keywords Essential oil · Food packaging · Plant extract · Shelf-life

P. Gumus (✉)
Faculty of Health Sciences, Department of Nutrition and Dietetics, Kilis 7 Aralik University, Kilis, Turkey

J. A. Malik et al. (eds.), *Food Process Engineering and Technology*,
https://doi.org/10.1007/978-981-99-6831-2_12

12.1 Introduction

Essential oils are obtained from many different aromatic plants and are highly volatile. Their constituents and extraction methods find implementation in the food industry as a protective tool to prolong the shelf-life of food. The term "essential" means that the oil includes the "essence of" the specific plant's fragrance. Natural plant oils have been used for various applications since early times (El Asbahani et al. 2015; Hamid et al. 2011; Herman et al. 2019; Rashidinejad and Jafari 2020). Essential oils are a rich source of numerous bioactive constituents, showcasing a wide range of beneficial properties such as antioxidative, antibacterial, antifungal, anticancer, antimicrobial, anti-inflammatory, antidiabetic, antiulcerogenic, antianxiety, and antitumoral activities. These properties make essential oils highly valuable not only in nutraceuticals, food, and cosmetic industries but also in the treatment of infectious diseases (El Asbahani et al. 2015; Hamid et al. 2011; Kumar et al. 2020; Miguel 2010; Morsy 2017; Nazzaro et al. 2020; Rashidinejad and Jafari 2020; Saad et al. 2013; Tongnuanchan and Benjakul 2014; Turek and Stintzing 2013). Essential oils are utilized for a variety of purposes, including perfumery, soap, and the cosmetic sector. They are also widely employed in the food and beverages sector as flavorings and preservation additives (Herman et al. 2019; Vintilă 2017).

Approximately 17,000 aromatic plants, belonging to angiospermic families like Zingiberaceae, Myrtaceae, Asteraceae, Lamiaceae, and Rutaceae, have been identified as sources of essential oils. These plants are known for their distinctive and unique essential oils, which have been commercialized and are widely used in the food industry.

Aromatic plants are among the best sources of essential oils, and these oils can be found in various parts of the plant, such as leaves, flowers, condiments, bark, seeds, peel, fruits, wood, stems, rhizomes, and roots.

Examples of essential oil sources include:

- Flowers (e.g., rose, mimosa, jasmine, lavender), tree blossoms, bulbs, and dried flower buds.
- Leaves (e.g., mint, eucalyptus, thyme, peppermint, savory, oregano).
- Leaves and stems (e.g., cinnamon, geranium, verbena).
- Grasses (lemongrass).
- Gum (frankincense).
- Bark (e.g., canella).
- Wood (e.g., cedar, pine).
- Roots (e.g., angelica, vetiver, valerian).
- Seeds (e.g., fennel, dill, coriander, and almond).
- Fruits (e.g., orange, anise, fennel).
- Peel (bergamot).
- Rhizomes (e.g., ginger, curcuma, galangal).
- Gums (e.g., myrrh, storax).

These diverse essential oils contribute to the aroma, flavor, and preservation of various food products (Calo et al. 2015; Dhifi et al. 2016; Hanif et al. 2019; Kumar et al. 2020; Moghaddam and Mehdizadeh 2017; Morsy 2017; Rashidinejad and Jafari 2020; Sadgrove and Jones 2015; Vintilă 2017).

The essential oil is a combination of low molecular weight components, which are responsible for its typical aroma. These oils are a mixture of volatile compounds, including phenolics, amines, terpenes, amides, alcohols, acids, esters, aldehydes, ketones, lactones, coumarins, and ethers. The components of essential oils can be categorized into two groups: terpene and aroma compounds (Calo et al. 2015; Hanif et al. 2019; Kumar et al. 2020; Morsy 2017; Vintilă 2017). Essential oils (EOs) are volatile oils extracted from plants that are rich in terpenes and aromatics. These oils can be categorized into four classes: terpenes, straight-chain compounds, phenylpropanoids, and other miscellaneous compounds (Morsy 2017; Setyaningsih and Siahaan 2018). Terpenes are the major chemical compounds found in essential oils and are composed of a few isoprene units. Terpenes are generally divided into different groups based on the number of carbon atoms: monoterpenes, sesquiterpenes, and diterpenes (Kumar et al. 2020; Morsy 2017). Essential oils are mixtures of hydrocarbons and their oxygenated derivatives, which consist of two different isoprenoid pathways (Sharifi-Rad et al. 2017). These oils are formed in the plant through two varied pathways: the mevalonate pathway and the non-mevalonate pathway. Terpenes are biogenerated by the mevalonate pathway (Dhifi et al. 2016; Hamid et al. 2011; Kumar et al. 2020).

The quantity of essential oil in plants is influenced by various factors, such as climatic, seasonal, and geographical conditions, as well as harvest time and extraction methods (Hamid et al. 2011; Saad et al. 2013).

Several extraction methods are preferred for the production of essential oils, including expression, fermentation, effleurage, steam distillation, and hydro distillation (Herman et al. 2019). These extraction processes may involve freeze drying, gas chromatography, rotary evaporation, steam distillation, and hydrolyzation (Herman et al. 2019).

The main methods of essential oil extraction include: maceration, cold pressing, solvent extraction, expression, high-pressure solvent extraction, enfleurage, defleurage, water distillation, steam distillation, water and steam distillation, turbo distillation, hydrodistillation, microwave-assisted distillation, microwave hydrodiffusion, microwave-assisted extraction, solvent-free microwave extraction, supercritical carbon dioxide extraction, ultrasonic extraction and phytonic process, controlled pressure drop process, supercritical fluid extraction, ultrasound-assisted extraction, headspace trapping techniques, solid phase micro-extraction, supercritical fluid extraction, simultaneous distillation extraction, microwave distillation, controlled instantaneous decomposition, thermomicro-distillation, microdistillation, membrane extraction (Fig. 12.1).

Freeze drying, rotary evaporation, hydrolyzation, and GC assays are also utilized in the extraction process of essential oils (Hanif et al. 2019; Herman et al. 2019; Moghaddam and Mehdizadeh 2017; Stratakos and Koidis 2016).

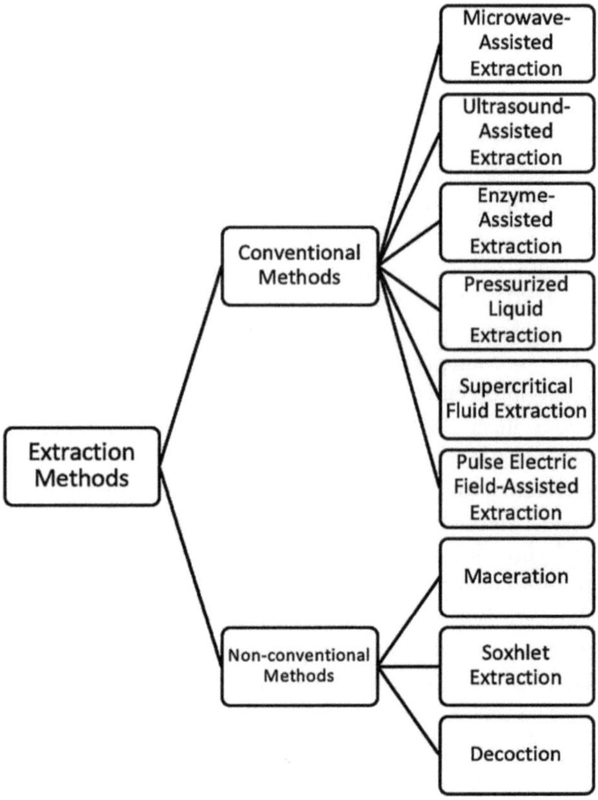

Fig. 12.1 Several methods for extracting essential oils form different plants. (Adapted from: Pai et al. 2022)

12.1.1 Plant Extracts

This section aims to provide information about the usage of plant extracts as antimicrobial and antioxidant agents in the food industry, as well as recent developments in different types of food packaging technologies. In recent years, plants and their extracts have been gaining popularity and have been extensively studied for their efficiencies in ensuring food safety and preservation (Calo et al. 2015; Christaki et al. 2021; Cordery et al. 2018). Plants have been used since ancient times as traditional medicine and as preservatives in foods, making them one of the oldest forms of healthcare known to mankind. Aromatic plants and their extracts also hold tremendous potential to become next-generation substances for enhancing human nutrition and health. The incorporation of plant extracts in food products and food packaging technologies offers a natural and sustainable approach to improve food safety and extend shelf life while minimizing the use of synthetic additives and chemicals (Christaki et al. 2012; Veiga et al. 2020).

Herbal extracts are considered to be one of the potential sources of bioactive compounds, such as phytochemicals, polyphenols, and carotenoids. These extracts are increasingly gaining significance as additives in the food industry due to their various beneficial attributes, including anticancer, antimicrobial, antioxidant, antifungal, anti-inflammatory properties, and their ability to scavenge free radicals (Altemimi et al. 2017; Christaki et al. 2012; Hashemi et al. 2020; Munekata et al. 2020; Nikmaram et al. 2018; Proestos 2020). For many years, plant extracts have been widely utilized for various purposes and in several industries, including food, medical, pharmaceutical, sanitary, aromatherapy, phytotherapy, perfumery, and cosmetic applications (Adeyinka and Richard 2015; Erkan et al. 2015; Regnier et al. 2012; Viji et al. 2017). They serve as next-generation eco-friendly bio-products that contribute to the production of high-quality foods and help to retard lipid oxidation in food during accelerated storage processes. By harnessing the potential of these plant extracts, the food industry can explore more sustainable and natural approaches to enhance food quality and safety (Godlewska et al. 2021; Inanç and Maskan 2012).

Plant extracts can be easily obtained from various types of plant materials, including herbs and spices (such as tea, oregano, cinnamon, mint, clove, ginger, sage, turmeric, garlic, anise, basil, and ginseng), flowers, seeds, bark, wood, buds, fruits (like dragon fruit, guarana, pomegranate, grapes, date, and kinnow), and roots and vegetables (such as broccoli, potato, drumstick, pumpkin, curry, nettle, cabbage, and onion). These extracts can be obtained through different extraction methods, such as fermentation, extraction, expression, and enfleurage. The efficiency of the extraction procedures depends on several factors, including the characteristics of the plant part used, the choice of solvent, the temperature, time, and pressure applied during the extraction process. These factors play a crucial role in determining the quantity and quality of the bioactive compounds present in the extracted plant materials, making them essential considerations for optimizing the extraction process in various applications, including food and pharmaceutical industries (Azmir et al. 2013; Cordery et al. 2018; Giannenas et al. 2020; Makhuvele et al. 2020; Shah et al. 2014; Viji et al. 2017).

Medicinal plants have been extensively used in folk medicine in many developing countries for maintaining good health. They play a crucial role in therapeutic purposes and the prevention of various chronic and infectious diseases, making them an essential part of traditional medicine (Giannenas et al. 2020; Lopes et al. 2020; Ullah et al. 2020). These plants contain not only primary metabolites, such as amino acids, fatty acids, carbohydrates, and organic acids, which are necessary for their survival, but also secondary metabolites, including phenolic compounds, flavonoids, steroids, resins, fatty acids, and gums. These secondary metabolites serve as a defense mechanism against pathogenic microorganisms and adverse environmental conditions and are believed to provide health benefits to human beings. The rich diversity of bioactive compounds present in medicinal plants makes them a valuable resource for the development of natural remedies and pharmaceutical products that can aid in treating various health conditions and promoting overall well-being (Akram et al. 2019; Benchaar et al. 2008; Ingle et al. 2017; Kapoor et al. 2015; Lopes et al. 2020; Makhuvele et al. 2020).

12.2 Scope of Herbal Extracts and Essential Oils for Extension of Shelf-Life of Packaged Foods

Food spoilage is described as any alteration in a food that leads to undesirable sensory perceptions and may be caused by various mechanisms, including microbial, chemical, and physical reactions. Several factors can affect food spoilage, such as moisture, light, temperature, and chemical reactions. One of the primary roles of food packaging is to slow down the natural processes that cause food deterioration during production, transport, processing, or storage. Packaging helps protect food from external factors that can accelerate spoilage, such as exposure to air, moisture, and light, as well as microbial contamination. By providing a barrier and maintaining a controlled environment, food packaging helps extend the shelf life of products and ensures that they remain safe and fresh for consumers (Sanches-Silva et al. 2014; Taormina 2021).

Shelf-life can be defined as the time elapsed after processing and packaging during which the food maintains a minimum level of quality sufficient for consumption. Two critical factors that determine the quality and shelf-life of a food product are lipid peroxidation and microbial contamination (Keshvari et al. 2013; Ukrainets et al. 2016). Lipid oxidation is a major cause of food deterioration, leading to off-odors, off-flavors, texture and color changes, and nutritional losses, which ultimately decrease the product's shelf-life and result in product loss. Some consequences of lipid oxidation include nutritional loss in food, reduced energy content, formation of off-flavors and odors, and changes in color. To maintain food quality and extend shelf-life, it is essential to implement appropriate food packaging methods and storage conditions that protect the product from lipid oxidation and microbial contamination. Effective packaging can significantly contribute to preserving the freshness, taste, and nutritional value of the food, ensuring a longer period of safe consumption (Sanches-Silva et al. 2014; Tian et al. 2013).

There are many factors impacting the quality of a food product: extrinsic factors (temperature, atmosphere, microorganisms, etc.) and intrinsic factors (nutrients, water activity, pH, etc.). Food packaging acts as a passive barrier, preserving contents from the external environment. The oxidative stability of a packaged food product can be influenced by its interaction with the external environment. By considering and controlling these factors, the food industry can extend shelf-life and ensure high-quality products for consumers (Koontz 2016; Taormina 2021).

Essential oils can be integrated into packaging as they have the ability to modify the matrix of packaging materials. They are highly volatile, sensitive to moisture, light, heat, and oxygen, and can be well-preserved through encapsulation (Morsy 2017; Tongnuanchan and Benjakul 2014; Vintilă 2017). Essential oils possess various bioactivities, including antimicrobial, antibacterial, anti-inflammatory, antifungal, anticancer, antidiabetic, and antianxiety activities. They also exhibit antioxidant activity, although their antioxidant content is lower compared to commercial antioxidants. By incorporating essential oils into packaging materials, it is possible to enhance the preservation of food products and provide added benefits, such as antimicrobial properties, which can contribute to extending the shelf-life of the packaged foods. Additionally, the use of essential oils may offer a more natural and sustainable alternative to synthetic additives in food packaging (Kumar et al. 2020; Morsy 2017; Setyaningsih and Siahaan 2018).

Many herbs, spices, plants, and fruits having antimicrobial and antioxidant compounds can be used in food packaging. Emergent terms of active and intelligent packaging technologies which are alternative to traditional preservative techniques, provide numerous innovative solutions for ensuring the quality, improvement food product safety, minimizing food losses and accomplishment desired characteristics of food. These new packaging methods preserve the food from environmental effects and external risks like humidity, light, temperature, odors, pathogens, microorganisms' shocks, dust, vibrations, physical damage, and compressive forces. By incorporating antimicrobial and antioxidant compounds from natural sources into active and intelligent packaging, it becomes possible to create more sustainable and effective ways of extending the shelf-life of food products and enhancing their safety during storage and distribution. These advancements in packaging technologies contribute to the overall efforts in reducing food waste and ensuring the delivery of safe and high-quality food to consumers (Ahmed et al. 2017; Akram et al. 2019; Han et al. 2018; Ribeiro-Santos et al. 2017).

Some studies have been focused on novel packaging technologies due to the new food packaging concept gaining interest among scientists. There are many advancements in packaging technology containing intelligent packaging and active packaging. There are also new trends to develop the stable properties of food packaging like barrier performance, mechanical strength, structural properties, and thermal stability. Sustainable (green) packaging may decrease the environmental effects of food packaging using biodegradable materials and plant extracts. Plant extracts that are an essential ingredient in biodegradable food packaging films have been used for obtaining desirable properties in the food industry due to their natural source and phytochemical features (Han et al. 2018; Mir et al. 2018). Plant extract process with vacuum packaging or modified atmospheric packaging is an effective preservation method and the incorporation of plant extracts into biodegradable packaging films is a promotive method for delaying oxidation, controlling foodborne pathogens and meeting demands of consumers, prevention the food deterioration, extending shelf-life and developing the safety and quality of food products (Han et al. 2018; Mir et al. 2018; Viji et al. 2017).

12.3 Conclusion

Plants have the potential to provide two types of oils: fixed oils, which include esters of glycerol and fatty acids, and essential oils, which are mixtures of volatile and semi-volatile organic compounds obtained from a single botanical source. Essential oils, found in various aromatic plants, are highly volatile. Aromatic plants and essential oils offer promising opportunities for economic and ecological development. Many herbs, spices, plants, and fruits with antimicrobial and antioxidant activities can be utilized in food packaging.

This study highlights the use of plant extracts, essential oils, their constituents, and extraction techniques in the food sector as a protective tool to prolong the shelf-life of food products. Future studies may explore recent developments in various food packaging technologies and novel potential applications to further extend the shelf-life of packaged foods. By leveraging the benefits of plant-based compounds in food packaging, the food industry can enhance food safety, reduce food waste, and improve the overall quality of packaged products.

References

Adeyinka A, Richard F (2015) Application of phytochemical extracts and essential oils in food products: a review. Int J Biotechnol Food Sci 3(3):31–35

Ahmed I, Lin H, Zou L, Brody AL, Li Z, Qazi IM, Pavase TR, Lv L (2017) A comprehensive review on the application of active packaging technologies to muscle foods. Food Control 82: 163–178

Akram MZ, Fırıncıoğlu SY, Jalal H, Doğan SC (2019) The use of essential oils in active food packaging: a review of recent studies. Turk J Agric Food Sci Technol 7(11):1799–1804

Altemimi A, Lakhssassi N, Baharlouei A, Watson DG, Lightfoot DA (2017) Phytochemicals: extraction, isolation, and identification of bioactive compounds from plant extracts. Plants 6(4):42–51

Azmir J, Zaidul ISM, Rahman MM, Sharif KM, Mohamed A, Sahena F, Jahurul MHA, Ghafoor K, Norulaini NAN, Omar AKM (2013) Techniques for extraction of bioactive compounds from plant materials: a review. J Food Eng 117(4):426–436

Benchaar C, Calsamiglia S, Chaves AV, Fraser GR, Colombatto D, McAllister TA, Beauchemin KA (2008) A review of plant-derived essential oils in ruminant nutrition and production. Anim Feed Sci Technol 145(1–4):209–228

Calo JR, Crandall PG, O'Bryan CA, Ricke SC (2015) Essential oils as antimicrobials in food systems–a review. Food control 54:111–119

Christaki E, Bonos E, Giannenas I, Florou-Paneri P (2012) Aromatic plants as a source of bioactive compounds. Agriculture 2(3):228–243

Christaki S, Moschakis T, Kyriakoudi A, Biliaderis CG, Mourtzinos I (2021) Recent advances in plant essential oils and extracts: delivery systems and potential uses as preservatives and antioxidants in cheese. Trends Food Sci Technol 116:264–278

Cordery A, Rao AP, Ravishankar S (2018) Antimicrobial activities of essential oils, plant extracts and their applications in foods: a review. J Agric Environ Sci 7(2):76–90

Dhifi W, Bellili S, Jazi S, Bahloul N, Mnif W (2016) Essential oils' chemical characterization and investigation of some biological activities: a critical review. Medicines 3(4):25–31

El Asbahani A, Miladi K, Badri W, Sala M, Addi EA, Casabianca H, El Mousadik A, Hartmann D, Jilale A, Renaud FNR, Elaissari A (2015) Essential oils: from extraction to encapsulation. Int J Pharm 483(1–2):220–243

Erkan N, Doğruyol H, Günlü A, Genç İY (2015) Use of natural preservatives in seafood: plant extracts, edible film and coating. Food Health 1(1):33–49

Giannenas I, Sidiropoulou E, Bonos E, Christaki E, Florou-Paneri P (2020) The history of herbs, medicinal and aromatic plants, and their extracts: past, current situation and future perspectives. In: Florou-Paneri P, Christaki E, Giannenas I (eds) Feed additives. Academic Press, Cambridge, pp 1–18

Godlewska K, Ronga D, Michalak I (2021) Plant extracts-importance in sustainable agriculture. Ital J Agron 16(2):e1851
Hamid AA, Aiyelaagbe OO, Usman LA (2011) Essential oils: its medicinal and pharmacological uses. Int J Curr Res 33(2):86–98
Han JW, Ruiz-Garcia L, Qian JP, Yang XT (2018) Food packaging: a comprehensive review and future trends. Compr Rev Food Sci Food Saf 17(4):860–877
Hanif MA, Nisar S, Khan GS, Mushtaq Z, Zubair M (2019) Essential oils. In: Malik S (ed) Essential oil research: trends in biosynthesis, analytics, industrial applications and biotechnological production. Springer, Cham, pp 3–17
Hashemi M, Daneshamooz S, Raeisi M, Jannat B, Taheri S, Noori SMA (2020) An overview on antioxidants activity of polysaccharide edible films and coatings contains essential oils and herb extracts in meat and meat products. Adv Anim Vet Sci 8(2):198–207
Herman RA, Ayepa E, Shittu S, Fometu SS, Wang J (2019) Essential oils and their applications: a review. Adv Nutr Food Sci 4(4):1–13
Inanç T, Maskan M (2012) The potential application of plant essential oils/extracts as natural preservatives in oils during processing: a review. J Food Sci Eng 2(1):1–9
Ingle KP, Deshmukh AG, Padole DA, Dudhare MS, Moharil MP, Khelurkar VC (2017) Phytochemicals: extraction methods, identification and detection of bioactive compounds from plant extracts. J Pharmacogn Phytochem 6(1):32–36
Kapoor A, Kaur G, Kaur R (2015) Antimicrobial activity of different herbal plants extracts: a review. World J Pharm Sci 4(7):422–459
Keshvari M, Asgary S, Jafarian-Dehkordi A, Najafi S, Ghoreyshi-Yazdi SM (2013) Preventive effect of cinnamon essential oil on lipid oxidation of vegetable oil. Arya Atheroscler 9(5): 280–286
Koontz JL (2016) Packaging technologies to control lipid oxidation. In: Hu M, Jacobsen C (eds) Oxidative stability and shelf-life of foods containing oils and fats. AOCS Press, Urbana, pp 479–517
Kumar A, Singh P, Gupta V, Prakash B (2020) Application of nanotechnology to boost the functional and preservative properties of essential oils. In: Prakash B (ed) Functional and preservative properties of phytochemicals. Academic Press, Cambridge, pp 241–267
Lopes TS, Fontoura PS, Oliveira A, Rizzo FA, Silveira S, Streck AF (2020) Use of plant extracts and essential oils in the control of bovine mastitis. Res Vet Sci 131:186–193
Makhuvele R, Naidu K, Gbashi S, Thipe VC, Adebo OA, Njobeh PB (2020) The use of plant extracts and their phytochemicals for control of toxigenic fungi and mycotoxins. Heliyon 6(10): e05291
Miguel MG (2010) Antioxidant and anti-inflammatory activities of essential oils: a short review. Molecules 15(12):9252–9287
Mir SA, Dar BN, Wani AA, Shah MA (2018) Effect of plant extracts on the techno-functional properties of biodegradable packaging films. Trends Food Sci Technol 80:141–154
Moghaddam M, Mehdizadeh L (2017) Chemistry of essential oils and factors influencing their constituents. In: Grumezescu AM, Holban AM (eds) Soft chemistry and food fermentation. Academic Press, Cambridge
Morsy NFS (2017) Chemical structure, quality indices and bioactivity of essential oil constituents. In: El-Shemy HA (ed) Active ingredients from aromatic and medicinal plants. IntechOpen, London, pp 175–206
Munekata PES, Rocchetti G, Pateiro M, Lucini L, Domínguez R, Lorenzo JM (2020) Addition of plant extracts to meat and meat products to extend shelf-life and health-promoting attributes: an overview. Curr Opin Food Sci 31:81–87
Nazzaro F, De Martino L, Fratianni F, De Feo V (2020) Essential oils from Mediterranean aromatic plants. In: Preedy VR, Watson RR (eds) The Mediterranean diet: an evidence-based approach. Academic Press, Cambridge, pp 555–564

Nikmaram N, Budaraju S, Barba FJ, Lorenzo JM, Cox RB, Mallikarjunan K, Roohinejad S (2018) Application of plant extracts to improve the shelf-life, nutritional and health-related properties of ready-to-eat meat products. Meat Sci 145:245–255

Pai S, Hebbar A, Selvaraj S (2022) A critical look at challenges and future scopes of bioactive compounds and their incorporations in the food, energy, and pharmaceutical sector. Environ Sci Pollut Res 29:35518–35541

Proestos C (2020) The benefits of plant extracts for human health. Foods 9(11):e1653

Rashidinejad A, Jafari SM (2020) Nanoencapsulation of bioactive food ingredients. In: Jafari SM (ed) Handbook of food nanotechnology. Academic Press, Cambridge, pp 279–344

Regnier T, Combrinck S, Du Plooy W (2012) Essential oils and other plant extracts as food preservatives. In: Bhat R, Alias AK, Paliyath G (eds) Progress in food preservation. Wiley, Chichester, pp 539–579

Ribeiro-Santos R, Andrade M, de Melo NR, Sanches-Silva A (2017) Use of essential oils in active food packaging: recent advances and future trends. Trends Food Sci Technol 61:132–140

Saad NY, Muller CD, Lobstein A (2013) Major bioactivities and mechanism of action of essential oils and their components. Flavour Fragr J 28(5):269–279

Sadgrove N, Jones G (2015) A contemporary introduction to essential oils: chemistry, bioactivity and prospects for Australian agriculture. Agriculture 5(1):48–102

Sanches-Silva A, Costa D, Albuquerque TG, Buonocore GG, Ramos F, Castilho MC, Machado AV, Costa HS (2014) Trends in the use of natural antioxidants in active food packaging: a review. Food Addit Contam Part A 31(3):374–395

Setyaningsih D, Siahaan DG (2018) The influence of essential oil addition to oxidative stability of palm-oil biodiesel. IOP Conf Ser Earth Environ Sci 209(1):e012007

Shah MA, Bosco SJD, Mir SA (2014) Plant extracts as natural antioxidants in meat and meat products. Meat Sci 98(1):21–33

Sharifi-Rad J, Sureda A, Tenore GC, Daglia M, Sharifi-Rad M, Valussi M, Tundis R, Sharifi-Rad M, Loizzo MR, Ademiluyi AO, Sharifi-Rad R, Ayatollahi SA, Iriti M (2017) Biological activities of essential oils: from plant chemoecology to traditional healing systems. Molecules 22(1):70–78

Stratakos AC, Koidis A (2016) Methods for extracting essential oils. In: Preedy VR (ed) Essential oils in food preservation, flavor and safety. Academic Press, Cambridge, pp 31–38

Taormina PJ (2021) Microbial growth and spoilage. In: Hardin MD, Taormina PJ (eds) Food safety and quality-based shelf-life of perishable foods. Springer, Cham, pp 41–69

Tian F, Decker EA, Goddard JM (2013) Controlling lipid oxidation of food by active packaging technologies. Food Funct 4(5):669–680

Tongnuanchan P, Benjakul S (2014) Essential oils: extraction, bioactivities, and their uses for food preservation. J Food Sci 79(7):R1231–R1249

Turek C, Stintzing FC (2013) Stability of essential oils: a review. Compr Rev Food Sci Food Saf 12(1):40–53

Ukrainets AI, Pasichniy VM, Zheludenko YV (2016) Antioxidant plant extracts in the meat processing industry. Biotechnol Acta 9(2):19–27

Ullah F, Ayaz M, Sadiq A, Ullah F, Hussain I, Shahid M, Yessimebkov Z, Adhikari-Devkota A, Devkota HP (2020) Potential role of plant extracts and phytochemicals against foodborne pathogens. Appl Sci 10(13):e4597

Veiga M, Costa EM, Silva S, Pintado M (2020) Impact of plant extracts upon human health: a review. Crit Rev Food Sci Nutr 60(5):873–886

Viji P, Venkateshwarlu G, Ravishankar CN, Gopal TS (2017) Role of plant extracts as natural additives in fish and fish products-a review. Fish Technol 54(3):145–154

Vintilă I (2017) Basic structure, nomenclature, classification and properties of organic compounds of essential oil. In: Hashemi SMB, Khaneghah AM, Santa-Ana AS (eds) Essential oils in food processing: chemistry, safety and applications. Wiley, Chichester, pp 173–190

Part IV
Potential of Nanomaterials in Food Packaging

Chapter 13
Engineered Nanomaterials in Food Packaging: Synthesis, Safety Issues, and Assessment

Jayasree T. Joshi, V. Harsha, Jobil J. Arakal, and Arya S. Krishnan

Contents

J. T. Joshi (✉) · V. Harsha
Agricultural and Food Engineering Department, Indian Institute of Technology Kharagpur, Kharagpur, West Bengal, India

J. J. Arakal
Department of Food Technology, Saintgits College of Engineering, Kottayam, Kerala, India

A. S. Krishnan
Department of Food Technology, Amal Jyothi College of Engineering, Kanjirappally, Koovappally, Kerala, India

J. A. Malik et al. (eds.), *Food Process Engineering and Technology*,
https://doi.org/10.1007/978-981-99-6831-2_13

Abstract Innovations in the development of nanotechnology have led to new opportunities in the food processing and packaging sectors. Engineered nanomaterials provide substantial advantages, such as: improving the barrier and physicomechanical characteristics of packaging materials; provide cost-effective packaging materials with better functionality, shelf-life, portability, and biodegradability. Nanoclay, silver, zinc oxide and titanium-based nanoparticles are often used in the packaging applications. In the synthesis of nanomaterials, bottom-up and top-down techniques are applied. Nanotechnology-based active and smart packaging materials release or absorb components into/from the food, thus providing information on the status of the food inside the package. With the introduction of nanotechnology in food packaging, a gap in understanding of nanoparticle behavior in food and environmental systems has arisen. The migration of nanoparticles from packaging materials into food, and the health hazards linked with it, are also major concerns. Hence, it is imperative to understand the toxicity, bioaccumulation, and long-term implications of engineered nanomaterials to the environment and public health. Common techniques for the assessment of engineered nanomaterials include: spectroscopy, microscopy, light scattering, chromatography, and electrophoretic techniques. The existing assessment techniques are challenged by particle size heterogeneity and aggregation, as well as nanoparticles (NPs) bio-uptake by different species. Research need be strengthened to develop standardized approaches for assessing nanoparticles in the complex media. The state-of-the-art in the synthesis of nanomaterials, safety issues and assessment approaches has been presented in this chapter.

Keywords Assessment · Engineered nanomaterials · Food packaging · Safety concerns · Synthesis

13.1 Introduction

Increased demand for new food items has prompted the food industry to seek new technologies to preserve its market dominance in the food processing sectors and to offer authentic, convenient, fresh, and delicious products with enhanced freshness and shelf-life (Ameta et al. 2020). Nanotechnology has emerged as a game-changer in the food processing industry, bringing about an industrial revolution in both developing and developed countries and making food processing investment profitable (Singh et al. 2017).

Figure 13.1 presents possible applications of nanotechnology in the food industry. Engineered nanomaterials (ENMs) are substances with particle sizes of 1–100 nm with distinct physicochemical properties from their original substance

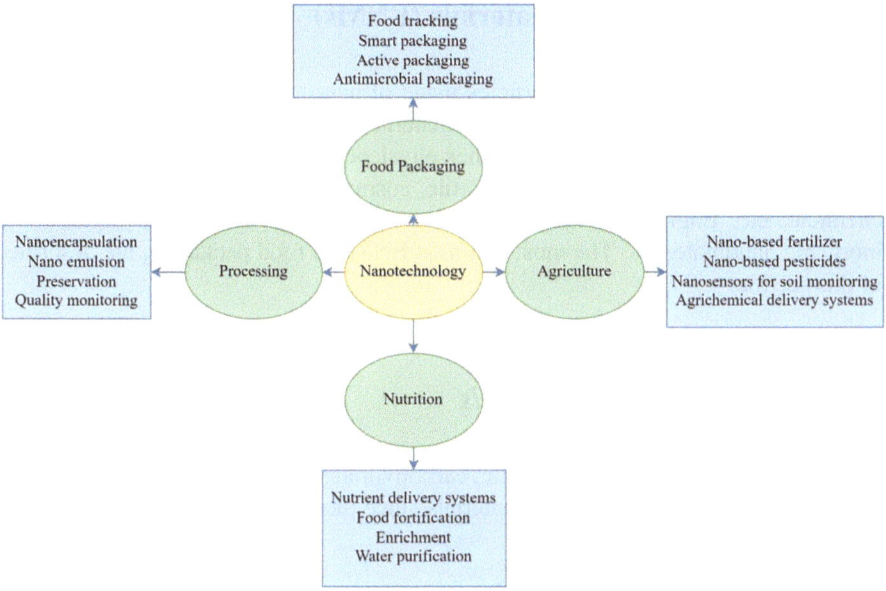

Fig. 13.1 Applications of nanotechnology in the food processing sectors

or their counterparts (FAO 2014). Their physicochemical characteristics are controlled by the quantum effects of their high surface area to mass ratio and extremely small size.

Advances in nanotechnology in the food industry are expected to multiply in next few years. Nanoscale polymers, nanoscale metal oxides (such as: iron and aluminum oxides, and titanium oxide), as well as polymeric nanocomposite materials are commonly used nanomaterials (NMs) in the electrical, construction, pharmaceutical, and medical sectors. ENMs are employed in photovoltaic cells, drug delivery nanovehicles, and immunological-sensing devices in the form of nanoshells, nanobranches, and nanocages (Kahru and Dubourguier 2010).

ENMs can enhance the quality and safety of foods. ENMs are exploited in food industry to improve the functional attributes of food products, to develop food delivery and monitoring systems, and to improve the physical-mechanical properties of packaging materials (Sanguansri and Augustin 2006; Weiss et al. 2006). Silver nanoparticles, nanoclay, nano-titanium dioxide, nano-zinc oxide, titanium nitride nanoparticles, and other engineered nanomaterials are used as functional additives in food packaging. Concerns for nanoparticle toxicity, long-term effects, and bioaccumulation have also been emerged. Owing to their extremely low concentration levels, detection and identification of ENMs are also challenging. The common techniques for the identification and characterization of ENMs are spectroscopy, microscopy, electrophoresis, and chromatography (Picó 2016).

This chapter provides an overview of applications of engineered nanomaterials in food packaging; as well as their synthesis and assessment techniques.

13.2 Engineered Nanomaterials (ENMs)

Engineered nanomaterials are particles made of nanoscale dimensions (1–100 nm) having distinctive characteristics. These materials have unique characteristics due to their size, shape, composition, and other physicochemical activities. These properties make promising applications in textile, cosmetics, biomedical, food, electrical, chemical, etc. Engineered nanomaterials are generally classified as organic and inorganic nanomaterials. The most common ENMs in food packaging are described in this section.

13.2.1 Organic Nanomaterials

Organic nanomaterials include protein, carbohydrates, and lipids, etc. These are less-toxic compared to inorganic nanomaterials since these are derived organically.

13.2.1.1 Lipid Nanomaterials (LNMs)

These nanomaterials have structures similar to nanoemulsions. They have a core and shell structure, in which the bioactive components, nutraceuticals, and lyophilic materials are encapsulated. The outer shell-layer contains protein, emulsifiers, polysaccharides, etc. The use of lipid-based nanomaterials in food packaging has the potential to extend shelf-life of foods. Mainly, these are used in edible packaging for food materials. LNM has less affinity towards water; therefore, they can resist moisture migration in edible films.

Nano lipids are lipid derivatives, such as: fatty acids, fat from vegetables and animals, including waxes in nano range dimensions. These include nano-emulsion, nano-liposomes, and polymeric nanomaterials. Xanthan gum can be incorporated with solid lipid nanomaterials to prepare the edible film for packaging or coating of fruits and vegetables (García et al. 2016). The inclusion of LNMs into polymer can improve the thermo-mechanical and moisture-barrier attributes of the polymer film.

Zambrano et al. (2013) incorporated solid-type LNMs into edible-coatings to increase the shelf-life of guava. They observed that lipid nanoparticles extended the storage time and preserved the fruit's quality. They also reported that excessive concentrations of solid-lipid nanoparticles resulted in physiological damage. The use of lipid-type nanomaterials (NMs) in the agricultural and food sectors has limitations because of insufficient information on safety and toxicity of NMs.

13.2.1.2 Carbohydrate-Based Nanomaterials

Carbohydrate-based nanomaterials are commonly available biopolymers in nature from renewable, and non-toxic sources. The various sources of nano-carbohydrates include starch, cellulose, and chitosan. The addition of these carbohydrates at the nano-level into the polymeric matrix form nanocomposites can improve the functional characteristic of the matrix. Starch nanoparticles have demonstrated improved properties in food packaging applications (Le Corre et al. 2010). Nano starch consists of amylose and amylopectin. Nano starch materials (also termed as starch nanocrystals) can be used as fillers to improve the thermo-mechanical properties (Dularia et al. 2019). Also, the addition of these materials makes better biodegradability and gas- and moisture barriers of packaging materials.

According to research studies, the structure and morphology of starch nanoparticles rely entirely on their botanical origin (Song et al. 2011). Nano starch materials were extracted from mung beans to develop composite films (Roy et al. 2020). The addition of these nanomaterials increased the mechanical and barrier characteristics of the coatings.

The nanoprecipitation process is used to manufacture cellulose, which is the most abundant natural polymer. It can be used as replacement for non-biodegradable materials. Nano cellulose offers good gas barrier properties due to formation of H-bonds, which make a strong network structure. Also, the inclusion of cellulose nanoparticles in a film improves the mechanical stability, tensile strength, and elastic modulus (Balakrishnan et al. 2018). Three forms of cellulose can be utilized for packaging, such as: cellulose nanocrystals (CNC), bacterial nano cellulose (BNC), and cellulose nanofibrils (CNF).

Sarwar et al. (2018) investigated physical, mechanical, and thermal properties of a polyvinyl alcohol nanocomposite film with nanocellulose and silver nanoparticles. The packaging films demonstrated strong thermal and mechanical qualities, and significant anti-bacterial activities against Gram-positive and Gram-negative bacteria.

Chitosan is the second most prevalent natural polymer next to cellulose, which may be found in the cell-walls of fungus, arthropod exoskeletons, and crustacean shells. The better antimicrobial properties, nontoxic benefits, and sufficient availability make its applications in the food packaging sector (Divya and Jisha 2018). Chitosan nanocrystals can be used as a biodegradable coating material to improve the shelf-life of fruits and vegetables. The addition of chitosan nanoparticles to PLA enhances moisture migration and oxygen barrier characteristics of a film.

13.2.1.3 Protein Nanomaterials

Protein-based nanomaterials are popular in variety of industries due to their bio-degradability, stability, and nontoxicity. These have applications in biomedical, drug delivery, bioimaging, biosensor, food packaging, etc. The primary protein structure is well-defined so that the protein-based nanomaterial can contribute to surface modifications. This property helps to make use of protein nanomaterials (such as: albumin, gelatin, sericin, collagen, keratin, and fibroin) for drug delivery. Because of the robust protein network, the integration of protein-based nanomaterials in packaging material can improve moisture barrier characteristics and mechanical strength (Zubair and Ullah 2020).

Plant-based (soy, zein, pea rice, wheat, lecithin) and animal-based (gelatin, silk, collagen, albumin, whey protein, casein, elastin) protein sources are available for the development of protein-based nanomaterials. Whey is a byproduct that is formed during the processing of cheese. Whey protein acts as a good biopolymer in edible packaging applications. Whey protein-based nanoparticles can be applied as an effective coating or biopolymer film in food packaging considering their high nutritional value and the ability to deliver active antimicrobial, anti-browning, and antioxidant components. These coatings can be effectively applied for fresh-cut fruits and vegetables, meat, and fishery products (Kandasamy et al. 2021).

Zein-based nanomaterials is another plant protein alternative for food packaging (Huang et al. 2018). The water vapor permeability of zein-based nanomaterial is comparably low compared to other biopolymer materials. The incorporation of zein-based nanomaterials with other organic and inorganic nanomaterials like isolates of whey protein and silver nanomaterials respectively can increase the mechanical and anti-microbial characteristics of polymer films.

Gelatin-based nanomaterials, which are made by partial hydrolysis of collagen, can also be effectively applied in the development of food containers. The addition of gelatin-based nano-materials enhanced the shelf-life of carrots, eggplants, orange, banana, and pineapple (Ramos et al. 2014).

Casein-based nanomaterials is another alternative, which has excellent gas barrier properties due to protein structure (Hydrogen bonding). Milk protein coatings can prevent oxidative degradation of sliced fruits and vegetables by inhibiting polyphenol oxidase (Porta et al. 2011). Edible films and coatings from casein-based nanomaterials can also retain variety of active substances that can improve the nutritional value of foods.

13.2.2 Inorganic Nanomaterials

Inorganic nanomaterials are metal or nonmetal compounds that are comparably stable materials. Commonly used inorganic nanomaterials include: silver, titanium dioxide, zinc oxide, and copper-based nanomaterials. The addition of these nano substances in polymer matrices can improve functional properties of the packaging materials and shelf-life of food materials.

13.2.2.1 Silver Nanomaterials (AgNMs)

Silver nanomaterials (AgNMs) are one of the commonly used inorganic nanomaterials. These are deployed in food processing, pharmaceutical, and healthcare industry. The physicochemical properties (optical, thermal, biological, and electrical) of silver nanomaterials have expanded their industrial applications. Silver nanomaterials have exhibited good antimicrobial properties against bacteria, viruses, yeast, and fungus (Martínez-Abad et al. 2012; Rai et al. 2009). The antimicrobial activity helps to keep safe and fresh food produced with enhanced shelf-life. Also due to this property, AgNMs are incorporated into both biodegradable and non-biodegradable matrices to develop better packaging materials. The inclusion of surface polishing of silver nanomaterials into packaging material makes it an active material for packaging applications (Toker et al. 2013).

The antimicrobial effect of nano-silver materials is higher compared to normal size silver materials because of their higher surface area that makes more interaction on the microbial cells (Emamhadi et al. 2016). It has also been observed that both insoluble and soluble forms of Ag are absorbed in the gastrointestinal system after consuming AgNMs (Marchioni et al. 2018). The maximum amount of silver that can migrate from packaging to food is 0.05 mg/l of water and 0.05 mg/kg in food (EFSA 2005).

13.2.2.2 Titanium Dioxide Nanomaterials (TiO_2NMs)

TiO_2 nanomaterials are non-toxic and thermo-stable metal nanomaterials (Huang et al. 2018). The widely used method for the production of titanium nanoparticles is sol-gel processing (Duncan 2011). The superior optical property widens the applications in different fields. They also have strong antibacterial characteristics, enabling widespread use in food packaging. In addition to its antibacterial properties, TiO_2 nanoparticles in packaging material can reduce the detrimental effects of UV radiation on food products (Duncan 2011). TiO_2 nanoparticles can act as oxygen scavengers under UV light, and can help to prevent food deterioration induced by oxygen (Xiao et al. 2004).

Packaging films containing TiO_2 can alter the structural, biochemical, and microbiological activities of the films. The coating of titanium dioxide nanomaterial to polypropylene film could suppress the growth of E. coli in lettuce (Chawengkijwanich and Hayata 2008). Compared to Gram-negative bacteria, Gram-positive bacteria are more resistant to TiO_2 nanoparticles (Xing et al. 2012). The inclusion of these nanomaterials with different substrates (such as: potato starch films, poly lactic acid (PLA)) and nano-fibers have improved the tensile property, and optical transparency of packaging films.

13.2.2.3 Zinc Oxide Nanomaterials (ZnONMs)

Zinc oxide is a non-metallic oxide with antimicrobial characteristics. Its excellent electrical and optical properties make it useful in multitude sectors, such as: electrical, medical, cosmetics, and material manufacturing. As ZnO-NMs are less expensive and less toxic than the silver nanoparticles, therefore they are employed in food packaging (Emamifar et al. 2010). They also act as an ultraviolet absorber, which helps to protect the food material that is sensitive to ultraviolet light.

The incorporation of zinc oxide nanomaterial into different packaging matrices improves their mechanical properties, and durability along with barrier and antimicrobial properties (Espitia et al.). According to researchers, adding zinc oxide nanoparticles to orange juice suppresses the growth of Lactobacillus plantarum (Emamifar et al. 2010). The use of these metal oxide nanoparticles in combination with PLA, HDPE, PVC, LLDPE, and other materials can improve the food packaging of various fruits and vegetables, and poultry products. These nanoparticles are also more effective in inhibiting the bacterial growth of E. coli, L. monocytogenes, S. aureus, Staphylococcus enteritids, and other food pathogens (da Silva et al. 2019).

13.2.2.4 Copper and Copper-Oxide Nanomaterials

Copper and copper-oxide nanomaterials have also applications in food packaging. These metal oxide nanoparticles also exhibit antimicrobial properties, making them efficient against bacteria, fungus, and viruses. Due to its lower reduction potential, copper nanomaterial oxidizes fast, creating copper ions with high mobility to interact with the cytoplasmic membrane (Conte et al. 2013; Llorens et al. 2012). These nanoparticles are also employed as water disinfectants, semiconductors, solar cells, and other uses in pharmaceuticals, chemical-engineering, and food industries. Copper nanoparticles combined with PLA can function as active packaging materials for dairy food products (Conte et al. 2013).

13.2.3 Biopolymer-Based Nanomaterials

Biopolymer nanomaterial consists of a bio-polymer material reinforced with organic or inorganic nanoparticles. The nanocomposite films produced from biopolymer-based nanomaterials exhibit excellent properties compared to native films. Renewable sources like protein and polysaccharides are widely used as organic matrices for the development of biopolymer nanocomposites (Borkotoky et al. 2019). The inclusion of these nanoparticles in packaging alters its characteristics and helps to improve the shelf-life of food materials, such as: dairy, fruits and vegetables, and meat products. The processing ability of biopolymers is low and their water barrier properties are also low. Therefore, the reinforced matrix should be selected to improve these properties. Polymer-based nanomaterials are polylactic acid (PLA), polyvinyl alcohol (PVA), Polyhydroxy alkenoates and related biopolymer blends.

PLA is prepared from lactic acid by condensation polymerization. It is considered safe to use for food applications by FDA (Food & Drug Administration). PLA possesses important characteristics, such as: renewability, crystallinity, biodegradability, and high mechanical strength. However, the brittleness of PLA polymers makes it difficult in the processing. Therefore, it is reinforced with inorganic nanomaterials (such as: silver, zinc oxide, and copper nanomaterials) to improve the rheological and thermo-mechanical properties of polymer layers (Dudkiewicz et al. 2011). Mulla et al. (2021) evaluated the effects of different PLA nanocomposites in food packaging. These authors observed that there was significant increase in the storage modulus, crystallinity, glass transition temperature, tensile strength, antibacterial activity, water-vapor transmission rate and permeability of oxygen in case of nanocomposites incorporated films.

PVA is prepared from polyvinyl acetate by hydrolysis; and it is non-toxic, recyclable, low-cost, and biodegradable. PVA has low permeability towards gasses, therefore it can act as a better barrier to oxygen. The structure of PVA is unstable, therefore reinforcement is required with other nanomaterials, such as: cellulose, antimicrobial agent, and plasticizers.

Polyhydroxy alkenoates are biodegradable hydrophobic polyester materials, which have applications in medical, cosmetic, packaging, tissue engineering, and other industrial sectors. One of the polyhydroxy alkanoates PHB (poly(3-hydroxy butyrate)) is used as a packaging film, but its crystallinity is high. Copolymer, PHBV (poly(3-hydroxy butyrate-co-3-hydroxy valerate)), also shows better handling characteristics with reduced crystallinity. The blend of PHB and PLA improves the ease of processing and packaging performance (Dudkiewicz et al. 2011).

13.3 Synthesis of Engineered Nanomaterials (ENMs)

13.3.1 Bottom-Up Approach

Structure building from atom to cluster to nanoparticles is a bottom-up or productive strategy. The most often utilized processing methods include: sol-gel method, chemical vapor deposition (CVD) method, spinning, pyrolysis, and biosynthesis. Sol-gel method is the most effective approach for the production of nanoparticles, particularly metal oxide nanoparticles. A chemical precursor of the nanoparticle is dissolved in water/ alcohol and is heated to form a gel in this process. Then, the gel is dried, pulverized, and calcined (Bokov et al. 2021).

In the chemical vapor deposition (CVD) technique, the precursors are allowed to disintegrate at high temperatures and in a vacuum chamber to form a thin film (Yanase et al. 2019). The spinning method involves spinning of the precursor solution in a revolving disc reactor. The spinning causes the particles or atoms to bind together, after which they are precipitated, dried, and stored. The attributes of synthesized nanoparticles will be determined by the operational factors (Hamouda et al. 2019), such as: feed flow rate, disc rotation speed, feeding position, disc surface properties, and liquid to precursor ratio.

Pyrolysis is the process of decomposing chemical precursors into various particles using proper heat treatment (Chaudhary et al. 2020). Nanoparticles may also be biosynthesized by employing bacteria, fungus, yeast, and plants (Ahmed et al. 2018; Nadagouda et al. 2014). When charged ions and cell-wall enzymes interact electrostatically, metallic ions are trapped on the surface of living cells. These metallic ions are then bio-reduced into their corresponding nanoparticles (Chaudhary et al. 2020).

13.3.2 Top-Down Approach

Mechanical milling, laser ablation, sputtering, electron beam lithography, and thermal breakdown are examples of top-down strategies, which minimize bulk materials to nanometric-scale components (Chaudhary et al. 2020). Nanoparticles synthesized using thermal breakdown processes involve breaking down the chemical bonds within the parent metals at various temperatures (Unni et al. 2017). The milling is a safe and energy-efficient method of producing nanoparticles of diverse sizes and dimensions. The parent material is milled into minute particles by breaking chemical bonds in this method (Wei et al. 2020).

Sputtering is a process in which energetic ions of gas or gaseous plasma impact a target material's surface, ejecting small particles from the surface (Hatakeyama et al. 2011). Laser ablation may be used to synthesize nanoparticles, such as: inorganic nanoparticles, carbon nanotubes, nanowires, and core-shell nanoparticles. This approach generates nanoparticles by nucleating and developing laser-vaporized species in a gas mixture. This method enables extremely fast vapor quenching of high-purity nanoparticles in the quantum size category of 10 nm (Chaudhary et al. 2020).

13.4 Packaging Applications of Engineered Nanomaterials (ENMs)

13.4.1 ENMs in Active Packaging

ENMs have turned into a noteworthy novel technology in the food packaging owing to its reduced weight, greater mechanical strength, increased heat resistance, and enhanced gas barrier qualities, UV radiation, volatiles from packaging films, and moisture. The effects and applications of nanoparticles in food packaging systems are being investigated (Costa 2010). Active nanoparticles embedded in polymer matrices can have two-fold advantage of improving the performance of food

Table 13.1 Recent studies on the application of ENMs in active packaging

Nanomaterial	Film material	Key insights	References
Laponite immobilized silver nanoparticles	Chitosan	Physicomechanical, water solubility, and OTR characteristics enhanced	Wu et al. (2018)
		Showed antimicrobial property to *S. aureus, A. niger, E. coli*, and *P. citrinum*	
Silver nanoparticles	Poly vinyl alcohol	Mechanical, optical, and water resistance properties enhanced	Mathew et al. (2019)
		Showed antimicrobial properties towards: *S. typhimurium and S. aureus*	
Silver–copper nanoparticles	Polylactide	Exhibited antimicrobial effects against Gram -positive and Gram-negative bacteria	Ahmed et al. (2018)
Titanium dioxide	Gluten and cellulose	Improved tensile strength, modulus of elasticity and water sensitivity observed	El-Wakil et al. (2015)
		Showed antimicrobial property to *Saccharomyces cerevisiae, E. coli and Staphylococcus aureus*	
Selenium-Silver nanoparticles	Furcellaran-gelatin	Improved mechanical, antimicrobial and barrier properties	Kandasamy et al. (2021)
		Showed antimicrobial property to *Staphylococcus aureus, and E. coli*	
Silver nanoparticle	Cellulose	Exhibited inhibitory effect to *E. coli* and *L. monocytogene*	Yu et al. (2016)
Silver-Copper nano particle	Fish gelatin	Improved mechanical and thermal properties, lower transparency	Arfat et al. (2017)
		Exhibited antimicrobial activity against Gram-positive and negative bacteria	

packaging material, as well as impart a value-addition (such as: antimicrobial, scavenger, antioxidant activity), thus promoting and prolonging the shelf-life of the end-product (Trovatti et al. 2012).

Active packaging utilizes nanomaterials that directly interact with food or the environment, providing better protection. ENMs in active packaging release or absorb substances from the food or food packaging headspace, thus extending the shelf-life. By integrating titanium nanoparticles into various polymers, Xiao et al. (2004) generated an oxygen scavenger film. The addition of oxygen scavengers prolonged the shelf-life of the package by enabling it to maintain extremely low oxygen levels. Table 13.1 summarizes some of the recent literature on the use of ENM in active packaging.

13.4.2 Role of ENMs in Intelligent Packaging

Intelligent packaging with nanosensors alerts customers about enzymatic reactions associated with molecular breakdown taking place in food, which renders the product unfit for human consumption. Gases and other enzymes can escape but cannot enter the package, extending the shelf-life of the product and reducing the use of chemical preservatives.

Smart packaging is a hybrid of active and intelligent packaging that simultaneously monitors and responds to changes in the product or surroundings (Vanderroost et al. 2014). The goal of ENM-based intelligent packaging is to utilize chemical or biosensors to assess the safety and quality of food as it moves from the producer to the consumer.

Sensory characteristics, leakages, microorganisms, gases, pH level, time, and temperature may be detected and analyzed using intelligent packaging. The features of intelligent packaging vary depending on the product being packed, such as medicines, beverages, food, and other home and health supplies (Kuswandi et al. 2011).

Intelligent packaging aids in the monitoring of a product along its supply chain, the management of the environment within and outside the box, and the communication of the product's status to its manufacturer, distributor, or consumer at any given time. Food contamination by microorganisms results in alterations in flavor, taste, and aroma, and also it can cause ill effects.

Another concern in packaged foods is the detection and mitigation of food adulterants (Sharma et al. 2017). Such challenges can be overcome by the use of various nano enabled sensors. Pathogenic microorganisms, contaminants, toxins, smell and taste are all detected by the sensor. The various applications of sensors in packaging are listed in Fig. 13.2.

13.4.3 Role of ENMs for Improving Antimicrobial Properties

The majority of ENMs have excellent antimicrobial characteristics. ENMs have been demonstrated to produce reactive oxygen species, which can impair enzymatic activity, damage cell organelles, and disrupt DNA synthesis (Yin et al. 2012). Nanoparticles also have a high positive zeta-potential, which makes it easier for them to interact with cell membranes, potentially disrupting the proper transmission pathway across the plasma membrane and resulting in cell death. Nanoparticles of titanium dioxide have also been shown to alter D-alanine metabolism in S. mutans (Kumar et al. 2017).

CuO-NPs have been shown to limit the enzymatic activity of nitrate reductase and nitrite reductase in Paracoccus denitrificans, resulting their inhibition (Čech-Barabaszová et al. 2020). Al-Naamani et al. (2016) studied the antibacterial properties of chitosan and chitosan-zinc oxide nanocomposite coatings on polyethylene

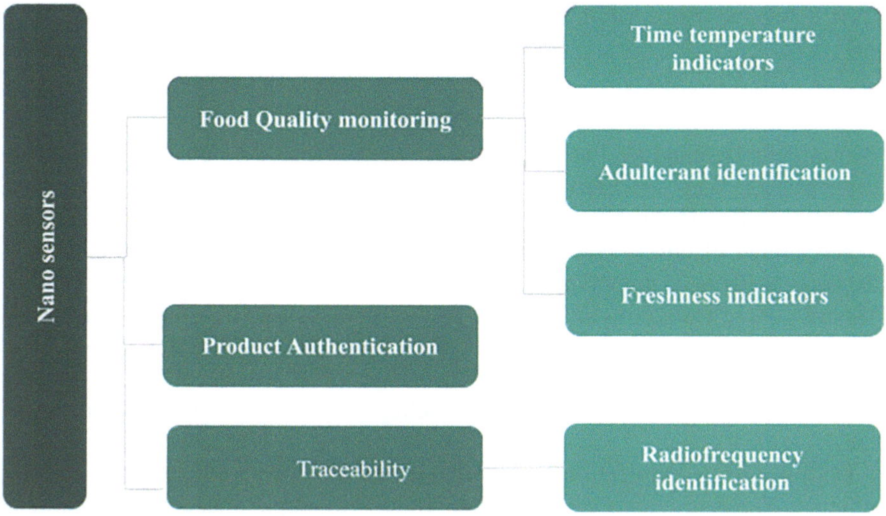

Fig. 13.2 Applications of nano-enabled smart sensors in food packaging

films. They observed that coating polyethylene with chitosan-ZnO nanocomposites could increase the antibacterial properties of the film. And they were able to obtain 99.9% inactivation of viable pathogenic microorganisms.

13.5 Safety Concerns

Nanomaterials in food packaging have mostly been employed to provide healthier, safer, and higher-quality products. However, there are safety concerns regarding the nanomaterial migration from packaging into the food matrix. The research on the feasibility and verification of the health effects are still underway. Nanomaterials of natural or synthetic origin in different geometries, size and textures are found in food matrices. They potentially alter food stability, digestibility, flow ability, microbial population, sensory and other organoleptic qualities (Chiralt et al. 2018).

Nanoscale particles are also produced during food processing activities, such as, size reduction and spray drying. Most analytical methods are sufficient to determine ENPs in a complex food system, whereas others need separation from the matrix to reduce interference from intrinsic nanoparticles. Literature reveals that ENMs modify their structure and composition in response to their surroundings (Picó 2016). In addition, interactions with other food constituents can alter their physicochemical properties. The possible risk of migration of ENMs from packaging material to food matrix is a serious issue (Kanmani and Rhim 2014). There is a scarcity of literature on potential consequences to human health and the environment. The research on bioconversion of ENMs after consumption is also limited. Therefore, there is a wide scope and urgent need for research in this area.

13.6 Assessment of ENMs (Fig. 13.3)

13.6.1 Microscopy

Microscopic techniques are commonly used for the characterization of nanoparticles, such as: Electron microscopy (EM), scanning probe and optical microscopy. EM is considered as the most promising technique in which a beam of accelerated electrons is used as the source of light (Barbacki 2007). Optical microscopy with a numerical aperture of one and wavelengths from 400 to 800 nm yield a resolution of 200 nm. The resolution of an electron microscope is substantially higher than that of a light microscope because the wavelength of an electron beam is much shorter than that of light. The wavelength of an electron microscope with a 100 kV acceleration voltage is 0.0037 nm (Dudkiewicz et al. 2011).

During electron interaction with particles, wavelengths are transmitted forming an image. The instruments are capable of producing sharper images of nanoparticles than a regular optical microscope (Tiede et al. 2008).

TEM is a useful tool in a variety of domains, not just nanotechnology. It has been used in material science, pharmaceutical research, medical applications, and semi-conductor research (Pakravan et al. 2012). The topological, structural, biochemical, and crystalline data of biopolymer frameworks can be obtained using TEMs. It is also being used to figure out how nanoparticles disperse in a lattice (Lloret et al. 2012).

By analyzing the damage induced by electron diffraction, TEM has shown to be a valuable tool in depicting core-shell and hollow nanoparticles and nanofibers. Although the working principles of both TEM and light microscope are the same, yet the major difference lies in the use of imaging sources. In TEM, high energy electron beams in the range of 80–300 KeV are used instead of a light, and high-resolution images are produced when the electron beams are passed through a thin layer of samples (Dudkiewicz et al. 2011).

When passing through the sample, electrons may get deflected and some parts may not be affected. The images are created by collecting emitted electrons using a camera. The darker portion of the image depicts places in the sample, where few or no electrons are transmitted; while the brighter portion depicts regions where more

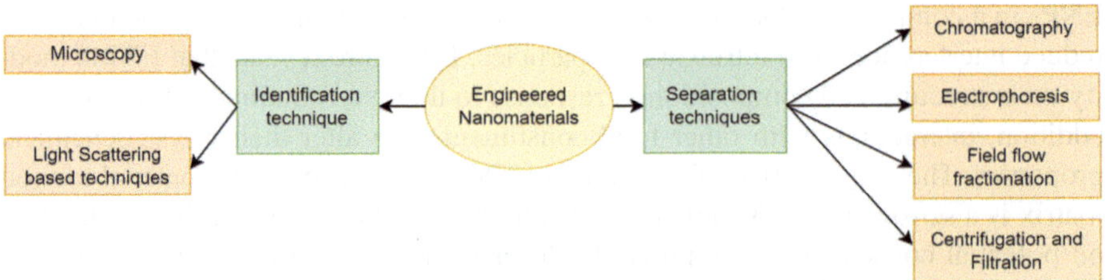

Fig. 13.3 Techniques for identifying and assessing the characteristics of ENMs

electrons are transmitted (Pakravan et al. 2012). Time-consuming and complicated sample preparation techniques are required for transmission electron microscopy (TEM).

Regular TEM models employ accelerated electron beams at high vacuum settings; however, certain materials are not viable under these circumstances. For the evaluation of such materials, cryo-TEM is used, in which the samples are quickly unclogged into a fluid solution and cooled to cryogenic temperatures (Frank 2013). Cryo-TEM has various advantages, such as: ability to analyze nanoparticles in liquid suspensions and investigate samples in vitreous ice (Kaushik et al. 2015; Majoinen et al. 2014). By using TEM, it is possible to determine elemental composition, size, shape and structure of nanoparticles (crystalline or amorphous). Although TEM can exhibit the morphology and crystalline structure, yet TEM chambers cannot be used for liquid samples; since the methodologies involved in the sample preparation like dehydration may lead to unwanted clumping of particles (Dudkiewicz et al. 2011).

EM imaging is done by examining sample particles with a high-energy electron beam, to study surface topography and other physical features of nanomaterials, such as: size, structure, dispersion, and agglomeration. Depending on the sample characteristics, resolutions up to 1 nm can be achieved with modern SEM imaging technique (Pang et al. 2013). However, organic samples need to be coated with an electrically conductive material, such as, metal or carbon to abstain from impacts of charging during imaging. SEM imaging secondary electrons are delivered due to the inelastic interactions of primary electrons and test particles, and are of low-energy contrasted with back-scattered electrons (Huang et al. 2015). Secondary electrons offer information on surface topography of test particles, whereas high-energy back-scattered electrons provide information on the elemental composition (Venkateshaiah et al. 2020). Number of detectors can be used to analyze secondary electrons, back-scattered electrons and electrons emitted from the surface of sample particles (Girão et al. 2017; Venkateshaiah et al. 2020).

Field Emission Scanning Electron Microscopy (FESEM) and Environmental Scanning Electron Microscopy (ESEM) are two variants of SEM, which can provide high resolution images compared with conventional SEM (Gupta et al. 2014). In FESEM, highly focused electron beams are produced with a field emission gun and this feature enables the development of high-resolution images, where information regarding elemental composition can be attained. FESEM does not require any sample preparation processes, such as: sample coating; and may operate at various temperatures. It can also function in a variety of gaseous environments allowing wet imaging of materials (Gupta et al. 2014; Yusof et al. 2014). TEM has a higher resolution than SEM, as TEM employs transmitted electrons to create an image, whereas SEM uses reflected electrons. As a consequence, TEM gives data on the sample's internal structure (such as: lattice parameters, morphology, and stress state), whereas SEM offers data alone on the sample's internal structure and morphology (Venkateshaiah et al. 2020).

Atomic Force Microscopy (AFM) provides basic information on surface characteristics with high spatial resolution (Huang et al. 2015). AFM is based on the basic principle of scanning probe electron microscopy (SPM), in which materials are studied using a sharp tip (which is typically a 3–6 microns long pyramid with a diameter of less than 10 Å). The sharp tip is attached at the free-end of a cantilever; and when the sample surfaces are scanned by the probe it gets deflected and the electrostatic forces are produced that are measured. Although AFM is suitable for all types of samples, not all sample variants produce the best quality results (Giessibl 2003; Santos and Castanho 2004).

13.6.2 Spectroscopy

Spectroscopic methods are used widely for nanoparticle characterization. Among the scattering techniques, light scattering techniques are more prominent and are classified into static light scattering (SLS), dynamic light scattering (DLS) and small angle neutron scattering (SANS) techniques (Tiede et al. 2008).

Dynamic scattering is also known as photon correlation spectroscopy and is widely used for the sizing of nanoparticles and to predict the state of agglomeration of nanomaterial in suspensions. However, DLS has several limitations, e.g., external interferences, such as: dust particles might alter the scattering intensity and hence the size of nanoparticles. Furthermore, as DLS is a quantitative approach, sizing data from compounds with different size categories are difficult to interpret (Dudkiewicz et al. 2011).

SLS, also known as Multi Angle Light Scattering (MALS), is a popular approach for determining physical properties of nanoparticles (such as: size and shape), and it may be used with other light scattering techniques, such as, DLS. SANS (small angle neutron scattering) may be utilized to characterize nanoparticles in both liquid and solid materials (Tiede et al. 2008).

Nanoparticle tracking analysis (NTA) is widely utilized for determining the hydrodynamic size of nanoparticles (Hou et al. 2018). In this technique, the motion of particles under Brownian movement are quantified with a video microscopy. The distance traveled by considerable amount of particles within a given time interval is measured and can be related to diameter data to obtain information on particle concentration and hydrodynamic size distribution. Nanoparticle tracking analysis (NTA) is extensively used in several nano toxicological studies to study size characterization.

Small angle X-ray scattering is another method to investigate the size distribution of macromolecular materials in monodisperse and polydisperse systems. For the analysis of elemental composition of nanoparticles, X-ray spectroscopy techniques based on fluorescence, absorption, and diffraction were applied (Tiede et al. 2008). X-ray absorption and emission spectroscopy are component specific methods that can give subjective data about metalloid species (Dudkiewicz et al. 2011; Huang et al. 2015).

13.6.3 Chromatography

Chromatographic techniques are tools for identifying and separating ENMs. Separation procedures are influenced by differences in molecular properties, such as: adsorption, partitioning, and affinity of the nanomaterials. The mixture to be separated is introduced in a stationary phase in chromatography. The stationary phase is slowly moved over by a mobile phase containing a suitable solvent. The components are then be transported individually based on their solvent solubility (Coskun 2016).

The widely employed chromatographic separation techniques for engineered nanomaterials include: size exclusion chromatography (SEC) and hydrodynamic chromatography (HDC). SEC has been used to analyze quantum dots, polystyrene nanoparticles, and carbon nanotubes because of its excellent separation efficiency (Huang et al. 2005; Krueger et al. 2005; Ziegler et al. 2005). Balthasar et al. (2005) used SEC to study the molecular weight distribution of antibody-modified gelatin nanoparticles and they observed that amount of solvent affects the separation quality. They also reported that nanoparticles produced with small amounts of solvent were more stable than those made with higher volumes.

Qiu et al. (2016) used a high-performance SEC in combination with a differential refractive index detector to separate and characterize starch nanoparticles; and they obtained a sample recovery rate of 95%.

Hydrodynamic chromatography (HDC) can be employed in the separation of nanomaterials. The chromatography column in HDC is filled with nonporous beads, which establish flow channels inside; and particles are separated by the flow velocity gradients surrounding the particle (Tiede et al. 2008). Helsper et al. (2013) used HDC to characterize liposome nanoparticles, and the chromatograms identified the particle sizes ranging from 20 to 350 nm, with retention time of around 0.05 min. Their studies also show that HDC can effectively separate silver, gold, and polystyrene nanoparticles with radii ranging from 10 to 60 nm from river water samples (Proulx and Wilkinson 2014).

13.6.4 Electrophoresis

Electrophoresis technique involves the movement of charged molecules in a solution in response to an electric field. In food analysis, gel electrophoresis and capillary electrophoresis are commonly used electrophoretic techniques.

Parak et al. (2003) used gel electrophoresis to separate nanogold conjugates, and they observed that smaller particles experience low-gel resistance and migrate more quickly. Hanauer et al. (2007) used gel electrophoresis technique to separate gold and silver nanoparticles of various sizes and shapes following the application of a charged polymer layer. They reported that nanorods with the highest aspect ratio were the slowest to respond to the electric field. For spherical particles, there was a noticeable trend of increased mobility with increasing size, but no such trend was observed for triangular-shaped particles. Capillary electrophoresis allows particles in different solutions to be separated according to their charge and size distribution.

Liu et al. (2005) employed capillary electrophoresis to separate silver nanoparticles. They found that the addition of an anionic surfactant (such as: sodium dodecyl sulphate) to the electrolyte will prevent the aggregation of silver particles during the separation process, which improves separation performance.

13.7 Conclusion

Nano-packaging of foods is a growing paradigm with a potential for improving shelf-life through its improved functional and barrier properties. Based on the foregoing discussion, it is concluded that nanotechnology-based solutions have enormous potential in the food industry, with the goal of improving conservation and quality. ENMs have been effective in active, smart, and intelligent packaging in different research studies. At the same time, further research is needed to understand the risks and long-term impacts of ENMs. Spectroscopy, microscopy, light scattering, chromatography, and electrophoretic methods are commonly used for evaluating engineered nanomaterials. Particle size heterogeneity and aggregation, as well as nanoparticle bio-uptake by diverse species, provide challenges to conventional evaluation approaches. The development of standardized methodologies, integrated approaches for analysis and quantification, hazardous effects, and bioaccumulation of ENMs require more investigation. Advances in these areas will be critical in acquiring a thorough understanding of nanoparticles and addressing related issues.

References

Ahmed AA, Hamzah H, Maaroof M (2018) Analyzing formation of silver nanoparticles from the filamentous fungus Fusarium oxysporum and their antimicrobial activity. Turk J Biol 42(1): 54–62

Al-Naamani L, Dobretsov S, Dutta J (2016) Chitosan-zinc oxide nanoparticle composite coating for active food packaging applications. Innov Food Sci Emerg Technol 38(part A):231–237

Ameta SK, Rai AK, Hiran D, Ameta R, Ameta SC (2020) Use of nanomaterials in food science. In: Ghorbanpour M, Bhargava P, Varma A, Choudhary DK (eds) Biogenic nano-particles and their use in agro-ecosystems. Springer, Singapore, pp 457–488

Arfat YA, Ahmed J, Hiremath N, Auras R, Joseph A (2017) Thermo-mechanical, rheological, structural and antimicrobial properties of bionanocomposite films based on fish skin gelatin and silver-copper nanoparticles. Food Hydrocoll 62:191–202

Balakrishnan P, Gopi S, Sreekala MS, Thomas S (2018) UV resistant transparent bionanocomposite films based on potato starch/cellulose for sustainable packaging. Starch 70(1–2):e1700139

Balthasar S, Michaelis K, Dinauer N, von Briesen H, Kreuter J, Langer K (2005) Preparation and characterization of antibody-modified gelatin nanoparticles as a drug carrier system for uptake in lymphocytes. Biomaterials 26(15):2723–2732

Barbacki A (ed) (2007) Mikroskopia elektronowa (Electron microscopy), 3rd edn. University of Technology Publishing House, Poznań, p 115

Bokov D, Turki A, Chupradit S, Suksatan W, Javed M, Shewael IH, Valiev GH, Kianfar E (2021) Nanomaterial by sol-gel method: synthesis and application. Adv Mater Sci Eng 2021:e5102014
Borkotoky SS, Ghosh T, Bhagabati P, Katiyar V (2019) Poly(lactic acid)/modified gum Arabic (MG) based microcellular composite foam: effect of MG on foam properties, thermal and crystallization behavior. Int J Biol Macromol 125:159–170
Čech-Barabaszová K, Holešová S, Bílý M, Hundáková M (2020) CuO and CuO/vermiculite based nanoparticles in antibacterial PVAc nanocomposites. J Inorg Organomet Polym Mater 30(10): 4218–4227
Chaudhary P, Fatima F, Kumar A (2020) Relevance of nanomaterials in food packaging and its advanced future prospects. J Inorg Organomet Polym Mater 30(12):5180–5192
Chawengkijwanich C, Hayata Y (2008) Development of TiO2 powder-coated food packaging film and its ability to inactivate Escherichia coli in vitro and in actual tests. Int J Food Microbiol 123(3):288–292
Chiralt A, Gonzales-Martinez C, Vargas M, Atares L (2018) Edible films and coatings from proteins. In: Yada RY (ed) Proteins in food processing, 2nd edn. Woodhead Publishing, London, pp 477–500
Conte A, Longano D, Costa C, Ditaranto N, Ancona A, Cioffi N, Scrocco C, Sabbatini L, Contò F, Del Nobile MA (2013) A novel preservation technique applied to Fiord latte cheese. Innov Food Sci Emerg Technol 19:158–165
Coskun O (2016) Separation techniques: chromatography. North Clin Istanbul 3(2):156–162
Costa C (2010) Nanocomposites to extend the shelf-life of ready to use fruit and vegetables. In: Proceedings of the 14th workshop on the developments in the Italian PhD research on food science technology and biotechnology, vol 18. University of Sassari, Oristano, 16 Sept 2009
Divya K, Jisha MS (2018) Chitosan nanoparticles preparation and applications. Environ Chem Lett 16:101–112
Dudkiewicz A, Tiede K, Loeschner K, Jensen LHS, Jensen E, Wierzbicki R, Molhave K (2011) Characterization of nanomaterials in food by electron microscopy. Trends Anal Chem 30(1): 28–43
Dularia C, Sinhmar A, Thory R, Pathera AK, Nain V (2019) Development of starch nanoparticles based composite films from non-conventional source-Water chestnut (Trapa bispinosa). Int J Biol Macromol 136:1161–1168
Duncan TV (2011) Applications of nanotechnology in food packaging and food safety: barrier materials, antimicrobials and sensors. J Colloid Interface Sci 363(1):1–24
EFSA (2005) Opinion of the scientific panel on food additives, flavorings, processing aids and materials in contact with food (AFC). Commission related to 7th list of substances for food contact materials. EFSA J 39:401–405
El-Wakil NA, Hassan EA, Abou-Zeid RE, Dufresne A (2015) Development of wheat gluten/ nanocellulose/titanium dioxide nanocomposites for active food packaging. Carbohydr Polym 124:337–346
Emamhadi MA, Sarafraz M, Akbari M, Fakhri Y, Linh NTT, Zuriaga E, Arias E, Peña C, Gavara R (2016) Properties of cellulose-starch nanocomposites obtained from bio-nanocomposite granules and thermoplastic films. Carbohydr Polym 151:913–921
Emamifar A, Kadivar M, Shahedi M, Soleimanian-Zad S (2010) Evaluation of nanocomposite packaging containing Ag and ZnO on shelf-life of fresh orange juice. Innov Food Sci Emerg Technol 11(4):742–748
FAO (2014) Report on the state of food and agriculture 2014 in brief. The State of Food and Agriculture (SOFA), pp 74–93. https://www.fao.org/3/i4036e/i4036e.pdf?
Frank J (2013) Story in a sample - the potential limitations of cryo-electron microscopy applied to molecular machines. Biopolymers 99(11):832–836
García CI, Hernández H, Quintanar D, Del Real L, de la Luz Zambrano M (2016) The evaluation of mechanical, thermal, optical and microstructural properties of edible films with solid lipid nanoparticles-xanthan gum stored at different temperatures and relative humidity. Food Bioproc Tech 9(10):1756–1768

Giessibl FJ (2003) Advances in atomic force microscopy. Rev Mod Phys 75(3):949–955

Girão AV, Caputo G, Ferro MC (2017) Application of scanning electron microscopy–energy dispersive X-ray spectroscopy (SEM-EDS). In: Rocha-Santos TA, Duarte AC (eds) Comprehensive analytical chemistry, vol 75. Elsevier, New York, pp 153–168

Gupta B, Tummalapalli M, Deopura BL, Alam MS (2014) Preparation and characterization of in-situ crosslinked pectin – gelatin hydrogels. Carbohydr Polym 106:312–318

Hamouda RA, Hussein MH, Abo-Elmagd RA, Bawazir SS (2019) Synthesis and biological characterization of silver nanoparticles derived from the cyanobacterium Oscillatoria limnetica. Sci Rep 9(1):1–17

Hanauer M, Pierrat S, Zins I, Lotz A, Sönnichsen C (2007) Separation of nanoparticles by gel electrophoresis according to size and shape. Nano Lett 7(9):2881–2885

Hatakeyama Y, Morita T, Takahashi S, Onishi K, Nishikawa K (2011) Synthesis of gold nanoparticles in liquid polyethylene glycol by sputter deposition and temperature effects on their size and shape. J Phys Chem 115(8):3279–3285

Helsper JP, Peters RJ, Brouwer L, Weigel S (2013) Characterization and quantification of liposome-type nanoparticles in a beverage matrix using hydrodynamic chromatography and MALDI-TOF mass spectrometry. Anal Bioanal Chem 405(4):1181–1189

Hou J, Ci H, Wang P, Wang C, Lv B, Miao L, You G (2018) Nanoparticle tracking analysis versus dynamic light scattering: case study on the effect of Ca2+ and alginate on the aggregation of cerium oxide nanoparticles. J Hazard Mater 360:319–328

Huang JY, Li X, Zhou W (2015) Safety assessment of nanocomposite for food packaging application. Trends Food Sci Technol 45(2):187–199

Huang XY, Mclean RS, Zheng M (2005) High-resolution length sorting and purification of DNA-wrapped carbon nanotubes by size-exclusion chromatography. Anal Chem 77(19): 6225–6228

Huang Y, Mei L, Chen X, Wang Q (2018) Recent developments in food packaging based on nanomaterials. Nanomaterials 8(10):830–835

Kahru A, Dubourguier HC (2010) From ecotoxicology to nanoecotoxicology. Toxicology 269(2-3):105–119

Kandasamy S, Yoo J, Yun J, Kang HB, Seol KH, Kim HW, Ham JS (2021) Application of whey protein-based edible films and coatings in food industries: an updated overview. Coatings 11(9): 1056–1061

Kanmani P, Rhim JW (2014) Physicochemical properties of gelatin/ silver nanoparticle antimicrobial composite films. Food Chem 148:162–169

Kaushik M, Basu K, Benoit C, Cirtiu CM, Vali H, Moores A (2015) Cellulose nanocrystals as chiral inducers: enantioselective catalysis and transmission electron microscopy 3-D characterization. J Am Chem Soc 137(19):6124–6127

Krueger KM, Al-Somali AM, Falkner JC, Colvin VL (2005) Characterization of nanocrystalline CdSe by size exclusion chromatography. Anal Chem 77(11):3511–3515

Kumar V, Guleria P, Mehta SK (2017) Nanosensors for food quality and safety assessment. Environ Chem Lett 15(2):165–177

Kuswandi B, Wicaksono Y, Abdullah A, Heng LY, Ahmad M (2011) Smart packaging: sensors for monitoring of food quality and safety. Sens Instrum Food Qual Saf 5(3-4):137–146

Le Corre D, Bras J, Dufresne A (2010) Starch nanoparticles: a review. Biomacromolecules 11(5): 1139–1153

Liu FK, Ko FH, Huang PW, Wu CH, Chu TC (2005) Studying the size/shape separation and optical properties of silver nanoparticles by capillary electrophoresis. J Chromatogr A 1062(1):139–145

Llorens A, Lloret E, Picouet PA, Trbojevich R, Fernandez A (2012) Metallic-based micro and nanocomposites in food contact materials and active food packaging. Trends Food Sci Technol 24(1):19–29

Lloret E, Picouet PA, Fernández A (2012) Matrix effects on the antimicrobial capacity of silver-based nanocomposite absorbing materials. LWT Food Sci Technol 49(2):333–338

Majoinen J, Haataja JS, Appelhans D, Lederer A, Olszewska A, Seitsonen J, Ikkala O (2014) Supracolloidal multivalent interactions and wrapping of dendronized glycopolymers on native cellulose nanocrystals. J Am Chem Soc 136(3):866–869

Marchioni M, Jouneau PH, Chevallet M, Michaud-Soret I, Deniaud A (2018) Silver nanoparticle fate in mammals: bridging in vitro and in vivo studies. Coord Chem Rev 364:118–136

Martínez-Abad A, Lagaron JM, Ocio MJ (2012) Development and characterization of silver-based antimicrobial ethylene–vinyl alcohol copolymer (EVOH) films for food-packaging applications. J Agric Food Chem 60(21):5350–5359

Mathew S, Snigdha S, Mathew J, Radhakrishnan EK (2019) Biodegradable and active nanocomposite pouches reinforced with silver nanoparticles for improved packaging of chicken sausages. Food Packag Shelf Life 19:155–166

Mulla MZ, Rahman MRT, Marcos B, Tiwari B, Pathania S (2021) Poly lactic acid (PLA) nanocomposites: effect of inorganic nanoparticles reinforcement on its performance and food packaging applications. Molecules 26(7):e1967

Nadagouda MN, Iyanna N, Lalley J, Han C, Dionysiou DD, Varma RS (2014) Synthesis of silver and gold nanoparticles using antioxidants from blackberry, blueberry, pomegranate, and turmeric extracts. ACS Sustain Chem Eng 2(7):1717–1723

Pakravan M, Heuzey MC, Ajji A (2012) Core–shell structured PEO - chitosan nanofibers by coaxial electrospinning. Biomacromolecules 13(2):412–421

Pang J, Liu X, Zhang X, Wu Y, Sun R (2013) Fabrication of cellulose film with enhanced mechanical properties in ionic liquid 1-allyl-3-methylimidaxolium chloride (AmimCl). Materials 6(4):1270–1284

Parak WJ, Pellegrino T, Micheel CM, Gerion D, Williams SC, Alivisatos AP (2003) Conformation of oligonucleotides attached to gold nanocrystals probed by gel electrophoresis. Nano Lett 3(1): 33–36

Picó Y (2016) Challenges in the determination of engineered nanomaterials in foods. Trends Anal Chem 84(part A):149–159

Porta R, Mariniello L, Di Pierro P, Sorrentino A, Giosafatto CVL (2011) Transglutaminase crosslinked pectin-and chitosan-based edible films: a review. Crit Rev Food Sci Nutr 51(3): 223–238

Proulx K, Wilkinson KJ (2014) Separation, detection and characterization of engineered nanoparticles in natural waters using hydrodynamic chromatography and multi-method detection (light scattering, analytical ultracentrifugation and single particle ICP-MS). Environ Chem 11(4):392–401

Qiu C, Yang J, Ge S, Chang R, Xiong L, Sun Q (2016) Preparation and characterization of size-controlled starch nanoparticles based on short linear chains from debranched waxy corn starch. LWT Food Sci Technol 74(2):303–310

Rai M, Yadav A, Gade A (2009) Silver nanoparticles as a new generation of antimicrobials. Biotechnol Adv 27(1):76–83

Ramos K, Ramos L, Cámara C, Gómez-Gómez MM (2014) Characterization and quantification of silver nanoparticles in nutraceuticals and beverages by asymmetric flow field flow fractionation coupled with inductively coupled plasma mass spectrometry. J Chromatogr A 1371:227–236

Roy K, Thory R, Sinhmar A, Pathera AK, Nain V (2020) Development and characterization of nano starch-based composite films from mung bean (Vigna radiata). Int J Biol Macromol 144:242–251

Sanguansri P, Augustin MA (2006) Nanoscale materials development from a food industry perspective. Trends Food Sci Technol 17(10):547–556

Santos NC, Castanho MA (2004) An overview of the biophysical applications of atomic force microscopy. Biophys Chem 107(2):133–149

Sarwar MS, Niazi MBK, Jahan Z, Ahmad T, Hussain A (2018) Preparation and characterization of PVA/nanocellulose/Ag nanocomposite films for antimicrobial food packaging. Carbohydr Polym 184:453–464

Sharma C, Dhiman R, Rokana N, Panwar H (2017) Nanotechnology: an untapped resource for food packaging. Front Microbiol 8:1735

da Silva BL, Abuçafy MP, Manaia EB, Junior JAO, Chiari-Andréo BG, Pietro RCR, Chiavacci LA (2019) Relationship between structure and antimicrobial activity of zinc oxide nanoparticles: an overview. Int J Nanomedicine 14:e9395

Singh T, Shukla S, Kumar P, Wahla V, Bajpai VK, Rather IA (2017) Application of nanotechnology in food science: perception and overview. Front Microbiol 8:1501

Song D, Thio YS, Deng Y (2011) Starch nanoparticle formation via reactive extrusion and related mechanism study. Carbohydr Polym 85(1):208–214

Tiede K, Boxall AB, Tear SP, Lewis J, David H, Hassellöv M (2008) Detection and characterization of engineered nanoparticles in food and the environment. Food Addit Contam 25(7):795–821

Toker RD, Kayaman-Apohan N, Kahraman MV (2013) UV-curable nano-silver containing polyurethane based organic/ inorganic hybrid coatings. Prog Org Coat 76(9):1243–1250

Trovatti E, Fernandes SC, Rubatat L, da Silva PD, Freire CS, Silvestre AJ, Neto CP (2012) Pullulan: nano-fibrillated cellulose composite films with improved thermal and mechanical properties. Compos Sci Technol 72(13):1556–1561

Unni M, Uhl AM, Savliwala S, Savitzky BH, Dhavalikar R, Garraud N, Arnold DP, Kourkoutis LF, Andrew JS, Rinaldi C (2017) Thermal decomposition synthesis of iron oxide nanoparticles with diminished magnetic dead layer by controlled addition of oxygen. ACS Nano 11(2):2284–2303

Vanderroost M, Ragaert P, Devlieghere F, De Meulenaer B (2014) Intelligent food packaging: the next generation. Trends Food Sci Technol 39(1):47–62

Venkateshaiah A, Padil VV, Nagalakshmaiah M, Waclawek S, Černík M, Varma RS (2020) Microscopic techniques for the analysis of micro and nanostructures of biopolymers and their derivatives. Polymers 12(3):512–516

Wei X, Wang X, Gao B, Zou W, Dong L (2020) Facile ball-milling synthesis of CuO/biochar nanocomposites for efficient removal of reactive red 120. ACS Omega 5(11):5748–5755

Weiss J, Takhistov P, McClements DJ (2006) Functional materials in food nanotechnology. J Food Sci 71(9):R107–R116

Wu Z, Huang X, Li YC, Xiao H, Wang X (2018) Novel chitosan films with laponite immobilized Ag nanoparticles for active food packaging. Carbohydr Polym 199:210–218

Xiao L, Green AN, Haque SA, Mills A, Durrant JR (2004) Light-driven oxygen scavenging by titania/polymer nanocomposite films. J Photochem Photobiol A Chem 162(2–3):253–259

Xing Y, Li X, Zhang L, Xu Q, Che Z, Li W, Li K (2012) Effect of TiO2 nanoparticles on the antibacterial and physical properties of polyethylene-based film. Prog Org Coat 73(2–3): 219–224

Yanase T, Miura T, Shiratori T, Weng M, Nagahama T, Shimada T (2019) Synthesis of carbon nanotubes by plasma-enhanced chemical vapor deposition using Fe1−xMnxO nanoparticles as catalysts: how does the catalytic activity of graphitization affect the yields and morphology? J Carbon Res C 5(3):46–51

Yin L, Colman BP, McGill BM, Wright JP, Bernhardt ES (2012) Effects of silver nanoparticle exposure on germination and early growth of eleven wetland plants. PloS One 7(10):e47674

Yu B, Liu T, Du Y, Luo Z, Zheng W, Chen T (2016) X-ray-responsive selenium nanoparticles for enhanced cancer chemo-radiotherapy. Colloids Surf B Biointerfaces 139:180–189

Yusof YM, Shukur MF, Illias HA, Kadir MFZ (2014) Conductivity and electrical properties of corn starch–chitosan blend biopolymer electrolyte incorporated with ammonium iodide. Phys Scripta 89(3):e035701

Zambrano ML, Mercado E, Ramirez P, Cornejo MA, Gutiérrez E, Quintanar D (2013) Use of solid lipid nanoparticles (SLNs) in edible coatings to increase guava (Psidium guajava L.) shelf-life. Food Res Int 51(2):946–953

Ziegler KJ, Schmidt DJ, Rauwald U, Shah KN, Flor EL, Hauge RH, Smalley RE (2005) Length-dependent extraction of single-walled carbon nanotubes. Nano Lett 5(12):2355–2359

Zubair M, Ullah A (2020) Recent advances in protein derived bionanocomposites for food packaging applications. Crit Rev Food Sci Nutr 60(3):406–434

Chapter 14
Nanomaterials in Foods: Recent Advances, Applications and Safety

Farhana Mehraj Allai, Khalid Gul, Z. R. Azaz Ahmad Azad, Insha Zahoor, Sadaf Nazir, and Arshied Manzoor

Contents

Abstract Nanotechnology has numerous promising food applications, such as: formulation of nanocapsules, nanofibers, nanocoatings, nanodispersions, nano-encapsulation, nanoemulsions, etc. Nanomaterials are being used due to their tunable physical, chemical and mechanical characteristics. Novel nanoparticles have been formulated for improving the safety and quality of foods. Shortage of food has evolved the idea of use of nanomaterials to preserve food so as to reach it globally. Food derived nanomaterials have significantly contributed in the development of nutraceutical and functional products. So, nanoparticles as well as nanomaterials have attracted several aspects and interest due to their unique characteristics. This chapter outlines the most recent trends in nanomaterials, its role at several levels of food industry, classification, potential application and the use of nanomaterials in food safety and toxicity.

F. M. Allai · Z. R. A. A. Azad (✉) · I. Zahoor · S. Nazir
Department of Post-Harvest Engineering and Technology, Faculty of Agricultural Sciences, Aligarh Muslim University, Aligarh, Uttar Pradesh, India

K. Gul
Department of Food Process Engineering, National Institute of Technology, Rourkela, Odisha, India

A. Manzoor
Department of Food Technology, School of Interdisciplinary Sciences & Technology, Jamia Hamdard, New Delhi, India

J. A. Malik et al. (eds.), *Food Process Engineering and Technology*,
https://doi.org/10.1007/978-981-99-6831-2_14

Keywords Delivery system · Food safety · Nanoparticles · Preservation · Packaging · Toxicity

14.1 Introduction

The rapid and increasing growth of nanotechnology during the last decades has found its way across the globe in almost all major sectors of industry. Among these major sectors, food is one of the promising domains that has revolutionized both traditional and commercial food products by using latest approaches such as, contact of nanoparticles with food (engineered nanoparticles), precision processing and handling, smart and active packaging, food additives, preservatives, encapsulation, formulation of novel products, nanosensors, food safety and security, etc. (Huang et al. 2010, Shukla 2012).

The word "nano" refers to the nanostructure material that should be something very tiny and small ranging between 1 to 100 nm (Alshammari et al. 2018). The intertwine of potential applications of food and nanoscience has led to a newer and beneficial field known as nanotechnology. The field of nanotechnology has significantly benefited food productivity and quality, development of functional food, developing and designing instruments to enhance food storage, quality, traceability, and shelf-life. In food processing, nanotechnology has helped in enhancement of bioavailability and absorption of essential micro as well as macro nutrients and other supplements with the use of nano-delivery technique (Sophiya and Thangjam 2018).

Another important aspect is food preservation, public acceptance and toxicity with novel nanoparticles that could bring risk fear among the consumers. The recognition of nanoscale chemicals together with food nanotechnology has attracted the attention of consumers regarding the potential toxicity.

Nano-based techniques are integral part of research and are facing various challenges in terms of cost economical operations to formulate safe edible products and non-toxic nano delivery systems. The application of this novel technology has gained the interest in food processing, preservation, storage and transportation in ensuring safety of food (Xiaojia et al. 2019). Also, several latest nano-based and innovative food packaging materials have been developed in association with nanosensors to check their beneficial effects on the environment and human health. Nanostructured materials are used for packaging of food, beverages, and health supplements, etc. Moreover, nanoscale ingredients also boost the color retention, aroma, flavor, increase of bioavailability of nutrients, extension of shelf-life, inhibition of microbial spoilage and enhance the durability of packaging materials (Dasgupta et al. 2015). Researchers have found that consumers accept most of nanoparticles in food packaging rather than in food matrices.

Nanotechnology as an emerging field has various potential applications as depicted in Table 14.1. It plays an important role in the nutraceutical and functional food sector. Encapsulation of certain components in nanostructured materials

Table 14.1 Potential application of nanotechnology

Agriculture	Food processing and engineering	Food packaging	Supplements
– Pesticides and other agro-chemicals of nano-formulated grade are used to enhance the efficacy in comparison to other formulations developed by conventional methods	– During processing and storage nano-based food additives have an application to improve the stability and bioavailability of nutrients present in the food, to increase the properties of the food product and to enhance the potency	– Detection of chemicals or food borne pathogens by attaching antibodies to fluorescent nanoparticles	– Nanosize powders to improve nutrient absorption
– For delivery of agrichemicals, pesticides, and fertilizers Nanocapsules are used	– Bioengineered nanocapsulated flavor enhancer	– Nano form materials covered with thin coating of metals such as metal oxides or aluminum in order to act as a barrier against the gases and vapors	– Cellulose nanocrystal composites as drug carriers
– Nanocapsules filled with tebuconazole and carbendazim are used as fungicide	– Nanofiltration in which nanopores can be used for the purification and softening of water	– Micro and nano size beads of polylactic and polycaprolactone incorporated into phase change materials to show enhanced thermal buffer capability of the packaging material	– For better stability, better absorption or targeted delivery nanoencapsulation of nutraceuticals is done
– For monitoring crop growth and soil conditions nanosensors are used	– In beverages synthetic amorphous silica (SAS) is used as a clarifying agent – SAS, calcium silicate, and dicalcium phosphate is also used in powdered food items as anticaking and free flow agent	– Nanocomposites of polymer nanoclay have beneficial effect in enhancing the thermal characteristics and tensile strength of packaging material	– Delivery of nutrients to cells more efficiently without affecting taste or color of food is done by coiled nanoparticles known as Nanocochleates
– Nanoparticles of inorganic metals and ashes of nanoscale have been developed for antifungal, antimicrobial and insecticidal properties	– Nanoparticles are used to bind and selectively remove the pathogens or any chemicals from food	– For plastic beer bottles, nanocrystals of silicate enriched with nanocomposite layers shows enhanced barrier characteristics	– For better absorption active molecules of vitamin sprays are dispersed into nanodroplets

(continued)

Table 14.1 (continued)

Agriculture	Food processing and engineering	Food packaging	Supplements
– Plant and animal pathogens are identified by nanosensors	– Iron of nano-structured form can prove to be beneficial for fortification of iron in certain foods	– Silicate nanoparticles are used for stronger, lighter and more heat-resistant films	
– Vaccines are delivered as Nanocapsules	– Smart delivery system of nano-based food additives for monitoring safety and quality of food	– Modified permeation behavior of foils	
– During storage pests are controlled by using NP loaded with garlic essential oil and coated with polyethylene glycol	– Nanotechnology is also used in diary industry like separation of milk proteins, standardization of milk and increasing the quality and safety of diary fluids	– Nano based food packaging material like biodegradable packaging, smart, active, intelligent and nanolaminates to protect food from microorganism and thus extends the shelf-life and quality of food	

(NSMs) can increase the solubility and bioavailability of several ingredients having nutritional characteristics like coloring agents, vitamins, minerals and thus during processing release of micro nutrients and other biologically active ingredients can take place. Nanofoods have two different forms: food packaging-nano-outside and food additives-nano-inside (Ranjan et al. 2014). Food packaging of nanoscale have significantly improved the shelf-life and quality of food product. In case of food additives, nanotechnologies have an application to influence the shelf-life, aroma, flavor and texture of food products.

14.2 Prospective and Current Applications of Nanomaterials in Food Sectors

In industrial applications, nanomaterials have gained significant importance in the food sector. These applications are usually associated with the improved characteristics as bioavailability or release of active substances. Nanotechnology can be employed at several stages during the production of food like processing, packaging and labeling, transportation, and quality checks during postproduction, etc. (Fig. 14.1). Nanoparticles are used generally in the form of nanocapsules, nanoemulsions, nanocomposites in food products. Nanotechnology is a vast field and has several classifications, each having specific functions (Table 14.2).

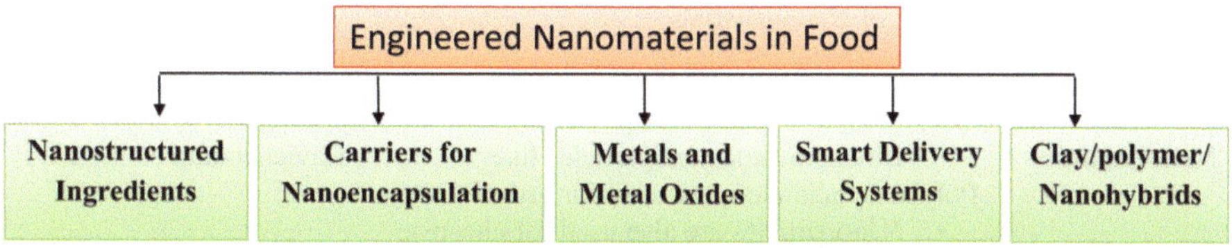

Fig. 14.1 Applications of nanotechnology at several stages of food production

14.3 Scope of Nanomaterials in Processing, Preservation, and Packaging of Foods

Nano sized particles are naturally present in the food matrices in the form of fats and carbohydrates (Peters et al. 2016). These ingredients can thus be used for the preparation of nanoemulsions, food grade polymers, and nanoencapsulates. Moreover, food industries and researchers have already identified various applications of nanotechnology in almost each and every sector of the food industry, starting from the field practices in agriculture (such as: vaccine delivery, identification of plant and animal pathogen, pesticides and fertilizers) till the product reaches to the consumer. In between, food processing, preservation and packaging are the most active areas in the food industry.

In food processing, nanotechnology has several functions such as incorporation of functional ingredients via nanoencapsulation into food, size reduction of food additives, in the form of nanoemulsions to develop aromas, flavor, consistency and texture to enhance the bioavailability of ingredients having nutritional benefits and thus making food products more healthy, hygienic and nutritive for end user (Chaudhry et al. 2017; Cushen et al. 2021).

Several nanocomposites were specifically used for packaging of food, such as: LDPE-low density polyethylene-silver/LDPE, CuO/LDPE, Zinc oxide/LDPE, polylactic acid, isotactic polypropylene, and gelatin. Many researchers have reported that polystyrene, chitosan and polyvinyl chloride as nanocomposite films are used to inactivated pathogens when Zinc oxide and Cu are bounded together. Other nanomaterials during the processing, preservation and packaging of food material are indicated in Table 14.3. In addition to various applications, food production chain has many stages:

- Delivery of agrochemicals
- Detection of pathogens by use of nanomaterials.
- Finally at any step of food chain, nanomaterials can be found as a result of infection, starting from pre-havest to postharvest where the use of nanoparticles is to enhance the effectiveness of the phytosanitary items.
- Moisture, temperature and time are monitored by biodegradable nanosensors.
- Preservation of food products is boosted.

Table 14.2 Classification of Nanomaterials

Type of Nanomaterials	Functions	References
Nanoparticles	• Biodegradable nanoparticles like Polylactic acid are often used for drug delivery • Nanoparticles are also used for encapsulation and delivery of micronutrients as well such as zinc, vitamins, iron and protein • For better efficiency polylactic acid is used in combination with polyethylene glycol and functional ingredient of nanoscale may be encapsulated and can be released at a specific site • Organic and inorganic nanomaterials especially silver nanoparticles, are used mostly due to its antimicrobial property	Ravichandran et al. (2011)
	• Gold nanoparticles are widely used as a detector or sensor • Nanoparticles of titanium dioxide are also recommended as food additive, disinfecting agent and flavor enhancer	Xiaojia et al. (2019)
Nanoemulsions	• Nanoemulsions have large surface area due to which it can interact with various biologically active components like lipase enzymes found in gastrointestinal tract that readily digests the tiny droplets of nanoemulsions	Zarif (2003)
	• Nanoemulsions are used for the formulation of flavored oils, sweeteners, and have antimicrobial property that are very effective on gram positive bacteria and thus can be used as disinfectant in case of packaging material	Wang et al. (2012)
	• Soybean and tributyl phosphate are used for the development of nanoemulsions. These nanoemulsions have then an application to reduce the growth of microorganisms and spoilage of food • It aids the uniform texture in ice cream and is used to improve the consistency, texture, aroma, and flavor in food product	Hogan et al. (2001) and Sanguansri and Augustin (2006)
Nano-Laminates	• Nanolaminates can be used in combination with edible films and coatings of various fruits and vegetables for packaging purpose • It also serves as carrier for various bioactive and functional components like antioxidants and antimicrobial agents • It improves the color, retains the flavor, and texture of food	Ranjan et al. (2014) and Sophiya and Thangjam (2018)
Nano-encapsulation	• During processing, packaging, storage and transportation nanoencapsulation increases the bioavailability, optimal hydration, quality and stability of the unstable food product	Assadpour and Jafari (2018)

(continued)

Table 14.2 (continued)

Type of Nanomaterials	Functions	References
	• It releases the bioactive compounds to the specific site at desired rate and time	Akhavan et al. (2018) and Katouzian et al. (2017)
	• Unwanted flavors in food are masked by encapsulation process.	Rezaeia et al. (2019)

- Processing: food additives like retention of color, enhancers of flavor, food supplements to enhance bioavailability of micronutrients, food structures-novel and latest like use of nanoemulsions to decrease fat content.
- Target delivery of nutraceuticals.

14.4 Nanomaterials: Research Advances in Foods

In twenty-first century, the major challenges across the globe are rapidly increasing growth of population, health of consumer, food supply, its safety and sustainability of bioresources. Latest techniques and novel approaches in science and technology are very much important for identifying the solutions to the above challenges like use of different nanomaterials as reported in Table 14.4. Nanotechnology is one of the recent techniques for adding nutritive ingredients to the food chain. Nanotechnology regulations are itself challenging as the word incorporates wide range of diverse technologies, materials and products with a broad spectrum of potential health benefits and risks (Marchant et al. 2009).

The recent advance in nanotechnology is dealt with deep insight and is forerunner of every researcher across the world. Nanotechnology is gaining its popularity tremendously because of its nanoscale structures or objects (Purohit et al. 2012). In finished product, fraction of nanoscale ingredients is used. It is the most common form, where the differences of the ingredients are made between the engineered nanomaterials and the manufactured nanomaterials. Engineered nanomaterials offers several benefits, such as: better physicochemical characteristics of material, surface charge, agglomeration state, good preservative properties, enhanced absorption capability of active compounds, and increased flavor or color retention. As the size of particle is reduced, the surface area per unit mass enhances in an indirect manner, i.e., inversely proportional. Surface area is increased by manipulating the porosity of nanoparticle. Large surface area can be achieved by formulating clusters or tiny particles to increase the surface to volume ratio of each component thus enhances the rate of dissolution, surface reactivity but modifies the toxicity profile.

Table 14.3 Nanoparticles in Food Industry

Domain	Nanomaterial	Function	References
Food processing	Nanoemulsions	– It can be used in sauces, processed meat, puddings, and mayonnaise to improve texture	Dongxiao and Geoffrey (2016)
	Nanoscale salt particles	– Less quantity of salt gives same taste to tongue receptor without compromising the taste	Dongxiao and Geoffrey (2016)
	Titanium dioxide	– Gives color to the food	Code of Federal Regulations (CFR) (2018)
	Nano sized micelles	– Used in milk protein casein to develop health promoting ingredient such as vitamin D_2	Semo et al. (2007)
	Nanoparticle encapsulated lycopene	– Used to enhance the bioavailability – Used in jam, jelly, paste sauce, tomato juice	Auweter et al. (1999)
	Nanotea enriched with nanoselenium	– Increases bioavailability and uptake of selenium	Ami et al. (2018)
	Silicon dioxide	– Flavor enhancer – Anticaking agent	European Commission (2008)
	Nanometric delivery system of essential oils- mixture of terpenes and d-limonene	– In fruit juices it is used to improve the antimicrobial activity	Donsì et al. (2011)
	Chitosan nanoparticle	– Increased microbial activity in fish finger	Abdou et al. (2012)
	Nanocapsulated oleoresin of paprika	Enhances sensory characteristics in chicken breast fillet	Yusop et al. (2012)
	PLGA nanoparticles	– Improves antimicrobial activity in meat products	Ravichandran et al. (2011)
Food Preservation	Gold based nanoparticle	– In milk, B1 aflatoxin can be detected.	Mao et al. (2006)
	Lipid based nanoencapsulation	– It is used widely to enhance the antioxidant activity by improving the solubility and bioavailability	Mozafari et al. 2006)
	Nano-silica with chitosan	– During prolonged storage it is used to preserve the fresh food.	Yu et al. (2012)
	-omega 3 nanoencapsulated	– Used in bread making – Acrylamide (degradation byproduct) was decreased strongly.	Gokmen et al. (2011)

(continued)

Table 14.3 (continued)

Domain	Nanomaterial	Function	References
Food Packaging	Nanoparticles and nanocomposites of silver	– Act as an antimicrobial – Zeolites- silver containing material are used for the disinfection purpose	US – FDA (2015)
	Active packaging coated with nano zinc oxide	– Improves shelf-life of fresh cut Fuji apples	Li et al. (2011)
	Nanoemulsions prepared from tributyl phosphate, nonionic or soybean	– Microbial growth is being checked and thus decreasing the spoilage of food.	Sanguansri and Augustin (2006)
	Zinc oxide nanoparticle	– Used in juice like orange and liquid egg albumin – Significantly decreases the yeast and mold counts, salmonella without affecting quality parameters.	Emamifar et al. (2011)
	Silver oxide	– Decreases growth of microorganisms in apples	Zhou et al. (2011)
	Magnesium oxide in combination with polylactic acid polymer	– It protects the material against the bacterial spoilage	Swaroop and Shukla (2018)
	Zerovalent iron particles	– It has an application as oxygen scavengers	Foltynowicz et al. (2017)
	Polyethylene with nano powder like kaolin, silver, anatase and rutile titanium dioxide	– Used to preserve fruits like strawberries	Yang et al. (2010)
	Silicate nanoparticles	– Limit flow of oxygen in packaging container and also check the moisture leakage so that product remains fresh for long time	Jones et al. (2008)

14.5 Safety and Characterization of Food Nanomaterials

Food nanoparticles have special characteristics including particle shape and size, surface area, charge, and topography, rate of dispersion, purity, solubility and reactivity etc. and have various advantages like (Akbari-Alavijeh et al. 2019; Assadpour and Jafari 2019a; Assadpour and Jafari 2019b; Assadpour and Jafari 2019c; Bata-Vidacs et al. 2013; Dima and Dima 2016; Esfanjani et al. 2017).

- Enhances the bioactive bioavailability.
- Microbial detection technique is incorporated with biofunctionalized nanomaterial that leads to real time, sensitive and quick mode of detection of cells of microorganisms.

Table 14.4 Nanomaterials in food research

Nanomaterial	Function	References
Aluminum oxide	– Optical sensing elements	Waterhouse et al. (2015)
Gold	– Textile coloration	Murdoch et al. (2011)
Iron	– Magnetic sensing when embedded in silicon dioxide	Leveneur et al. (2011)
Nano barcodes	– Authenticity of food, antimicrobial agent for packaging of food.	Dongxiao and Geoffrey (2016)
Nano filteration	– Used in purification of sugar, whey processing, water, juice clarification and concentration	Salehi 2014, Van Der Bruggen et al. (2003)
Nanoparticles based on starch or protein	– Delivery of bioactive substances to enhance efficacy – Encapsulating materials	Dongxiao and Geoffrey (2016)
Nanoparticles of aluminum oxide, silver and zinc oxide	– In silk worm it used to improve the grasserie and rice weevil disease caused due to *B. Mori* nuclear polyhedrosis virus (Baculovirus BmNPV) and *S. Oryzae* respectively	Goswami et al. (2010)
Nanosensors	– Smart sensing chips for safety analysis	Dongxiao and Geoffrey (2016)
Nickel oxide	– Co-catalyst	Chen et al. (2015)
Silver	– Antimicrobial agent – When meat starts to spoil, the chemical reaction between the hydrogen sulfide and silver nano layer takes place and thus changes the clolour of the label.	Smolander et al. (2004)
Titanium dioxide	– Coating of TiO_2 is used for sterilization of surfaces of food product or drinking water. – Whitening agent like candy and dairy	Chen et al. (2015)
Wheat straw or hulls of soy bean	– Electrochemical, mechanical and thermal properties which are further used for biosensors.	Dongxiao and Geoffrey (2016)
Zinc oxide	– UV sensing	Allen et al. (2011)

- Nanomaterial based biosensors have been made for detection of species of bacteria, formulation of microbial inhibition tests, and formation of bacterial biofilm.
- Nanomaterial in conjugation with biomolecules forms the bases for the manufacturing of novel nano based biorecognition devices.
- Nanomaterials also give protection to the bioactive compounds against the external parameters like light, temperature and oxygen.
- Passage of food through gastrointestinal tract is attributed by the different physicochemical parameters such as ionic strength, enzymes and pH.

- The addition of nanomaterials as nutraceuticals and an essential ingredient in the food matrices without affecting the sensory and physical properties.

Nanoscale materials are present in some traditional food items like mayonnaise contains fat globules and proteins are found in milk, but the application of these nanoparticles in food stuffs, water and agriculture may pose serious issue to human health

Nanoparticles have remarkable applications and benefits in the food industry, but it is a need of an hour to recognize the consumers concern towards the food safety and toxicity. Nanoparticles enter into the food system that influences the human health negatively by contaminating the environment and buildup of toxic compounds in the food chain. If consumers are directly exposed to nanoparticles associated with the matrix of food then it causes serious issue to the consumer's health (Klaine et al. 2008).

In food industry contaminants, pathogens and microbiological hazards are of prime concern. Silver nanoparticles are used in the food chain to kill or inactivate the microbes without adding any sort of deterioration to the food products (Handford et al. 2014). Moreover, nanomaterial - based assays and nanosensors can be used to diagnose the pathogens and contaminants present in the food that were caused due to the nanoparticles (Duncan 2011). There are few studies that have shown the presence of nanoparticles in food causing toxicity by analyzing food ingredients used during processing, preservation and packaging. Most importantly, nanoscale food additives can directly come in contact with different organs of our body that may result in increased level of risk depending on the quantity and concentration of additives used in food.

Food safety is a major public health issue across the globe. The main aim of the food safety is to ensure that the foods that are being consumed should not produce any harmful effect on the consumer health. During processing, packaging, storing and distribution the food must be protected from any sort of biological, chemical and physical hazards. Recent techniques in nanotechnology have drastically changed the applications being used in food industry for processing, safety and security, enhancing the shelf-life, decreasing the packaging waste and increasing the nutraceutical value (Wesley et al. 2014). In addition, nanoparticles including carbon nanotubes, quantum dots, metal nanoparticles, and other nanomaterials can be used for the manufacturing of nanosensors for analysis and detection of microorganisms and other tests for application of food safety as shown in Table 14.5.

14.6 Conclusion

Food grade nanomaterials have promising applications in food processing sector including as food additives, protection of nutraceuticals, enhancement of their bioavailability, production of smart packaging, and development of biosensors for diagnoses of microbes, pathogens and toxins. Public awareness about the benefits,

Table 14.5 Nanoparticles for detection of microbes

Nanoparticles	Function	References
Gold nanoparticles	*Salmonella enterica serotype Typhi*	Vikesland and Wigginton (2010)
Graphene oxide gold nanoparticles	– Detects *E. Coli* bacterial cell – Detects *Cronobacter sakazakii* bacterium that is poisonous to infants as it can be found in formula milk powder	Shukla et al. (2018)
MgO NPs	– It is used for the treatment of heavy metal infected waste water and bacterial infection	Cai et al. (2017)
Optical detection platform in collaboration with gold nanoparticles	– In pork samples it is used to detect *salmonella*	Oh et al. (2017)
Palladium nanoparticles in association with lateral flow immune strip	– Is used against *Klebsiella* that allows visualization and binding with specific bacteria	Tominaga (2018)
Quantum dot	*E. Coli O* 157:H7	Hahn et al. (2008)
Silver nanocolloids are used in surface enhanced Raman scattering	– Used to detect bacteria	Jarvis and Goodacre (2004), Kahraman et al. (2008)
Titanium dioxide	– To reduce some both gram positive and negative bacteria and few viruses through the formation of hydroxyl free peroxide	Zan et al. (2007)

challenges, and opportunities that comes in collaboration with this novel technique will bring better acceptance of the emerging technology. Despite the enhancing opportunities, the prime concern is about the health of a consumer that certainly consumes the food containing nanoparticles, which could make an entry to pass into the human tissues, resulting in the accumulation of contaminants and toxic substances that produces unintentional effects to the health of human body. However, the use of several nanoparticles having bioengineered characteristics poses harmful effects on the human health and the environment. At the same time, the intervention of nanomaterials in food industry has brought several changes in food processing, preservation and packaging. These modifications and changes have considerably enhanced the shelf-life of foods with reduced rate of food spoilage and deterioration. Knowing the properties of the nanoparticles in foods is an essential aspect for producers as they significantly influence the food safety and quality. That is the reason why more and more scientific research has been done in order to formulate the new, latest and novel techniques for detecting and controlling the negative aspect of nanomaterials.

References

Abdou ES, Osheba AS, Sorour MA (2012) Effect of chitosan and chitosan-nanoparticles as active coating on microbiological characteristics of fish fingers. Int J Appl Sci Technol 2(7):158–169

Akbari-Alavijeh S, Shaddel R, Jafari SM (2019) Nanostructures of chitosan for encapsulation of food ingredients. In: Jafari SM (ed) Biopolymer nanostructures for food encapsulation purposes, vol 1. Academic, San Diego, CA, pp 381–418

Akhavan S, Assadpour E, Katouzian I, Jafari SM (2018) Lipid nano scale cargos for the protection and delivery of food bioactive ingredients and nutraceuticals. Trends in Food Sci Technol 74: 132–146

Allen MW, Zemlyanov DY, Waterhouse GIN, Metson JB, Veal TD, McConville CF, Durbin SM (2011) Polarity effects in the x-ray photoemission of ZnO and other wurtzite semiconductors. Appl Phys Lett 98(10):101906

Alshammari FH, Jong-Hun H, Byung-Chun K, Irfan AR (2018) The intertwine of nanotechnology with the food industry. Saudi J Biol Sci 25(1):27–30

Ami P, Falguni P, Nihir S, Chandraprakash K (2018) Application of nanotechnology in the food industry: present status and future prospects. In: Grumezescu AM, Holban AM (eds) Handbook of food bioengineering: impact of nanoscience in the food industry (chapter 1), vol 12. Elsevier, New York, ny, pp 1–27

Assadpour E, Jafari SM (2018) A systematic review on nanoencapsulation of food bioactive ingredients and nutraceuticals by various nanocarriers. Crit Rev Food Sci Nutr 59(19): 3129–3151

Assadpour E, Jafari SM (2019a) An overview of biopolymer nanostructures for encapsulation of food ingredients. In: Jafari SM (ed) Biopolymer nanostructures for food encapsulation purposes, vol 1. Academic, San Diego, CA, pp 1–35

Assadpour E, Jafari SM (2019b) A systematic review on nanoencapsulation of food bioactive ingredients and nutraceuticals by various nanocarriers. Crit Rev Food Sci Nutr 59(19): 3129–3151

Assadpour E, Jafari SM (2019c) Nanoencapsulation: techniques and developments for food applications. In: Rubio AL, Rovira MJF, Sanz MM, Gomez-Mascaraque LG (eds) Nanomaterials for food applications (Chapter 3). Elsevier, New York, pp 35–61

Auweter H, Bohn H, Haberkorn H, Horn D, Luddecke E, Rauschenberger V (1999) Production of carotenoid preparations in the form of cold water-dispersible powders, and the use of the novel carotenoid preparations. US patent no. 5,968,251, USA

Bata-Vidacs I, Adanyi N, Beczner J, Farkas J, Szekacs A (2013) Nanotechnology and microbial food safety. In: Mendez-Vilas A (ed) Microbial pathogens and strategies for combating them: science, technology and education, vol 1. FormaTex Research Center Publisher, Badajoz, pp 155–169

Cai Y, Li C, Wu D, Wang W, Tan F, Wang X, Wong PK, Qiao X (2017) Highly active MgO nanoparticles for simultaneous bacterial inactivation and heavy metal removal from aqueous solution. Chem Eng J 312:158–166

Chaudhry Q, Castle L, Watkins R (2017) Nanotechnologies in food: what, why and how? In: Chaudhry Q, Castle L, Watkins R (eds) Nanotechnologies in food, 2nd edn. RCS Publication, London, pp 1–19

Chen WT, Chan A, Jovic V, Sun-Waterhouse D, Murai K, Idriss H, Waterhouse GIN (2015) Effect of the TiO2 crystallite size TiO2 polymorph and test conditions on the photo-oxidation rate of aqueous methylene blue. Top Catal 58(2–3):85–102

Chen WT, Chan A, Sun-Waterhouse D, Moriga T, Idriss H, Waterhouse GIN (2015) Ni/TiO2: a promising low-cost photocatalytic system for solar H2 production from ethanol-water mixtures. J Catal 326:43–53

Code of Federal Regulations (CFR) (2018) Electronic code of federal regulations. Food and drugs (part 73). The United States office of the Federal Register (OFR). US Government Publishing Office, Washington, DC

Cushen M, Kerry J, Morris M, Cruz-Romero M, Cummins E (2021) Nanotechnologies in the food industry—recent developments, risks and regulation. Trends in Food Sci Technol 24(1):30–46

Dasgupta N, Ranjan S, Mundekkad D, Ramalingam C, Shanker R, Kumar A (2015) Nanotechnology in agro-food: from field to plate. Food Res Int 69:381–400

Dima S, Dima C (2016) Protection of bioactive compounds. In: Apetrei C (ed) Frontiers in bioactive compounds: natural sources, physicochemical characterization and applications. Bentham Science Publishers, Sharjah, pp 255–301

Dongxiao S-W, Geoffrey W (2016) Recent advances in the application of nanomaterials and nanotechnology in food research. In: Grumezescu AM (ed) Novel approaches of nanotechnology in food: nanotechnology in the Agri-food industry, vol 1. Elsevier, New York, NY, pp 21–66

Donsì F, Annunziata M, Sessa M, Ferrari G (2011) Nanoencapsulation of essential oils to enhance their antimicrobial activity in foods. LWT 44(9):1908–1914

Duncan TV (2011) Application of nanotechnology in food packaging and food safety: barrier materials, antimicrobials and sensors. J Colloid Interface Sci 363(1):1–24

Emamifar A, Kadivar M, Shahedi M, Soleimanian-Zad S (2011) Effect of nanocomposite packaging containing ag and ZnO on inactivation of lactobacillus plantarum in orange juice. Food Control 22(3–4):408–413

Esfanjani AF, Jafari SM, Assadpour E (2017) Preparation of a multiple emulsion based on pectin-whey protein complex for encapsulation of saffron extract nano droplets. Food Chem 221:1962–1969

European Commission (2008) Regulation (EC) 1333/2008 of the European parliament and of the council of 16 December 2008 on food additives. European Commission, Brussels, p 18. https://eur-lex.europa.eu/LexUriServ/LexUriServ.do?uri=OJ:L:2008:354:0016:0033:en:PDF

Foltynowicz Z, Bardenshtein A, Seangerlaub S, Antvorskov H, Kozak W (2017) Nanoscale, zerovalent iron particles for application as oxygen scavenger in food packaging. Food Packag Shelf Life 11:74–83

Gokmen V, Mogol B, Lumaga R, Fogliano V, Kaplun Z, Shimoni E (2011) Development of functional bread containing nanoencapsulated omega-3 fatty acids. J Food Eng 105(4):585–591

Goswami A, Roy I, Sengupta S, Debnath N (2010) Novel applications of solid and liquid formulations of nanoparticles against insect pests and pathogens. Thin Solid Films 519(3): 1252–1257

Hahn MA, Keng PC, Krauss TD (2008) Flow cytometric analysis to detect pathogens in bacterial cell mixtures using semiconductor quantum dots. Anal Chem 80(3):864–872

Handford CE, Dean M, Henchion M, Spence M, Elliot CT, Campbell K (2014) Implications of nanotechnology for the Agri-food industry: opportunities, benefits and risks. Trends in Food Sci Technol 40(2):226–241

Hogan SA, McNamee BF, O'Riordan ED, O'Sullivan M (2001) Microencapsulating properties of sodium caseinate. J Agric Food Chem 49(4):1934–1938

Huang Q, Yu H, Ru Q (2010) Bioavailability and delivery of nutraceuticals using nanotechnology. J Food Sci 75(1):R50–R57

Jarvis RM, Goodacre R (2004) Discrimination of bacteria using surface enhanced Raman spectroscopy (SERS). Anal Chem 76(1):40–47

Jones N, Ray B, Ranjit KT, Manna AC (2008) Antibacterial activity of ZnO nanoparticle suspensions on a broad spectrum of microorganisms. FEMS Microbiol Lett 279(1):71–76

Kahraman M, Yazıcı MM, Sahin F, Culha M (2008) Convective assembly of bacteria for surface-enhanced Raman scattering. Langmuir 24(3):894–901

Katouzian I, Esfanjani AF, Jafari SM, Akhavan S (2017) Formulation and application of a new generation of lipid nano-carriers for the food bioactive ingredients. Trends in food. Sci Technol 68(Suppl. C):14–25

Klaine SJ, Alvarez PJJ, Batley GE, Fernandes TF, Handy RD, Lyon DY, Mahendra S, Mclaughlin MJ, Lead JR (2008) Nanomaterials in the environment: behavior, fate, bioavailability, and effects. Environ Toxicol Chem 27(9):1825–1851

Leveneur J, Waterhouse GIN, Kennedy J, Metson JB, Mitchell DRG (2011) Nucleation and growth of Fe nanoparticles in SiO2: a TEM, XPS, and Fe L-edge XANES investigation. J Phys Chem C 115(43):20978–20985

Li M, Zhu L, Lin D (2011) Toxicity of ZnO nanoparticles to Escherichia coli: mechanism and the influence of medium components. Environ Sci Technol 45(5):1977–1983

Mao X, Huang J, Leung MF, Du Z, Ma L, Huang Z, Gu L (2006) Novel coreshell nanoparticles and their application in high-capacity immobilization of enzymes. Appl Biochem Biotechnol 135(3):229–239

Marchant GE, Sylvester DJ, Abbott KW (2009) What does the history of technology regulation teach us about nano oversight? J Law Med Ethics 37(4):724–731

Mozafari MR, Flanagan J, Matia-Merino L, Singh H (2006) Recent trends in the lipid-based nanoencapsulation of antioxidants and their role in food. J Sci Food Agric 86(13):2038–2045

Murdoch M, Waterhouse GIN, Nadeem MA, Metson JB, Keane MA, Howe RF, Llorca J, Idriss H (2011) The effect of gold loading and particle size on photocatalytic hydrogen production from ethanol over au/TiO2 nanoparticles. Nat Chem 3(6):489–492

Oh SY, Heo NS, Shukla S, Cho HY, Vilian ATE, Kim J, Lee SY, Han YK, Yoo SM, Huh YS (2017) Development of gold nanoparticle aptamer-based LSPR sensing chips for the rapid detection of salmonella typhimurium in pork meat. Sci Rep 7(1):10130

Peters RJ, Bouwmeester H, Gottardo S, Amenta V, Arena M, Brandhoff P (2016) Nanomaterials for products and application in agriculture, feed and food. Trends Food Sci Technol 54:155–164

Purohit K, Khitoliya P, Purohit R (2012) Recent advances in nanotechnology. Int J Sci Eng Res Technol 1(11):1–11

Ranjan S, Dasgupta N, Chakraborty AR, Samuel SM, Ramalingam C, Shankar R, Kumar A (2014) Nanoscience and nanotechnologies in food industries, opportunities and research trends. J Nanopart Res 16(6):2464

Ravichandran M, Hettiarachchy NS, Ganesh V, Ricke SC, Surendra S (2011) Enhancement of antimicrobial activities of naturally occurring phenolic compounds by nanoscale delivery against listeria monocytogenes, Escherichia coli O157, H7 and salmonella typhimurium in broth and chicken meat system. J Food Safety 31(4):462–471

Rezaeia A, Fathib M, Jafari SM (2019) Nanoencapsulation of hydrophobic and low-soluble food bioactive compounds within different nanocarriers. Food Hydrocoll 88:146–162

Salehi F (2014) Current and future applications for nanofiltration technology in the food processing. Food Bioprod Process 92(2):161–177

Sanguansri P, Augustin MA (2006) Nanoscale materials development—a food industry perspective. Trends Food Sci Technol 17(10):547–556

Semo E, Kesselman W, Danino D, Livney YD (2007) Casein micelle as a natural nano-capsular vehicle for nutraceuticals. Food Hydrocoll 21(6):936–942

Shukla K (2012) Nanotechnology and emerging trends in dairy foods, the inside story to food additives and ingredients. Int J Nanosci Nanotechnol 1(1):41–58

Shukla S, Haldorai Y, Bajpai VK, Rengaraj A, Hwang SK, Song X, Kim M, Huh YS, Han YK (2018) Electrochemical coupled immune-sensing platform based on graphene oxide/gold nanocomposite for sensitive detection of Cronobacter sakazakii in powdered infant formula. Biosens Bioelectron 109:139–149

Smolander M, Hurme E, Koivisto M, Kivinen S (2004) PCT international patent application WO 2004/102185 A1. FAO, Geneva

Sophiya DL, Thangjam R (2018) Progress and challenges of nanotechnology in food engineering. In: Grumezescu AM, Holban AM (eds) Handbook of food bioengineering: impact of nanoscience in the food industry, vol 12. Academic, New York, NY, pp 87–112

Swaroop C, Shukla M (2018) Nano-magnesium oxide reinforced polylactic acid biofilms for food packaging applications. Int J Biol Macromol 113:729–736

Tominaga T (2018) Rapid detection of Klebsiella pneumoniae, Klebsiella oxytoca, Raoultella ornithinolytica and other related bacteria in food by lateral-flow test strip immunoassays. J Microbiol Methods 147:43–49

US – FDA (2015) Environmental decision memo for food contact notification No. 1569. Division of Biotechnology and GRAS Notice Review (HFS-255). FDA - Office of Food Additive Safety

Van Der Bruggen B, Vandecasteele C, Van Gestel T, Doyen W, Leysen R (2003) A review of pressure-driven membrane processes in wastewater treatment and drinking water production. Environ Prog 22(1):46–56

Vikesland PJ, Wigginton KR (2010) Nanomaterial enabled biosensors for pathogen monitoring- a review. Environ Sci Technol 44(10):3656–3669

Wang Y, Zhang Q, Zhang CL, Li P (2012) Characterization and cooperative antimicrobial properties of chitosan/nano—ZnO composite nanofibrous membranes. Food Chem 132(1): 419–427

Waterhouse GIN, Chen WT, Chan A, Jin H, Sun-Waterhouse D, Cowie B (2015) Structural, optical and catalytic support properties of gold-Al2O3 inverse opals. J Phys Chem C 119(12): 6647–6659

Wesley SJ, Raja P, Sunder Raj AA, Tiroutchelvamae D (2014) Review on nanotechnology applications in food packaging and safety. Int J Eng Res Technol 3:645–651

Xiaojia H, Hua D, Huey-min H (2019) The current application of nanotechnology in food and agriculture. J Food Drug Anal 27(1):1–21

Yang FM, Li HM, Li F, Xin ZH, Zhao LY, Zheng YH, Hu QH (2010) Effect of nano-packaging on preservation quality of fresh strawberry (Fragaria ananassa Duch. cv Fengxiang) during storage at 4oC. J Food Sci 75(3):236–240

Yu YW, Zhang SY, Ren YZ, Li H, Zhang XN, Di JH (2012) Jujube preservation using chitosan film with nano-silicon dioxide. J Food Eng 113(3):408–414

Yusop SM, O'Sullivan MG, Preub M, Weber H, Kerry JF, Kerry JP (2012) Assessment of nanoparticle paprika oleoresin on marinating performance and sensory acceptance of poultry meat. LWT 46(1):349–355

Zan L, Fa W, Peng TP, Gong ZK (2007) Photocatalysis effect of nanometer TiO2 and TiO2-coated ceramic plate on hepatitis B virus. J Photochem Photobiol 86(2):165–169

Zarif L (2003) Nano-cochleate cylinders for oral & parenteral delivery of drugs. J Liposome Res 13(1):109–110

Zhou L, Lv S, He G, He Q, Shi BI (2011) Effect of PE/Ag2O nanopackaging on the quality of apple slices. J Food Qual 34(3):171–176

Part V
Nanotechnology Applications in Food Technology

Chapter 15
Scope of Nanotechnology in Food Science and Food Engineering

Rajni Gautam and Nidhi Gaur

Contents

Abstract Nanotechnology (NT) has huge potential for providing new horizons and innovations in nanomedicine, nanoelectronics, nanosensing, nanomagnetics, etc. Nanotechnology has affected our lives. Nanotechnology has turned out to be a cardinal role player spanning from development, processing and packaging of foods to shelf-life extension. In this chapter, introduction of nanotechnology with recent advancements and applications of NT in food science/food industry are discussed along with risk assessment for use of nanomaterials at in food engineering and food products.

Keywords Food packaging · Food processing · Nanoemulsions · Nanoencapsulation · Nanomaterials · Nanosensors · Nanotechnology

R. Gautam (✉) · N. Gaur
School of Basic and Applied Sciences, K.R. Mangalam University, Gurgaon, Haryana, India
e-mail: rajni.gautam@krmangalam.edu.in

J. A. Malik et al. (eds.), *Food Process Engineering and Technology*,
https://doi.org/10.1007/978-981-99-6831-2_15

15.1 Introduction

Among all basic amenities of life, food is of utmost importance and hence agriculture equipped with latest technological advancements is the major factor for economic growth for most nations across the world. With ever rising population, one of the key challenges is to fulfil the increasing demands of masses leading to new techniques to enhance food productivity. Nanotechnology has played key role in the development, processing and packaging of food products leading to new products with improved taste, texture and increased shelf-life. Broadly, the application at nanoscale can be classified in two groups: nanoscale ingredients and nanosensors.

The nanoscale food ingredients involve food processing (Chellaram et al. 2014), e.g., nanostructures are used as additives, antimicrobial agents, fillers to increase the shelf-life of packaging material (Abbas et al. 2009). Nanomaterials are classified as a material having at least one dimension less than 100 nm in diameter. Figure 15.1 shows the nanoscale and typical nanomaterials.

Food nanosensing (Duncan 2011) strives to achieve improvised quality of food and safety evaluation. Major properties of materials at nanoscale (such as: greater strength, improved optical properties, superconductive nature and modified antimicrobial potential) have been widely investigated to improvise foods. In addition to these, various advances like microelectromechanical systems, DNA microarrays and microfluidics will further improvise the food processing for rapid sampling of contaminants, bioseparation of proteins, and delivery of nutrients.

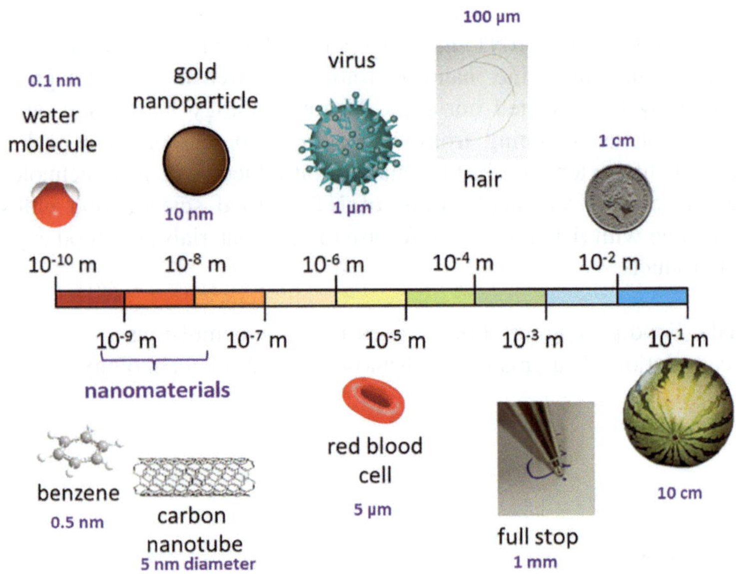

Fig. 15.1 Nanoscale and typical nanomaterials (*Source*: ChemBAM n.d.)

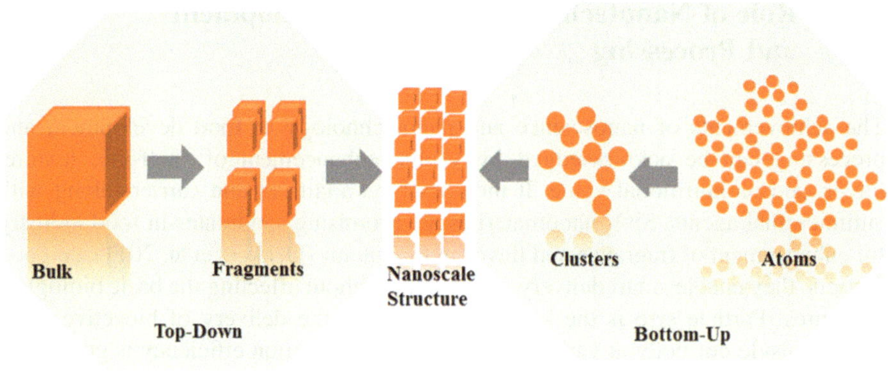

Fig. 15.2 Fabrication techniques of nanostructured materials (reprinted with permission from Gautam et al. 2022)

Nanomaterials have many superior properties over the bulk materials, e.g., larger surface area, high porosity, improved physical/chemical/mechanical properties. This difference is due to larger surface area for the same mass of material in bulk form. This makes the nanomaterials more chemically reactive. Also, quantum effects start appearing at this scale leading to different properties, such as: optical, electrical and magnetic properties.

Nanomaterials can be grouped as one dimensional (1D), two dimensional (2D) and three dimensional (3D). Example of 1D nanomaterial is a very thin surface coating. Nanowires and nanotubes come under 2D nanomaterials. Nanoparticles (NPs) refer to 3D nanomaterials. Nanotechnology is related to synthesis, characterization and applications of these nanomaterials by controlled manipulation of shape and size.

The basic techniques to synthesize nanomaterials are top-down and bottom-up (Fig. 15.2). Top-Down approach means going from higher to lower dimension, e.g., milling process in which nanostructured material is obtained by mechanical crushing of bulk material. Photolithography is also important Top-Down approach for nanostructured material fabrication. On the contrary, Bottom-up strategy refers to building up of nanostructures from the assembly of basic units, e.g., sputtering, Physical Vapor Deposition (PVD) and Chemical Vapor Deposition (CVD) and Sol-Gel Method.

This chapter focuses on recent advancements and contributions of nanotechnology (NT) in the food sector along with assessment of risks and health complications linked with the consumption of nanomaterials (NMs) in food items. The chapter includes: (1) nanotechnological advancements for food processing are discussed; (2) nanosensors used for detection of food quality; and (3) recent trends in food packaging. The chapter also concludes with the discussion of safety issues.

15.2 Role of Nanotechnology in Food Development and Processing

The achievements of nanoscience and nanotechnology in food development and processing can be acknowledged by visible enhancement of shelf-life, texture, flavor, and its nutritional value. It includes food additives and carriers along with antimicrobial agents. SiO_2 nanomaterials are promising candidates in food industry for enhancement of fragrance and flavor in food items (Dekkers et al. 2011). As food carriers, they enable smart delivery of nutrients without affecting the basic biological structures. Particle size is the key factor affecting the delivery of bioactive compounds inside our body at various sites; because absorption efficiency is greater for submicron particles. The main challenge is to deliver the required amount of compound at the fixed destination/target place along with maintenance of required levels for required period of time (Ezhilarasi et al. 2013). Therefore, role of nanotechnology can be summarized into formation of encapsulation, biopolymer matrices, and emulsions to offer efficient delivery systems.

15.2.1 Nanoencapsulation

To boost the food taste, techniques of nanoencapsulation are used to upgrade the release of flavor and to improve its retention for maintenance of culinary balance (Saifullah et al. 2019). Nanocapsules are used to carry out nanoencapsulation (Quintanilla-Carvajal et al. 2010) and to deliver the desired component. In addition, nanocapsules also help in preserving food by entrapping the unwanted components like odor (Assadpour and Mahdi-Jafari 2019).

The benefits of nanoencapsulation includes wide range from ease of handling, improved stability, enhanced flavor, moisture and controlled release driven by pH, to improved efficacy. Nanoemulsion keeps the texture and food taste intact and provides the basic benefit of protection and release of the required component at the desired place even in adverse situations. The degradation of bioactive components can be reduced by nanoencapsulation by acting and even providing a shield against exchange of gas and moisture even when the food is unpacked (Singh 2016). Nanocapsules have applications for delivering vaccines, fertilizers and pesticides to the plants, as food nutrients (such as: vitamin and minerals), and in enhancing the nutritional content of the food (Jampilek et al. 2019).

In our body, nanocapsules help in the assimilation of lipids, carbohydrates, and proteins, which are present in the food that is otherwise difficult to achieve owing to their low water solubility. Moreover, they enable these bioactive compounds to resist the acidic conditions prevalent in the intestine. Nanocapsules transport the food nutrients through the intestinal tract and also being subcellular in size they increase the solubility.

The primarily methods for preparation of nanocapsules are (Rong et al. 2011): nanoprecipitation method, emulsion-diffusion method, double emulsification method, emulsion-coacervation method, polymer coating method, and layer-by-layer method. Encapsulation of flavonoids and vitamins is suitably done through polymeric nanoparticles. Nanoliposomes (Hamadou et al. 2020) can play crucial role in controlling the delivery of enzymes, nutrients, vitamins, antimicrobials within the system. Yang et al. (2015) indicated that ferritin - trapped rutin can be made with enhanced stability by using ferritin nanocages compared with free-rutin. A biologically important plant pigment anthocyanins (with biological importance) is also used after its nanoencapsulation (Zhang et al. 2014). One major advantage of encapsulation lies in the immense health benefits in the form of edible capsules (Koo et al. 2005).

Now-a-days, nanoencapsulation of probiotics (Centurion et al. 2021) for designing of vaccines has gained importance due to their remarkable ability to deliver them to the gastrointestinal region in an efficient manner and in regulating the immune response. The examples of nanoencapsulation are: nanoparticles based on starch are used to deliver lipid bodies (Pradhan et al. 2015); Archaeosomes for antioxidants (Pradhan et al. 2015); and slow degradation of milk by using nanoencapsulating *α-tocopherol* in fat droplets (Relkin et al. 2008). Figure 15.3 shows common polymers that can be employed for encapsulation.

In food industry, some of the commercialized food products using nanoencapsulation by different companies are (Pradhan et al. 2015; Yalçinöz and Erçelebi 2018):

- Canola active oil used in nanoencapsulation of fortified *phytosterols*;
- Color Emulsion for *Beta-carotenal*, paprika nanoemulsions;
- Fortified Fruit Juice encapsulates fortified vitamin, *lycopene*;
- Nano-nutraceuticals to nanoencapsulation of nanoclusters to enhance the flavor of shake without adding sugar; and
- Plant-based lipid *Nano Resveratrol*.

15.2.2 Nanoemulsions

Nanoemulsions are used in dressing salad, flavored oils, sweeteners, and many other processed foods. They aid in the release of several flavors using various stimuli, such as: pH, heat, ultrasonic waves, etc. (Azrini et al. 2019). In addition to release, maintaining the flavor and prevention of degradation due to oxidation reactions is an added benefit. They can be formulated using high energy (i.e.: high pressure homogenization, ultrasound method and ultra-speed liquid coaxial jets) and low energy techniques (i.e. membrane emulsification, spontaneous emulsification, solvent displacement, emulsion inversion point, and phase inversion point) (Azrini et al. 2019).

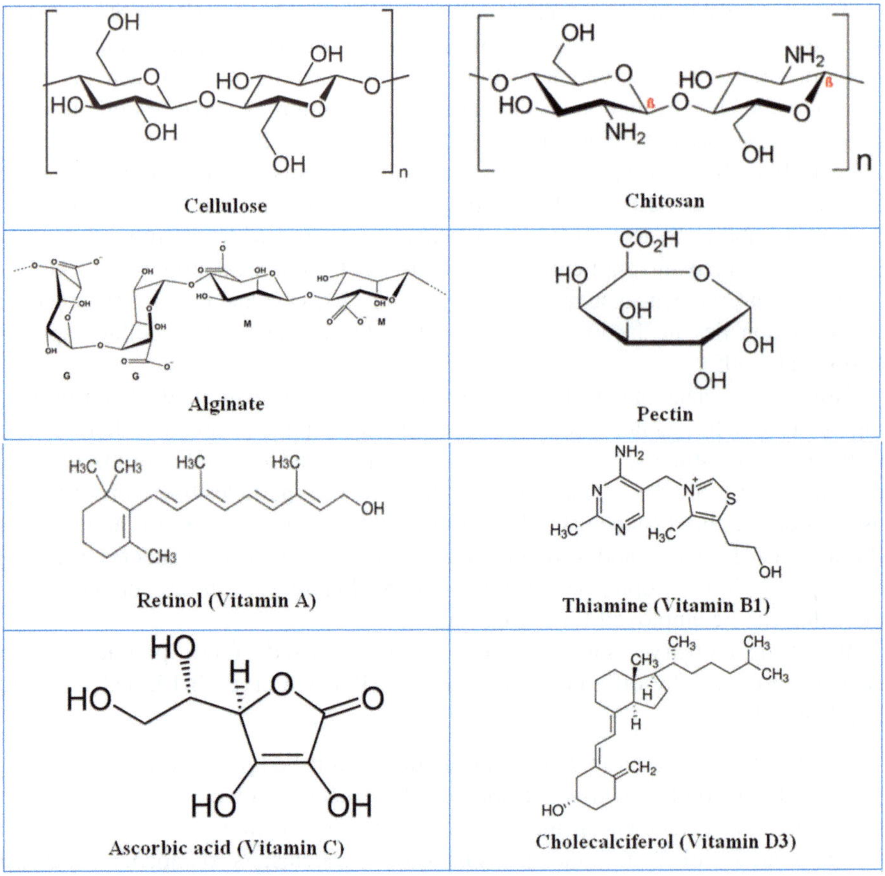

Fig. 15.3 Common biopolymers used in encapsulation

The nanoemulsions can also be produced by the diffusion process involving aqueous phase. The position of the lipophilic component in the nanoemulsion is controlled by surface activity, coefficient of oil-water partition, solubility, and melting point, etc. Nanoemulsion formation can lead to encapsulation of many lipophilic components, such as: flaxseed oil, *citral*, *tributyrin*, *β-carotene*, *capsaicin*, *coenzyme - Q* and many vitamins, which are oil soluble (Aswathanarayan and Vittal 2019).

Nanoemulsion offers more stability as compared to the conventional emulsions. Because of their ability to be digested easily and absorbed by the digestive enzymes due to larger surface area and small size, nanoemulsions are in high demand. Also being smaller in size and increased solubility, they can easily permeate the mucous layer in small intestine (Macierzanka et al. 2020). Nanoemulsions are also beneficial in improving the texture of the food product (e.g.: improvement in texture of ice cream by incorporating proteins (such as: egg, milk, and vegetable protein) or carbohydrates (such as: starch, pectin, alginate)) (Aswathanarayan and Vittal 2019). In addition, creaming and sedimentation can be reduced by using weighting agents, such as, Brominated vegetable oil. Availability of nutrients is greatly improved by the use of nanoemulsions (Chanamai and McClements 2000).

Hydrolyzed milk proteins like *α-lactalbumin* have emerged as an efficient carrier of drugs, nutrients, and supplements (Layman et al. 2018; Lin et al. 2021). For delivery of hydrophobic nutraceuticals, Casein micelles are commonly used (Sadiq et al. 2021) that show productive antimicrobial activity and effectiveness against Gram-positive organisms (Yang et al. 2014) enabling their use in food decontamination applications. Nanoemulsions also play cardinal role in stunted microbial growth (Javadzadeh and Azharshekoufeh 2017).

Self-assembled nanoemulsions can maintain the essence and aroma of the functional compounds by prohibiting enzyme growth and protecting from other deteriorating factors like heat, oxidation, and pH variation (Gupta et al. 2016). Vitamins (A, D and E3), *β-carotene*, and *lycopene* (Zhang et al. 2016) have been encapsulated by the self-assembled nanoemulsions. Nanoemulsions have gained popularity as phytochemicals (such as: *Carotenoids* and *polyphenols*), which are highly beneficial for lowering blood pressure, anticancer agents, regulating digestive system, strengthening the immune system, regulating cholesterol and sugar levels, and as antioxidants. However, shortage of bioavailability of phytochemicals has always been a challenge for the manufacturers. Nanoemulsions have brought a breakthrough in solving this problem as they can increase the solubility of the phytochemicals using efficient delivery systems. Bioavailability of phytochemicals can be increased due to smaller lipid size (Aqil et al. 2013).

15.3 Role of Nanosensing in Food Science and Food Engineering

Nanosensors can help to detect unwanted organisms (i.e.: pathogens) that grow in the food products (Kumar et al. 2020). This helps in controlling the food quality by detecting the spoilage by detecting the change in the color or smell of the food. The device contains a sensor to detect the variation in physical parameters, such as: heat, light, humidity, gas or any chemical. This change is converted into an electrical signal for data processing (Bouwmeester et al. 2009). Nanosensors are important to control the quality of food products and timely alerting the customers on the food safety status. This is done by detecting pesticides (Christopher et al. 2020), pathogens (Carlson et al. 2018), and toxins (Girigoswami et al. 2021) that may be present in the food products.

15.3.1 Gas Sensors

Gas sensors are generally employed to detect the presence of harmful gases like hydrogen, ammonia, etc. (Pereira et al. 2021). These nanosensors are preferred over conventional sensors because of their increased sensitivity and other novel attributes. Metals (such as: Palladium and gold (Korotcenkov and Cho 2017)) are generally employed in fabricating gas sensors. Conducting polymers can also be employed to make gas sensors.

Having conducting particles embedded in the insulating polymer matrix, the electrical, optical, and magnetic properties of conducting polymers are exploited and presence of the gas is recorded on conducting polymer - based sensor (Pirsa 2017). Nanosensors incorporating carbon black and polyaniline (Arshak et al. 2007) are employed for foodborne pathogen detection. With color changes, these sensors can alert on the quality of a food item without the requirement of sending it to the laboratory. Gas sensors and temperature-time indicators are commonly used sensors in food packaging. Other types include electronic noses, nano-test strips, array biosensors, and nanocantilevers. Behaving like a nose, Electronic-nose sensor uses different chemical sensors linked to data processing system (Tan and Xu 2020) to detect variety of odorants and chemicals on the food surface. Electronic tongue sensors are also based on similar principle and produce color when encountered with spoiled food material indicating that food is unfit for consumption (Baldwin et al. 2011).

15.3.2 Biosensors

There is a lot of literature available on nanobiosensors for pathogen detection (Helmke and Minerick 2006; Subramanian et al. 2006). Nanostructures (such as: thin films, nanofibers, nanorods and carbon nanotubes) have been investigated for biosensing applications (Nikolelis and Nikoleli 2018). There are two main units of any biosensor: bioreceptor and transducer (Fig. 15.4).

Examples of bioreceptors are antibody, DNA strain, enzyme, protein that are accountable for chemical or physical change. Examples of transducers are optical transducer, electrochemical detector or magnetic detectors (Shirzadfar and Khanahmadi 2018), which convert this change in chemical or physical property into an electrical signal. These physical or chemical changes occur when the analyte comes in contact of bioreceptor that is further converted into voltage or current signals to be read by an electronic device. Biosensors are safe devices to detect biological substances present in the food and have become an integral part of food industry (Shirzadfar and Khanahmadi 2018).

Optical immunosensors based on thin film are used for sensing of microbial cells. Gold - based nanoparticles are helpful for detection of toxins present in milk, such as: *aflatoxin B1* (Akgönüllü et al. 2021). Carbon nanotube - based biosensors

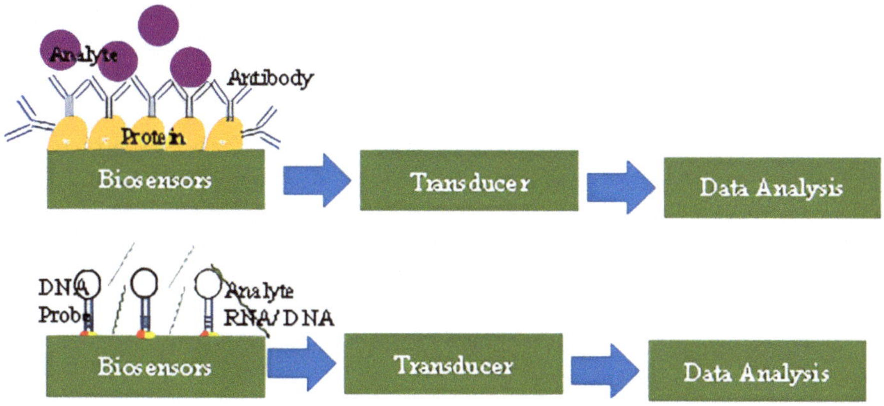

Fig. 15.4 Components of a biosensing module (reprinted with permission from Gautam et al. 2022)

have gained importance due to their simple nature, low-cost, easy and rapid detection of microorganisms and other toxins. Carbon nanotube - based sensors generally work on the principle of change in the conductivity when toxins or foreign antibodies are attached to nanotubes (Wang et al. 2009). Therefore, nanotechnology plays an indispensable role to track, trace and monitor the quality of food by detection of pesticides (Nile et al. 2020), pathogens and toxins (Stephen Inbaraj and Chen 2016).

15.4 Role of Nanotechnology in Food Packaging

One of the motivating factors for extensive research on nanotechnology in food packaging is to reduce the plastic consumption and to look for a better replacement having improved antimicrobial/antioxidant, thermal, and mechanical properties and which can be recycled. A good packaging material is a necessary requirement to maintain the standard of the food product. It must possess gas and moisture permeability also being strong and biodegradable (Couch et al. 2016). Nanotechnology-based smart food packaging has enhanced mechanical strength, improved antimicrobial or antioxidant properties, better barrier properties along with detection of safety status by employing nanosensors for microbes detection (Mustafa and Andreescu 2020).

Organic compounds (such as: essential oils and acids) have been examined by researchers to explore their use as antimicrobial packaging (Gálvez et al. 2007). However, because of their sensitivity to physical conditions like temperature and pressure, these organic compounds may be a misfit to food processing steps. The inorganic nanoparticles show better stability under extreme conditions and enhanced antibacterial activity and hence are becoming popular in packaging industry. With active antimicrobial packaging, any growth of microbes or unwanted microorganisms on food surfaces may be curbed (Aziz and Karboune 2018). Silver, copper, chitosan along with metal oxide nanoparticles (such as: Titanium oxide and Zinc oxide) are used as advanced packaging materials (Kumar and Anthony 2016).

New and advanced nanomaterials are being regularly developed for the applications in food industry to improve their packaging properties, i.e., mechanical strength, moisture and gas permeability, and heat resistance, etc. Polymer nanocomposites are also emerging to be a potential candidate in food packaging industry because of these properties (Cheikh et al. 2022).

15.4.1 Nanocomposites

Recently, the use of nanocomposites has dramatically increased for the purpose of food packaging. They consist of polymers and nanoparticles (Sinha and Okamoto 2003). They impart highly flexible chemical functionality and are best suitable for creating high barrier properties (Jordan et al. 2005). Prime role of nanocomposites lies in increasing the life span of a food item and keeping them fresh for long times. For example, the cans of carbon dioxide (CO_2) dissolved beverages if layered by nanocomposites may efficiently bring down leakage of CO_2 (Jordan et al. 2005).

Nanoclay is an important nanocomposite used as gas barrier. It is naturally occurring, stable and inexpensive aluminum silicate, also called as *phyllosilicate* (Bordes et al. 2009). Food shelf-life can also be extended by using nanocomposite and nanolaminates as they provide a shield from adverse thermal and mechanical conditions. Nano - based polymers have come up as cost effective packaging materials (Sarfraz et al. 2021). To provide fire resistance, make it lighter and stronger, the nanoscale fillers (for example: clay and silicate nanoplatelets, silica (SiO_2) chitin or chitosan) are mixed with the polymer matrix (Badar et al. 2021). Nanocomposite films with antimicrobial properties with fillers offer many advantages due to their integrated structure and barrier properties (Jamróz et al. 2019). Nanocomposite - based on nanoclays can be classified as: *intercalated* nanocomposites and *exfoliated* nanocomposites (Liu et al. 2006):

- *Intercalated* nanocomposites are the multi-layered polymeric structures created by the insertion of polymer chains into clay region (Camargo et al. 2009).
- Exfoliated nanocomposites are the result of randomly dispersed clay layers in the polymer matrix.

Some of the commercial nanoclays are Aegis, Imperm, and Durethan (Bumbudsanpharoke and Ko 2019). They have improved properties, such as: biodegradable property, low density, and transparent nature. For example, the barrier property of clay is markedly improved by Aegis, thereby preserving CO_2 in the carbonated beverages (Mohammad et al. 2022). *Polyamide* in the form of *Durethan* can improve the strength of paperboard containers for juice storage (Basumatary et al. 2022). *Nanocor*, another nanoclay based polymer, is used to manufacture plastic beverage bottles (Sujithra and Manikkandan 2019). Other packaging applications include nanocomposites having antimicrobial properties like Silver coated nanocomposites, *Zinc oxide* and *pediocin* embedded nanocomposite films (de Dicastillo et al. 2020).

Bionanocomposites are generally made from biodegradable and eco-friendly green polymers (such as: *starch* and *cellulose* derivatives) that are being explored as better layering materials for the packaging purpose (García-Guzmán et al. 2022). Guard IN Fresh is one of the commercially available nanocomposites which helps in preventing leakage of *ethylene* gas (Yucel 2016) and used for vegetables and fruits ripening. Nanocomposites have additional advantage of being biodegradable and environment friendly. Top Screen DS13 being water based is an example of easily recyclable nanocomposite (Vasile 2018).

NanoCeram PAC is another nanocomposite based coating material which absorbs components causing foul odor and bad taste (Pradhan et al. 2015). Nanoclays with enzymes incorporated in it also prove to be a good food packaging material (Bumbudsanpharoke and Ko 2019). Although it is not widely followed as enzymes are vulnerable to many factors like temperature, yet with unfavorable pH and presence of proteases with increased surface area and better transfer properties, they are still one of the widely used materials in the food packaging industry.

15.4.2 Edible Coatings

Edible films have marked a new revolution in the food packaging industry. Food items are covered in form of a layer of edible material, which acts as a barrier to food degrading factors. Nanolamination is an important commercially used methods to prepare edible films in the food sector. It comprises of multilayers with each layer having nano-dimensions (i.e. 1–100 nm) (Shit and Shah 2014). These edible coatings are present on food items like biscuits, chocolates, fruits, vegetables, meats, and other baked items, etc.; and provide protection from gas and moisture. Along with improvised textural properties, they add to the flavor and nutrients along with better antimicrobial properties.

Polysaccharides, *proteins*, and *lipids* are being used to prepare edible nanolaminates. Each of these has different ability to prevent spoilage; for example, protection against oxygen is better managed by polysaccharide and protein-based films while significant protection from moisture is provided by lipid-based nanolaminates, but they have reduced mechanical strength. Polysaccharides, proteins and lipids do not give all the essential benefits in an edible coating. To improvise, newer additives, e.g., polyols are under investigation. The coating can be done by either spraying the desired substances or dipping them into solutions (Shit and Shah 2014). The extent of adsorption depends on the nature of adsorbing surface and food. Different nanolaminate layers are formed by using several adsorbing substances like *polyelectrolytes* (*proteins and polysaccharides*), and *lipids*.

15.5 Safety Issues

There are many risk factors related to use of nanomaterials in food sector. The safety concern mainly arises because of their possible penetration into the food from packaging material and its consequent impact on consumer's health (Bradley et al. 2011). Even for those substances, which are considered safe, the risks of its nano-counterparts are hard to be ignored and needs further study and attention because of the difference in physical and chemical properties at this level.

Because of their small size, nanomaterials show increased risk of accumulation within the body. For example, it has been documented that human lung cells are adversely affected by silica nanoparticles (used as anti-caking agent) (Athinarayanan et al. 2014). In general, inorganic nanoparticles are considered nondigestible and organic nanoparticles are digestible. Both types affect the human body in a different way. Minerals and metals compose the nondigestible inorganic nanoparticles whereas lipids, proteins, and carbohydrates are digestible organic nanoparticles (McClements and Xiao 2017). In addition to this, because of different structure and arrangement there is different digestion and absorption rate, which may further decide the fate of nanoemulsion.

Extensive research is being carried out in this field to explore the possible implications of nanoemulsions on human body. Because of high absorption rate of nanoemulsions, they are readily adsorbed and can lead to accumulation in the biological system, which may be toxic for the body. Thus, high consumption of nanoemulsions is dangerous for health due to involvement of chemical substances (Wooster et al. 2017).

Emulsifiers are not effective in making nanoemulsions in their natural forms and if consumed at a very high level they affect adversely. Amongst different nanoemulsions, the ones with lipid base are best suited above all other conventional nanoemulsions but high lipid content may lead to cardiovascular disease and obesity in humans.

Normal working of cells gets disrupted as the nanoemulsions may attack the immune system by attaching to the cellular receptors and confounding them. Also, nanoparticles lead to degradation of proteins resulting in the disruption of working of cells. For example, Silver nanoparticles adversely affect the human system. They reduce the ATP content influencing lung fibroblast, leading to chromosomal abnormality and even cancer. The nanoparticles may cross the boundaries of the cells due to smaller size resulting in oxidative damage of cells and formation of free radicals. Nanomaterials used in Food packaging like carbon nanotubes migrate into food leading to toxicity in the human skin and lungs (Bradley et al. 2011).

To study the transport of nanoparticles in the food, a model was proposed by Cushen et al. incorporating silver and copper nanofillers (Cushen et al. 2014). One of the striking observations included the action of nanofiller percentage in the nanocomposites driving migration apart from particle size, or contact time, or temperature (Mahler et al. 2012). Further to ensure product quality, some standards for commercial products must be laid down by regulatory authorities considering environmental regulations and safety. The disposal waste of nanomaterials by industries also poses challenge for the natural ecosystems. Their accumulation in the soil or water may cause disruption in the biotic components of the region. Nanoparticles get dissolved in the water bodies and proving toxic to the animal and plant life (Matranga and Corsi 2012). In turn, this toxicity enters the food chain and subsequently the food web. The benthic species are affected if these nanoparticles get settled on the marine bed. They also result in stunted plant growth.

15.6 Conclusion

In this chapter, all the important contributions and achievements of Nanotechnology in the field of food science and engineering during last 10 years are discussed in detail. This is an attempt to summarize the role of nanomaterials in food sector by using nanosensors and smart food packaging. Nanotechnology has enabled food products with improved texture and taste along with long shelf-life. Nanocapsules have applications in healthcare by providing improved delivery of treatment drugs and necessary nutrients inside human body with numerous health benefits. New smart biodegradable packaging and edible coatings as packaging materials provide ecological benefits and show novel prospects for use in future. This chapter also documents the risks and harmful effects linked with the usage of nanomaterials in food products.

Acknowledgements Authors express their gratitude to K.R. Mangalam University for the necessary financial assistance.

References

Abbas KA, Saleh AM, Mohamed A, Mohd-Azhan N (2009) The recent advances in nanotechnology and its applications in food processing: a review. J Food Agri Environ 7(3):14–17

Akgönüllü S, Yavuz H, Denizli A (2021) Development of gold nanoparticles decorated molecularly imprinted–based plasmonic sensor for the detection of aflatoxin M1 in milk samples. Chemosensors 9(12):9120363

Aqil F, Munagala R, Jeyabalan J, Vadhanam MV (2013) Bioavailability of phytochemicals and its enhancement by drug delivery systems. Cancer Lett 334(1):32–38

Arshak K, Adley C, Moore E, Cunniffe C, Campion M, Harris J (2007) Characterisation of polymer nanocomposite sensors for quantification of bacterial cultures. Sensors Actuators B Chem 126(1):6–10

Assadpour E, Mahdi-Jafari S (2019) A systematic review on nanoencapsulation of food bioactive ingredients and nutraceuticals by various nanocarriers. Crit Rev Food Sci Nutr 59:3129–3151

Aswathanarayan JB, Vittal RR (2019) Nanoemulsions and their potential applications in food industry. Front Sustainable Food Syst 3:00095

Athinarayanan J, Periasamy VS, Alsaif MA, Al-Warthan AA, Alshatwi AA (2014) Presence of nanosilica (E551) in commercial food products: TNF-mediated oxidative stress and altered cell cycle progression in human lung fibroblast cells. Cell Biol Toxicol 30:89–100

Aziz M, Karboune S (2018) Natural antimicrobial/antioxidant agents in meat and poultry products as well as fruits and vegetables: a review. Crit Rev Food Sci Nutr 58(3):1194256

Azrini N, Azmi N, Elgharbawy AAM, Motlagh SR, Samsudin N, Salleh HM (2019) Nanoemulsions: factory for food, pharmaceutical and cosmetics. Food Processes 7(9):617–623

Badar R, Zahid AA, Yar M (eds) (2021) Chitin-based nanomaterials micro and nano technologies biopolymeric nanomaterials. Elsevier, New York, NY, pp 249–275

Baldwin EA, Bai J, Plotto A, Dea S (2011) Electronic noses and tongues: applications for the food and pharmaceutical industries. Sensors 11:04744–04766

Basumatary IB, Mukherjee A, Katiyar V, Kumar S (2022) Biopolymer-based nanocomposite films and coatings: recent advances in shelf-life improvement of fruits and vegetables. Crit Rev Food Sci Nutr 62:1912–1935

Bordes P, Pollet E, Avérous L (2009) Nano-biocomposites: Biodegradable polyester/nanoclay systems. Prog Polym Sci 34(2):125–155

Bouwmeester H, Dekkers S, Noordam MY, Hagens WI, Bulder AS, de Heer C, ten Voorde SECG, Wijnhoven SWP, Marvin HJP, Sips AJAM (2009) Review of health safety aspects of nanotechnologies in food production. Regul Toxicol Pharmacol 53(1):52–62

Bradley EL, Castle L, Chaudhry Q (2011) Applications of nanomaterials in food packaging with a consideration of opportunities for developing countries. Trends Food Sci Technol 22(11): 604–610

Bumbudsanpharoke N, Ko S (2019) Nanoclays in food and beverage packaging. J Nanomater 2019: 8927167

Camargo PHC, Satyanarayana KG, Wypych F (2009) Nanocomposites: synthesis, structure, properties and new application opportunities. Mater Res 12(1):1–39

Carlson K, Misra M, Mohanty S (2018) Developments in micro- and nanotechnology for foodborne pathogen detection. Foodborne Pathog Dis 15(1):16–25

Centurion F, Basit AW, Liu J, Gaisford S, Rahim MA, Kalantar-Zadeh K (2021) Nanoencapsulation for probiotic delivery. ACS Nano 15(12):18653–18660

Chanamai R, McClements DJ (2000) Impact of weighting agents and sucrose on gravitational separation of beverage emulsions. J Agric Food Chem 48(11):5561–5565

Cheikh D, Majdoub H, Darder M (2022) An overview of clay-polymer nanocomposites containing bioactive compounds for food packaging applications. Appl Clay Sci 216:106335

Chellaram C, Murugaboopathi G, John AA, Sivakumar R, Ganesan S, Krithika S, Priya G (2014) Significance of nanotechnology in food industry. APCBEE Procedia 8:109–113

Chem BAM (n.d.) Nanotechnology. https://chembam.com/definitions/nanotechnology/. Accessed 3 Aug 2023

Christopher FC, Kumar PS, Christopher FJ, Joshiba GJ, Madhesh P (2020) Recent advancements in rapid analysis of pesticides using nano biosensors: a present and future perspective. J Clean Prod 269:122356

Couch LM, Wien M, Brown JL, Davidson P (2016) Food nanotechnology: proposed uses, safety concerns and regulations. Agro Food Ind Hi Tech 27(1):36–40

Cushen M, Kerry J, Morris M, Cruz-Romero M, Cummins E (2014) Evaluation and simulation of silver and copper nanoparticle migration from polyethylene nanocomposites to food and an associated exposure assessment. J Agric Food Chem 62(6):1403–1411

de Dicastillo CL, Vidal CP, Falcó I, Sánchez G, Márquez P, Escrig J (2020) Antimicrobial bilayer nanocomposites based on the incorporation of as-synthetized hollow zinc oxide nanotubes. Nanomaterials 10(3):0503

Dekkers S, Krystek P, Peters RJB, Lankveld DPK, Bokkers BGH, Van Hoeven-Arentzen PH, Bouwmeester H, Oomen AG (2011) Presence and risks of nanosilica in food products. Nanotoxicology 5(3):393–405

Duncan TV (2011) Applications of nanotechnology in food packaging and food safety: barrier materials, antimicrobials and sensors. J Colloid Interface Sci 363(1):17–21

Ezhilarasi PN, Karthik P, Chhanwal N, Anandharamakrishnan C (2013) Nanoencapsulation techniques for food bioactive components: a review. Food Bioprocess Technol 6(3):628–647

Gálvez A, Abriouel H, López RL, Omar NB (2007) Bacteriocin-based strategies for food biopreservation. Int J Food Microbiol 120(1–2):51–70
García-Guzmán L, Cabrera-Barjas G, Soria-Hernández CG, Castaño J, Guadarrama-Lezama AY, Rodríguez S (2022) Progress in starch-based materials for food packaging applications. Polysaccharides 3(1):136–177
Gautam R, Abhilasha, Kumari N (2022) Nanosensor technology and their applications: an overview. Intl J Nanoparticles 14(2–4):238–259
Girigoswami A, Ghosh MM, Pallavi P, Ramesh S, Girigoswami K (2021) Nanotechnology in detection of food toxins—focus on the dairy products. Biointerface Res Appl Chem 11(6): 14155–14172
Gupta S, Khan S, Muzafar M, Kushwaha M, Yadav AK, Gupta AP (2016) Encapsulation: entrapping essential oil/flavors/aromas in food. In: Grumezescu AM (ed) Encapsulations, vol 2. Academic, San Diego, CA, pp 229–268
Hamadou HA, Huang WC, Xue C, Mao X (2020) Formulation of vitamin C encapsulation in marine phospholipids nanoliposomes: characterization and stability evaluation during long term storage. LWT 127:109439
Helmke BP, Minerick AR (2006) Designing a nano-interface in a microfluidic chip to probe living cells: challenges and perspectives. PNAS 103(17):6419–6424
Jampilek J, Kos J, Kralova K (2019) Potential of nanomaterial applications in dietary supplements and foods for special medical purposes. Nanomaterials 9(2):296
Jamróz E, Kulawik P, Kopel P (2019) The effect of nanofillers on the functional properties of biopolymer-based films: a review. Polymers 11(4):675
Javadzadeh Y, Azharshekoufeh L (2017) Therapeutic nanostructures for dermal and transdermal drug delivery. In: Grumezescu AM (ed) Nano-and microscale drug delivery systems: design and fabrication. Elsevier, New York, NY, pp 131–146
Jordan J, Jacob KI, Tannenbaum R, Sharaf MA, Jasiuk I (2005) Experimental trends in polymer nanocomposites—a review. Mater Sci Eng A 393(1–2):44–48
Koo OM, Rubinstein I, Onyuksel H (2005) Role of nanotechnology in targeted drug delivery and imaging: a concise review. Nanomedicine 1(3):4–11
Korotcenkov G, Cho BK (2017) Metal oxide composites in conductometric gas sensors: achievements and challenges. Sensors Actuators B Chem 244:182–210
Kumar VV, Anthony SP (2016) Antimicrobial studies of metal and metal oxide nanoparticles. In: Grumezescu AM (ed) Surface chemistry of nanobiomaterials: applications of nanobiomaterials. Elsevier Inc, New York, NY, pp 265–300
Kumar H, Kuča K, Bhatia SK, Saini K, Kaushal A, Verma R, Bhalla TC, Kumar D (2020) Applications of nanotechnology in biosensor-based detection of foodborne pathogens. Sensors (Basel) 20(7):1966
Layman DK, Lönnerdal B, Fernstrom JD (2018) Applications for a-lactalbumin in human nutrition. Nutr Rev 76(6):4–11
Lin T, Meletharayil G, Kapoor R, Abbaspourrad A (2021) Bioactives in bovine milk: chemistry, technology, and applications. Nutr Rev 79(Suppl 2):48–69
Liu J, Boo WJ, Clearfield A, Sue HJ (2006) Intercalation and exfoliation: a review on morphology of polymer nanocomposites reinforced by inorganic layer structures. Mater Manuf Process 21(2):143–151
Macierzanka A, Ménard O, Dupont D, Gutkowski K, Staroń R, Krupa L (2020) Colloidal transport of lipid digesta in human and porcine small intestinal mucus. Food Res Int 138(Part A):109752
Mahler GJ, Esch MB, Tako E, Southard TL, Archer SD, Glahn RP, Shuler ML (2012) Oral exposure to polystyrene nanoparticles affects iron absorption. Nat Nanotechnol 7(4):3–11
Matranga V, Corsi I (2012) Toxic effects of engineered nanoparticles in the marine environment: model organisms and molecular approaches. Mar Environ Res 76:32–40
McClements DJ, Xiao H (2017) Is nano safe in foods? Establishing the factors impacting the gastrointestinal fate and toxicity of organic and inorganic food-grade nanoparticles. NPJ Sci Food 1(6):41538

Mohammad ZH, Ahmad F, Ibrahim SA, Zaidi S (2022) Application of nanotechnology in different aspects of the food industry. Discover Food 2(12):44187

Mustafa F, Andreescu S (2020) Nanotechnology-based approaches for food sensing and packaging applications. RSC Adv 10(33):19309–19336

Nikolelis DP, Nikoleli GP (eds) (2018) Nanotechnology and biosensors, Ist edn. Elsevier, Amsterdam, p 470

Nile SH, Baskar V, Selvaraj D, Nile A, Xiao J, Kai G (2020) Nanotechnologies in food science: applications, recent trends, and future perspectives. NanoMicro Lett 12:45

Pereira PFM, de Sousa Picciani PH, Calado V, Tonon RV (2021) Electrical gas sensors for meat freshness assessment and quality monitoring: a review. Trends Food Sci Technol 118:36–44

Pirsa S (2017) Chemi-resistive gas sensors based on conducting polymers. In: Khosrow-Pour M (ed) Materials science and engineering: concepts, methodologies, tools, and applications, vol 3. IGI Global Publishing Company, Hershey, PA, pp 543–574

Pradhan N, Singh S, Ojha N, Shrivastava A, Barla A, Rai V, Bose S (2015) Facets of nanotechnology as seen in food processing, packaging, and preservation industry. Biomed Res Int 2015: 365672

Quintanilla-Carvajal MX, Camacho-Díaz BH, Meraz-Torres LS, Chanona-Pérez JJ, Alamilla-Beltrán L, Jimenéz-Aparicio A, Gutiérrez-López GF (2010) Nanoencapsulation: a new trend in food engineering processing. Food Eng Rev 2(1):39–50

Relkin P, Yung JM, Kalnin D, Ollivon M (2008) Structural behaviour of lipid droplets in protein-stabilized nano-emulsions and stability of α-tocopherol. Food Biophysics 3(2):163–168

Rong X, Xie Y, Hao X, Chen T, Wang Y, Liu Y (2011) Applications of polymeric nanocapsules in the field of drug delivery systems. Curr Drug Discov Technol 8(3):173–187

Sadiq U, Gill H, Chandrapala J (2021) Casein micelles as an emerging delivery system for bioactive food components. Foods 10(8):1965

Saifullah M, Shishir MRI, Ferdowsi R, Tanver MR, Van Vuong Q (2019) Micro and nano encapsulation, retention and controlled release of flavor and aroma compounds: a critical review. Trends Food Sci Technol 86:230–251

Sarfraz J, Gulin-Sarfraz T, Nilsen-Nygaard J, Pettersen MK (2021) Nanocomposites for food packaging applications: an overview. Nanomaterials 11(1):10

Shirzadfar H, Khanahmadi M (2018) Application of biomarkers and biosensors to detect and track pathogenic agents. Inte J Anal Tech 4(1):1–5

Shit SC, Shah PM (2014) Edible polymers: challenges and opportunities. J Polym 2014:427259

Singh H (2016) Nanotechnology applications in functional foods; opportunities and challenges. Prev Nutr Food Sci 21(1):1–8

Sinha S, Okamoto M (2003) Polymer/layered silicate nanocomposites: a review from preparation to processing. Prog Polym Sci 28(11):1539–1641

Stephen Inbaraj B, Chen BH (2016) Nanomaterial-based sensors for detection of foodborne bacterial pathogens and toxins as well as pork adulteration in meat products. J Food Drug Anal 24(1):15–28

Subramanian A, Irudayaraj J, Ryan T (2006) A mixed self-assembled monolayer-based surface plasmon immunosensor for detection of E. coli O157:H7. Biosens Bioelectron 21(7):7–11

Sujithra S, Manikkandan TR (2019) Application of nanotechnology in packaging of foods: a review. Int J Chem Tech Res 12(4):7–14

Tan J, Xu J (2020) Applications of electronic nose (e-nose) and electronic tongue (e-tongue) in food quality-related properties determination: a review. Artif Intell Agric 4:104–115

Vasile C (2018) Polymeric nanocomposites and nanocoatings for food packaging: a review. Materials 11(10):1834

Wang L, Chen W, Xu D, Shim BS, Zhu Y, Sun F, Liu L, Peng C, Jin Z, Xu C, Kotov NA (2009) Simple, rapid, sensitive, and versatile SWNT-paper sensor for environmental toxin detection competitive with ELISA. Nano Lett 9(12):4147–4152

Wooster TJ, Moore SC, Chen W, Andrews H, Addepalli R, Seymour RB, Osborne SA (2017) Biological fate of food nanoemulsions and the nutrients they carry-internalisation, transport and cytotoxicity of edible nanoemulsions in Caco-2 intestinal cells. RSC Adv 7(64):40053–40066

Yalçinöz Ş, Erçelebi E (2018) Potential applications of nano-emulsions in the food systems: an update. Mater Res Express 5(6):166–170

Yang SC, Lin CH, Sung CT, Fang JY (2014) Antibacterial activities of bacteriocins: application in foods and pharmaceuticals. Front Microbiol 5:241

Yang R, Zhou Z, Sun G, Gao Y, Xu J, Strappe P, Blanchard C, Cheng Y, Ding X (2015) Synthesis of homogeneous protein-stabilized rutin nanodispersions by reversible assembly of soybean (Glycine max) seed ferritin. RSC Adv 5(40):31533–31540

Yucel U (2016) Intelligent packaging. In: Gerlis R (ed) Reference module in food science. Elsevier, New York, NY. https://doi.org/10.1016/B978-0-08-100596-5.03212-1

Zhang T, Lv C, Chen L, Bai G, Zhao G, Xu C (2014) Encapsulation of anthocyanin molecules within a ferritin nanocage increases their stability and cell uptake efficiency. Food Res Int 62: 183–192

Zhang R, Zhang Z, Kumosani T, Khoja S, Abualnaja KO, McClements DJ (2016) Encapsulation of β-carotene in nanoemulsion-based delivery systems formed by spontaneous emulsification: influence of lipid composition on stability and bioaccessibility. Food Biophysics 11(2):154–164

Chapter 16
Scope of Nanoencapsulation for Delivery of Functional Food Ingredients

Sonia Mor, Navdeep Nain, Anamika Das, Anu Kumari, and Vini Swarup

Contents

Abstract Nanotechnology has numerous prospects for the creation of novel food systems, such as: functional foods, nutraceuticals, bioactive compounds, pharmaceutical foods, etc. Functional foods can prevent and reduce incidence of specific diseases and can promote our health. Nonetheless, their utilization has been

S. Mor (✉) · A. Das · A. Kumari · V. Swarup
Warner College of Dairy Technology, Sam Higginbottom University of Agriculture, Technology & Sciences (SHUATS), Prayagraj, Uttar Pradesh, India
e-mail: sonia.mor@shiats.edu.in; anu.kumari@shiats.edu.in

N. Nain
University Institute of Engineering and Technology, Kurukshetra University, Kurukshetra, Haryana, India

J. A. Malik et al. (eds.), *Food Process Engineering and Technology*,
https://doi.org/10.1007/978-981-99-6831-2_16

restricted because of their low dissolvability, dependability, bioavailability, and target particularity, and aftereffects when utilized in enormous dosages. In fact, nanoparticles can make a variety of functional components more soluble and stable, can improve their absorption, and prevent quick degradation in our body, and can extend their duration of circulation. Moreover, these nanoparticles (NPs) display high differential take-up effectiveness in the target cells (or tissues) compared with normal cells (or tissues), in this way bringing about diminished poisonousness.

Keywords Bioactives · Bioavailability · Delivery · Functional ingredients · Nanotechnology

16.1 Introduction

Nanoparticles (NPs) are particulate substances that have one dimension <100 nm. In comparison to their respective detritus at larger dimensions, the nanoparticles exhibit distinct physical, chemical, and organic characteristics, due to several factors, such as: an appreciably larger surface area relative to the volume, quicker reactivity or balancing in a chemical reaction, improved mechanical strength, etc.

Analysts have created novel nanoencapsulation systems for wellbeing and controlled/centered arrival of various bioactive ingredients, especially doctor prescribed medications and food nutraceuticals, so that these might be enough brought into definitions and can bring about the best bioavailability. Stakeholders are becoming more concerned about food quality and safety due to rising global food demand and security. In particular, effective methods for long-term food preservation are urgently needed. Numerous components in functional foods have low water solubility, are vulnerable to chemical deterioration, and are incompatible with the food matrix in which they are used. Limited oral bioavailability, unpleasant sensory characteristics, and subpar release profiles are other problems. These restrictions can be overwhelmed, and the bioavailability can be expanded through nanoencapsulation procedures.

Encapsulation is a method that creates capsules by packing tiny bits of core elements into a wall material. Various bioactive components (such as: nutraceuticals, antioxidants, enzymes, polyphenols, and micronutrients etc.) can be safeguarded using this method (Gouin 2004). Size of all nanoparticles are different from each other such as, nanocapsules and nanospheres, which are colloidal-sized particles, have diameter in range of 10–1000 nm, microcapsules are particles with diameters ranging from 3 to 800 nm (Ezhilarasi et al. 2013). By using one or more techniques (such as: homogenization, extrusion, ultrasonication, nanoemulsification, nanoprecipitation, nanocomplexation, and supercritical fluid technology), the bioactive substances are entrapped inside the nanoparticles to create nanoencapsules. Since the delivery of bioactives is inversely related to particle size, these nanosized particles are superior to microparticles (Kulkarni and Feng 2013).

Nanoparticles significantly enhance the encapsulated molecule's solubility and bio-adhesive characteristics, extending its gastrointestinal time and increasing its bioavailability. For instance, particles smaller than 200 nm can bind to intestinal epithelial cells or M-cells, diffuse into enterocytes, and then travel to the targeted area through the circulatory system.

Additionally, nanoparticles can permit the regulated release of bioactive to target site. The disagreeable flavor of the bioactives is also covered up by nanoencapsulation, protecting the reinforced vehicle's natural qualities. As a result, use nano delivery technologies in the dairy and food industries is beneficial. Examples of nanofortification are:

- Canola Active oil from Shemen has been fortified with nanoencapsulated phytosterols;
- Daily boost is a fortified beverage with nanoencapsulated vitamins;
- Fortified fruit juices are fortified with nanoencapsulated vitamins, lycopene, thiamine, and iron;
- LycoVit is fortified with nanoencapsulated lycopene from Baden Aniline and Soda Factory (BASF);
- Nova

To create nanoparticles laden with bioactive substances, various fabrication techniques have been examined and investigated. The release mechanisms from the nanoencapsules on ingestion are established by the solubility characteristics of the bioactive substances and the characteristics of the nanoshells into which they are placed. The *in vivo* bioavailability and functioning of nanoparticles are typically controlled by their size, hydrophobicity, and charge of the active molecule.

The nanoparticles'ability to function and survive in gastrointestinal transit environments is one of the biggest obstacles in their manufacturing. In the food sector, encapsulation has been utilized to supply several bioactive substances (carotenoids, polyphenols, and omega-fatty acids, etc.) that are responsive to environmental variables (Fathi et al. 2014).

The most popular biocompatible and environmentally friendly nanoparticles are covered in this chapter, including solid lipid nanoparticles, niosomes, nanostructured lipid carriers, liposomes, and emulsions. Their development and scope for incorporation in food ingredients has also been presented.

16.2 Encapsulation Techniques

The particle size of any bioactive component directly influences how well it is delivered to different target locations in our body (Kawashima 2001). The cerebral blood barrier and intestinal epithelium are hypothesized to be more permeable due to nanoscale dimensions (Moutinho et al. 2012). Compared to microencapsulation, nanoencapsulation offers a better potential to enhance the bioavailability, handling the release, and to allow precise aiming of the bioactive substances (Mozafari et al.

2006). Protein and peptide nanoencapsulation are desirable methods for improving protein and peptide stability and maintaining its activity until they reach their target site. Therefore, encapsulating compounds in the micro dimension is appealing.

16.2.1 Nanogels

In drug and food designing, one of the most adaptable conveyance frameworks is nanogels, especially for biomacromolecules and hydrophobic mixtures (Zhang et al. 2016). Physical or chemical cross-linking of monomers or polymers can create three-dimensional cross-linked networks with submicron widths (Abaee et al. 2017; Mokhtari et al. 2017). Physical cross-connecting occurs commonly between polymer precursors, whereas chemical cross-connecting occurs between monomers or polymer antecedents (Zhang et al. 2016). The nanogels' unique property of cross-linking allows them to swell rather than dissolve (Oh et al. 2007). With variable pore diameters and tuned pore topologies, nanogels can create porous structures with high-level loading capacity (Gokmen and Du Prez 2012).

In addition to their advantages due to their nanosize, Both hydrophilic and hydrophobic nanogels, also known as nano hydrogels and nano organogels, can collect both water-soluble and water-insoluble bioactive molecules. Many polysaccharides, including chitosan, heparin, alginate, pullulan, pectin, hyaluronic acid, chondroitin sulphate, starch, cellulose, cyclodextrin, and dextran, have been used to make nanogels, according to Debele et al. (2016). Additionally, review on hydrogels made from soy and whey proteins for the delivery of bioactives was published by Abaee et al. (2017). Biopolymer-based nanogels can be employed to transport many bioactive substances, particularly hydrophobic substances.

16.2.2 Nanoemulsions

Nanoemulsions are suitable delivery mechanisms for bioactive compounds that are hydrophobic with low solubility. Due to their improved water solubility and capacity to prevent bioactive components from degrading, they are potential nanocarriers for a variety of chemicals. Formulated nanoemulsions can be used as-is or dried to create powder forms utilizing several techniques, including freeze drying and spray drying.

The stability and bioavailability of nanoemulsions have been the subject of numerous investigations (Jafari et al. 2015). Oil-in-water (O/W) nanoemulsions are employed for the blending of lipophilic bioactive substances. However, water-in-oil (W/O) emulsions can also be utilized to encapsulate hydrophilic substances, such as, the majority of polyphenols and vitamins (Mohammadi et al. 2016). These emulsions typically protect the lipophilic substances because the oil phase contains dietary oils.

Food specialists have recently begun using double emulsions to nanoencapsulate bioactive ingredients in food (Alfaro et al. 2015; Mehrnia et al. 2016; Mohammadi et al. 2016). Alfaro et al. (2015) created frozen yoghurt enhanced with a sodium caseniate-based nanoemulsion of purple rice bran, which is high in oryzanol. The fortified yoghurt was found to be comparable in viscosity and hardness to the unfortified yoghurt, with the observed emulsion droplet size ranging from 150 to 300 nm.

Commercial milk with nanoemulsified curcumin droplets was able to stop the oxidation of milk's lipids and showed potent radical scavenging abilities (Joung et al. 2016). The reduced droplet size of the vitamin D3-loaded nanoemulsion was made possible by using sodium caseinate as a stabilizer by Tipchuwong et al. (2017). Over the course of 56 days, the vitamin stability of fortified low-fat ice cream with the nanoform of vitamin D3 was monitored. The nanoemulsification significantly improved the vitamin stability.

16.2.3 Nanofibers

The consistent creation of nonwoven nanofibers with widths of only a couple of nanometers is made conceivable by a noteworthy innovation called electrospinning, which could be used to move bioactive food parts (Esfanjani and Jafari 2016). Pumping a solution (polymer or biopolymer) from a spinneret toward a ready collector is the electrospinning process, which is simple and reliable for producing ultrathin electrospuns. At long last, nonwoven ultrafine strands are gathered onto the assortment plate (Ghorani and Tucker 2015). The created nanoscale fibre mat exhibits excellent properties, such as a high surface area to volume proportion and adequate porosity, which make them suitable as carriers of bioactive food components and nutraceuticals to the extent that the particles included could boost functionality because of the ultrathin fibres (Katouzian and Jafari 2016). Emulsion electrospinning can produce centre shell structures that are better at binding hydrophobic mixes than blend electrospinning can produce nanofibers. In addition, emulsion electrospinning rather than coaxial electrospinning may still result in the bioactive compounds losing their activity due to the interface tension between the water and oil stages (Nikmaram et al. 2017).

16.2.4 Coacervation

One or more polyelectrolytes are phase separated from solution during the coacervation process, and then a freshly formed coacervate phase is placed over the bioactive component. To reinforce the coacervate, a hydrocolloid shell can also be cross-connected using an appropriate chemical or enzymatic cross-linker (such as: transglutaminase or glutaraldehyde) (Zuidam and Shimoni 2010). If only one type of polymer is employed, the process is classified as simple coacervation, and if two or more types are utilized, it is classified as complex coacervation.

The type of biopolymer (including its molar mass, charge, and flexibility), pH, concentration, ionic strength, ratio of biopolymers, etc. influence how strongly the biopolymers interact and what kind of complex is formed (Turgeon et al. 2003). Hydrogen bonds, hydrophobic contacts, and electrostatic interactions between biopolymers with opposing charges can all have a substantial impact on how the complex forms. Utilizing polyanion tripolyphosphate (TPP) as coacervation cross-connecting agent, the incubation method can be used to encapsulate bovine serum albumin (BSA) in chitosan because of the possibility of controlled release based on exceptionally high constituent loading (up to 99%) as well as the possibilities of controlled release depending on temperature, mechanical stress, or sustained release (Gan and Wang 2007; Gouin 2004).

The size of BSA-loaded chitosan-TPP nanoparticles ranged from 200–580 nm. Using a transmission electronic microscope (TEM), the structural alteration of BSA-loaded nanoparticles was investigated, which revealed that the particles swelled and degraded with time. Additionally, they showed that the polyionic coacervation process may be used to control over the effectiveness of protein encapsulation and the resulting release profile.

Capsaicin was encapsulated by Jincheng et al. (2010) utilizing a sophisticated coacervation technique and additional drying in a vacuum oven. To enrich fruit juices with omega fatty acids, scientists developed a multilayer stable nanocoacervative system using gum Arabic and sodium caseinate to deliver the bioactive components EPA and DHA (Ilyasoglu and El 2014). The nanoparticles contained 50 mg of EPA and DHA and had a mean size of 232 nm. Such a stable nanosystem offers considerable potential to PUFA-enrich or fortify dairy products, reducing the likelihood of oxidation. Ahmadi et al. (2015) created functional yoghurt with encapsulated CoQ loaded into lactoglobulin and gum Arabic coacervation complex in response to health advantages of coenzyme Q10 (CoQ).

16.2.5 Nanosponges

Because they are solid, porous particles with nanocavities in their structure (either internal or external cavities), nanosponges, as cutting-edge delivery systems, can entrap the bioactives. Because they can transport significant quantities of different chemicals, these nanocarriers are superior to other nanocarriers. However, by improving their solubility, nanosponges can boost the controlled release of insoluble and inadequately water-soluble bioactive compounds (Tejashri et al. 2013). They can also trap them at very low concentrations to remove metallic ions and organic pollutants from aqueous solutions and to clean the environment (up to ppb) (Trotta 2011). Additionally, nanosponges can be utilised to mask unpleasant odours, turn liquid elements into solid powders, and offer a sustained and protracted release for delicate and biodegradable ingredients. These additional applications could enhance material attributes (Trotta 2011).

16.2.6 Nanoprecipitation

The process of nanoprecipitation (solvent displacement) is based on the idea that when an organic phase (solvent mix) holding a polymer and a bioactive compound is added to an aqueous phase, the polymer automatically collapses to form microparticles or nanoparticles that can encapsulate bioactive compounds in the organic phase. The main requirements for this approach are that the solvent should be dispersed and that the polymer and bioactive component both can be dissolved in the solvent. The variations of this approach include: flash precipitation, two-step precipitation, and ouzo effect.

Biodegradable polymers (such polylactocaprone (PCL), polylactide (PLA), polylactide-co-glicolide (PLGA), polyalkylcyanoacrylate (PACA), etc.) are often used in the nanoprecipitation process. The utilization of an organic solvent during the preparation process is the principal drawback, which may restrict its application unless the organic solvent is entirely flushed from the system. Noronha et al. (2013) created PCL nanocapsules that contained tocopherol. With a low PDI of 0.112 and a negative zeta potential of −30 mV, PCL has been shown to be a nontoxic, biodegradable, and biocompatible polyester that can encapsulate tocopherol at a rate of 99% while maintaining good stability. Similarly, Yallapu et al. (2010) used the nanoprecipitation technique to encapsulate curcumin in PLGA while using PEG-5000 as a stabilizer. Curcumin nanoparticles had improved cellular absorption and superior bioactivity in both *in vitro* and *in vivo* tests.

Gelatin and Tween 20 were used as stabilizers while carotene (a chemical of relevance to the dairy industry) was encapsulated into PLA and PLGA nanoparticles. After redispersion of the lyophilized material, it was shown that the initial nanoparticle size of 80 nm had not considerably changed the average particle size (Ribeiro et al. 2008). Accordingly, based on the physicochemical characteristics of the fortificant vehicle, the carotene nanoparticles can be utilized either in their original liquid form or in solid form (obtained following lyophilization) for the fortification of dairy products.

16.2.7 Nanocochelates

Phosphatidylserine, negatively charged lipid, calcium, and divalent cation, are the main components of nanocochelates, which are well-defined, compact, and durable lipid-based carriers (Zarif et al. 2000). They are comprised of a solid, continuous lipid bilayer sheet that has been spirally wound up, with little to no internal aqueous gap. They are multilayered cigar-shaped structures.

Due to their unique structure and properties, nanocochelates are extraordinary possibility for oral and foundational conveyance of numerous bioactives. They can move hydrophobic, amphiphilic, negatively or positively charged biocompounds as well as those that are hydrophobic.

16.2.8 Supercritical Fluid Technique

Since the supercritical fluid has properties that lay between those of fluids and gases, such as low consistency and high diffusion rates, it can be used to encapsulate molecules that are sensitive to heat, such as taste or aroma components, polyunsaturated fatty acids, sterols, etc. In this process, a supercritical fluid is used to dissolve the polymer and bioactive component. The supercritical fluid spray-dries and the solute particles with bioactive chemical precipitate after the solution is blasted via a jet. This method is quick, easy, runs at extremely low temperatures, and requires the least amount of organic solvent. Although it has a excellent deal of potential for widespread use in nanoencapsulation, the applications are restricted since the polymer and bioactive ingredients must be completely dissolved in the supercritical fluid. Recently, techniques such supercritical antisolvent precipitation, gas antisolvent, and fast expansion from supercritical fluid have been used.

Through a supercritical antisolvent technique, Fernández-Ponce et al. (2015) created co-precipitated quercetin nanoparticles utilizing ethyl cellulose as a polymer and ethyl acetate as a non-toxic solvent. Utilizing surfactants like Tween 80, Lutrol F68, Solutol HS15, and sodium lauryl sulphate, researchers created phytosterol-loaded nanoparticles (size >500 nm) with improved stability (Türk and Lietzow 2004).

16.2.9 SLNs and Nanostructured Lipid Carriers

The innovative nanoparticulate solid lipid nanoparticles (SLNs) were created to extend duration of release while shielding the active ingredients from unfavorable chemical processes including oxidation (Müller et al. 2000). To contain the bioactive components in their lipid matrix, SLNs feature lipid droplets that are completely crystalline and structured (Weiss et al. 2008). They are either made of a single solid lipid or a mixture of several substances. The origins of solid matrix of SLN are the O/W (oil/water) emulsions, where a hard solid phase replaces the liquid lipid/oil phase.

Later, nanostructured lipid carriers (NLCs) were developed to nanoencapsulation industry to solve the drawbacks of SLNs. Maximizing loading volume and reducing bioactive chemical leakage were the main objectives. They are formed by mixing solid and liquid lipids with the bioactive components (inner phase) in water (outer phase). The lipid mixture present in NLCs exhibits a slower polymorphic transition and a low crystallinity index. The internal phase of the NLCs' composition can therefore improve the loading capacity, encapsulation effectiveness, and bioavailability of drugs that are nanoencapsulated in comparison to SLN (Müller et al. 2002).

16.2.10 Nanoliposomes

The nanoscale variant of liposomes (called nanoliposomes) can encapsulate hydrophilic, hydrophobic, and amphiphilic molecules (Akhavan et al. 2018; Ghorbanzade et al. 2017). The lipid/water interface can contain amphiphilic chemicals, hydrophilic substances can be added to the inside, and hydrophobic substances can be contained within hydrophobic bilayers. Marsanasco et al. (2015) discovered that chocolate milk supplemented with liposomes comprising omega-3 fatty acids, vit. E, and folic acid was more popular than non-fortified milk. The coating of positively charged whey protein isolate enhanced the stability of liposomes. The longer shelf-life and reversal of osmotic effects, when salts or sugars are present, are resulted in decreased sensitivity to low pH during GIT transit (Frenzel and Steffen-Heins 2015).

Niosomes are vesicles created when non-ionic surfactants self-assemble in aqueous fluids to form closed bilayer structures (Uchegbu and Vyas 1998). Similar to liposomes, interactions between water molecules and surfactants lead to their production. They serve as a useful vehicle for the delivery of hydrophobic bioactives, which associate with the lipid bilayer, and hydrophilic bioactives, which are contained in the interior aqueous compartment, due to the presence of bilayer structure (Kumar and Rajeshwarrao 2011). Niosomes are less expensive than liposomes, are biodegradable, and immune-neutral by nature. Non-ionic surfactants and stabilizers are main building blocks of niosomes. As opposed to their anionic, cationic, or amphoteric cousins, nonionic surfactants have higher stability, compatibility, and toxicity, making them the most frequently utilised surfactants in the creation of vesicles (Marianecci et al. 2014).

16.3 Nanoencapsulation of Different Food Ingredients

Through particular methods, the field of food bioactive components being nanoencapsulated has grown significantly during the previous few decades. These additives should be positioned inside the nanocarriers so that a regulated release could be accomplished to provide bioactive elements and nutraceuticals to the cells and safeguard them from harmful ecological and food process aspects. Numerous food additives, such as phenolic composites, antioxidants, food colourants, antimicrobial compounds, essential oils, mineral, essential fatty acids, fish oil, and vitamins, can be nanoencapsulated using various techniques.

16.3.1 Nanoencapsulation of Phenolic Contents and Antioxidants

Numerous research have examined the anti-oxidant, anti-allergenic, anti-inflammatory, anti-atherogenic, anti-thrombotic, and anti-microbial properties of phenolic compounds in food products and other chemical compositions. Due to their amazing ability to enhance the life-span of food products while favourably affecting their sensory or nutritional aspects, antioxidants that chelate free radicals are frequently utilised as ingredients in food compositions.

Free radicals can be reduced and eliminated by enzyme-based antioxidants. Superoxide dismutase (SOD), catalase (CAT), glutathione peroxidase, and glutathione reductase are examples of such enzymes. Carotenoids, vitamin C, vitamin E, plant phenolic compounds, lipoic acid, and glutathione are examples of non-enzymatic antioxidants (GSH). Phenolic compounds are the main subclass of non-enzymatic antioxidants. The process of nanoencapsulation can boost the solubility, absorption, and penetration of phenolic compounds and antioxidants into the body as well as their protection within food compositions. Antioxidants and phenolic elements can be nanoencapsulated using a variety of methods (Akhavan et al. 2018).

According to Esfanjani and Jafari (2017), phenolics can be nanoencapsulated using a complicated emulsion system, such as multiple nanoemulsions. Chen et al. (2015) demonstrated that curcumin may load phenolic chemicals and antioxidants into nano-liposomes produced using thin film and dynamic high-pressure microfluidization. This type of inclusion demonstrated adequate bioavailability, good water solubility, and physicochemical instability when compared to free curcumin.

16.3.2 Essential Fatty Acids and Fish Oil Nano-Encapsulation

The best and simplest unit of lipids is the fatty acid, which is frequently divided into saturated and unsaturated fatty acids. The three omega-3 fatty acids that are most important for humans are linolenic acid, eicosapentaenoic acid (EPA), and docosahexaenoic acid (DHA) as human bodies does not synthesize of such fatty acids due to the lack of specific desaturase enzymes. The finest source of EPA and DHA is fish oil. However, plant oils, seeds, and nuts are the typical sources of linolenic acids. Fish oil and essential fatty acid oxidation is especially hazardous since it not only alters the flavour of the oils but also creates hazardous chemicals (Ghorbanzade et al. 2017).

By employing nanoemulsions as encapsulation structures, fish oil and essential fatty acids have been protected and their release has been managed. When fish oils and essential fatty acids are contained, nanoemulsion is utilised to restrict and control their release. Nanoliposomes are used to strengthen the stability against oxidation of important fatty acids and fish oils, as well as to increase their absorption (Akhavan et al. 2018).

16.3.3 Nanoencapsulation of Vitamins

Vitamins are crucial nutrients for humans because they have an impact on a variety of biochemical processes that occur within the body. There are 13 different vitamins named, including folate (B9), cyanocobalamin, thiamin (B1), riboflavin (B2), nicotinic acid (B3), pyridoxine (B6), and pantothenic acid (B5). Carotenoids and the lipophilic Vit. A, E, K, and D provide evidence of vitamin A's beneficial effects. Typically, leaching, which includes washing, blanching, and cooking processes, results in the loss of water-soluble vitamins. These processes play essential roles in physiological reactions. Additionally, biopolymeric nanocarriers have been successfully used to nanoencapsulate vitamins. Approximately 65% of vitamin C was successfully encapsulated (Akhavan et al. 2018).

Pezeshky et al. (2016) employed thin-film-hydration sonication to create nanoliposomes with vitamin A palmitate infused into them. The ability of vitamin A palmitate to be encapsulated was diminished at high cholesterol levels. The size of the nanoparticles was close to 76 nm, and the encapsulation rate was raised to 15.8% by using a lecithin-cholesterol mixture (50/10 mg).

16.3.4 Nano-Encapsulation of Minerals

The most significant bioactive ingredients concerned with cellular metabolism and organic processes are minerals. These micronutrients cannot be synthesized by plants or animals. However, plant roots can commonly absorb these nutritional ions that are present in the soil naturally. Fruits and vegetables are the primary sources of minerals (Gharibzahedi and Jafari 2017). Many minerals are susceptible to oxidative deterioration; as a result, this is crucial to safeguard them in fortified/supplemented food products throughout their lifespan. The difficult situations involving the transport of minerals and other bioactive substances in food applications can be successfully handled by nano-encapsulation.

Iron deficient rats responded more significantly to ferrous glycinate-containing nanoliposomes made by Ding et al. (2011) from egg phosphatidylcholine than to the two free Fe-forms.

16.3.5 Nano-Encapsulation of Food Flavors

Food systems depend on flavors as essential ingredients to make the food appropriate for consumers and capable of satisfying them. Some of food flavours are, allylpyrazine in roasted nuts, 2-isobutyl-3-methoxypyrazine in green pepper, aldehydes in fruits, terpenoids in citrus and piney, phenolics in smoked goods, methoxypyrazines in vegetables, etc.

How taste composition can affect the creation and characteristics of nanoemulsions is an important topic. Lower fold oils are prone to droplet growth while being stored, and the growth rate accelerates with increasing storage time. The development of complexes between amylose and volatile compounds was investigated by Putseys et al. (2009) as an essential quality of natural nanocarriers. It is known that starch can be combined with volatile substances, particularly tiny nonpolar molecules, to create inclusion complexes.

16.4 Conclusion

For the safe and targeted release of nutraceuticals and food-active substances, various nanocarriers have recently been created. Such nanocarriers can be added to dietary ingredients to increase a nutrient's bioavailability. A few of these are producible utilising high-stress homogenizers and sonication tools, while electrospinning/spraying can best shape some nanocarriers. Using these useful bioactive substances and nanoencapsulation techniques, the food sector should aim to increase the quality and lifespan of food items. Researchers should use advantageous bioactive ingredients and nanoencapsulation technology to increase the quality and life expectancy of food products. Increasing the oxidative stability of edible oils and foods that are high in oil, reducing the growth of infectious organisms in food, enhancing the texture and sensory qualities of finished food products are a few examples. These are only a few of the issues that food-based bioactive nanoparticles may successfully resolve. More thorough research is required to address the issue of both the protection of nanomaterials and nanocarriers inside human bodies, which is of the utmost importance to both consumers and governing bodies.

References

Abaee A, Mohammadian M, Jafari SM (2017) Whey and soy protein-based hydrogels and nanohydrogels as bioactive delivery systems. Trends Food Sci Technol 70:69–81

Ahmadi N, Nasirpour A, Sheikhzeinodin M, Keramat J (2015) Microencapsulation of ubiquinone using complex coacervation for functional yoghurt. Food Sci Biotechnol 24:895–904

Akhavan S, Assadpour E, Katouzian I, Jafari SM (2018) Lipid nanoscale cargos for the protection and delivery of food bioactive ingredients and nutraceuticals. Trends Food Sci Technol 74:132–146

Alfaro L, Hayes D, Boeneke C, Xu Z, Bankston D, Bechtel PJ, Sathivel S (2015) Physical properties of a frozen yogurt fortified with a nanoemulsion containing purple rice bran oil. LWT 62(2):1184–1191

Chen X, Zou LQ, Niu J, Liu W, Peng SF, Liu CM (2015) The stability, sustained release and cellular antioxidant activity of curcumin nanoliposomes. Molecules 20(8):14293–14311

Debele TA, Mekuria SL, Tsai HC (2016) Polysaccharide based nanogels in the drug delivery system: application as the carrier of pharmaceutical agents. Mater Sci Eng C 68:964–981

Ding BM, Zhang XM, Xia SQ (2011) Effectiveness of treatment of iron-deficiency anemia in rats with ferrous glycinate nanoliposomes. J Food Sci Biotechnol 30(1):49–54

Esfanjani AF, Jafari SM (2016) Biopolymer nanoparticles and natural nano-carriers for nano-encapsulation of phenolic compounds. Colloids Surf B: Biointerfaces 146:532–543

Esfanjani AF, Jafari SM (2017) Nanoencapsulation of phenolic compounds and antioxidants. In: Jafri SM (ed) Nanoencapsulation of food bioactive ingredients: principles and applications, vol 1. Academic, Cambridge, Massachusetts, pp 63–101

Ezhilarasi PN, Karthik P, Chhanwal N, Anandharamakrishnan C (2013) Nanoencapsulation techniques for food bioactive components: a review. Food Bioprocess Technol 6(3):628–647

Fathi M, Martin A, McClements DJ (2014) Nanoencapsulation of food ingredients using carbohydrate-based delivery systems. Trends Food Sci Technol 39(1):18–39

Fernández-Ponce MT, Masmoudi Y, Djerafi R, Casas L, Mantell C, de La Ossa EM, Badens E (2015) Particle design applied to quercetin using supercritical anti-solvent techniques. J Supercrit Fluids 105:119–127

Frenzel M, Steffen-Heins A (2015) Whey protein coating increases bilayer rigidity and stability of liposomes in food-like matrices. Food Chem 173:1090–1099

Gan Q, Wang T (2007) Chitosan nanoparticle as protein delivery carrier: systematic examination of fabrication conditions for efficient loading and release. Colloids Surf B: Biointerfaces 59(1): 24–34

Gharibzahedi SMT, Jafari SM (2017) The importance of minerals in human nutrition: bioavailability, food fortification, processing effects and nanoencapsulation. Trends Food Sci Technol 62: 119–132

Ghorani B, Tucker N (2015) Fundamentals of electrospinning as a novel delivery vehicle for bioactive compounds in food nanotechnology. Food Hydrocoll 51:227–240

Ghorbanzade T, Jafari SM, Akhavan S, Hadavi R (2017) Nanoencapsulation of fish oil in nano-liposomes and its application in fortification of yogurt. Food Chem 216:146–152

Gokmen MT, Du Prez FE (2012) Porous polymer particles—a comprehensive guide to synthesis, characterization, functionalization and applications. Prog Polym Sci 37(3):365–405

Gouin S (2004) Microencapsulation: industrial appraisal of existing technologies and trends. Trends Food Sci Technol 15(7–8):330–347

Ilyasoglu H, El SN (2014) Nanoencapsulation of EPA/DHA with sodium caseinate–gum arabic complex and its usage in the enrichment of fruit juice. LWT 56(2):461–468

Jafari SM, Fathi M, Mandala I (2015) Emerging product formation. In: Galanakis CM (ed) Food waste recovery. Academic, San Diego, CA, pp 293–317

Jincheng W, Xiaoyu Z, Sihao C (2010) Preparation and properties of nanocapsulated capsaicin by complex coacervation method. Chem Eng Commun 197(7):919–933

Joung HJ, Choi MJ, Kim JT, Park SH, Park HJ, Shin GH (2016) Development of food-grade curcumin nanoemulsion and its potential application to food beverage system: antioxidant property and in vitro digestion. J Food Sci 81(3):N745–N753

Katouzian I, Jafari SM (2016) Nano-encapsulation as a promising approach for targeted delivery and controlled release of vitamins. Trends Food Sci Technol 53:34–48

Kawashima Y (2001) Nanoparticulate systems for improved drug delivery. Adv Drug Deliv Rev 47:1–2

Kulkarni SA, Feng SS (2013) Effects of particle size and surface modification on cellular uptake and biodistribution of polymeric nanoparticles for drug delivery. Pharm Res 30(10):2512–2522
Kumar GP, Rajeshwarrao P (2011) Nonionic surfactant vesicular systems for effective drug delivery—an overview. Acta Pharm Sin B 1(4):208–219
Marianecci C, Di Marzio L, Rinaldi F, Celia C, Paolino D, Carafa M (2014) Niosomes from 80s to present: the state of the art. Adv Colloid Interf Sci 205:187–206
Marsanasco M, Márquez AL, Wagner JR, Chiaramoni NS, Alonso SDV (2015) Bioactive compounds as functional food ingredients: characterization in model system and sensory evaluation in chocolate milk. J Food Eng 166:55–63
Mehrnia MA, Jafari SM, Makhmal-Zadeh BS, Maghsoudlou Y (2016) Crocin loaded nano-emulsions: factors affecting emulsion properties in spontaneous emulsification. Int J Biol Macromol 84:261–267
Mohammadi A, Jafari SM, Assadpour E, Esfanjani AF (2016) Nano-encapsulation of olive leaf phenolic compounds through WPC–pectin complexes and evaluating their release rate. Int J Biol Macromol 82:816–822
Mokhtari S, Jafari SM, Assadpour E (2017) Development of a nutraceutical nano-delivery system through emulsification/internal gelation of alginate. Food Chem 229:286–295
Moutinho CG, Matos CM, Teixeira JA, Balcao VM (2012) Nanocarrier possibilities for functional targeting of bioactive peptides and proteins: state-of-the-art. J Drug Target 20(2):114–141
Mozafari MR, Flanagan J, Matia-Merino L, Awati A, Omri A, Suntres ZE, Singh H (2006) Recent trends in the lipid-based nanoencapsulation of antioxidants and their role in foods. J Sci Food Agric 86(13):2038–2045
Müller RH, Mäder K, Gohla S (2000) Solid lipid nanoparticles (SLN) for controlled drug delivery–a review of the state of the art. Eur J Pharm Biopharm 50(1):161–177
Müller RH, Radtke M, Wissing SA (2002) Nanostructured lipid matrices for improved microencapsulation of drugs. Int J Pharm 242(1–2):121–128
Nikmaram N, Roohinejad S, Hashemi S, Koubaa M, Barba FJ, Abbaspourrad A, Greiner R (2017) Emulsion-based systems for fabrication of electrospun nanofibers: food, pharmaceutical and biomedical applications. RSC Adv 7(46):28951–28964
Noronha CM, Granada AF, Carvalho SM, Lino RC, Maciel MV, Barreto PLM (2013) Optimization of α-tocopherol loaded nanocapsules by the nanoprecipitation method. Ind Crop Prod 50:896–903
Oh JK, Siegwart DJ, Matyjaszewski K (2007) Synthesis and biodegradation of nanogels as delivery carriers for carbohydrate drugs. Biomacromolecules 8(11):3326–3331
Pezeshky A, Ghanbarzadeh B, Hamishehkar H, Moghadam M, Babazadeh A (2016) Vitamin a palmitate-bearing nanoliposomes: preparation and characterization. Food Biosci 13:49–55
Putseys JA, Derde LJ, Lamberts L, Goesaert H, Delcour JA (2009) Production of tailor-made short chain amylose–lipid complexes using varying reaction conditions. Carbohydr Polym 78(4): 854–861
Ribeiro HS, Chu BS, Ichikawa S, Nakajima M (2008) Preparation of nano-dispersions containing β-carotene by solvent displacement method. Food Hydrocoll 22(1):12–17
Tejashri G, Amrita B, Darshana J (2013) Cyclodextrin based nanosponges for pharmaceutical use: a review. Acta Pharma 63(3):335–358
Tipchuwong N, Chatraporn C, Ngamchuachit P, Tansawat R (2017) Increasing retention of vitamin D3 in vitamin D3 fortified ice cream with milk protein emulsifier. Int Dairy J 74:74–79
Trotta F (2011) Cyclodextrin nanosponges and their applications. In: Bilensoy E (ed) Cyclodextrins in pharmaceutics, cosmetics, and biomedicine: current and future industrial applications. Wiley, Hoboken, NJ, pp 323–342
Turgeon SL, Beaulieu M, Schmitt C, Sanchez C (2003) Protein–polysaccharide interactions: phase-ordering kinetics, thermodynamic and structural aspects. Curr Opin Colloid Interface Sci 8(4–5):401–414
Türk M, Lietzow R (2004) Stabilized nanoparticles of phytosterol by rapid expansion from supercritical solution into aqueous solution. AAPS PharmSciTech 5(4):36–45

Uchegbu IF, Vyas SP (1998) Non-ionic surfactant-based vesicles (niosomes) in drug delivery. Int J Pharm 172(1–2):33–70

Weiss J, Decker EA, McClements DJ, Kristbergsson K, Helgason T, Awad T (2008) Solid lipid nanoparticles as delivery systems for bioactive food components. Food Biophysics 3(2): 146–154

Yallapu MM, Gupta BK, Jaggi M, Chauhan SC (2010) Fabrication of curcumin encapsulated PLGA nanoparticles for improved therapeutic effects in metastatic cancer cells. J Colloid Interface Sci 351(1):19–29

Zarif L, Graybill JR, Perlin D, Mannino RJ (2000) Cochleates: new lipid-based drug delivery system. J Liposome Res 10(4):523–538

Zhang H, Zhai Y, Wang J, Zhai G (2016) New progress and prospects: the application of nanogel in drug delivery. Mat Sci Eng 60:560–568

Zuidam NJ, Shimoni E (2010) Overview of microencapsulates for use in food products or processes and methods to make them. In: Zuidam NJ, Nedovíc VA (eds) Encapsulation technologies for active Food ingredients and food processing. Springer, New York, pp 3–29

Chapter 17
Scope of Nanotechnology for Sustainable Production of Nutritive Foods

Annika Durve Gupta

Contents

Abstract To meet the nutritional requirements of the world's fast rising residents is a pressing necessity. Pest infestations, microbiological assaults, natural calamities, reduced soil value, and nutrient accessibility are factors that can cause crop damage in conventional farming. To address these difficulties, more creative technologies have been developed. Nanotechnology has aided the revolution technology for use in agriculture, which has the ability to change the current agrarian system while also ensuring food security. As a result, nanoparticles are emerging as potential tools to

A. D. Gupta (✉)
Department of Biotechnology, B.K. Birla College of Arts, Science and Commerce (Autonomous), University of Mumbai, Mumbai, Maharashtra, India

J. A. Malik et al. (eds.), *Food Process Engineering and Technology*,
https://doi.org/10.1007/978-981-99-6831-2_17

change existing agriculture techniques. For plant health management and soil enhancement, a wide range of nanoparticle-based formulations have been studied, including nano-sized herbicides, insecticides, fertilizers, fungicides, and sensors. Food processing, food packing, functional food production, food safety, detection of foodborne microbes, and delayed life-span of food and/or food products are just some of the nanotechnological applications that have been developed in retort to the rising need for nanoparticles in food science. The promise of nanoparticles in the food commerce is explored in this chapter, with the goal of supplying clients with safe, infection-free food and assuring customer approval of food with increased functional properties.

Keywords Food nutrition · Food preservation · Food safety · Nano sensors · Nano fertilizers · Nano-processed food products · Nanoparticles

17.1 Introduction

Sustainable nutrition is a global concern, as people around the world fight to deal with increasing incidence of chronic illnesses and nutrient shortages while also conserving scarce natural resources. Sustainable nutrition entails making healthy, nutrient-packed foods more available, inexpensive, and ethnically appropriate, while simultaneously protecting the environment and supporting local communities.

Health-promoting elements including phenolic substances, minerals, and vitamins are progressively being combined into food and beverages to produce functional foods that are especially developed to enhance human health and welfare. However, incorporating these nutraceuticals into foods may be a challenging task due to their unpleasant flavors, low solubility, lack of chemical stability or low bioavailability. Encapsulating bioactive components in nanoparticle-based delivery systems is a common solution to this issue.

The bioavailability of encapsulated bioactive chemicals frequently rises when the size of the particles containing them is decreased due to their quick absorption, capacity to penetrate the mucus barrier, or immediate reception by cells. Nanoparticles can be created to travel through specific regions of the GI tract before releasing their payload at a predetermined location, perhaps improving their health advantages. Nutraceutical-loaded nanoparticles can be made using natural nanocarriers, specialized equipment, lipid formulations, biopolymer nanoparticles (NPs), and other methods. The five categories are based on the main mechanisms or chemicals to manufacture nanoparticles. Low birth weight, protein energy undernourishment, arresting height, reduced weight for height, and underweight in kids, chronic energy deficiency in adults, micronutrient deficiencies, and diet-related noncommunicable diseases are public nutritional concerns for public health throughout the world.

Some developing countries are undergoing rapid socioeconomic, demographic, dietary, and health-care transitions. While these countries persist to face concerns (such as: poverty, starvation, and communicable diseases), they are also dealing with obesity and an increase in the prevalence of noncommunicable diseases, such as, high blood pressure and diabetes, particularly in metropolitan areas.

To deal with the challenge of malnutrition in any society, a thorough understanding of all facets of appropriate nutrition is required. Malnutrition is mostly caused by an inadequate/unbalanced intake of foods/nutrients. The WHO (World Health Organization) provides a recommended dietary allowance (RDA) of nutrients for each individual and each age group. This RDA focuses on nutrients and is technically oriented. Foods provide not only nutrition, but also a variety of non-nutrient phytochemicals for health benefits.

Food-based strategies are important routes to reach a good nutritional status because of the nutrients that are received buy us from the food. Malnourishment is typically related with a pattern of calorie and nutrient consumption rather than a lack of food. A healthy way of living demands adequate and suitable nutrition. When combined with regular exercise, a balanced meal can help retain a healthy body, decrease the threat of chronic illness (such as: heart diseases), and encourage general health.

Food must suit to the physiological needs of a person in terms of safety, excellence, and amount, as well as be socially and culturally adequate. "*Food security occurs when all people have physical, social, and economic access to sufficient, safe, and nutritious food that fits their dietary needs and food preferences for an active and healthy life* (International Institute for Population Sciences (IIPS) and Macro International 2007)" and it emphasizes availability, accessibility, and absorption or utilization of foods. Absorption determines nutrition outcomes. Nutritional security is defined as the ingestion of an adequate amount of energy, minerals protein, and vitamins, always (Fig. 17.1).

Sustainable nutrition is a worldwide challenge, as people fight to deal with the increased frequency of chronic diseases and dietary shortages while simultaneously conserving precious natural resources. Sustainable nutrition comprises increasing the availability, affordability, and cultural appropriateness of healthful, nutrient-dense foods while also protecting the environment and strengthening local communities. Healthy eating habits can help shield against the devastation caused by hunger and noncommunicable diseases such as, heart ailments, diabetes, malignancy and stroke.

While part of animal-sourced food production in climate change is being investigated, restricting these sources of high-quality protein is likely to have an unintended negative impact on human health, worsening inequalities and undernutrition, particularly among children, which can have life-long consequences. When compared to the huge carbon emissions caused by fossil fuel transportation, removing dairy cows from regenerative agricultural systems, for example, would have a minor impact on greenhouse gas emissions while greatly limiting the availability of some critical dietary minerals, such as calcium and vitamin D, which are already underutilized in the American diet.

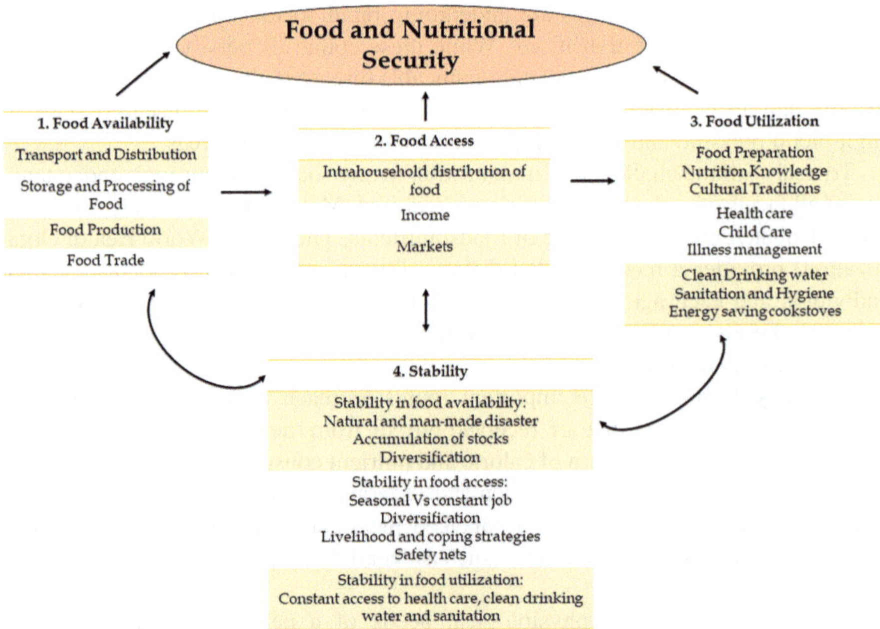

Fig. 17.1 General definition of a food security

Excellent quality protein, calcium, vitamins (such as: B6, B12, C, D and A), niacin, riboflavin, selenium, zinc, iodine and potassium are all found in milk. Because scientific data supports the role of milk and dairy foods in global nutrition and health, access to nutrient-dense foods like dairy foods may exacerbate present nutrient gaps and the developing malnutrition epidemic.

Dairy is important for nutrition at all stages of life, but it is especially important for children's nutrition, which is why at least 160 million children worldwide get and benefit from school milk programs today. This contribution is critical for all children, but especially for those who live in marginalized and poor communities with limited access to nutritional foods. Animal and plant diets should be considered as complementary rather than competing entities in terms of nutritional, social, economic, and environmental benefits. Rather than eliminating animal-sourced goods, the focus should be on prioritizing and supporting long-term solutions that make healthy foods (such as: milk and cheese, fresh fruits and vegetables, whole grains, and high-quality animal and plant-based proteins) more affordable and accessible.

Although this method promotes nutrition equity and healthier people, yet it is not an easy answer based on the following statement (Burlingame 2012): "*Sustainable diets have reduced environmental repercussions and contribute to nutrition and food security as well as a healthful life for existing and impending generations. Sustainable diets are ecologically and environmentally friendly, acceptable in many cultures, widely available, cost-effective, and cheap, nutritionally sufficient, reliable, and nutritious, & they maximize human and natural resources*".

This chapter explores how food-grade nanoparticles can be utilized to improve the functionality of nutraceuticals in functional foods and beverages.

17.2 Sustainable Food Systems Versus Healthy Diets

Despite a wealth of scientific proof connecting diet to human wellbeing and environmental sustainability, no commonly accepted criteria for sustainable food production and healthy diets have existed in the past. The EAT-Lancet Commission (37 experts in 16 countries: operating for human wellbeing, agriculture, economics, nutrition, political science, and environmental sustainability) analysed 2019 data and set international scientific targets for food systems that define a "safe operating range" (Burlingame 2012). These objectives are focused on two key areas that have an impact on us and the environment:

17.2.1 Target 1: Healthy Diets

The Commission characterizes a "planetary well-being diet," with utilization limits for every nutrition type, in view of significant concentrate on food varieties, dietary examples, and wellbeing suggestions. Notwithstanding its name, this is an adaptable dietary example that incorporates a low to direct measure of fish and poultry, as well as no or very little red meat, handled meat, refined grains, added sugar, and boring vegetables. As indicated by the Commission, inescapable reception of this eating example would have significant medical advantages, remembering a huge decrease for complete mortality.

17.2.2 Target 2: Sustainable Food Production

Because current food products contribute to climate alteration, biodiversity damage, pollution, and unsustainable water and land use changes, the Commission creates a set of guidelines for worldwide food products to follow in order to "reduce the threat of irreversible and potentially disastrous shifts in the Earth system." These restrictions are connected to climate alteration (based on greenhouse gas releases), land system alteration (based on agricultural activities), brackish water consumption, biodiversity damage (based on extinction rate), and the nitrogen & phosphate cycle (grounded on the operation of these diseases). It will need a significant effort to transition to a sustainable food system capable of delivering nutritious diets to an approximate 10 billion people by 2050.

The Commission emphasized that "statistics are stable and effective sufficient to compel on the spot action, and put-off will growth the opportunity of catastrophic,

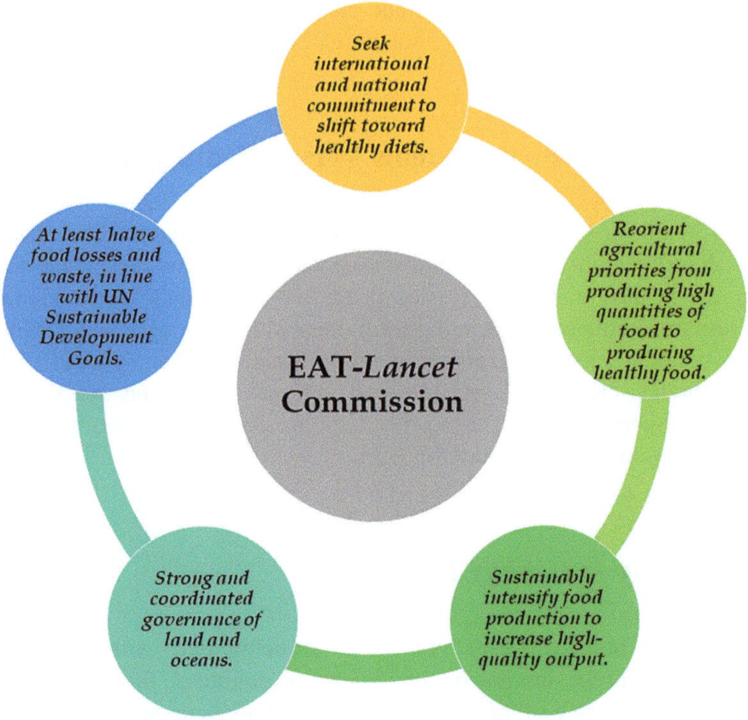

Fig. 17.2 Great food transformation: five broad initiatives for nationwide, district, town, and neighborhood change

even disastrous, consequences". Fortunately, their studies confirmed that this transition became potential through a mixture of full-size nutritional shifts closer to mostly plant-primarily based totally consuming models, remarkable discounts in meals deficiencies and waste, and big advances in meals manufacturing practices. Of course, such a "Great Food Transformation" will necessitate a multi-sector, multistage endeavor influenced via way of means of medical goals. To get started, the Commission proposed five most important goals for national, regional, and city transformation (Fig. 17.2).

17.3 Synthesis of Nanoparticles

Nanotechnology is a cutting-edge technology that aims to manufacture materials with a variety of forms as well as to improve food quality at the molecular stage. The nanoparticles range from 1 to 100 nm (Loghman et al. 2012; Ognik et al. 2016). When a substance is converted from its natural state to nanoparticles, its chemical and physical characteristics are altered. These modifications can improve

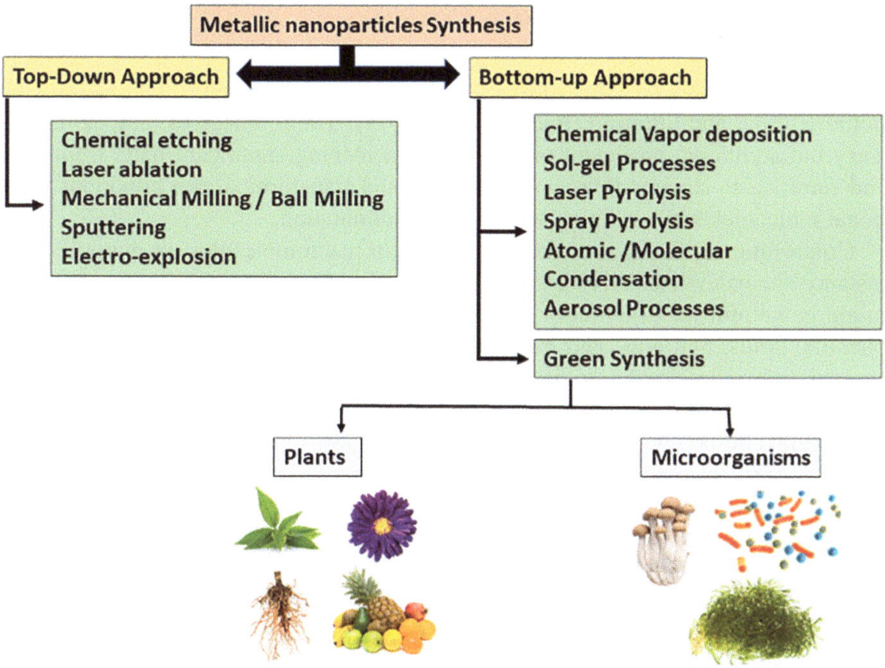

Fig. 17.3 Numerous methods for creation of nanoparticles

manufacture, transport, detection, treating, and food safety (Mohapatra et al. 2014; Otles and Yalcin 2008). Nanomaterials are not novel, but they have been employed to protect human and animal health in a variety of fields, including diagnostics, nutrition, the chemical production, biotechnology, drug delivery and therapeutics, biomedicine and vaccinations (Blecher et al. 2011; Iqbal et al. 2016; Lambe et al. 2016; Stephen et al. 2015; Yang et al. 2016). Figure 17.3 depicts various methods for synthesis of nanoparticles.

17.4 Use of Nanomaterials for Sustainable Food Production

Assuming humankind is to take care of a bigger and more well-off worldwide populace, it should increase food creation and productivity (Willett 2019). It should create more food, diminish food squander, and circulate food even more equally. Simultaneously, it is fundamental to do as such without harming the climate for people in the future. The following are few examples of how nanotechnology can add for the advancement of new maintainable and biologically cordial food source.

17.4.1 Valorization of Waste Streams

Nanotechnology can possibly further develop processes like centrifugation, gravitational settling, and filtration (Bacchin et al. 2011). These processes may be used to carry out additional proficient procedures of screening significant parts from food and rural waste streams that can be cleaned and transformed into important functional molecules while limiting waste and contamination.

Controlling colloidal associations between macromolecules or particles, for instance, at nanometer sizes can be used to change their screening state, which might be helpful for isolating or suspending them. Proteins, polysaccharides, polyphenols, lipids, nutrients, and minerals, for example, can be extracted from waste streams using these technologies. Waste from horticulture or food production can also be used to make functional nanoparticles. For example, cellulose can be extracted from food waste, wood or cotton, and transformed into useable nanofibers with a variety of uses in the food and additional activities.

Combining nanocellulose to food varieties, for instance, may lessen the pace of fat or starch breakdown due to gastric juices, maybe working on human wellbeing by bringing down glucose or cholesterol spikes in the wake of eating (DeLoid et al. 2018). For instance, adding nanocellulose to food varieties can decrease the pace of starch or lipid processing under gastrointestinal circumstances, which could help human wellbeing by alleviating glucose or lipid concentration after food intake. Nanocellulose is also being utilized to reduce the calorie thickness of fatty food sources, such as dressings, sauces, bread shop goods, and meat items, by re-enacting some of the enticing tactile and mouthfeel attributes traditionally associated with fats. Nanocellulose can be utilized as a key component in food packing components to give more eco-friendly and practical options to plastics. In non-food applications, the capability of dried cellulose nanofibers to soak up and store huge amount of liquids (e.g., blood and urine) is being utilized to enhance diapers, tampons, and pads (McClements 2019).

Biochar is an additional nanostructured substance that has huge promise as a useful material in culinary as well as non-food functions (Barrow 2012). It is created by burning low-oxygen farming waste. Since the material created is plentiful in carbon and minerals, it tends to be used as a manure on farming harvests. Besides, its application might give extra advantages by improving soil quality and limiting ozone depleting substance outflows (CO_2) into the climate. Nanotechnology can likewise be utilized to make another age of networks able to do productively eliminating synthetics from waste streams (McClements 2019). For example, nanofilters can be made by electrospinning regular or engineered polymer for make mats of caught nanofibers (Soares et al. 2018). These nanofilters can be designed to be really proficient and particular at disengaging specific atoms or particulate matter from arrangements because of their customized creation, high unambiguous surface region, high viewpoint proportion, and controlled pore size. A nanofilter of this kind is utilized to clean water and confine significant particles from squander streams.

17.4.2 Boosting Action of Fertilizers and Pesticides

Nanotechnology is likewise being utilized to work on the viability of farming herbicides and composts. It is being utilized, specifically, to make nanoparticle-based conveyance procedures to expand the adequacy of pesticides and manures while lessening their negative ecological ramifications (McClements 2019).

17.4.2.1 Nanopesticides

Pests like Microbes, bugs, and insects can destroy lot of horticultural fruits and animal products made by the food supply network. Thus, lessening these misfortunes is essential for the food supply increment. Pesticides should be applied cautiously to help agrarian yields and forestall crop misfortunes, yet extreme use can harm water, land assets, and air, along with the food crops which they are intended to treat. High degree of pesticide used can cause persistent issues like malignant growth, disease, and asthma. Pests (like microbes, bugs, and rodents) waste lot of farming harvests and animal items. Subsequently, diminishing these misfortunes is basic to supply increment the effectiveness and manageability of the food. Pesticide concentration and effectiveness can be improved by immobilizing pesticides in nanoparticles whose arrangement, size, shape, and surface properties can be customized to explicit applications (Fig. 17.4) (McClements 2019).

Copper-based nanoparticles, for instance, have been demonstrated to be more proficient than conventional copper-based pesticides at shielding watermelon from root contagious diseases (Elmer et al. 2018).

Other inorganic nanoparticles (like aluminum, silicon, silver, zinc, iron, and zinc) have likewise been demonstrated to be successful bug sprays against an assortment of pests on different yields (Shang et al. 2019). The nano size of nanomaterial-based pesticides is habitually connected to their great adequacy. These small particles can infiltrate more effectively through the openings in the stems, leaves, and roots, which are generally home to harmful microorganisms. Moreover, the surface highlights of nanoparticles, like charge and hydrophobicity, can be changed to assist them with sticking more safely to establish surfaces, decreasing their vulnerability to draining into the ecosystem.

Pesticide-loaded nanoparticle surfaces, for instance, can be made cationic with the end goal that they are electrically drawn to plant or pest anionic surfaces. At last, more effective pesticides can be made by catching polymeric microgels having nanoparticles loaded with pesticide inside them: the transfer rate can then be balanced by varying the microgels' pore size, and the delivery can be set off by utilizing microgels whose properties can explicit natural changes like enzyme activity, temperature, humidity, or pH.

There has been a lot of attention in making nanopesticides, which are plant-based to cure crops as opposed to inorganic ones, basically in light of the fact that they might be less risky to the climate and human wellbeing (McClements 2019).

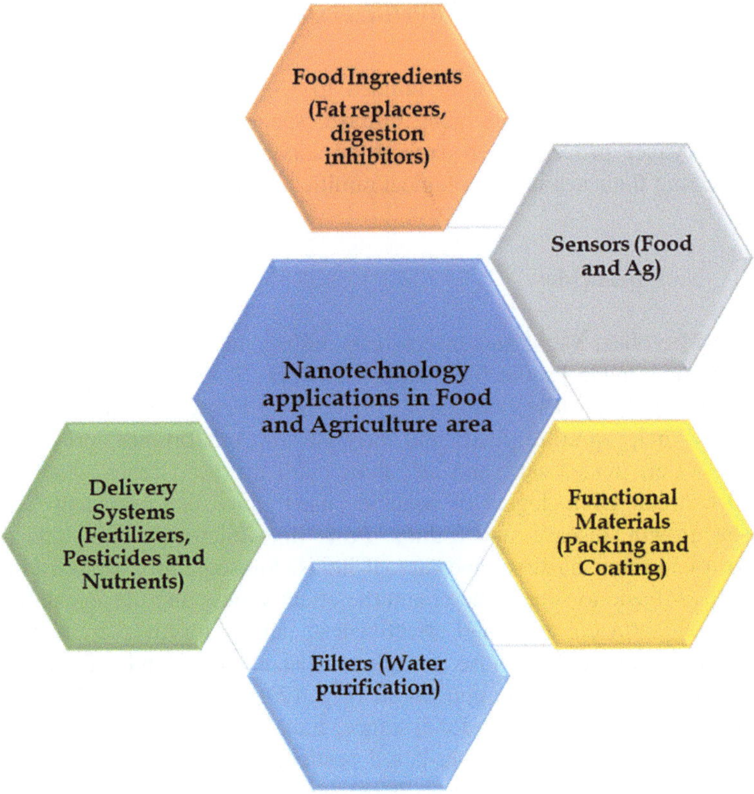

Fig. 17.4 Applications of nanotechnology in food and agriculture sectors

Secondary metabolites are normally delivered by plants, some of which have solid antibacterial action or other natural compounds that might keep bugs from harming crops, like offensive smells, severe preferences, or substances harmful to specific species. Plant secondary metabolites can be separated and utilized as normal pesticides.

Essential oils obtained from different plants are the most completely researched normal pesticides for this cause (Rao et al. 2019). Their power can be enormously expanded by changing over them into oil-in-water nanoemulsions, which will cause tiny essential oil droplets in water. To shield plants from bugs, these rejuvenating balm nanoemulsions can be showered on them or added to the dirt. Numerous different phytochemicals, like flavonoids, phenolics, and alkaloids, have been displayed to have huge antibacterial action (Zaynab et al. 2018), thus could be utilized to create plant-based nanopesticides. It could be feasible to decrease the utilization of engineered pesticides in the future by creating more successful plant-based other options (Fig. 17.5).

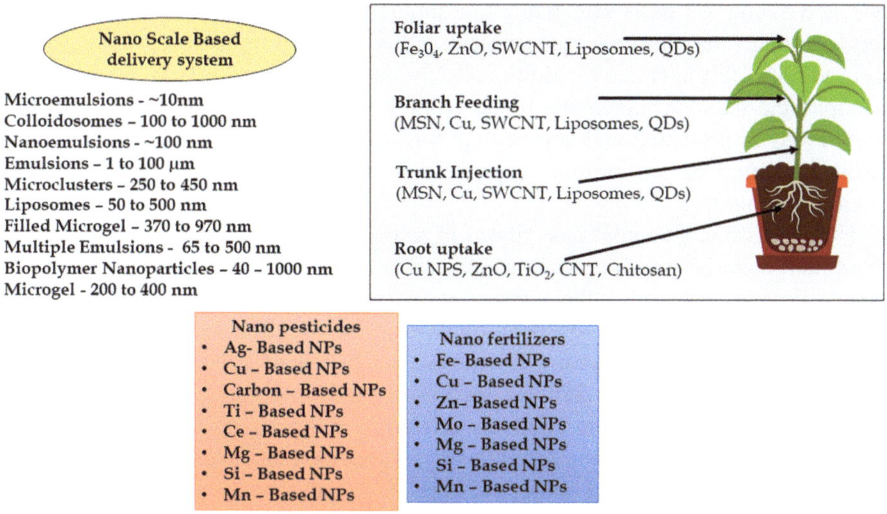

Fig. 17.5 Various forms of nanomaterials for the production of nanofertilizers and nanopesticides along with the NP interaction with the respective part of a plant

17.4.2.2 Nanofertilizers

Farming harvests require a great deal of water and supplements to flourish and remain sound, which are not generally available in nature. Therefore, ranchers should apply adequate measures of normal or manufactured fertilizers to harvests to guarantee significant yields, good plants, and decreased waste. Fertilizers as often as possible contain both macronutrients (like N, P, K) and micronutrients (e.g., Cu, Mn, Zn). Fertilizers, similar to pesticides, should be utilized with caution to refrain from them adversely affecting the climate by contaminating water supplies and soils, e.g., eutrophication. Nanotechnology innovations, like those used to further develop pesticide viability, can be used to increment fertilizer adequacy (McClements 2019). Nanofertilizers can be made by epitomizing supplements into little particles. The practical viability of these nanofertilizers can then be adjusted by changing the production, size, shape, or surface properties of the nanoparticles. As indicated by different investigations, nanofertilizers act uniquely in contrast to normal fertilizers because of their little size (Ma et al. 2018).

Nanofertilizers, for instance, have been shown to be more viable than regular fertilizers in supporting plant development, limiting misfortune, and working on better dietary quality (Fig. 17.5). Nonetheless, some nanofertilizers have been professed to have the contrary impact when applied to explicit farming harvests. Accordingly, it is essential to upgrade the synthesis and size of nanoparticles utilized in definite applications. For an assortment of reasons, nanofertilizers might find lasting success than regular fertilizers (Ma et al. 2018). First off, their nano size assists them with going through the pores in crop' stems, roots, or leaves, permitting supplements to arrive at within the plants. Second, nanofertilizer surfaces can be tweaked to connect more effectively to establish surfaces, decreasing the amount of fertilizer that filters into the environment and contaminates the ecosystem. At long last, various fertilizers and additionally pesticides can be mixed in a solitary nanoparticle-based arrangement. In the event that nanopesticides and nanofertilizers are viewed as protected and compelling, business reasonability will be significant. Ordinary composts are minimal expense substances that are normally used from country to country.

Numerous nanoformulations created in research centers, then again, incorporate exorbitant materials or procedures which are hard to cheaply scale up. It ought to be noted, in any case, that the greater expense of a nanoformulation might be counterbalanced by the investment funds acknowledged by supporting agrarian efficiency and limiting harvest misfortunes. For instance, it has been asserted that consuming an additional a $26 per section of land on a nanoparticle-based compound expanded watermelon yields by around $4600 per section of land (Dimkpa and Bindraban 2018). The numerous strategies expected to make the up-and-coming age of nanopesticides and nanofertilizers have proactively been talked about (Rodrigues et al. 2017).

The researchers accentuate the need of life-cycle investigation to decide the way of nanoparticles move from soils to plants to bugs to insects to people and the ecosystem. They likewise talk about the requirement for better comprehension of the expected pessimistic effects of nanoparticles on harvests, individuals, and the climate. There is a particular need to all the more likely comprehend how long nanoparticles stay in the ecosystem and to decide their conceivable effect on plant and soil microbiomes.

17.4.3 Nanomaterial-Aided Precision Agriculture

Enhanced harvests, reduced damages, and improved nutritive value of agricultural crops can be obtained from horticultural harvests by intently checking their development, wellbeing state, and nourishing necessities as they age and proper after care (McClements 2019). Robotized automated systems, for instance, can be utilized to apply an advanced portion of water, fertilizers, or pesticides to plants, or computerized hardware can reap the plants after they completely grow up. To work successfully, these automated systems require exact data about the situation with the plants, which is subject to sensor innovation accessibility (Wang et al. 2016).

Nanotechnology is being utilized to foster another age of sensors fit for giving information on plant development, wellbeing, and healthful prerequisites. Analysts are growing minimal expense compact detecting gadgets that ranch laborers can use to catch ongoing information on agrarian yield, soil, and ecological components (McClements 2019).

In these gadgets, nanomaterials are regularly utilized to further develop awareness and selectivity. Samples of plant tissue or soil is taken, blended in with water, squashed, and saved in a little sample tubes. Whenever a test contains a selected

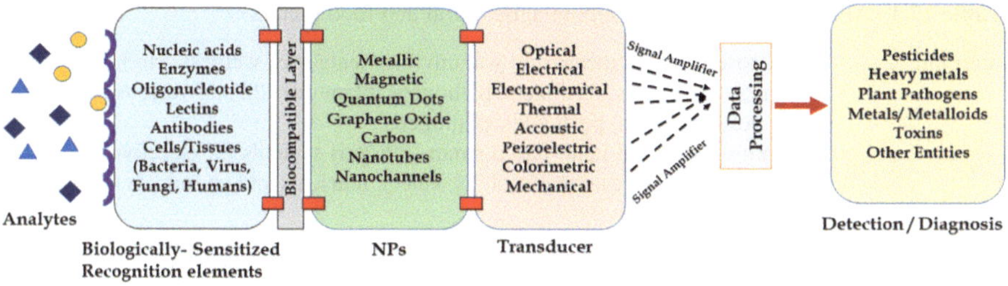

Fig. 17.6 Use of nanomaterials for production of nanosensors

compound, like a marker of plant development, nourishment, or well-being, these instruments produce quantitative outcomes, (for example, an electronic sign or signal response). The farmer can then settle on an educated choice on the most proficient method to deal with the yields in view of this information.

Remote sensors, for example, those that recognize the development of electromagnetic radiation (like noticeable or infrared radiation) by plants utilizing gadgets mounted on farm haulers, reapers, drones, planes, or satellites, can offer important data at times. Another age of nano-empowered sensors that will be put in soil or harvests and constantly send information to a distant PC is being created (Atzberger 2013). The use of this kind of sensor would empower better control of horticultural harvest yield.

Regardless, for the up-and-coming age of sensors to be generally utilized in the field, they should be affordable, dependable, reliable, and safe. Comparative sensors could be utilized to distinguish rural issues like pathogenic microorganisms or dangerous substances, consequently upgrading sanitation. Furthermore, earlier pest detection would permit farmers to apply insecticides to their crops before they were harmed, reducing losses & enhancing sustainability. Agricultural plants have been proven to internalize carbon nanotube-based sensors that communicate signals to a smartphone when the plants become contaminated with pesticides, demonstrating the technology's potential (Omanović and Maksimovic 2016; Wong et al. 2017). Nano-empowered sensors of this sort are supposed to be utilized in accuracy farming from now on. Regardless, it is basic to guarantee that the sensors don't taint the yields and that they are economical, trustworthy, and robust (Fig. 17.6 and Table 17.1).

17.4.4 Nanomaterial-Based Functional Materials

Nanotechnology is additionally being utilized to make another age of additional material which is practical and harmless to the ecosystem, like engineered polymers (plastics), that are well-known to be naturally harmful (McClements 2019). A significant measure of exploration has been led on the improvement of

Table 17.1 Applications of nanosensors in agricultural and food sectors

Farming	• Nanosensors for measuring soil environments (pH, wetness, etc.), pesticides, herbicides, pesticides, fertilizers, pathogens, plant growth, and various pesticides, herbicides, fertilizers, pesticides, pathogens • Nanosensors and nano-based smart transport systems for effective use of agricultural natural reserves (such as water), nutrients and chemicals through precision agriculture • Nanoparticles for delivering growth hormones or DNA to plants in the field • Nanochips for preservation and tracking of identities • Nanocapsules for supplying pesticides, herbicides, fertilizers and vaccines • An aptasensor for measuring pesticides and pesticides (e.g.: horate, acetamiprid, and isocarbofos, e.g., cocaine, oxytetracycline, tetracycline, kanamycin) • Adapter sensors for heavy metal detection (Hg^{2+}, As^{3+}, Cu^{2+}, etc.)
Food treating	• Nanoencapsulated flavor enhancer • Nanotubes and nanoparticles as gelling agents & thickeners • Nanocapsule injection of plant steroids to substitute cholesterol in meat • Selective binding of contaminants or pathogens from food and nanoparticles for removal of OTA, fumonisin B1)
Food packing	• DNA biochips for identifying pathogens and identifying the presence of various types of dangerous bacteria and fungi that damage meat, fish, or fruits • Wearable nanosensors for detecting compounds, microbes and poisons in food • Embedded in packaging items to detect chemicals distributed when food is damage, detecting tastes (for example, to detect bitterness, sweetness, saltiness, umami, sourness) or nose (for example, of wine) • Detection of ethylene by electromechanical nanosensor • Nanosensors used as labels or coatings improve food packaging integrity by detecting leaks, indicating temperature changes, and ensuring microbial safety • An aptasensor for detecting microbial cells (e.g.: *Salmonella, E. Coli, listeria monocytogenes*), drugs (e.g.: Cocaine, oxytetracycline, tetracycline, kanamycin) • Aptasensors for heavy metal detection (Hg^{2+}, As^{3+}, Cu^{2+}, etc.)
Food transfer	• examining environmental situations during distribution and storing with nanosensors • Nanosensors for tracing and examining of products in transit and storage. This is important for products with short shelf-life • Aptasensor to detect microbial cells • Smart sensor technology that monitors the condition of grains, milk products, fruits and vegetables in the storage area to identify the cause and kind of spoilage (e.g.: *Salmonella, E. Coli, listeria*)
Nourishment	• Foods containing nanocapsules that provide nutrients • Nanospirals (50 nm coiled nanoparticles) for more efficient delivery of nutrients (vitamins, lycopene, omega fatty acids, etc.) to cells while preserving the color and flavor of food

eco-friendly nanostructured packaging fabrics and coverings for agrarian and food wares specifically (Fig. 17.7) (Youssef and El-Sayed 2018).

These materials are every now and again produced using regular polymers like proteins and polysaccharides. Besides, natural or inorganic nanoparticles might be inserted in the polymer framework to further develop useful properties like exterior,

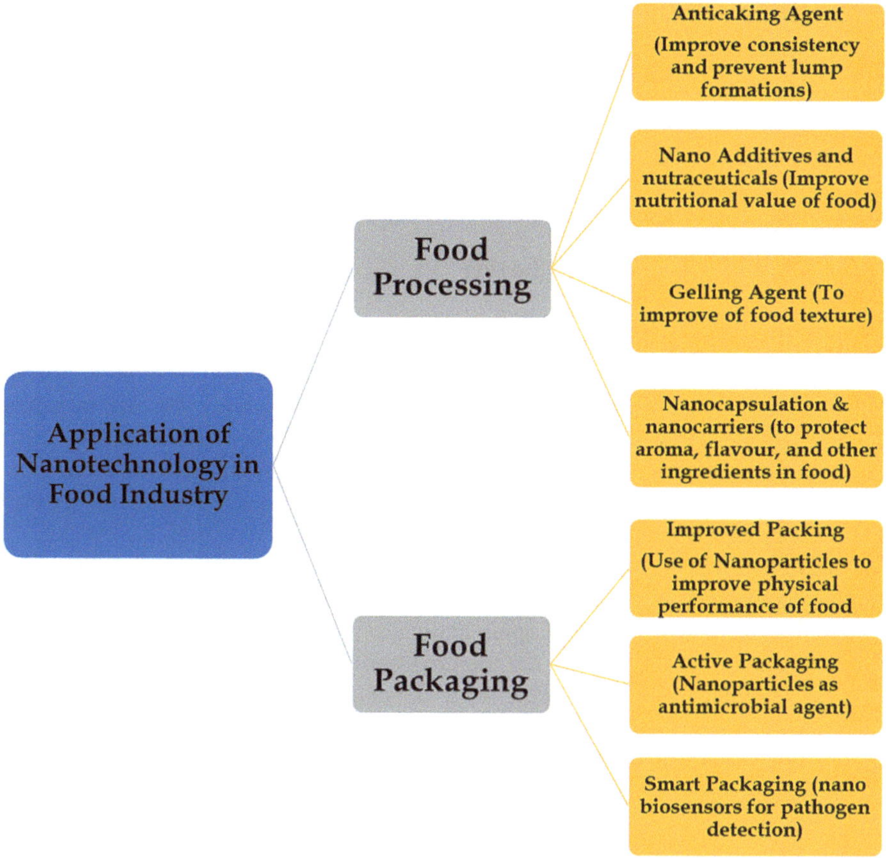

Fig. 17.7 Applications of nanomaterials in food packaging and processing

mechanistic strength, or obstruction characteristics, as well as antibacterial or anti-oxidant action. They may likewise contain normal substances, for example, anthocyanins, which go about as regular sensors and give data about the newness or nature of the packed food. A full comprehension of the underlying association and collaborations of the different components at the nanoscale is expected for the production of better packaging materials. This information can be utilized to upgrade their practical presentation. Further it will be essential to guarantee that such packaging materials can be efficiently manufactured at a sensible expense while holding their optimal utilitarian elements under certifiable circumstances.

17.5 Role of Nanomaterials in our Health

Investigators are also using nanotechnology to improve the security and well-being of foods (McClements 2019). This segment highlights variety of prospective applications of nanotechnology in this field.

17.5.1 Enhancing Food Safety

Food spoilage from damage and harmful microbes results in food waste and builds the hazard of foodborne sickness and death. Customarily, microbial tainting of food sources was overseen through handling or stockpiling systems, for example, heat processing, lack of hydration, freezing, or pH change. For this point, synthetic additives like benzoic acid, natural acids, nitrates, and sulfur dioxide have as of late been used. In any event, when microorganisms promote antibacterial activity, many synthetic compounds become obsolete.

Furthermore, in response to customer demand for "clean label" products, the food industry is trying to substitute synthetic preservatives with more natural ones. There has been a lot of interest in developing antimicrobials which are plant-based that are not only effective but also safe for the environment. As a result, investigators are utilizing nanotechnology to produce a novel group of inorganic and natural nanoparticles that may be used as efficient antibacterial specialists in food. Inorganic nanoparticles such as gold, titanium, silver, and copper have been shown to exhibit antibacterial characteristics (Wang et al. 2017). These nanoparticles interact with microorganism cell, causing infinitesimal holes causing cell lysis. Moreover, some inorganic nanoparticles produce Reactive oxygen species (ROS), which obliterate microorganisms' cell structures, like DNA, RNA, proteins, or phospholipids.

Despite their extraordinary antibacterial properties, inorganic nanoparticles are not especially name amicable, and there is a lot of worry about their expected harmfulness. Accordingly, numerous researchers are utilizing nanotechnology to make antimicrobials that are completely made out of regular natural substances.

Antimicrobial nanoemulsions got from essential oils, like thyme, cloves, lemon, cinnamon, peppermint, orange, and garlic, are quite compelling (Rao et al. 2019). Plants produce essential oils to safeguard themselves from insects like microbes, parasites, yeasts, and bugs, as recently expressed. To endure the numerous bugs to which they have been uncovered over their reality, these oils contain a complicated mix of variety of antibacterial synthetics. In antimicrobic nanoemulsions, these oils can kill microbes by disturbing cell films and slowing down significant metabolic cycles. These nanoemulsions can possibly decrease bacterial infections in new harvests, for example, alfa alfa seeds, radish seeds, and mung beans (Landry et al. 2015). These natural antimicrobics may eventually substitute several of the synthetic antimicrobials presently utilized in the food business.

Harvard University has contrived an original methodology for the advancement of antimicrobial medicines using NT (nanotechnology) (Huang et al. 2019). Utilizing an electrospray/ionization strategy, pure water can be transformed into a fog of nanometer-sized beads containing receptive oxygen species ("Engineered Water Nanostructures," or EWNS). ROS are unquestionably powerful antimicrobial specialists that can hurt microorganisms when they come into contact with them, as recently showed. These oils can kill microorganisms by breaking cell films and slowing down significant metabolic cycles in antimicrobial nanoemulsions. These nanoemulsions can possibly decrease bacterial growth in new food, for example, alfa-alfa seeds, radish seeds, and mung beans (Landry et al. 2015). To kill microbes in the environment or on food, they can be blended into food wrapping supplies. However long they are protected to eat, they can be utilized as utilitarian parts in dinners and drinks. One potential trouble with consuming antimicrobic nanoparticles in food is that they might go through the GIT and upset the stomach microbiota, possibly harming human wellbeing (Lamas et al. 2020). Therefore, it is fundamental to guarantee the wellbeing of the new age of nano-empowered nanoparticles.

17.5.2 *Variation in Absorption of Macronutrients*

It has been contended that the new huge expansion in corpulence, obesity, and diabetes is the consequence of expanded utilization of ultra-processed food sources (McClements 2019). The human gastrointestinal system advanced to be phenomenally productive at getting supplements from the food varieties accessible to our forebears, which were often whole or minimally processed foods. Sodas, organic product, breads, juices, treats, snacks, pastas, and cakes are highly processed, annihilating their original structure. Therefore, the chemicals in our GI lot quickly breakdown these food sources, bringing about an increase in glucose or cholesterol levels as well as a decline in satiety reaction. Accordingly, highly-processed food varieties might disrupt our metabolic and hormonal systems, expanding our chance of overeating and constant infection (McClements 2019).

Accordingly, numerous scientists are investigating ways of lessening macronutrient assimilation inside the GIT. Nanotechnology is one methodology being explored for this reason. As previously stated, the incorporation of nanocellulose into diets has been shown to lower the rate and amount of macronutrient digestion (DeLoid et al. 2018). Several physiological and physicochemical tools have been proposed to explain this anomaly, including the confinement of bile salts, enzymes, or calcium to nanofibers; nanofiber gelation or thickening of gastrointestinal liquids, which slows mixing and diffusion processes; and the development of nanofiber coatings around macronutrients, which reduces the action of digestive juices. Other edible nanoparticles, such as nanochitin, have been shown to inhibit macronutrient absorption (Zhou et al. 2020). Thus, these nanomaterials might be utilized as practical parts to diminish macronutrient assimilation in handled food sources, so helping with the battle against overweightness, diabetes, and obesity. Regardless, more exploration is expected to ensure that they are protected and have no bad wellbeing influences. Late examination, for instance, has uncovered that nanochitin can restrict the bioaccessibility of oil dissolvable nutrients (Zhou et al. 2020).

17.5.3 Enhancing Bioavailability of Bioactive Compounds

Food handling can be improved by combining bioactive compounds with essential nutritive properties, like vitamins, minerals, or nutraceuticals (McClements 2019). Nanotechnology is particularly well suited to improving food substance and increasing bioavailability bioactive ingredient orally. The bioavailability can be explained as the percentage of a substance that is absorbed in an active state into the systemic circulation (Nile et al. 2020).

A few novel nanotech-based procedures are being developed to expand the biological availability of bioactive synthetic compounds in food varieties: excipient & delivery systems as shown in Fig. 17.8 (McClements 2019). Bioactive-stacked palatable nanoparticles are used in delivery systems to bring bioactive mixtures into the GIT so epithelial cells can ingest them. The delivery frameworks for lipid nanoparticles are intended to break and create nanostructures (mixed micelles) inside the GI liquids, consequently helping bioactive transport and assimilation. Excipient systems are foods that are created to boost the bioactive compounds availability in natural foods taken with them, like vegetables & fruits. They do as such by damaging the GIT and establishing a condition that expands the bioaccessibility, steadiness, or potentially assimilation of the bioactive synthetic compounds. *In vitro* examinations, for instance, have shown that when carrot carotenoids are combined with excipient nanoemulsions, their bioaccessibility is essentially supported (Zhang et al. 2016; Youssef and El-Sayed 2018).

Excipient nanoemulsions could in this way give the premise to novel age of dressings, creams, or sauces pointed toward supporting the biological availability of bioactive parts in vegetables & fruits, ingested simultaneously. In the two instances, drop size influences the pace of lipid absorption and bioactive biological availability,

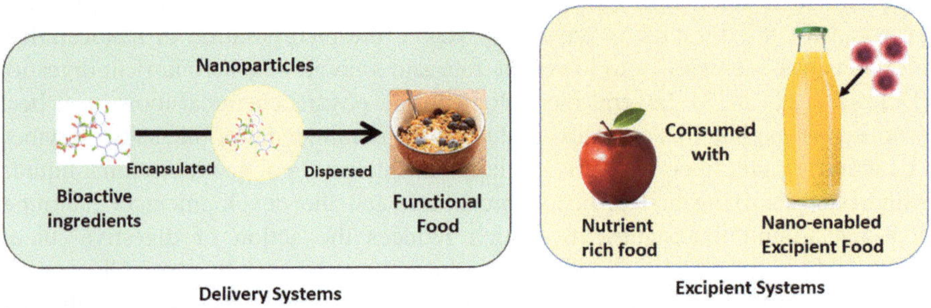

Fig. 17.8 Nanotechnology for boosting the bioavailability of nutrients and nutraceuticals by utilizing transport methods or excipient methods

Table 17.2 Nanotechniques to capture and transport of functional materials (Singh et al. 2017)

Nanomaterial technique	Characteristic features	Examples	References
Edible coatings	To keep fresh goods fresh for a lengthy period of time	Cellulose nanocrystal: Entrapped gelatin-based edible coatings; Coatings of chitosan/ nanosilica; Nano-SiO_2 based chitosan film; Coatings of alginate/ lysozyme nanolaminate;	Fakhouri et al. (2014) and Medeiros et al. (2014), Shi et al. (2013) and Yu et al. (2012)
Hydrogels	It is easily packed in capsules, preserves pharmaceuticals from harsh environments, and allows them to be delivered in reaction to environmental cues like temperature and pH	Hydrogels made of protein	Qui and Park (2001)
Inorganic NPs	They have excellent encapsulation capability & hard exteriors that enable for regulated functionalization	Mesoporous silica nanoparticles	Tang et al. (2012)
Liposomes	Because a liposome contains water-based solution within a hydrophobic membrane, it can be utilized to deliver hydrophilic or hydrophobic molecules	Liposomes having cationic lipid altered with an acid-labile Polymer hyper-branched Poly(glycidol) (HPG)	Yoshizaki et al. (2014)
Nano-emulsions	• Improved resistance to gravitational separation & droplet aggregation; • Improved optical clearness; and • Improved bioavailability orally	b-carotene-based nanoemulsion	Kong et al. (2011) and Yuan et al. (2008)
Polymeric micelles	Water-insoluble chemicals are dissolved in the hydrophobic core, resulting in elevated solubility and minimal toxicity	PEO-b-PCL (poly (ethylene glycol) block-poly (caprolactone)) polymeric micelles	Ma et al. (2008)

prompting an expansion in the surface region of the lipid stage presented to lipase in the GI liquids (Fig. 17.8 and Table 17.2).

17.5.4 Regulated or Directed GI Delivery

A wide scope of bioactive compounds that might be advantageous to human wellbeing should be conveyed in their entire to explicit areas inside the GIT (McClements 2017, World Health Organization (WHO) 2009). Stomach enzymes (like lactase or lipase) can be carried to the small intestine in people with lacks protein, like lactose intolerance or pancreatitis.

Probiotic microbes, likewise, should be delivered to the colon to be important to the stomach microbiota. However, acids, digestive enzymes, and bile salts frequently destroy enzymes and probiotics in the upper GIT before they reach their intended destination (Fig. 17.9). As a result, these delicate bioactive compounds must be encapsulated and protected under certain GIT conditions before being released at the necessary region of action. Nanostructured particles, like biopolymer-based nanoparticles or microgels, are frequently used to accomplish this (McClements

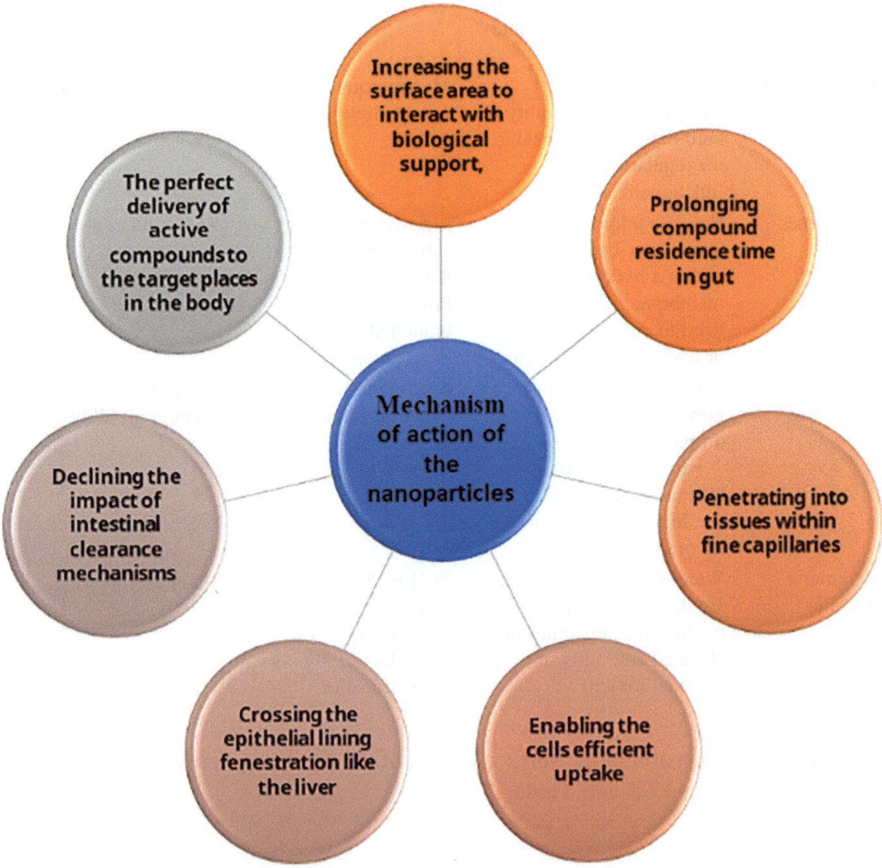

Fig. 17.9 The mechanism of action of the nanoparticles

2017). These biopolymer particles' width and pore size can be adjusted to limit the passage of GI contents into them. Co-encapsulation of antiacids with the bioactive ingredients helps to protect them from degrading in the extremely acidic gastric secretions.

17.5.5 Nano-Enabled Sensing

Food characteristics are often tested using modern analytical tools like chromatography, mass spectrometry, or spectroscopy. However, nanotechnology is presently being exploited to create a novel group of small sensing devices capable of delivering information instantly and economically with no need for specialized knowledge (Wang and Duncan 2017). Food manufacturers, wholesalers, merchants, and customers may now instantly check the condition of beverage and food commodities. By embedding nano-enabled sensors into food packing substances or linking them with mobile phones, consumers may instantly detect the qualitative qualities of their foods. In theory, these sensors may offer information regarding food makeup, flavor, value, and security. Several nanotechnological approaches are being employed to accomplish this target. Nano-enabled sensors, for instance, have been designed to detect the quantities of synthetic colors, poisons and pesticides in meals using Raman spectroscopy techniques where detection of inelastic scattering from metallic nanoparticles is done (Zheng and He 2014).

17.6 Safety Issues

Beside the gigantic advantages of nanotechnology to the food business, there are sure wellbeing issues encompassing nanoparticles. Many have investigated the security of nanomaterials, zeroing in on the prospect of nanoparticles getting deposited from wrapping components in the food and its impact on purchaser wellbeing (Bradley et al. 2011; Jain et al. 2016).

Albeit a substance is GRAS (for the most part viewed as protected); extra examination is expected to decide the risk of its nano structures since the properties of physiochemical nature in nanostates vary emphatically from those in macrostates. Besides, the little size of the nano-materials may build the chance of accumulation inside body tissues (Savolainen et al. 2010). When presented to silica nanoparticles utilized as against solidifying specialists, for instance, human lung cells can be hurt (Athinarayanan et al. 2014; Sahu et al. 2008).

Disintegration is impacted by various boundaries, including molecule surface shape, focus, surface energy, collection, and adsorption. Cushen and associates fostered a model to dissect molecule relocation from food bundling. They concentrated on Ag and Cu removal from nanocomposites and found that the level of nanofiller in the nano-composites was quite possibly the best basic component driving the movement, offsetting molecule size, temperature, and contact length. Since each nanomaterial is particular, poisonousness will absolutely be evaluated dependent upon the situation (Mahler et al. 2012; Cushen et al. 2014). Administrative bodies should lay out some business item principles to guarantee item quality, wellbeing and security prerequisites, and natural requirements.

17.7 Conclusion

Since the utilization of nanometer-scale structures in the food business has extended in prevalence lately, there has been a large interest and effective work seen in this area. Nanobiotechnology-based devices and components are increasing and extra effective as the innovation moves along. Its purposes in food hustling and sanitation are notable. Moreover, encouraging outcomes have been accounted for in food safeguarding utilizing nanomaterials, wherever they could shield the edible products from dampness, oils, scents, off-flavors, and gases. They are proficient transporters of bioactive substances to explicit tissues. Despite the fact that nanotechnology progresses are making new ways consistently, there are as yet innumerable challenges and chances to work on current innovation, as well as questions with respect to the consequences of nanotechnology that should be directed to calm purchaser anxieties. While managing the development of nanotechnology in food systems, straightforwardness of security issues and natural effect ought to be focused on, and henceforth commanded testing of nano food varieties before they are delivered to the market is required.

References

Athinarayanan J, Periasamy VS, Alsaif MA, Al-Warthan AA, Alshatwi AA (2014) Presence of nanosilica (E551) in commercial food products: TNF-mediated oxidative stress and altered cell cycle progression in human lung fibroblast cells. Cell Biol Toxicol 30:89–100

Atzberger C (2013) Advances in remote sensing of agriculture: context description, existing operational monitoring systems and major information needs. Remote Sens 5:949–981

Bacchin P, Marty A, Duru P, Meireles M, Aimar P (2011) Colloidal surface interactions and membrane fouling: investigations at pore scale. Adv Colloid Interf Sci 164:2–11

Barrow CJ (2012) Biochar: potential for countering land degradation and for improving agriculture. Appl Geogr 34:21–28

Blecher K, Nasir A, Friedman A (2011) The growing role of nanotechnology in combating infectious disease. Virulence 2:395–401

Bradley EL, Castle L, Chaudhry Q (2011) Applications of nanomaterials in food packaging with a consideration of opportunities for developing countries. Trends Food Sci Technol 22:603–610

Burlingame B (2012) Sustainable diets and biodiversity—directions and solutions for policy research and action. In: proceedings of the international scientific symposium biodiversity and sustainable diets united against hunger. FAO, Rome, p 61

Cushen M, Kerry J, Morris M, Cruz-Romero M, Cummins E (2014) Evaluation and simulation of silver and copper nanoparticle migration from polyethylene nanocomposites to food and an associated exposure assessment. J Agric Food Chem 62:1403–1411

DeLoid GM, Sohal IS, Lorente LR, Molina RM, Pyrgiotakis G, Stevanovic A, Demokritou P (2018) Reducing intestinal digestion and absorption of fat using a nature-derived biopolymer: interference of triglyceride hydrolysis by nanocellulose. ACS Nano 12:6469–6479

Dimkpa CO, Bindraban PS (2018) Nanofertilizers: new products for the industry? J Agric Food Chem 66:6462–6473

Elmer W, De La Torre-Roche R, Pagano L, Majumdar S, Zuverza-Mena N, Dimkpa CO, Gardea-Torresdey J, White JC (2018) Effect of metalloid and metal oxide nanoparticles on fusarium wilt of watermelon. Plant Dis 102:1394–1401

Fakhouri FM, Casari ACA, Mariano M, Yamashita F, Mei LI, Soldi V (2014) Effect of a gelatin-based edible coating containing cellulose nanocrystals (CNC) on the quality and nutrient retention of fresh strawberries during storage. In: Proceedings of the IOP conference series: materials science and engineering—2nd international conference on structural Nano composites, nanostructure, Madrid, vol 64. IOP Publishing, Madrid, p 012024. https://doi.org/10.1088/1757-899X/64/1/012024

Huang RZ, Vaze N, Soorneedi A, Moore MD, Xue YL, Bello D, Demokritou P (2019) Inactivation of hand hygiene-related pathogens using engineered water nanostructures. ACS Sustain Chem Eng 7:19761–19769

International Institute for Population Sciences (IIPS), Macro International (2007) National Family Health Survey (NFHS-3) for 2005/2006, India, vol I. International Institute for Population Sciences, Mumbai, pp 1–168

Iqbal HM, Villalba A, Khandia R, Munjal A, Dhama K (2016) Recent trends in nanotechnology-based drugs and formulations for targeted therapeutic delivery. Recent Pat Inflam Allergy Drug Discov 10:18722. https://doi.org/10.2174/1872213X10666161213162823

Jain A, Shivendu R, Nandita D, Chidambaram R (2016) Nanomaterials in food and agriculture: an overview on their safety concerns and regulatory issues. Crit Rev Food Sci Nutr 1160363:297–317. https://doi.org/10.1080/10408398.2016.1160363

Kong M, Chen XG, Kweon DK, Park HJ (2011) Investigations on skin permeation of hyaluronic acid-based nanoemulsion as transdermal carrier. Carbohydr Polym 86:837–843. https://doi.org/10.1016/j.carbpol.2011.05.027

Lamas BG, Breyner NM, Houdeau E (2020) Impacts of foodborne inorganic nanoparticles on the gut microbiota-immune axis: potential consequences for host health. Part Fibre Toxicol 17:19–23

Lambe U, Minakshi P, Brar B, Guray M, Ranjan K (2016) Nano diagnostics: a new frontier for veterinary and medical sciences. J Exp Biol Agric Sci 4:307–320

Landry KS, Micheli S, McClements DJ, McLandsborough L (2015) Effectiveness of a spontaneous carvacrol nanoemulsion against salmonella enterica Enteritidis and Escherichia coli O157:H7 on contaminated broccoli and radish seeds. Food Microbiol 51:10–17

Loghman A, Iraj SH, Naghi DA, Pejman M (2012) Histopathologic and apoptotic effect of nanosilver in liver of broiler chickens. Afr J Biotechnol 11:6207–6211

Ma ZS, Haddadi A, Molavi O, Lavasanifar A, Lai R, Samuel J (2008) Micelles of poly(ethylene oxide)-b-poly(epsilon caprolactone) as vehicles for the solubilization, stabilization, and controlled delivery of curcumin. J Biomed Mater Res A 86:300–310

Ma CX, White JC, Zhao J, Zhao Q, Xing BS (2018) Uptake of engineered nanoparticles by food crops: characterization, mechanisms, and implications. Annu Rev Food Sci Technol 9:129–153

Mahler GJ, Esch MB, Tako E, Southard TL, Archer SD, Glahn RP (2012) Oral exposure to polystyrene nanoparticles affects iron absorption. Nat Nanotech 7:264–271

McClements DJ (2017) Designing biopolymer microgels to encapsulate, protect and deliver bioactive components: physicochemical aspects. Adv Colloid Interf Sci 240:31–59

McClements DJ (2019) Future foods: how modern science is transforming the way we eat. Springer, Cham, pp 1–35

Medeiros BGS, Souza MP, Pinheiro AC, Bourbon AI, Cerqueira MA, Vicente AA (2014) Physical characterization of an alginate/lysozyme nano-laminate coating and its evaluation on 'Coalho' cheese shelf-life. Food Bioproc Technol 7:1088–1098

Mohapatra P, Swain RK, Mishra SK, Behera T, Swain P (2014) Effects of dietary nano-selenium supplementation on the performance of layer grower birds. Asian J Anim Vet Adv 9:641–652

Nile SH, Baskar V, Selvaraj D, Nile A, Xiao JB, Kai GY (2020) Nanotechnologies in food science: applications, recent trends, and future perspectives. Nanomicro Lett 12:45–53

Ognik K, Stepniowska A, Cholewinska E, Kozowski K (2016) The effect of administration of copper nanoparticles to chickens in drinking water on estimated intestinal absorption of iron, zinc and calcium. Poult Sci 95:2045–2051

Omanović E, Maksimovic M (2016) Nanosensors applications in agriculture and food industry. Bull Chem Technol Bosnia and Herzegovina 47:59–70

Otles S, Yalcin B (2008) Smart food packaging. Elektroniczne Czasopismo Naukowe z Dziedziny Logistyki (Electronic Scientific Journal in the field of Logistics) 4:1–7

Qui Y, Park K (2001) Environment sensitive-hydrogels for drug delivery. Adv Drug Devl Rev 53: 321–339

Rao JJ, Chen BC, McClements DJ (2019) Improving the efficacy of essential oils as antimicrobials in foods: mechanisms of action. Annu Rev Food Sci Technol 10(1):365–387

Rodrigues SM, Demokritou P, Dokoozlian N, Hendren CO, Karn B, Lowry GV (2017) Nanotechnology for sustainable food production: promising opportunities and scientific challenges. Environ Sci Nano 4:767–781

Sahu A, Bora U, Kasoju N, Goswami P (2008) Synthesis of novel biodegradable and self-assembling methoxy poly(ethylene glycol)-palmitate nanocarrier for curcumin delivery to cancer cells. Acta Biomater 4:1752–1761

Savolainen K, Pylkkänen L, Norppa H, Falck G, Lindberg H, Tuomi T (2010) Nanotechnologies, engineered nanomaterials and occupational health and safety: review. Saf Sci 6:1–7

Shang YF, Hasan MK, Ahammed GJ, Li MQ, Yin HQ, Zhou J (2019) Applications of nanotechnology in plant growth and crop protection: review. Molecules 24(14):2558. https://doi.org/10.3390/molecules24142558

Shi S, Wang W, Liu L, Wu S, Wei Y, Li W (2013) Effect of chitosan/nano-silica coating on the physicochemical characteristics of longan fruit under ambient temperature. J Food Eng 118: 125–131

Singh T, Shukla S, Kumar P, Wahla V, Bajpai VK, Rather IA (2017) Applications of nanotechnology in food science: perception and overview. Front Microbiol 8:01501

Soares RMDS, Siqueira NM, Prabhakaram MP, Ramakrishna S (2018) Electrospinning and electrospray of bio-based and natural polymers for biomaterials development. Mater Sci Eng C 92:969–982

Stephen BJ, Jain M, Dhama K, Singh SV, Datta M (2015) Nanotechnologybased therapeutics, drug delivery mechanisms and vaccination approaches for countering Mycobacterium avium subspecies paratuberculosis (MAP) associated diseases. Asian J Anim Vet Adv 10:830–842

Tang F, Li L, Chen D (2012) Mesoporous silica nanoparticles: synthesis, biocompatibility and drug delivery. Adv Mater 24:1504–1534

Wang Y, Duncan TV (2017) Nanoscale sensors for assuring the safety of food products. Curr Opin Biotechnol 44:74–86

Wang LL, Hu C, Shao LQ (2017) The antimicrobial activity of nanoparticles: present situation and prospects for the future. Int J Nanomedicine 12:1227–1249

Wang P, Lombi E, Zhao FJ, Kopittke PM (2016) Nanotechnology: a new opportunity in plant sciences. Trends Plant Sci 21:699–712

Willett W (2019) Food in the Anthropocene: the EAT−lancet commission on healthy diets from sustainable food systems. Lancet 393:447–492

Wong MH, Giraldo JP, Strano MS (2017) Nitroaromatic detection and infrared communication from wild-type plants using plant nano bionics. Nat Mater 16(2):264–272

World Health Organization (WHO) (2009) Report infant and young child feeding model chapter for textbooks for medical students and allied health professionals. WHO, Geneva, p 58. https://www.who.int/nutrition/publications/infantfeeding/9789241597494.pdf
Yang L, Li W, Kirberger M, Liao W, Ren J (2016) Design of nanomaterial-based systems for novel vaccine development. Biomater Sci 4:785–802
Yoshizaki Y, Yuba E, Sakaguchi N, Koiwai K, Harada A, Kono K (2014) Potentiation of pH-sensitive polymer-modified liposomes with cationic lipid inclusion as antigen delivery carriers for cancer immunotherapy. Biomaterials 35:8186–8196
Youssef AM, El-Sayed SM (2018) Bionanocomposites materials for food packaging applications: concepts and future outlook. Carbohydr Polym 193:19–27
Yu Y, Zhang S, Ren Y, Li H, Zhang X, Di J (2012) Jujube preservation using chitosan film with nano-silicon dioxide. J Food Eng 113:408–414
Yuan Y, Gao Y, Zhao J, Mao L (2008) Characterization and stability evaluation of beta-carotene nanoemulsions prepared by high-pressure homogenization under various emulsifying conditions. Food Res Int 41:61–68
Zaynab M, Fatima M, Abbas S, Sharif Y, Umair M, Bahadar K (2018) Role of secondary metabolites in plant defense against pathogens. Microb Pathog 124:198–202
Zhang RJ, Zhang ZP, McClements DJ (2016) Impact of lipid content on the ability of excipient emulsions to increase carotenoid bioaccessibility from natural sources (raw and cooked carrots). Food Biophys 11:71–80
Zheng JK, He LL (2014) Surface-enhanced Raman spectroscopy for the chemical analysis of food. Compr Rev Food Sci Food Saf 13:317–328
Zhou HL, Tan YB, Lv SS, McClements DJ (2020) Nanochitin-stabilized Pickering emulsions: influence of nanochitin on lipid digestibility and vitamin bioaccessibility. Food Hydrocoll 106:105878

Part VI
Food Science and Nutritional Research for Human Health

Chapter 18
Bioaccessibility and Bioavailability of Phenolics in Grapes and Grape Products

Müzeyyen Berkel Kaşıkçı and Neriman Bağdatlıoğlu

Contents

Abstract Grape is very popular fruit since ancient times. Grapes can be utilized to produce a variety of products including grape juice, raisin, wine, grape molasses and churchkhela. Flavonoids, stilbenes and anthocyanins are the most abundant phenolics found in grapes. Phenolic compounds are antioxidants that have been linked to many health benefits, including lower blood pressure and cardiovascular disease risk, cancer protection, anti-atherosclerotic affects and anti-tumor properties. The bioaccessibility and bioavailability of phenolics is crucial in terms of exerting health benefits. The bioaccessibility and bioavailability of phenolics are affected by on a variety of factors including initial concentration of phenolics, gastrointestinal conditions, food matrix, and interindividual differences. The purpose of this chapter is to present and discuss bioaccessibility and bioavailability studies on grape and grape products.

Keywords Bioaccessibility · Bioavailability · Grape · Grape products · Grape juice · Phenolic compounds

---

M. B. Kaşıkçı (✉) · N. Bağdatlıoğlu
Department of Food Engineering, Faculty of Engineering, Manisa Celal Bayar University, Manisa, Turkey
e-mail: muzeyyen.berkel@cbu.edu.tr

J. A. Malik et al. (eds.), *Food Process Engineering and Technology*,
https://doi.org/10.1007/978-981-99-6831-2_18

18.1 Introduction

Grapes are one of the most popular foods around the world. Besides, grapes are really healthy fruits attributed to their phenolic content. Several products can be manufactured from grape, for instance, grape juice, raisin, grape wine, grape molasses (pekmez) and leather (pestil). Phenolic compounds are antioxidants and these compounds have many health benefits such as reducing blood pressure and cardiovascular disease risk, cancer-protective effects, anti-atherosclerotic effects and antitumoral effects. Phenolic compounds need to be bioaccessible and bioavailable in order to execute their health effects (Lingua et al. 2019; Mandalari et al. 2010). The aim of this chapter was to collect and compare studies on bioaccessibility and bioavailability of phenolics in grape and grape products.

18.2 Phenolics in Grape and Grape Products

Grapes have several phenolic compounds as shown in Table 18.1. Class and amount of phenolic compounds vary due to grape varieties and growth conditions. Stilbenes and flavonols are most abundant phenolics in grapes, and red grapes are rich in anthocyanins (Lingua et al. 2018, 2019).

The phenolic compounds are stilbenes and flavonols found in white grapes (Chardonay) from most to least are coutaric acid (51.7 mg/kg fresh weight (FW)), caftaric acid (37.5 mg/kg FW), isoquercetin epicatechin, fertaric acid, gallic acid

Table 18.1 Phenolic compounds in grapes

White grape	Red grape
Coutaric acid	Catechin
Caftaric acid	Peonidin-3-*O*-glucoside
Isoquercetin	Epicatechin
Epicatechin	Quercetin-3-*O*-glucoronide
Fertaric acid	Epicatechin-3-*O*-gallate I
Gallic acid glucoside	(Epi)-catechin-(epi)catechingallate II
Catechin	Procyanidin dimer
Epicatechin gallate	Caftaric acid
Quercetin-3-glucuronide	(Epi)-catechin-(epi)catechingallate I
Kaempferol-3-glucoside	Cyanidin-3-*O*-glucoside
Astilbin	Kaempferol-3-*O*-glucuronide
Isorhamnetin-3-glucoside	Peonidin-3-*O*-(6″-*p*-coumaroyl)-glucoside
Quercetin	Isorhamnetin-3-glucoside
Kaempferol	Delphinidin-3-*O*-glucoside
	Quercetin-3-*O*-rutinoside
	Petonidin-3-*O*-glucoside
	Gallic acid hexose

glucoside, catechin, quercetin-3-glucuronide, epicatechin gallate, kaempferol-3-glucoside, astilbin, isorhamnetin-3-glucoside, kaempferol and quercetin (Lingua et al. 2019). Another white grape variety (Viognier) has same phenolics, which of coutaric acid and caftaric acid has highest amounts (Lingua et al. 2018).

Red grapes had higher amounts of phenolic compounds than white grapes (Lingua et al. 2018, 2019). Red grapes contain isoquercetin (494.36 mg/kg FW), isorhamnetin-3-glucoside (417.07 mg/kg FW), malvidin-3-glucoside (334.63 mg/kg FW), syringetin-3-glucoside (183.25 mg/kg FW), malvidin-3-acetylglucoside (166.35 mg/kg FW), epicatechin gallate, malvidin-3-coumaroylglucoside, peonidin-3-glucoside and kaempferol-3-glucoside in descending order of amounts (Lingua et al. 2018).

Various products can be manufactured from grapes such as grape juice, wine, raisins, leather (pestil), and molasses (pekmez). Processes during manufacturing of these products can change quantity and quality of phenolics (Celep et al. 2015; He et al. 2016; Kamiloglu and Capanoglu 2014; Lingua et al. 2018, 2019; Mohamedshah et al. 2020).

The major phenolics of grape juice were caffeoyl-tartaric acid epicatechin, proanthocyanidin and protocatechuic-glucoside. High-pressure homogenization process (HPHP) and thermal treatment (at 80 °C, 30 min and 90 °C, 30 s) increased caffeoyl-tartaric acid and proanthocyanidin content in grape juice (He et al. 2016).

White wines (from Chardonay and Viognier grapes) had also similar phenolics with these grapes, but in different quantities. Gallic acid, caftaric acid, catechin, epicatechin, coutaric acid, fertaric acid, caffeic acid, ethyl gallate, astilbin, quercetin-3-glucuronide and quercetin contents in white wine from Chardonay grapes were 0.58, 2.37, 0.30, 0.13, 5.96, 2.81, 4.80, 2.75, 0.10, 0.15 and 0.11 mg/L, respectively. White wine from Viognier grapes contain gallic acid (0.55 mg/L), caftaric acid (2.45 mg/L), catechin (0.43 mg/L), epicatechin (0.15 mg/L), coutaric acid (3.05 mg/L), fertaric acid (3.29 mg/L), caffeic acid (1.91 mg/L), ethyl gallate (2.88 mg/L), astilbin (0.34 mg/L), quercetin-3-glucuronide (0.25 mg/L) and quercetin (0.11 mg/L). Some phenolics found in grapes such as gallic acid-glucoside, epicatechin gallate, isoquercetin, kaempferol-3-glucoside, isorhamnetin-3-glucoside and kaempferol disappeared after wine making process (Lingua et al. 2019).

18.3 Bioaccessibility and Bioavailability

Bioaccessibility, which describes the released proportions of compounds from a complex food matrix in the gastrointestinal system, is the first step of bioavailability. Bioaccessibility and bioavailability mainly depend on food matrix, initial concentration of phenolic compound, food processing and gastrointestinal conditions (Mandalari et al. 2013; Minekus et al. 2014; Mohamedshah et al. 2020). Bioavailability analyses of phenolic compounds are really important analyses, but unfortunately these analyses are time consuming, difficult and expensive. For bioavailability analyses of phenolic compounds, detailed research is necessary. Bioaccessibility analyses of phenolic compounds are easier and cheaper than bioavailability analyses of these compounds. Ethics committee approval is not required in bioaccessibility analyses, but it is necessary in clinical and *in vivo* analyses (Kamiloglu and Capanoglu 2014; Lingua et al. 2019; Minekus et al. 2014).

18.4 Bioaccessibility of Phenolics in Grapes and Grape Products

Bioaccessibility and bioavailability of grape phenolics are influenced by grape variety, matrix and processes. Various processes can affect bioaccessibility and bioavailability of phenolics positively or negatively.

Bioaccessibility of white grape phenolics are shown in Table 18.2. *In vitro* digestion decreased total phenolic content (TPC), and bioaccessibility of total phenolics (TP) was 18% (Lingua et al. 2019) and 9.5% (Zoubiri et al. 2019) in white grape. Unlike this, in Niagara grapes (white grape), flavanols, hydroxybenzoic acids, hydroxycinnamic acids increased after *in vitro* digestion (Gomes et al. 2019). Bioaccessibility of gallic acid-glucoside, kaempferol-3-glucoside, quercetin-3-glucuronide and isoquercetin was 73%, 19.5%, 4.3% and 1.2%, respectively in white grape (Lingua et al. 2019). Bioaccessibility range of phenolic acids, flavonols, and flavan-3-ols in Niagara grape (white grape) was 21–67%, 0–3% and 64–585%, respectively (Mohamedshah et al. 2020). Bioaccessibility of flavan-3-ols, which more than 100%, probably means that bound flavan-3-ols are releasing from the food matrix and/or other phenolics are hydrolyzing to flavan-3-ols during *in vitro* digestion. Gallic acid, caftaric acid, catechin, epicatechin, coutaric acid, fertaric acid, epicatechin gallate, caffeic acid, ethyl gallate, astilbin, isorhamnetin-3-glucoside, quercetin and kaempferol disappeared after *in vitro* digestion in white grape (Lingua et al. 2019).

On the question of red grape, bioaccessibility of TP and total flavonoids (TF) was 13% and 8.3% (Zoubiri et al. 2019), respectively (Table 18.3). *In vitro* digestion did not change flavanols, but decreased flavonols and anthocyanins in Bordo grape (red grape) (Gomes et al. 2019). Anthocyanin bioaccessibility of Concord grape (red grape) varied between 14% and 39%. Anthocyanin bioaccessibility values of red seedless, red seeded and black seedless grapes were 34–59%, 20–44% and 51–85%, respectively. Bioaccessibility range of phenolic acids in Concord grape, red seedless, red seeded, black seedless and green seedless were 9–85%, 7–26%, 19–26%, 39–142% and 44–62%, respectively. Bioaccessibility range of stilbenes in red seeded and black seedless were 0–19% and 5–14%, respectively. Bioaccessibility range of flavonols in Concord grape, Niagara grape, red seedless, red seeded, black seedless and green seedless were 2–8%, 64–585%, 23–44%, 15–30%, 50–153% and 20–116%, respectively. Bioaccessibility values of flavan-3-ols in Concord grape, red seedless, red seeded, black seedless and green seedless were 0–2%, 4–11%, 2–4%, 8–33% and 2–35%, respectively (Mohamedshah et al. 2020).

Table 18.2 Bioaccessibility (%) of phenolics in white grapes

Grape variety	TP	TF	Flavonols	Flavan-3-ols	PA	Isoquercetin	Digestion method	References
White	18%					1.2%	Celep et al. (2015)	Lingua et al. (2019)
Dabouki	9.5%	7%					Minekus et al. (2014)	Zoubiri et al. (2019)
Niagara			64–585%	0–3%	21–67%		Kean et al. (2008)	Mohamedshah et al. (2020)

TP total phenolics, *TF* total flavonoids, *PA* Phenolic acids

Table 18.3 Bioaccessibility (%) of phenolics in red grapes

Grape variety	TP	TF	Antho	Flavonols	Flavan-3-ols	PA	Malvidin-3-glc	Isoquercetin	Digestion method	References
Cardinal	13%	8.3%							Minekus et al. (2014)	Zoubiri et al. (2019)
Concord			9–52%	2–8%	0–2%	9–52%			Kean et al. (2008)	Mohamedshah et al. (2020)
Syrah							9%	4%	Celep et al. (2015)	Lingua et al. (2018)

TP total phenolics, *TF* total flavonoids, *Antho* anthocyanins, *PA* phenolic acids, *malvidin-3-glc* malvidin-3-glucoside

In the case of grape juice, anthocyanin bioaccessibility of Concord juice (red grape juice) was between 86% and 135%. Bioaccessibility range of phenolic acids in Concord juice, Niagara juice (white grape juice) and Niagara juice with SO_2 were 70–85%, 39–48% and 42–66%, respectively (Mohamedshah et al. 2020). Bioaccessibility range of stilbenes in Concord juice, Niagara juice and Niagara juice with SO_2 were 44–63%, 38–52% and 25–53%. Bioaccessibility range of flavonols in Concord juice, Niagara juice and Niagara juice with SO_2 were 36–43%, 37–43% and 22–34%, respectively. Bioaccessibility values of flavan-3-ols in Concord juice, Niagara juice and Niagara juice with SO_2 were 75–95%, 63–101% and 48–76%, respectively (Mohamedshah et al. 2020). This research indicated that in spite of grape juices had lower initial phenolic content than grapes; bioaccessibility of phenolics in grape juices generally were higher than grapes.

Regards to grape juice sediment, bioaccessibility of trans-resveratrol (121%), vanillic acid (76%), gallic acid (76%), caffeic acid (72%), caftaric acid (68%), ferulic acid (68%) catechin (50%) and epicatechin (48%) from Bordo grape juice sediment (red grape juice sediment) were high. Bioaccessibility of vanillic acid, caftaric acid, epicatechin, ferulic acid, caffeic acid and catechin in Isabel grape juice sediment (red grape juice sediment) was 71%, 65%, 57%, 51%, 13% and 2%, respectively (da Silva Haas et al. 2019).

As far as process effects on bioaccessibility of phenolics in grape juices is considered, high pressure homogenization processing (HPHP) increased the bioaccessibility of caffeoyl-tartaric acid and did not change bioaccessibility of epicatechin, proanthocyanidin and protocatechuic-glucoside in grape juice. Thermal treatments (at 80 °C, 30 min and 90 °C, 30 s) increased bioaccessibility of TP in grape juice. Bioaccessibility values of caffeoyl-tartaric acid and proanthocyanidin increased after thermal treatments, while bioaccessibility values of epicatechin and protocatechuic glucoside decreased after thermal treatments (at 80 and 90 °C) (He et al. 2016).

He et al. (2016) also examined the effect of milk matrix such as whole milk, skimmed milk and soy milk on the bioaccessibility of phenolics in grape juice. All milk matrix improved total phenolic bioaccessibility in grape juice by 26.6–31.1%. After *in vitro* digestion, the ABTS antioxidant activity of soy milk added grape juice and control grape juice increased.

Regarding wines, most of phenolic compounds found in white wine such as gallic acid, caftaric acid, catechin, epicatechin, coutaric acid, epicatechin gallate, caffeic acid, astilbin, quercetin-3-glucuronide, isoquercetin, kaempferol-3-glucoside were absent after *in vitro* digestion. Bioaccessibility values of ethyl gallate, quercetin and coutaric acid were 79%, 55% and 51%, respectively in white wine (Lingua et al. 2019). TPC in Papazkarası red wine decreased from 19.64 to 4.69 mg gallic acid equivalents (GAE)/g dry extract after *in vitro* digestion. Gallic acid, caffeic acid and *p*-coumaric acid also decreased after *in vitro* digestion in Papazkarası red wine. Vanillin, rutin and quercetin disappeared after *in vitro* digestion in Papazkarası red wine (Celep et al. 2015).

Bioaccessibility of TP and TF for white dried grape was 24.9% and 41.1%, respectively. TP and TF bioaccessibility of red dried grape was 27.9% and 44.2%, respectively. Recovery (%) of ABTS, CUPRAC, FRAP, and DPPH antioxidant activity after *in vitro* digestion for white dried grape and red dried grape was in the range of 12.3–34.4% and 10.3–34.4%, respectively (Zoubiri et al. 2019).

As regards grape molasses (pekmez) and grape leather (pestil), total proanthocyanidin content of molasses (pekmez) decreased after *in vitro* digestion. Bioaccessibility range of total proanthocyanidins was 12–50% and 4–8% in grape molasses (pekmez) and grape leather (pestil), respectively. Bioaccessibility range of total flavonoids was 10–17% and 7–9% for grape molasses (pekmez) and grape leather (pestil), respectively (Kamiloglu and Capanoglu 2014).

In the case of grape jam, bioaccessibility of TP and TF for white grape jam was 25% and 14.9%, respectively. Bioaccessibility of TP and TF for red grape jam was 28.6% and 40.3%, respectively. TF bioaccessibility of red grape jam was higher than white grape jam while there were no statistical differences between TF bioaccessibilities of grapes (Zoubiri et al. 2019).

Turning to fortification studies, in Marinelli's research, spaghetti was enriched with red grape marc flour. *In vitro* digestion increased TPC from 2.11 to 5.53 mg GAE/g dry weight (dw) in spaghetti containing red grape marc flour. Total anthocyanins decreased from 0.084 to 0.037 mg GAE/g dw in spaghetti containing red grape marc flour after *in vitro* digestion (Marinelli et al. 2018). In another fortification study, breads fortified with grape pomace powder digested, and bioaccessibility range of anthocyanins, flavones and phenolic acids were 20.2–28%, 11.9–17.1% and 17–23.3%, respectively (Rocchetti et al. 2021).

18.5 Bioavailability of Phenolics in Grapes and Grape Products

Consuming grape phenolics has been linked to a number of health benefits, but their effects on the body can only be experienced if bioavailable amounts get to the target organs where action is necessitated (Castello et al. 2018; Iglesias-Carres et al. 2019; Lutz et al. 2014; Stalmach et al. 2012). However, phenolic compounds generally has low bioavailability values, due to low solubility, bound form in the food matrix and host-related factors such as age, genetics, enzymatic activity, colonic microflora (Cosme et al. 2020; Zhao et al. 2020) (Table 18.4, 18.5, 18.6, 18.7, 18.8, and 18.9).

There is limited research on bioavailability of phenolics in grape and grape products (Table 18.5). In these studies, red grape juice, red grape pomace drink and grape seed extracts were studied in humans and/or animals (Castello et al. 2018; Chen et al. 2017; Iglesias-Carres et al. 2019; Lutz et al. 2014; Stalmach et al. 2012). It has been predicted that approximately 95% of absorbed phenolics are modified to their conjugates. These products are secreted into the blood and then excreted by the kidneys. Nevertheless, more than 75% of phenolics consumed are not found in urine because they were not absorbed; these phenolics were probably excreted in bile or metabolized in the colon by the colonic microflora. Colonic products may be

Table 18.4 Bioaccessibility of phenolics in grape juice

Grape juice	Antho (%)	PA (%)	Stilbenes (%)	Flavonols (%)	Flavan-3-ols (%)	Digestion method	Reference
White (Niagara)		39–48	38–52	37–43	63–101	Kean et al. (2008)	Mohamedshah et al. (2020)
SO_2 white (SO_2 Niagara)		42–66	25–53	22–34	48–76		
Red (Concord)	86–135	70–85	44–63	36–43&	75–95		

Antho anthocyanins, *PA* phenolic acids

absorbed into the vein and pass to the circulatory system. To sum up, high levels of phenolics in plasma probably come to mean that these phenolics have high bioavailabilities (Lutz et al. 2014).

In a clinical study, it is found that 40% of the phenolics in Concord grape juice were recovered in the ileal effluent after being consumed by participants with ileostomies, indicating that 40% of phenolics will go to the large intestine in healthy subjects with a healthy colon. Hydroxycinnamate tartarate ester had highest recoveries, which was between 63% and 93%, amongst phenolic compounds. Recoveries of anthocyanins were between 13% and 34%, malvidin has the highest recovery amongst other anthocyanins. Although sulphated and methylated epicatechins were absent in Concord grape juice, they were found in ileal fluid. Recovery of total procyanidins and total flavonol glucosides in ileal fluid was 27% and 38%, respectively (Stalmach et al. 2012).

Concord grape juice was also consumed by healthy volunteers with a healthy colon to discover peak plasma concentrations (C_{max}) and urinary excretion, which were indicators of bioavailability. One nanomole per liter for petunidin-3-*O*-glucoside and 355 nmol/L for dihydrocoumaric acid were peak plasma concentrations (C_{max}). Urinary excretion ranged from 0.26% for total anthocyanins to 24% for hydroxycinnamate tartarate esters metabolites. These findings of the Concord grape juice study showed that anthocyanins had lower bioavailability values, while hydroxycinnamate metabolites had higher. Hydroxycinnamate metabolites were absorbed in small intestine and also in colon, with the colonic microflora appearing to be able to take responsibility for hydrogenation of the hydroxycinnamate (Stalmach et al. 2012).

In addition, bioavailability of Autumn Royal grape juice (red grape juice) was studied in humans. After uptake of Autumn Royal grape juice by subjects, gallic acid and catechin appeared in plasma. Recovery in plasma differed from 0.06% (30 min) to 0.24% (180 min) for gallic acid, while it varied from 0.32% (30 min) to 0.71% (180 min) for catechin. In 24-h urine samples, only gallic acid and catechin were found amongst phenolics. Because phenolics are often excreted within the first 24 h after ingestion, the excretion in 24-h urine is relative to the quantities consumed the day before. Despite the fact that catechin and gallic acid are poorly absorbed, these

Table 18.5 Effect of different processes on phenolics bioaccessibility of grape juice

Process	TP (%)	Caffeoyl-tartaric acid (%)	Epicatechin (%)	Proanthocyanidin (%)	Protocatechuic-glucoside (%)	Digestion method	Reference
Control	99	16	2.9	1.3	2.5	Wootton-Beard et al. (2011)	He et al. (2016)
HPHP	100	18	2.9	1.4	2.4		
T80	132	29	2.1	2.9	2.4		
T90	125	29	2.1	3.4	2.4		

HPHP high pressure homogenization processing, *T80* thermal treatment at 80 °C for 30 min, *T90* thermal treatment at 90 °C for 30 s, *TP* total phenolics

Table 18.6 Bioaccessibility of phenolics in wine

Wine	CA	EA	Q	Malv-3-glc	Catechin	Q-3-glcr	Epi	Syr-3-glc	Malvidin-3-acglc	Ast	Digestion method	References
White (Chardonay and viognier mix)	51%	79%	55%		NC	NC	NC			NC	Celep et al. (2015)	Lingua et al. (2019)
Red (Syrah)	·		NC	5%	22%	3%	14%	12%	7%	7%	Hu et al. (2013)	Lingua et al. (2018)

CA coutaric acid, *EA* ethyl gallate, *Q* quercetin, *Malv-3-glc* malvidin-3-glucoside, *Q-3-glcr* quercetin-3-glucuronide, *Epi* epicatechin, *Syr-3-glc* syringetin-3-glucoside, *Malvidin-3-acglc* malvidin-3-acetylglucoside, *Ast* astilbin, *NC* not calculated, because the relevant phenolic compound disappeared after *in vitro* digestion

Table 18.7 Bioaccessibility of phenolics in raisins

Raisin	TP (%)	TF (%)	Digestion method	Reference
White	25	41	Minekus et al. (2014)	Zoubiri et al. (2019)
Red	28	44		

TP total phenolics, *TF* total flavonoids

Table 18.8 Bioaccessibility of phenolics in grape jam and pekmez (grape molasses)

Product	TP (%)	TF (%)	Total proanthocyanidins	Digestion method	References
White grape jam	25	15		Minekus et al. (2014)	Zoubiri et al. (2019)
Red grape jam	29	40			
Pekmez (grape molasses)	21	17	50%	McDougall et al. (2005)	Kamiloglu and Capanoglu (2014)

TP total phenolics, *TF* total flavonoid

Table 18.9 Bioaccessibility of phenolics in grape pestil (grape leather)

TP	TF	Total proanthocyanidins	Digestion method	Reference
50%	9%	8%	Minekus et al. (2014)	Zoubiri et al. (2019)

TP total phenolics, *TF* total flavonoids

phenolics are bioavailable and may donate to the alleged health benefits of Autumn Royal grape juice consumption (Lutz et al. 2014).

In another clinical study, red grape pomace drink was ingested by healthy subjects. Phenolic metabolites were discovered in urine and plasma, following ingestion. The primary metabolites were phenyl-γ-valerolactones, hydroxybenzoic acids, simple phenols, hydroxyphenylpropionic acids, hydroxycinnamates, and (epi)catechin phase II conjugates. Both urine and plasma samples showed high inter-individual variability. In this clinical study, peak plasma concentrations (C_{max}) ranged from 2.9 nmol/L for catechol-sulphate to 1171.2 nmol/L for 5-(4′--Hydroxyphenyl)-γ-valerolactone-3′-glucuronide (Castello et al. 2018).

In the case of *in vivo* studies, red grape cultivar (organic and conventional) and rat exposure to various light schedules each day had an impact on the bioavailability of these grape polyphenols. The findings of this study implied that grapes consumed outside of their typical season or cultivation type (organic or conventional) affect phenolics bioavailability. Rats given organic red grapes had greater total serum metabolite concentration regardless of photoperiod exposure, while rats given conventional and organic red grapes had different serum metabolite profiles regardless of photoperiod exposure. Grape phenolics were more bioavailable in the rats subjected to 6 h of light per day, which simulated winter conditions (Iglesias-Carres et al. 2019; Stalmach et al. 2006).

In an *in vivo* study, the bioavailability of grape phenolics was discovered in non-diabetic and diabetic rats. Standardized grape polyphenol (SGP) mixture involving grape seed extract, Concord grape juice and resveratrol were administered to all rats. Bioavailability of methylated flavan-3-ol, resveratrol and quercetin metabolites was significantly higher in non-diabetic rats than diabetic rats. Plasma levels of all phenolics were higher in non-diabetic rats. Higher urinary excretion for all analyzed phenolic compounds was observed in diabetic rats than non-diabetic rats. Urinary excretion levels of all phenolic compounds were conversely associated to plasma levels and were most likely an important contributor to the diminished plasma levels. It was showed that diabetes could change bioavailability of phenolic compounds (Chen et al. 2017).

According to the limited clinical and *in vivo* studies, phenolic compounds in red grape products have low bioavailability values. Interindividual differences, exposure to sun light, diseases such as diabetes can alter bioavailability of grape phenolics.

18.6 Conclusion

In conclusion, grape and grape products have bioaccessible and bioavailable phenolic compounds in different percentages. Grape juice has highest phenolic bioaccessibilities amongst grape and grape products. There is limited research on bioavailability of phenolic compounds in grapes and grape products. These studies focused on red grape juice, red grape pomace drink and red grape polyphenols, all of which had low bioavailability of phenolics, but still contained bioavailable phenolic compounds. Various factors such as food matrix, process conditions, individual related factors, diseases, even seasons can modify bioaccessibility and bioavailability of grape phenolics. Further research is needed to understand bioavailability and absorption mechanisms of phenolics in grape and grape products.

References

Castello F, Costabile G, Bresciani L, Tassotti M, Naviglio D, Luongo D, Ciciola P, Vitale M, Vetrani C, Galaverna G, Brighenti F, Giacco R, del Rio D, Mena P (2018) Bioavailability and pharmacokinetic profile of grape pomace phenolic compounds in humans. Arch Biochem Biophys 646:1–9

Celep E, Charehsaz M, Akyüz S, Acar ET, Yesilada E (2015) Effect of in vitro gastrointestinal digestion on the bioavailability of phenolic components and the antioxidant potentials of some Turkish fruit wines. Food Res Int 78:209–215

Chen TY, Ferruzzi MG, Wu QL, Simon JE, Talcott ST, Wang J, Ho L, Todd G, Cooper B, Pasinetti GM, Janle EM (2017) Influence of diabetes on plasma pharmacokinetics and brain bioavailability of grape polyphenols and their phase II metabolites in the Zucker diabetic fatty rat. Mol Nutr Food Res 61(10):1700111

Cosme P, Rodríguez AB, Espino J, Garrido M (2020) Plant phenolics: bioavailability as a key determinant of their potential health-promoting applications. Antioxidants 9(12):1263

da Silva Haas IC, Toaldo IM, Gomes TM, Luna AS, de Gois JS, Bordignon-Luiz MT (2019) Polyphenolic profile, macro- and micro-elements in bioaccessible fractions of grape juice sediment using in vitro gastrointestinal simulation. Food Biosci 27:66–74

Gomes TM, Toaldo IM, Haas ICS, Burin VM, Caliari V, Luna AS, de Gois JS, Bordignon-Luiz MT (2019) Differential contribution of grape peel, pulp, and seed to bioaccessibility of micronutrients and major polyphenolic compounds of red and white grapes through simulated human digestion. J Funct Foods 52:699–708

He Z, Tao Y, Zeng M, Zhang S, Tao G, Qin F, Chen J (2016) High pressure homogenization processing, thermal treatment and milk matrix affect in vitro bioaccessibility of phenolics in apple, grape and orange juice to different extents. Food Chem 200:107–116

Hu JL, Nie SP, Min FF, Xie MY (2013) Artificial simulated saliva, gastric and intestinal digestion of polysaccharide from the seeds of Plantago asiatica L. Carbohydr Polym 92(2):1143–1150

Iglesias-Carres L, Mas-Capdevila A, Bravo FI, Arola L, Muguerza B, Arola-Arnal A (2019) Exposure of Fischer 344 rats to distinct photoperiods influences the bioavailability of red grape polyphenols. J Photochem Photobiol B 199:111623

Kamiloglu S, Capanoglu E (2014) In vitro gastrointestinal digestion of polyphenols from different molasses (pekmez) and leather (pestil) varieties. Int J Food Sci Technol 49(4):1027–1039

Kean EG, Hamaker BR, Ferruzzi MG (2008) Carotenoid bioaccessibility from whole grain and degermed maize meal products. J Agric Food Chem 56(21):9918–9926

Lingua MS, Theumer MG, Kruzynski P, Wunderlin DA, Baroni MV (2019) Bioaccessibility of polyphenols and antioxidant properties of the white grape by simulated digestion and Caco-2 cell assays: comparative study with its winemaking product. Food Res Int 122:496–505

Lingua MS, Wunderlin DA, Baroni MV (2018) Effect of simulated digestion on the phenolic components of red grapes and their corresponding wines. J Funct Foods 44:86–94

Lutz M, Castro E, García L, Henríquez C (2014) Bioavailability of phenolic compounds in grape juice cv. Autumn Royal. CYTA J Food 12(1):48–54

Mandalari G, Tomaino A, Rich GT, Lo Curto R, Arcoraci T, Martorana M, Bisignano C, Saija A, Parker ML, Waldron KW, Wickham MSJ (2010) Polyphenol and nutrient release from skin of almonds during simulated human digestion. Food Chem 122(4):1083–1088

Mandalari G, Bisignano C, Filocamo A, Chessa S, Sarò M, Torre G, Faulks RM, Dugo P (2013) Bioaccessibility of pistachio polyphenols, xanthophylls, and tocopherols during simulated human digestion. Nutrition 29(1):338–344

Marinelli V, Padalino L, Conte A, Del Nobile MA, Briviba K (2018) Red grape marc flour as food ingredient in durum wheat spaghetti: nutritional evaluation and bioaccessibility of bioactive compounds. Food Sci Technol Res 24(6):1093–1100

McDougall GJ, Fyffe S, Dobson P, Stewart D (2005) Anthocyanins from red wine—their stability under simulated gastrointestinal digestion. Phytochemistry 66(21):2540–2548

Minekus M, Alminger M, Alvito P, Ballance S, Bohn T, Bourlieu C, Carrière F, Boutrou R, Corredig M, Dupont D, Dufour C, Egger L, Golding M, Karakaya S, Kirkhus B, Le Feunteun S, Lesmes U, Macierzanka A, Mackie A, Marze S, McClements DJ, Menard O, Recio I, Santos CN, Singh RP, Vegarud GE, Wickham MSJ, Weitschies W, Brodkorb A (2014) A standardised static in vitro digestion method suitable for food—an international consensus. Food Funct 5(6): 1113–1124

Mohamedshah Z, Chadwick-Corbin S, Wightman JD, Ferruzzi MG (2020) Comparative assessment of phenolic bioaccessibility from 100% grape juice and whole grapes. Food Funct 11(7): 6433–6445

Rocchetti G, Rizzi C, Cervini M, Rainero G, Bianchi F, Giuberti G, Lucini L, Simonato B (2021) Impact of grape pomace powder on the phenolic bioaccessibility and on in vitro starch digestibility of wheat based bread. Foods 10(3):507

Stalmach A, Mullen W, Nagai C, Crozier A (2006) On-line HPLC analysis of the antioxidant activity of phenolic compounds in brewed, paper-filtered coffee. Braz J Plant Physiol 18(1): 253–262

Stalmach A, Edwards CA, Wightman JD, Crozier A (2012) Gastrointestinal stability and bioavailability of (poly)phenolic compounds following ingestion of Concord grape juice by humans. Mol Nutr Food Res 56(3):497–509

Wootton-Beard PC, Moran A, Ryan L (2011) Stability of the total antioxidant capacity and total polyphenol content of 23 commercially available vegetable juices before and after in vitro digestion measured by FRAP, DPPH, ABTS and Folin-Ciocalteu methods. Food Res Int 44(1): 217–224

Zhao D, Simon JE, Wu Q (2020) A critical review on grape polyphenols for neuroprotection: strategies to enhance bioefficacy. Crit Rev Food Sci Nutr 60(4):597–625

Zoubiri L, Bakir S, Barkat M, Carrillo C, Capanoglu E (2019) Changes in the phenolic profile, antioxidant capacity and in vitro bioaccessibility of two Algerian grape varieties, Cardinal and Dabouki (Sabel), during the production of traditional sun-dried raisins and homemade jam. J Berry Res 9(4):563–574

Chapter 19
Recent Developments for Formulation of Infant Foods

Ameeta Salaria, Shalini Arora, Rita Mehla, Tarun Pal Singh, and Anuj

Contents

A. Salaria
Sher-e-Kashmir University of Agriculture Sciences and Technology, Jammu, Jammu and Kashmir, India

S. Arora (✉) · Anuj
Department of Dairy Technology, College of Dairy Science and Technology, Lala Lajpat Rai University of Veterinary and Animal Sciences, Hisar, Haryana, India

R. Mehla
Dairy Chemistry Department, National Dairy Research Institute, Karnal, Haryana, India

T. P. Singh
Goat Products Technology Laboratory, ICAR-Central Institute for Research on Goats, Farah, Uttar Pradesh, India

J. A. Malik et al. (eds.), *Food Process Engineering and Technology*,
https://doi.org/10.1007/978-981-99-6831-2_19

Abstract The development of infants needs an adequate amount of high-quality nutrition. Breastfeeding is the healthiest nutritional source for neonates throughout the initial months before weaning. Healthcare practitioners recommend that mothers utilize a breast milk substitute (BMS) if breast milk is unavailable. Infant formula is a breast-milk substitute that can meet all of the infant's dietary requirements from birth to the introduction of supplemental feeding. These formulas are designed to provide adequate nutrition for infants. Still, some studies have shown the risks associated with infant formula consumption. The health issues include increased gastrointestinal diseases, respiratory infections, altered intellectual development, etc. Nutritionists and researchers are attempting to find ways to produce safe infant formula. Generally, different types of infant formulas, depending upon the added components and based upon cow milk, whey protein, elemental formula or amino-acid-based formulae, lactose-free formulae, etc., are available on the market. The current approaches and methods in developing infant formulae include supplementing the fat globular membrane, high-pressure pasteurization, and synthesis of human milk oligosaccharides *in vitro,* etc. This review paper will focus on various approaches to developing and formulating infant food. It further helps to disseminate current information to scientists and nutritionists about developing nutritious, risk-free infant food.

Keywords Human milk · Infant formula · Plant-based formulae · Formulation approaches

19.1 Introduction

Milk is nature's perfect diet for new-borns and developing children, except in rare cases where the infant becomes lactose intolerant. Practically, a mother's milk is desirable and essential from the moment the baby is born. It is recommended that the baby must be breast-fed until grows up. Human milk owing to the type and concentration of numerous vitamins and amino acids tends to be an optimal meal for babies under 6 months. Human milk provides essential nutrients to the infant at a critical developmental stage and protects the baby from disease. Mother's milk contains a variety of bioactive substances, cells, and hormones that are essential for immune response development and regulation as well as the treatment of late metabolic disorders such as diabetes and obesity (Ramiro-Cortijo et al. 2020). Breast-feeding has declined worldwide due to the widespread availability and commercial marketing of ready-to-use infant formulas. As a response, the WHO has adopted an International Code of Marketing of Breast Milk Substitutes, urging developing nations to breast-feed their children. According to CODEX guidelines, infant formula is a breast-milk substitute specially prepared to suit the nutritional needs of babies during the first months of life until proper complementary feeding is introduced. In cases of breast-feeding failure or insufficient milk syndrome (World Health Organization, and Food, Agriculture Organization of the United Nations 2004), the formula may be required for low birth weight/premature delivery infants. They may also aid in lowering environmental pollutant intake if the mother is exposed to high amounts of such pollutants.

Reducing the danger of transmittable infection from the mother's milk to newborns is a critical issue, such as mastitis, which can cause food poisoning, stomach pains, diarrhoea, and weight loss in the suckling infant. The use of antibiotics to treat the infection may necessitate the newborn to be placed on formula until the mother recovers. Drug consumption substantially impacts breast milk, and some substances, such as heroin, can lead to addiction in the newborn. Therefore, this is an area where the use of formulas should be explored (Packard 1982). The prime objective of any IMF is to impart comprehensive benefits contained in Human milk and thus act as a similar indistinguishable substitute for it. However, research comparing never and partially-breastfed newborns and young children to exclusively breastfed infants and young children has found various evidence of formula-feeding's adverse health effects. For the child, this means a higher risk of diarrhoea, respiratory infection, malocclusion, and all-cause mortality, as well as a higher chance of obesity and type-2 diabetes (Victora et al. 2016).

19.2 Bioactive in Human Milk

Breast milk is better for human infants than bovine milk because the content of milk varies based on the species from which it is derived, allowing species-specific juveniles to grow and develop at the proper rate. Indeed, experts believe that a child under the age of 1 should not be given conventional cow's milk, goat's milk, condensed milk, dry milk, evaporated milk, or any other type of milk. This is due to changes in milk composition identified *via* research over the last decade. While both cow's milk and human milk contain roughly the same quantity of water, the amounts of carbohydrates, protein, fat, vitamins, and minerals in each differ dramatically. Human milk has 87% water, 1% protein, 4% fat, and 7% carbohydrate (1–2.4% oligosaccharides) (Fig. 19.1). It is also high in minerals and vitamins (calcium, phosphorus, magnesium, potassium, sodium etc.) (Boquien 2018). It contains valuable bioactive components. Human milk lipids are essential owing to their unique content and nutritional and bioactive qualities. Human milk contains lipids that assist in babies' proper growth and development. The complexity of lipids in human milk

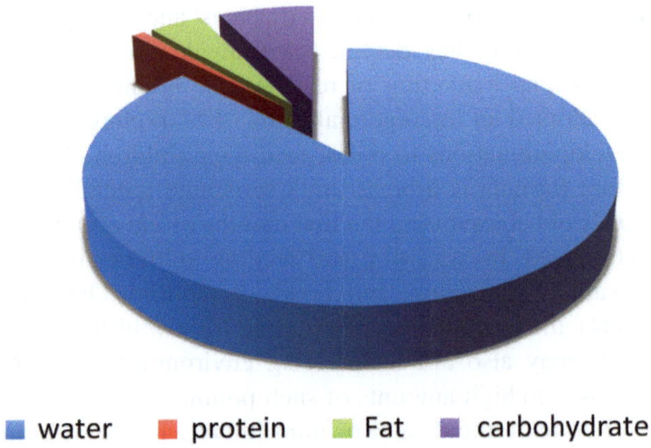

Fig. 19.1 Human milk composition

has been related to neonatal health outcomes such as optimal growth, neurocognitive nervous system development, inflammation and infection risk evaluation, and a reduced risk of adult metabolic and cardiovascular problems (Ramiro-Cortijo et al. 2020). Human milk house a distinct lipid profile of numerous fatty acids and a significant proportion of essential fat compared to bovine milk. It also has a higher linoleic and oleic acid content and decreased hypercholesterolemic activity (Kelishadi et al. 2012).

Human milk contains bioactive compounds such as medium chain monoglycerides, lactoferrin, lysozyme, immunoglobulin, cytokines, and oligosaccharides above the prime macro and micro nutrients (Yi and Kim 2021). These bio-actives protect against infectious bacteria or viruses and save from gastrointestinal or respiratory conditions (Pérez-Escalante et al. 2022). Furthermore, their involvement was already related to anti-inflammation activity and positive immune system modulation via various neurotransmitter systems. So far, approximately 200 various types of HMO have been characterized, demonstrating a fascinating and valuable impact on overall health (Bering 2018).

19.3 Human Milk Substitute (Infant Formula)

Researchers have been challenged to create a baby formula with the same nutritional content as breast milk. The majority of baby formula is made from cow's milk. However, compared to human breast milk, cow's milk has larger concentrations of protein, minerals, and fat. As a result, cow milk has to be skimmed and diluted to more nearly mimic the content of human breast milk (Cook 1989; Koletzko et al. 2005). Vegetable oils, vitamins, minerals, and iron are added to baby formula made from cow's milk for the majority of healthy full-term newborns.. Iron, nucleotides, and mixes of fats obtained from plants or animals that promote health are typical supplements. Infant formulae made with bovine milk are divided into term and preterm variants. The term contains a 1.5 g/100 mL protein content and 68 kcal/100 mL calorie value (Martin et al. 2016). These formulae include the amino acids cysteine and taurine to support the growth of the cardiovascular and neurological systems (Skolnick et al. 2020). Another category is the preterm formula, created for premature newborns with more effective energy content (80 kcal/100 mL). These formulations contain many proteins and minerals (Bourlieu et al. 2015). Additionally, soy milk is utilized in the newborn formula as a substitute for cow or human milk. It is the best option for newborns with congenital lactase deficiency in toddlers with food allergies. Food containing soy should be consumed after 6 months of age (Fiocchi et al. 2010). Because soy-based formula contains phytoestrogens, its uses are constrained by the worry that the newborn may be harmed (Adgent et al. 2012; Upson et al. 2015). There are three types of newborn diets based on cow and soy milk: powdered, liquid, and ready-to-feed infant formula (Martin et al. 2016). Figure 19.2 summarizes the information on human milk and new-born formula.

19.4 Apprehensions in Consumption of Infant Formula

19.4.1 Lactose Intolerance

Lactose (0.9–7.2%) is a significant carbohydrate in human milk (Jenness 1979), the greatest compared to other animal milk (50% more than bovine milk) (Le Huërou-Luron et al. 2010). In formula milk, lactose is supplied as the predominant carbohydrate source. Some alternative carbohydrates, such as corn syrup-based solids (CSS), have been utilized as a lactose substitute to avoid lactose intolerance (Anderson et al. 2022). Lactose intolerance (LI) requires lactose malabsorption (LM), which is a prerequisite. Undigested lactose comes into touch with the intestinal flora in cases of LM (primary or secondary). Short chain fatty acids (SCFA) and hydrogen (H_2), as well as carbon dioxide (CO_2), methane (CH_4), and carbon monoxide (CO), are produced when bacteria break down lactose (Misselwitz et al. 2019).

However, lactose-free formula intake may also have ill effects on infants' health (Schaafsma 2008). To assess lactose intolerance in baby formula and substitute the syrup with maize solids containing lactose compared to breast milk, a randomised, double-blind experiment was conducted. Additionally, certain newborn formulas contain fructose in a variety of forms, such as sucrose, which can cause sudden liver failure in newborns with inherited lactose intolerance who have received an incorrect diagnosis (Li et al. 2018).

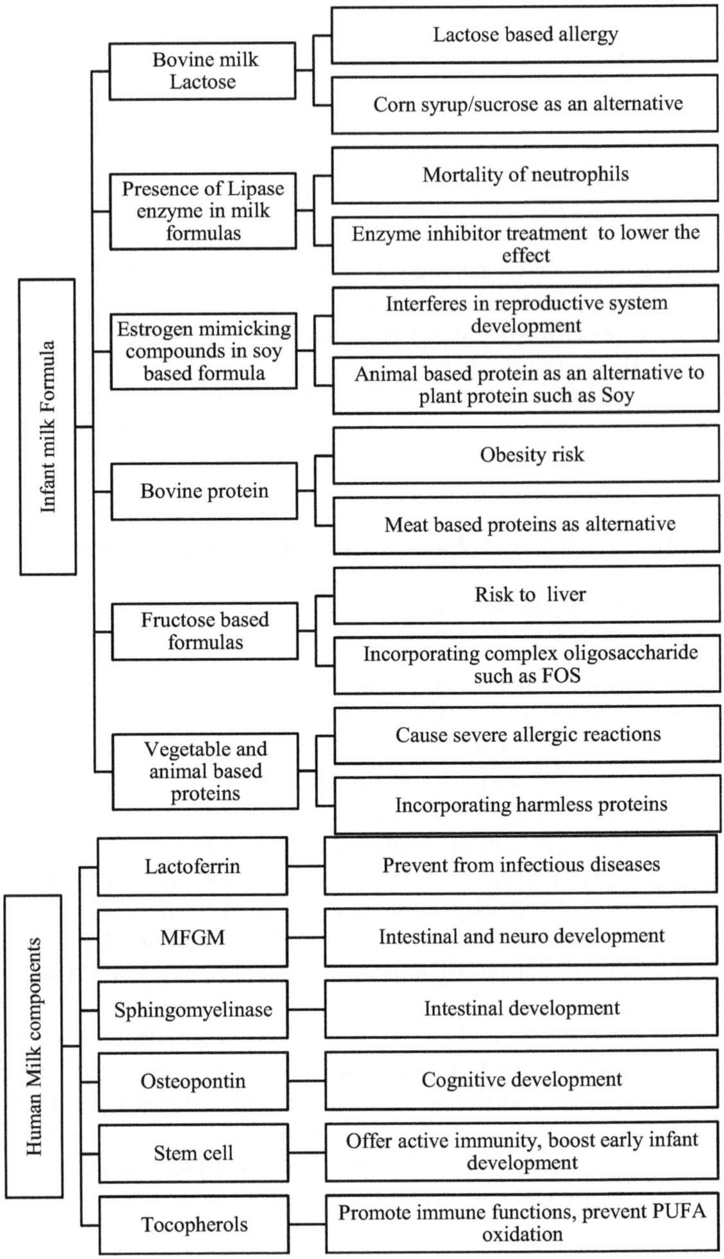

Fig. 19.2 Comparison of infant formula over human milk on infants heath. (***Adapted and Modified from*:** Hussain et al. 2021)

19.4.2 Effect on Growth and Cognitive Development

There are a number of results that represent early brain development and cognitive performance that are favourably correlated with breastfeeding. Infants who are exclusively breastfed have more white matter and subcortical grey matter in their brains than children who are formula-fed, according to studies using brain imaging. However, particularly in research involving human participants, it can be challenging to separate the effects of nutrition in breast milk from other confounding variables that impact brain development. The pigment lutein and the natural form of tocopherol, both of which are specifically deposited in the brain, are nutrients found in human breast milk. While most newborn formulae contain the synthetic version of tocopherol as a supplement, most infant formulas lack lutein, which is the main carotenoid in breast milk (Liu et al. 2019). New-borns fed mother milk were compared to infants fed formula in another combined analysis of six trials, and there was an observation that infants with mother feed had higher IQ scores than those on formula-feed (Dewey 2001).

Another study reported lower IQ among formula-fed newborns than in breastfed babies (Rossiter et al. 2015). Furthermore, Promotion of Breastfeeding Intervention Trial (PROBIT) investigations recently adding LCPUFA to a formula did not affect neurodevelopment compared to a control formula (Le Huërou-Luron et al. 2010).

19.4.3 Gastrointestinal Functions and Allergic Diseases

The most common gastrointestinal disorder-related cause of mortality in preterm neonates is necrotizing enterocolitis (NEC) (Niño et al. 2016). In the second to third week of life, necrotizing enterocolitis often develops. Prematurity, low birth weight, and formula feeding have been identified as the main risk factors among those that have been studied (Afzal et al. 2017). Higher microbial diversity is frequently thought to be advantageous. Childhood asthma has been linked to decreased gut microbial diversity, according to a recent research (Abrahamsson et al. 2014). Infants who are breastfed may be protected from developing asthma and allergy diseases, while infants who are given formula have lower gut bacterial diversity than breastfed infants (Oddy 2017). In recent research, newborns who were given infant milk were on the verge of catching respiratory tract infections more than those fed mothers. Similarly, infants nursed for more than 6 months have a low risk of disease in their first year (Mendonça et al. 2017).

19.4.4 Effects on Reproductive Function

Phytoestrogens are compounds having estrogenic action that are obtained from plants. The isoflavone family of phytoestrogens are present in baby formulae made from soy. Infants who drink soy-based formula every day consume between 4 and 11 mg of phytoestrogens per kilogram of body weight (Setchell et al. 1997). Some infant formulas have soy as their principal ingredient; this soy-based formula exposes newborns to an estrogen-like chemical in the first 6 months of life. According to recent research by Adgent et al. (2018), Children who were fed soy formula tended to enlarged vaginal cells and a minor reduction in uterine volume compared to children who consumed cow milk. On the other hand, hormonal concentration and nip bud diameter did not show any such changes. Furthermore, steroid hormone signalling is yet additional crucial development of the reproductive system (Ho et al. 2017). Babies that consume soy formula respond estrogenically in their vaginal tissues (Adgent et al. 2018).

19.4.5 The Essential Role of Infant Formulas: Necessity Amidst Potential Menaces

Babies are utterly reliant on their mothers for nutrition. However, nursing is not always recommended considering the impact on the health of the breastfeeding mother (AAP 2012; WHO 2010) or the baby's health (Seoane et al. 2020). When all other options for continuing breast-feeding have been exhausted, international scientific medical associations urge that infant milk formulae must be utilized (AAP 2012; Pediatria 2018). Infant formulas required as a supplement or substitute for human milk are the optimum choice for child growth relative to other uncooked alternatives altered to deliver appropriate nutrients. Compared to other uncooked food sources, infant formulae used as a supplement or replacement for human milk are the most incredible option for child growth since they can be modified to offer adequate nutrients. Even the most eager mother with TB, a heart condition, epilepsy, or another condition may be discouraged from nursing by her primary care physician (Wayler and Kein 1965). When asked why they could not finish their planned breastfeeding time, more than half of the women cited lactation issues, concerns about their children's weight gain, illnesses in the mothers or babies that required medicine, and issues with milk production (Odom et al. 2013). Some women must rely on the bottle because they are employed, must complete their education, or are raising adopted children (U.S. 1972). Iron deficiency (Bathla and Arora 2022; Duncan et al. 1985; Pisacane et al. 1995), vitamin D deficiency (Kreiter et al. 2000), and contact with the contaminants in the environment are the issues that require the most attention. Up to 30% of breast-fed newborns develop iron deficiency anaemia, and greater than 60% of anaemic infants have iron shortage by the time they are 12 months old. Iron deficiency is about twofold as common in breast-fed newborns (Pisacane et al. 1995). However, the cause is unknown. There has also been an increase in cases of dietary rickets in breast-fed newborns, predominantly in northern regions (Kreiter et al. 2000). This is unlikely to be a concern with sun exposure. Infants born to mothers with a deficiency of vitamin D and those without sunlight are at an elevated risk of rickets. The AAP and CPS recommend dosing most breastfed newborns a 200 Dose of vitamin D up to 2 months of age (C.P.S. 1998; Wagner and Greer 2008). Breast-fed newborns may often encounter environmental toxins (such as lead and polychlorinated biphenyls), prescribed and non-prescribed medicines, and dreadful infections carried by the carrier mother (such as the Human Immunodeficiency Virus). In the recent Covid crisis, formula milk has emerged as a new life saver for kids who have lost or have infected mothers. When another source is required, modern formulas are a lifesaver for busy paediatricians. Because of the increased risk of breast-feeding, formulas are occasionally needed for the hour. The possible dangers of formula feeding should be weighed against these advantages and risks, not the least of which is the chance of infectious organisms like contaminated water contaminating the formula (AIDS and United Nations Children's 1992; Humphrey and Iliff 2001; Mbori-Ngacha et al. 2001).

19.5 The Worldwide Market for Infant Formulas and Its Growth

According to the UN, the global population of 7.2 billion people will grow by 1 billion over the next 12 years, achieving 9.6 billion by 2050. The international market for baby nutritional supplements was estimated at 71.40 billion U.S. dollars in 2018 and is expected to hit 98.90 billion U.S dollars in 2024, with an annual growth rate (CAGR) of 5.58% (https://www.mordorintelligence.com/). The $50 billion infant formula business is predicted to be the fastest-growing packaged food category over the next 5 years, with annual increases exceeding 7%. Some industry experts foresee yearly growth of 8–9% even faster. The industry is not only rapidly increasing, but it also appears to be immune to global downturns (Rollins et al. 2016). This massive expansion is principally determined by expanding marketplaces in the Asia Pacific, which contribute to massive international revenues. Asian and South Asian countries are crucial marketplaces for baby food formula selling, responsible for nearly all projected growth through 2020 (Euromonitor 2016). With 46% of the global market share in 2015, China is the largest and fastest-growing market for infant formula. The world's most significant population of children between the ages of 0 and 4 is found in Southeast Asia, where rising urbanization and high birth rates are two factors driving the increase of milk formula sales.

Many countries, including China, are one of several nations with low rates of exclusive breastfeeding. Many parents assume milk-based human imitations to be a superior and better substitute to mother milk due to aggressive advertising (Chi

2017). The United States is the second-largest market for milk formulas after Asia. Due to limited maternity leave and the lack of legal prohibitions on advertising breast milk substitutes, mothers in the United States face difficulties while trying to breastfeed their babies. The baby milk industry is heavily concentrated among six important worldwide companies and their subsidiary businesses, in addition to being a vast and enormously lucrative market: Nestlé, Danone, Mead Johnson (previously RB), Abbott, Friesl and Campina, and Heinz. Nestle was the market leader, with a huge global footprint and a market share of about one-fifth (22%) of the global market. Due to its dominance in the fastest-growing market regions, such as Latin America (50%) and the Middle East and Africa (30% sales), Nestlé's 2015 success was driven by these regions. By 2019, the milk formula market will be anticipated to reach over 70 billion USD (Euromonitor 2016).

There are three types of baby milk formula available: (1) powder particle: The most basic form of baby formula, which needs to be reconstituted with water before ingesting; (2) fluid: A densely packed liquid that has to be mixed thoroughly with equal quantities of water; and (3) fully prepared: The far costlier version of formula milk, with no reconstitution required.

19.6 Innovative Advances in the Formulation of Infant Formula: Exploring Recent Approaches

Nowadays, many researchers and manufacturers are working on incorporating beneficial substances suggested by the Codex Alimentarius Commission (CODEX 2007). Numerous variants and types of formula milk are available in the market today. The main objective of each is to support the infant's growth and developmental requirements and make a safe choice of Human milk in case of exigencies. To fulfill these requirements, today's formula milk incorporates novel human milk bio actives and other active compounds from different sources with claimed health and functional benefits. Also, novel processing methods are in demand to prevent damage to essential functional compounds from conventional heating methods. Replacing conventional heat processing should also ensure safety concerns over multiple pathogens and their pathogenicity.

19.6.1 Luetin Supplementation

In recent years, researchers focused on yet another element of human milk, the lutein which belongs to a group of carotenes bearing structural resemblance with β carotene but no vitamin A activity so far. Lutein's biological and physiological effects are linked to its role in the development and function of baby eyes (Hammond et al. 2001; Landrum and Bone 2001). These findings demonstrate that lutein is essential for preserving visual function. However, because humans cannot produce these carotenoids in the body, they must be consumed through food or supplements (Eggersdorfer and Wyss 2018). Thus, due to this fact, these essential nutrients should be added to the diets of babies and toddlers. Whereas adults and older children meet their daily supply of lutein and zeaxanthin from vegetables, the newborn's primary source of lutein and zeaxanthin is breast milk, especially during the first few months after birth (Zimmer and Hammond 2007). Lutein levels of mother's milk vary from 3 to 232 g/L as determined by the breastfeeding period and native habitat. These variances are primarily determined by the breastfeeding woman's nourishment (Commission 2007; Jewell et al. 2004). Compared to the lutein-fortified formula (0.7–9.7 nmol/g fat), the lutein concentration of human milk varies depending on the breastfeeding stage, from 0.42 to 9.98 nmol/g fat. Depending on the stage of breastfeeding and the household, breast milk can contain anywhere between 3 and 232 g/L of lutein. Contrary to breast milk, infant formulas only include minute levels of lutein. Lutein is not often added to formulas and is not naturally found in infant formulae. HPLC analyses yield results between 0.07 and 0.13 g/L (Canfield et al. 2003). Lutein is a widely available carotenoid in the developing brain, contributing to over 50% of the carotenoid composition. Lutein values were significantly higher than other carotenoids throughout four main brain areas (Vishwanathan et al. 2014); this may be related to better learning capabilities. The cerebellum and the frontal, occipital, and temporal cortices contain greater than two-thirds of the carotenoids, including lutein (Eggersdorfer and Wyss 2018). In another study, lutein, in conjunction with DHA and choline, improves recognition memory results in newborns up to 5 months of age (Cheatham and Sheppard 2015). According to recent study, lutein containing baby milk is harmless and helps infants' neurocognitive growth (Capeding et al. 2010; Kon et al. 2014; Mackey et al. 2013; Zielińska-Pukos et al. 2017). Currently, lutein is found in a variety of newborn formulae. However, more study on lutein is needed to confirm the results in new-born's regarding its effect on growth rates (Kon et al. 2014).

19.6.2 Glycoproteins: Lactoferrin and Osteopontin

19.6.2.1 Lactoferrin

It is a multimodal protein source occurring in healthy breastmilk with bacteriostatic and bactericidal properties and has been linked to immunological performance, high iron absorption, and suppression of cancer (Lönnerdal 2009). Lactoferrin concentration varies in colostrum, but it accounts for a significant proportion of protein content in mothers' milk (Lönnerdal 2014; O'Callaghan et al. 2011). Because it has significantly less protein, adding lactoferrin from a bovine source to the formula is costly. Furthermore, sensitivity to heat processing makes it vital to maintain during infant formula manufacture (O'Callaghan et al. 2011). Lactoferrin added commercially exhibits bioactive characteristics similar to human lactoferrin. It was discovered that when lactoferrin was added to the newborn formula, the prevalence of respiratory diseases dropped while hematocrit increased (Lönnerdal et al. 2011; O'Callaghan et al. 2011). Lactoferrin is currently a widely available component approved for use in newborn formulas (Wakabayashi et al. 2018). Many preclinical types of research have revealed that lactoferrin-enriched formula improves memory and learning capabilities in some of animal investigations (Chen et al. 2015; van de Looij et al. 2014). It is a safe component found in a limited number of commonly available newborn formulas (Manzoni 2016; O'Callaghan et al. 2011). Furthermore, increased lactoferrin concentration in baby formula products remains unknown (Wakabayashi et al. 2018). But the ideal dosage of enhanced lactoferrin in infant formula products is unknown (Manzoni 2016).

19.6.2.2 Osteopontin

Osteopontin, like lactoferrin, is a protein with multiple functions involved in various biological operations, including bone formation, biomineralization, immunoregulatory roles, and cell growth. Osteopontin is a highly acidic phosphorylated glycoprotein that undergoes posttranslational modification (Sodek et al. 2000). Although glycoprotein was found in bone, milk has the highest concentrations. According to commercial enzyme-linked immunosorbent assays (ELISA), human colostrum has high osteopontin levels, while mature milk has lower levels (Schack et al. 2009a). Compared to human milk, the osteopontin in bovine milk is substantially low, with just negligible levels present in infant formula. In terms of immunological activity, osteopontin can bind directly to various bacterial strains, enhancing macrophage phagocytosis (Schack et al. 2009b). This protective mechanism may play an important role in illness prevention in breastfed newborns. Like lactoferrin, osteopontin is resistant to in vivo stomach digestion, and all animal species have this feature. After being clinically validated for safety and efficacy in vitro and in vivo clinical models, commercially available bovine osteopontin was added to baby formula. Compared to the control formula group, which did not include osteopontin, an randomized controlled trial, with 240 babies was carried out in China found a novel formula containing osteopontin (LacprodanOPN-10®) at 130 mg/L leads to a decreased incidence of fever and a proinflammatory immune response (Lönnerdal et al. 2016). Donovan et al. (2014) found that feeding bovine osteopontin-containing formula to rhesus monkeys affected the expression of many regulatory genes. Furthermore, this shift resulted in an expression level similar to that of breast-fed infants. According to these studies, osteopontin may have health benefits.

19.6.3 Bovine Milk Fat Globule Membrane/Bovine Milk Fat Supplementation in Infant Formula

Several studies have concluded that infants benefit from oral supplementation with bovine milk fat globule membranes (MFGMs). Bovine milk fat is enclosed in a complex structure known as MFGM. It comprises a triple phospholipid cholesterol layer with proteins and glycoproteins inserted. MFGM acts as a stabilizer in milk, reducing emulsion deterioration and coalescence. The positive actions of MFGM in newborns aid in intestinal and neurodevelopment and lower diarrhoea morbidity (He et al. 2017). Similarly, incorporating bovine lipids in infant formula has been attempted in conjunction with plant lipids to create a formula similar to breast milk in terms of fatty acid makeup (Cebo 2012). This membrane contains numerous physiologically active components, many of which have antimicrobial properties (Hernell et al. 2016; Lee et al. 2018), such as gangliosides (Rueda 2007), oligosaccharides (Triantis et al. 2018), and the glycoproteins butyrophilin, lactadherin, and mucin (Liao et al. 2011; Spitsberg 2005). MFGM of both human and bovine origin have identical polar lipid groups but unique polar lipid compounds and share equivalent compositional phospholipids (Fontecha et al. 2020). A comparison of human and bovine MFGM found that both contain comparable numbers of proteins. Mucin 1, Xanthine oxidase, CD36, butyrophylline, adipophylline, lactaderine, and fatty acid-binding protein are the primary bovine MFGM proteins (Cavaletto et al. 2008; Manoni et al. 2020). Furthermore, cow's milk whey proteins, particularly beta-lactoglobulin, lactoferrin, and immunoglobulin, have been linked to MFGM (Almeida et al. 2013; Fontecha et al. 2020; Timby et al. 2017). When released from the lactating mammary gland's epithelial cells, the milk fat globule membrane (MFGM) encloses the triglyceride-rich core of the milk fat globule. This membrane contains various physiologically active components (Hernell et al. 2016; Lee et al. 2018), many of which have antibacterial properties, such as gangliosides (Rueda 2007), oligosaccharides (Triantis et al. 2018), and the glycoproteins butyrophilin, lactadherin, and mucin (Liao et al. 2011; Spitsberg 2005). Standard formulae are free of MFGM since bovine milk fat is replaced with oils of plant origin throughout the manufacturing process (Timby et al. 2017). Bovine milk fractions enhanced in MFGM are now commercially accessible, and infant formulae containing MFGM have been introduced in several countries (Almeida et al. 2021). A rising number of clinical trials using MFGM supplementation for newborns or children show that it has a good effect on both neural development (Gurnida et al. 2012; Timby et al. 2014) and immune defence (Timby et al. 2015; Zavaleta et al. 2011). According to the investigations, the components of MFGM were primarily responsible for these effects (Hernell et al. 2016; Nieto-Ruiz et al. 2019).

Timby et al. (2014) observed that children who were given investigational reduced calorie and protein-based newborn milk supplemented with MFGM of bovine origin performed better in cognitive growth and development than the conventional formula administered. Timby et al. (2015) also investigated the effect of MFGM incorporation in formula milk on respiratory and other infections in children less than 12 months of age. The findings demonstrated that MFGM supplementation has immune-enhancing effects and reduces disease occurrence. Thus, these results support the significance of MFGM enrichment in combatting infection among infants. There is substantial evidence that MGFM encompasses several beneficial components that offer neural development, with considerable differences relative to newborns with formula feeding (Gurnida et al. 2012; Tanaka et al. 2013). Bovine MFGM fraction has comprehensive antimicrobial action aiding in preventing bacterial and viral diarrhoea (Zavaleta et al. 2011). They discovered in their clinical study that giving infants whey protein concentrate dietary supplementation with MFGM lowered the risk of an episode of dysentery and its prevalence. Infant studies are continuing (for example, the Timbly study at Ume University in Sweden, which began in 2008), and a few fortifiers incorporating MFGM in addition to IF are currently commercially available (for example, Lacprodan® MFGM10)® from Arla Foods. The dairy industry has not yet adopted it due to no pilot-scale production studies and is still limited to lab-scale (Dewettinck et al. 2008). Furthermore, there is no restriction regarding its inclusion in newborn formulae. Before introducing new innovative IF products, precise data about safety and effectiveness derived from (human) investigations is required, as with other new advances in this industry.

19.6.4 Inclusion of Plant Proteins in Infant Formulas

Plant proteins that partially replace dairy proteins significantly impact neonatal formula. The proteins authorized for the first Infant feeds (0–6 months) are protein from cow, goat, soy, or hydrolyzed protein from rice, according to appropriate European Regulation (E.U. 2017). Furthermore, the requirement for animal-based proteins is projected to expand by 2050, owing primarily to population growth and the rise of middle-class populations in developing nations (Egbert and Payne 2009; Thomson 2007). As a result, finding other sources with nutritional and physiological properties similar to human milk is necessary. In this regard, there is a growing interest in using plant-based proteins as partial substitutes for proteins of animal origin in meals (Ainis et al. 2017). Because of their high nutritional quality, strong technical and functional qualities (Barac et al. 2012), and low price, legume proteins, such as quinoa, quinoa (Venlet et al. 2021), pea, faba bean (Le Roux et al. 2020a; b; Roux et al. 2020), lentil (Alonso-Miravalles et al. 2021) potato (Roux et al. 2020; Schuh et al. 2019), and chickpea (Malunga et al. 2014; ULLOA et al. 1988), have been proposed as potentially viable protein sources for newborn formulae and may be a viable alternative to animal-derived proteins (Ainis et al. 2017; Alves and Tavares 2019; Chihi et al. 2016). Pea proteins (Pisum sativum) are becoming a viable alternative to soy protein products due to their higher vital amino acid composition (Chihi et al. 2016) and relatively decent digestibility (O'Kane et al. 2005). Furthermore, this protein can also bind fat and water, emulsify, and gel (Sandberg 2011). However, usage of plant-based protein must strictly conform with Regulation EU 2016/127, and consider novel foods, as new procedures for processing to provide protein isolates may be needed, possibly requiring EU authorization as a novel food (Turck et al. 2016). Because the incorporation of new protein sources in infant formula depends on some "standard" properties, like solubility, emulsification, and stability, as well as nutrient content depending on the needs of infants, most research on new plant-based infant formulas has mainly focused on physicochemical and functional properties, as well as digestibility (E.U. 2017; Floris et al. 2010). Lentil proteins, can be substituted in bovine infant formulae owing to its fairly good amino acid content and higher protein values and may cater to the physicochemical and nutritional requirements of new-borns The author and co-worker (Alonso-Miravalles et al. 2021) stated in a recent study that legume proteins are an excellent substitute protein sources for newborn formulae from a physicochemical and nutritional standpoint. Chickpeas, another legume, stands out as a promising alternative protein in infant formulae Researchers found that for infants older than 6 months, chickpea-based formulas might replace cow's milk-based formulas in the future after evaluating the nutritional quality of such formulations (Malunga et al. 2014; ULLOA et al. 1988). In malnourished children with lactose intolerance, chickpeas were an excellent milk replacement, either alone or in combination with methionine (Kalach et al. 2019). Due to its high-quality protein content in a few experiments, which included the identification of two allergens, quinoa has been suggested as a substitute source of cow's milk proteins in follow-up formulations (Hildebrand et al. 2021). It seems that quinoa is becoming more well-liked as a gluten-free source of protein for various customers, including new mothers. However, some questions have been raised over the quantity of saponins present, which might have hampered nutritional absorption in the intestine (Ballegaard et al. 2021). A new patent application [WO2018050705] describes baby formula for babies with cow's milk allergies (Schuh et al. 2019). According to Regulation EU No 609/2013, plant-based baby formulae based on proteins other than soy and hydrolyzed rice are presently not approved (E.U. 2013); however, in the most recent Regulation EU 2016/127, it is suggested that in order to make sure product development and innovation, other additives not covered by the specific needs of the Regulation should be permitted, provided their applicability for infant feed (E.U. 2017). Some toddler formulations based on pea, rice, buckwheat, and almond have recently been offered to suit the nutritional needs of toddlers (Vandenplas et al. 2021). Almond protein includes all essential amino acids and is easily digestible, although sulphur amino acids (methionine + cysteine) seem to be the first limiting amino acid in total seed proteins, followed by lysine and threonine (Ahrens et al. 2005). Research in infants aged 5–9 months with confirmed CMA fed almond formula revealed similar growth to those receiving soy-based formula or the CMP-eHF. Secondary sensitization was not found in the fortified almond formula group as in the other two groups (Salpietro et al. 2005). Buckwheat is not just a great source of protein in terms of amount; it is also a complete protein comprising all nine essential amino acids that are not produced naturally and must be supplied to individuals via food (Ikeda 2002). Whereas buckwheat is suggested as one of the first solid meals to introduce in allergic children, mostly in Japan and Korea

(Takahashi et al. 1998), and is a common nutrient in baby food products, it is rarely utilized in Western nations where only findings are documented (Sammut et al. 2011).

The protein supply significantly impacts IF qualities, and process parameters must be tailored to each formulation to achieve IF quality criteria. Before including any of these plant-based proteins in infant food formulations, safety must be evaluated.

19.6.5 Goat Milk-Based Oligosaccharides as Potential Additives

Infant formula made from cow's milk is widespread, but goat's milk-based dairy products are becoming increasingly popular since they are regarded to be closer to human milk than cow's milk (Urashima et al. 2011) and have lower amounts of -S1 casein (Ceballos et al. 2009). Goat milk is also said to have considerable lactoferrin *N*-glycan similarity with human milk (Le Parc et al. 2014). HMO's complexity makes it nearly tough to replicate in newborn formulae (Moro et al. 2005). To address this issue, current infant formulae are frequently supplemented with fructooligosaccharides (FOS) and galactooligosaccharides (GOS) (Oozeer et al. 2013). However, their usefulness in providing health advantages comparable to breast milk has yet to be demonstrated (Holma et al. 2002). Goats' milk oligosaccharides, particularly sialylated and fucosylated oligosaccharides, may prevent pathogen adhesion from gutting epithelial cells by acting as soluble analogs of host receptor molecules and/or by altering the articulation of such structures. This is because gut cell surface carbohydrate groups and these substances share similar structural characteristics.

Additionally, it has been shown that oligosaccharides influence gut pathogen protection by enhancing immunological activation (Triantis et al. 2018). These results suggest that the oligosaccharides in newborn formula made with goat milk have excellent prebiotic and anti-infection properties and may protect the child from gastrointestinal diseases. This need more study to be confirmed. Leong et al. (2019) discovered 14 measurable oligosaccharides in goats' milk-based newborn formula at stages 1 and 2. This was similar to the oligosaccharide content of fresh goat milk. Five oligosaccharides had structural similarities to those identified in human milk. Bifidobacteria and lactobacilli grew much faster due to these oligosaccharides, but *S. typhimurium* and *E. coli* NCTC 10418 were less likely to adhere to cell lines. According to these studies, oligosaccharides, which are often included in infant formula made with goat milk, have strong prebiotic and anti-pathogen adherence properties, which may be beneficial for gut health.

19.6.6 PUFA and Oligosaccharides Addition to Infant Formula

DHA and ARA are required for neonatal growth and development (Koletzko et al. 2001). Until recently, infant formulae solely contained their predecessors, LA and ALA. The inclusion of LA and ALA to formulae is standard and must be included in all newborn formulas on the suggestion of ESPGHAN. The addition of fatty acids in the formula is vital for child health, especially for breastfed newborns (Koletzko et al. 2001).

Similarly, bacterial substrates in infant formula aid digestive and gastrointestinal function (Bertelsen et al. 2016). Prebiotics were not included in infant formulas made with bovine milk Gibson and Roberfroid (1995) demonstrated in favour of its use to alter the microbiota in the colon. At present, about seven food-grade oligosaccharides have permitted baby formulae usage. These mainly include fructooligosaccharides (FOS), lactulose (LOS), galactooligosaccharides (GOS), polydextrose (PDX), and inulin (Vandenplas et al. 2015). Numerous clinical trials have shown the advantages of oligosaccharide mixtures in infant formulas. Two significant commercially available prebiotics often added to baby formula are galactooligosaccharides (GOS) and polydextrose (PDX). Newborn formula with goat milk as an additive contains naturally occurring oligosaccharides and has strong prebiotic and gut modulation capacities, indicating that they may benefit infant gut health.

19.6.7 Synthesis of Human Milk Oligosaccharides

One of the most significant differences between human breast and cow milk is the presence of HMOs in human milk, which are nonexistent in bovine milk and infant formula (Bode 2012). Human milk oligosaccharides (HMOs) are an essential and valuable element of human milk (Bode 2012). HMOs are a complex mixture of over 200 non-digestible and non-nutritional carbohydrates in human breast milk. The most frequent HMOs are fructosylated oligosaccharides (HMOs), which have been shown to improve health by enhancing bifidobacterial development and preventing toxin and pathogen binding (Bode 2012). *Bifidobacterium longum subsp. Bifidobacterium infantis* grows well on a medium supplemented with HMOs, including 2′-FL, as the sole carbohydrate source (Asakuma et al. 2011; Bunesova et al. 2016; LoCascio et al. 2007; Marcobal et al. 2010). L-fucose and 2′-FL promote brain growth (Matthies et al. 1996). Enzymes and other altered microorganisms were used to make HMOs in the lab. With such a 20% conversion ratio, 6-Sialyllactose (Dominant Sialylated Human Milk Oligosaccharide) was effectively synthesized by Guo et al. (2018) using a Novel exo-Sialidase enzyme from *Bacteroides fragilis* NCTC9343 to successfully synthesize 6-Sialyllactose (Dominant Sialylated Human Milk Oligosaccharide) with such a 20% conversion ratio. In another investigation,

Yu et al. (2018) prepared (HMOs) oligosaccharide 2′-fucosyl lactose utilizing metabolically modified *Saccharomyces cerevisiae* via a salvaging mechanism. Lee et al. (2012) showed enzymatic fucosylation in a batch fermentation using modified E. coli. These recombinant methods have much potential for removing contaminants that are good for health and enhancing the nutritional profile of baby formula. HMOs were previously exclusively found at various amounts in human milk, but they are now available in certain commercial baby formulas, making the new availability of synthetic 20-FL significant. Furthermore, the US Food and Drug Administration has no reservations about the proposed addition of baby formula since specific 20-FL components have been declared to be Generally Recognized as Safe (GRAS) (U.S. 2018). Internationally, the European Union has approved the use of 20-FL in infant formula (E.U. 2017). Similarly, countries that reference the European authorization will safely use 20-FL in infant formula.

19.6.8 Gangliosides

Human milk contains gangliosides, with the two main molecular species being monosialodihexosylganglioside and disialoganglioside GD3 (GM3). In the early stages of breastfeeding, the overall ganglioside level is most significant and mostly made up of GD3. Whereas breast-feeding, ganglioside GM3 becomes more common while ganglioside GD3 concentration decreases (Palmano et al. 2015). It has been shown that giving GD3 to infants enhances brain growth. When 2–3 mg of ganglioside were added to 100 g of newborn formula, infants aged 2–8 weeks scored better on the Griffiths scales for Hand and Eye Coordination, Performance, and General IQ (Zheng et al. 2019). There is justification for using gangliosides in the mixture. However, despite the bioactive's safety, tolerability, and effectiveness, only a few studies have been done on human neonates.

19.6.9 MicroRNA

Human milk contains bioactive components called microRNAs (miRNAs), identified in human milk cells, skim milk, and lipid fractions (Alsaweed et al. 2016). Exosomes are membrane vesicles that may take up these brief single-stranded noncoding molecules (19–24 nt), which may be conveyed kinetically in vivo. Exosomes serve as miRNA barrier mechanisms that stop GI tract breakdown. These little noncoding molecules (19–24 nt) are swept up by membrane vesicles and regulate immunological responses, cellular expression, proliferation, differentiation, and death. Exosomes, membrane vesicles, keep miRNA from degrading in the GI tract (Lukasik et al. 2017). Human bodily fluids such as saliva, plasma, urine, and tears contain a variety of miRNAs (Weber et al. 2010). Human milk contains bioactive substances called microRNAs (miRNAs). They are non-coding RNA molecules that control how genes are expressed throughout several aspects of the cell cycle, including cell division, differentiation, cell death, and immune response. Although it has not been verified, in silico approaches suggest that dietary factors affect miRNA expression in milk (Zhang et al. 2018). Several available miRNAs, which move freely throughout the body and are found in human milk, are important for a baby's growth (Mosca and Giannì 2017; Golan-Gerstl et al. 2017). Bovine milk contains miRNA in an exosome container, which has been shown to be bioavailable to human tissue using an in vitro CaCo-2 cell model, a human colon cancer cell line (s). Bovine milk also contains human milk (Zempleni et al. 2017). This study shows that human milk miRNAs may enter both phenotypically healthy and cancerous cells, affecting biological processes (Golan-Gerstl et al. 2017).

Cow milk samples contain some miRNAs with identical nucleotide sequences to human milk, but no direct evidence of equivalent effects has been observed (Zempleni et al. 2017). In contrast, milk-based infant formula exhibited a roughly twofold drop in miRNA-148a-3p expression. When milk was pasteurized and homogenized, the expression of other abundant miRNAs decreased. MiR-29b promotes osteoblast formation, whereas miR-200c reduces cancer risk by inhibiting epithelial-to-mesenchymal transition (Korpal et al. 2008). Processing procedures considerably influenced both miRNAs in the skim milk component.

Similarly, a mouse model saw no intestinal absorption of miRNAs in the digestive process. A few miRNAs enter the systemic circulation with slightly elevated serum levels (Title et al. 2015). As a result, the stability of several endogenous miRNAs remains unknown, and it is unclear if their inclusion in the baby formula should be explored. Furthermore, little study has been done to examine how biological miRNAs in infant formula affect the health and development of newborns.

Many newborn formulae do not currently contain human or bovine milk miRNA (Golan-Gerstl et al. 2017). This could be related to miRNA degradation caused by milk homogenization and cooking (Zempleni et al. 2017). Future studies in this field might examine whether cow milk miRNAs are immune to processing processes in developing newborn formulae and whether human breast milk miRNA has physiologic effects that differ from cow miRNA. Regardless of other factors, miRNA analysis has the potential to act as a beneficial tool in quality checks, primarily in milk.

19.6.10 High-Pressure Pasteurization

Most heat treatments, including UHT application, may influence the nutritional profile of infant formula, including vitamins D, A, and B, as well as protein. So as an alternative, some novel technologies like HPP can be suitably utilized. There has been considerable development in the inactivation of microorganisms by HPP in recent years, and numerous researchers illustrated the potential use of HPP treatment as an alternative to pasteurization, particularly in milk and some other food products

(Daryaei et al. 2016). Ultra-High Temperature (UHT) treatment (135–145 °C) is often used to process infant formula, precisely to manufacture sterilized products. On the other hand, UHT damages the nutritional profile of infant formula, including vitamins D, A, and B, as well as protein. HPP-treated milk has better solubility during the defragmentation of casein micelles (Goyal et al. 2013). HPP (600 MPa/5 min) at 40 °C was employed to manufacture a hypoallergenic and reduced protein newborn formula. The denatured α-β-Lg and significant retention of α-la result in the manufacture of a less allergenic formula utilizing HPP (Wazed and Farid 2019).

Similarly, a substantial study has been done on High-Pressure Processing (HPP), a non-thermal approach, for pasteurizing dairy and non-dairy products (Devi et al. 2013; Torres and Velazquez 2005). In a previous study, the impact of HPP on the retention of LF in milk and other media was examined. It was discovered that the source, the iron saturation level of the LF, and the treatment medium all affected the variance in LF retention (Masschalck et al. 2001; Mayayo et al. 2014; Mazri et al. 2012). A little study has been conducted so far regarding the effectiveness of HPP on maintaining LF levels in IMF formulation, despite some information on other treatment media. A recent study investigated the effect of HPP on microbial load and furan levels in infant meals made from vegetables (Kultur et al. 2017). The use of HPP resulted in a considerable reduction in microbial counts. Furan content in the heat-processed vegetable-based baby formula was 5.84 ng/g. Furan levels were determined to be below the detection limit (1.0 ng/g) in all HPP-exposed samples. HPP was a viable heat processing option for infant diets. HPP was a viable option for heat-processing infant diets (Kultur et al. 2017). Commercially available HPP-treated newborn foods from various companies include wholesome and nutritious blends, boosters, and gluten-free items (Javed et al. 2021). HPP foods are made up of purééd vegetables and fruits that can be used as primary feeding constituents for newborns and have a more excellent market value in terms of convenience and simplicity than fresh items (Al-Ghamdi et al. 2020).

19.6.11 Microwave and Ohmic Heating in Infant Foods

Conventional heating methods, such as pasteurization, are used to inhibit the growth of pathogens in the baby formula; nevertheless, traditional heating techniques have some adverse effects, such as the production of hazardous compounds such as hydroxymethylfurfural (HMF) and the destruction of certain bioactive substances that are undesirable for the nutritional profile of the formula (Roux et al. 2016). The formation of Maillard reaction products, which can result in product darkening and the synthesis of 5-hydroxymethylfurfural (HMF), an antinutrient agent, is a classic example (Jaeger et al. 2010; Sakr and Liu 2014). Because pureed baby foods, infant foods, or human milk are babies' sole food sources, such products must have appropriate nutritional content to sustain good dietary health and a healthy growth rate (Rodríguez 2001). Ohmic and microwave heating are newer technologies that can reduce damages encountered in conventional heat treatment and achieve better nutritional profiles in infant formula milk. Ohmic and microwave cooking are the most convenient options for preparing pathogen-free newborn feeds (Pires et al. 2020). The quick application allows for the eradication of microorganisms in a shorter period, limiting the negative impacts on sensory characteristics and nutrient losses (Evrendilek et al. 2012). Hradecky et al. (2017) reported samples that underwent conventional sterilisation included higher oxidative metabolism and Maillard breakdown product-related chemicals than those that underwent ohmic heating. In a pilot study, baby formula was concurrently subjected to steam injection UHT therapy and OH treatment (T0 50 °C–Tf 120 °C) (Roux et al. 2016). The chemical markers evaluated (furosine, FAST index, carboxymethyl lysine) showed a comparable quantity of the Maillard process in the products produced with the two treatments when time/temperature parameters varied from 120 to 140 °C from 0.4 up to 155 s were utilised. Mesías et al. (2016) found that ohmic heating had no effect on the total amino acid content and, consequently, did not modify the protein quality of the product. However, OH retained vitamin C better. The degree of heat denaturation of whey proteins was similarly comparable. Finally, they showed that the nutritional integrity of protein in vegetable baby meals might be effectively maintained using ohmic treatment as an alternative to conventional sterilizing.

The FDA has now approved microwave technology (MW) for commercial product sterilization, and is often used for cooking meals. One of the difficulties in the procedure is that, in contrast to water baths, current radiation heats objects without first passing through food (Martins et al. 2019). The main problem with MW heating is the uneven temperature distribution that causes hot and cold spots to appear all over the product. This may affect the items' sensory properties and microbiological safety (Chandrasekaran et al. 2013; Vadivambal 2010). Therefore, it must undergo rigorous testing before being used commercially in any baby meal. The best sterilization technique must be chosen by considering the variables that affect food safety, such as product variety, dietary impact, sensory quality, etc.

19.6.12 Replacing Dairy Proteins with Meat Protein

Conventionally proteins of dairy origin are used in infant milk formulations (Rudloff and Lönnerdal 1992). It has been observed that higher protein levels in such products may be the leading factor in fat deposition in neonates (Luque et al. 2015). It is owing to higher protein levels occurring in bovine milk formulations. In a recent study, higher animal protein consumption was associated with more significant physical growth and body weight in breastfed new-borns, but not with excessive adiposity gain as with dairy protein (Tang and Krebs 2014). Another study discovered that lamb meat-based formula was very well accepted as a relatively safe formula for new-borns who are sensitive to bovine milk hydrolysate formula (Weisselberg et al. 1996). In another study it was observed that the home-made meat-based formula HMMBF is an effective oligoantigenic meal for the cure of food-related atopic dermatitis its further spread (Cantani 2006). In terms of bioavailability of minerals, Fe bioavailability was greater than 80% in meat and chicken-based samples and roughly 20% in banana-based samples for vegetable- and meat-based baby diets (do Nascimento da Silva et al. 2013). Thus, including meat proteins may be another alternative to dairy, but its safety must be thoroughly examined.

19.7 Development of Secure Infant Formula for Optimal Well-Being

Development of safe formula at the industrial level is subject to a number of limitations and difficulties. While attempts are being made to use novel additives based on animal and vegetable components to lessen the risks associated with newborn milks, there are still substantial challenges to overcome.

Mother's milk naturally contain various bioactive components all of which are necessary for a baby to grow and flourish. These elements are essential to the baby's growth and general wellness. However, it is difficult to replicate these bioactive ingredients in baby formula on an industrial scale. Even if microbes are capable of producing these materials on a small scale, expanding production to satisfy commercial demand is extremely difficult.

Alpha-linolenic acid (ALA) and arachidonic acid (ARA), two important fatty acids, have also showed promise in improving a newborn's cognitive ability. However, due to their expensive cost, these fatty acids aren't frequently used in new-born formulae. To make these fatty acids more widely available, it is crucial to find affordable sources or alternative production processes (Hussain et al. 2023).

The industrial manufacture of safe infant formula must also ensure regulatory compliance. Infant formula must adhere to the rules and regulations set out by regulatory organisations like the FDA in order to be safe, nutritionally adequate, and correctly labelled. It might be difficult to meet these conditions while developing the formula and needs close attention to detail.

Despite these limitations, there is still potential for the development of safe infant formula on a large scale because to continued advances in nutrition and creative approaches. With the ultimate objective of developing safe and nutritionally superior newborn formulae, researchers and producers continue to investigate novel strategies and technologies to address these issues.

19.8 Conclusion

Infant formula development and formulation have improved significantly in recent years in order to enhance its nutritional profile, safety, and general quality. These methods seek to guarantee newborns' optimum growth and development while offering them a practical substitute for breast milk. Scientists have made great progress in their understanding of the components of breast milk and have been able to apply this expertise to the development of newborn formula. To closely resemble the composition of breast milk, this entails optimising the mix of proteins, carbs, lipids, vitamins, minerals, and other crucial components. In addition, improvements in ingredient sourcing and processing methods have made it possible to include nutrients that were previously challenging to include.

One strategy is the inclusion of advantageous substances such hormones, lutein, MicroRNA, milk fat globule membranes (MFGM), lactoferrin and osteopontin, human milk oligosaccharides (HMOs), and other compounds. In order to improve the safety and quality of infant formula, modern processing methods including high-pressure pasteurisation, microwave and ohmic heating, and protein separation have also been used.

Recent efforts have concentrated on lowering the cost of infant formula while adding ingredients like MFGM, HMOs, and human milk biogenesis. However, producing and incorporating human milk components into infant formula is the most problem. Formula manufacturers still face significant challenges in accurately reproducing the intricate makeup of human breast milk and properly combining these components.

Researchers are working diligently to develop infant formula that more nearly reflects the properties and advantages of human breast milk via technological advancement. These developments might lead to better overall health outcomes for newborns who need to be fed with the formula.

References

AAP (2012) Breastfeeding and the use of human milk. Pediatrics 129(3):e827–e841. https://doi.org/10.1542/peds.2011-3552

Abrahamsson TR, Jakobsson HE, Andersson AF, Björkstén B, Engstrand L, Jenmalm MC (2014) Low gut microbiota diversity in early infancy precedes asthma at school age. Clin Exp Allergy 44(6):842–850. https://doi.org/10.1111/cea.12253

Adgent MA, Daniels JL, Rogan WJ et al (2012) Early-life soy exposure and age at menarche. Paediatr Perinat Epidemiol 26(2):163–175. https://doi.org/10.1111/j.1365-3016.2011.01244.x

Adgent MA, Umbach DM, Zemel BS et al (2018) A longitudinal study of estrogen-responsive tissues and hormone concentrations in infants fed soy formula. J Clin Endocrinol Metab 103(5): 1899–1909. https://doi.org/10.1210/jc.2017-02249

Afzal B, Elberson V, McLaughlin C, Kumar VH (2017) Early onset necrotizing enterocolitis (NEC) in premature twins. J Neonatal Perinatal Med 10(1):109–112. https://doi.org/10.3233/npm-1616

Ahrens S, Venkatachalam M, Mistry AM, Lapsley K, Sathe SK (2005) Almond (Prunus dulcis L.) protein quality. Plant Foods Hum Nutr 60(3):123–128. https://doi.org/10.1007/s11130-005-6840-2

AIDS WHOGPo, United Nations Children's F (1992) Consensus statement from the WHO/UNICEF Consultation on HIV transmission and breast-feeding, Geneva, 30 April-1 May 1992. World Health Organization, Geneva

Ainis WN, Ersch C, Ipsen R (2017) Partial replacement of whey proteins by rapeseed proteins in heat-induced gelled systems: effect of pH. Food Hydrocoll 77:397–406

Al-Ghamdi S, Sonar CR, Patel J, Albahr Z, Sablani SS (2020) High pressure-assisted thermal sterilization of low-acid fruit and vegetable purees: microbial safety, nutrient, quality, and packaging evaluation. Food Control 114:107233. https://doi.org/10.1016/j.foodcont.2020.107233

Almeida C, Conte Junior C, de Oliveira SA, Alvares T (2013) Proteína do soro do leite: composição e suas propriedades funcionais/whey protein: composition and functional properties. Enciclopédia Biosfera 9:1840–1854

Almeida CC, Mendonça Pereira BF, Leandro KC, Costa MP, Spisso BF, Conte-Junior CA (2021) Bioactive compounds in infant formula and their effects on infant nutrition and health: a systematic literature review. Int J Food Sci 2021:8850080. https://doi.org/10.1155/2021/8850080

Alonso-Miravalles L, Barone G, Waldron D et al (2021) Formulation, pilot-scale preparation, physicochemical characterization and digestibility of a lentil protein-based model infant formula powder. J Sci Food Agric 102(12):5044–5054. https://doi.org/10.1002/jsfa.11199

Alsaweed M, Lai CT, Hartmann PE, Geddes DT, Kakulas F (2016) Human milk miRNAs primarily originate from the mammary gland resulting in unique miRNA profiles of fractionated milk. Sci Rep 6(1):20680. https://doi.org/10.1038/srep20680

Alves AC, Tavares GM (2019) Mixing animal and plant proteins: is this a way to improve protein techno-functionalities? Food Hydrocoll 97:105171. https://doi.org/10.1016/j.foodhyd.2019.06.016

Anderson CE, Whaley SE, Goran MI (2022) Lactose-reduced infant formula with corn syrup solids and obesity risk among participants in the Special Supplemental Nutrition Program for Women, Infants, and Children (WIC). Am J Clin Nutr 116(4):1002–1009. https://doi.org/10.1093/ajcn/nqac173

Asakuma S, Hatakeyama E, Urashima T et al (2011) Physiology of consumption of human milk oligosaccharides by infant gut-associated bifidobacteria. J Biol Chem 286(40):34583–34592. https://doi.org/10.1074/jbc.M111.248138

Ballegaard A-SR, Larsen JM, Rasmussen PH, Untersmayr E, Pilegaard K, Bøgh KL (2021) Quinoa (Chenopodium quinoa Willd.) seeds increase intestinal protein uptake. Mol Nutr Food Res 65(13):2100102. https://doi.org/10.1002/mnfr.202100102

Barac M, Cabrilo S, Stanojevic S et al (2012) Functional properties of protein hydrolysates from pea (Pisum sativum, L) seeds. Int J Food Sci Technol 47(7):1457–1467. https://doi.org/10.1111/j.1365-2621.2012.02993.x

Bathla S, Arora S (2022) Prevalence and approaches to manage iron deficiency anemia (IDA). Crit Rev Food Sci Nutr 62(32):8815–8828. https://doi.org/10.1080/10408398.2021.1935442

Bering SB (2018) Human milk oligosaccharides to prevent gut dysfunction and necrotizing enterocolitis in preterm neonates. Nutrients 10(10):1461. https://doi.org/10.3390/nu10101461

Bertelsen RJ, Jensen ET, Ringel-Kulka T (2016) Use of probiotics and prebiotics in infant feeding. Best Pract Res Clin Gastroenterol 30(1):39–48. https://doi.org/10.1016/j.bpg.2016.01.001

Bode L (2012) Human milk oligosaccharides: every baby needs a sugar mama. Glycobiology 22(9):1147–1162. https://doi.org/10.1093/glycob/cws074

Boquien CY (2018) Human milk: an ideal food for nutrition of preterm newborn. Front Pediatr 6:295. https://doi.org/10.3389/fped.2018.00295

Bourlieu C, Bouzerzour K, Ferret-Bernard S et al (2015) Infant formula interface and fat source impact on neonatal digestion and gut microbiota. Eur J Lipid Sci Technol 117(10):1500–1512. https://doi.org/10.1002/ejlt.201500025

Bunesova V, Lacroix C, Schwab C (2016) Fucosyllactose and L-fucose utilization of infant Bifidobacterium longum and Bifidobacterium kashiwanohense. BMC Microbiol 16(1):248. https://doi.org/10.1186/s12866-016-0867-4

C.P.S. (1998) Canadian Paediatric Society, dietitians of Canada and Health Canada. Nutrition for healthy term infants. Minister of Public Works and Government Services, Ottawa

Canfield LM, Clandinin MT, Davies DP et al (2003) Multinational study of major breast milk carotenoids of healthy mothers. Eur J Nutr 42(3):133–141. https://doi.org/10.1007/s00394-003-0403-9

Cantani A (2006) A home-made meat-based formula for feeding atopic babies: a study in 51 children. Eur Rev Med Pharmacol Sci 10(2):61–68

Capeding R, Gepanayao CP, Calimon N et al (2010) Lutein-fortified infant formula fed to healthy term infants: evaluation of growth effects and safety. Nutr J 9(1):22. https://doi.org/10.1186/1475-2891-9-22

Cavaletto M, Giuffrida MG, Conti A (2008) Milk fat globule membrane components—a proteomic approach. In: Bösze Z (ed) Bioactive components of milk. Springer, New York, pp 129–141

Ceballos LS, Morales ER, de la Torre AG, Castro JD, Martínez LP, Sampelayo MRS (2009) Composition of goat and cow milk produced under similar conditions and analyzed by identical methodology. J Food Compos Anal 22(4):322–329. https://doi.org/10.1016/j.jfca.2008.10.020

Cebo C (2012) Milk fat globule membrane proteomics: a 'snapshot' of mammary epithelial cell biology. Food Technol Biotechnol 50:306–314

Chandrasekaran S, Ramanathan S, Basak T (2013) Microwave food processing—a review. Food Res Int 52(1):243–261. https://doi.org/10.1016/j.foodres.2013.02.033

Cheatham CL, Sheppard KW (2015) Synergistic effects of human milk nutrients in the support of infant recognition memory: an observational study. Nutrients 7(11):9079–9095. https://doi.org/10.3390/nu7115452

Chen Y, Zheng Z, Zhu X et al (2015) Lactoferrin promotes early neurodevelopment and cognition in postnatal piglets by upregulating the BDNF signaling pathway and polysialylation. Mol Neurobiol 52(1):256–269. https://doi.org/10.1007/s12035-014-8856-9

Chi L (2017) Why is Asia demanding so much baby formula? BBC. http://www.bbc.co.uk/news/business-38851056

Chihi M-L, Mession J-L, Sok N, Saurel R (2016) Heat-induced soluble protein aggregates from mixed pea globulins and β-lactoglobulin. J Agric Food Chem 64(13):2780–2791. https://doi.org/10.1021/acs.jafc.6b00087

CODEX (2007) Codex Alimentarius Commission. Procedural manual, 17th ed.

Commission CA (2007)

Cook DA (1989) Nutrient levels in infant formulas: technical considerations. J Nutr 119(12 Suppl):1773–1777. https://doi.org/10.1093/jn/119.suppl_12.1773; discussion 1777–8

Daryaei H, Yousef AE, Balasubramaniam VM (2016) Microbiological aspects of high-pressure processing of food: inactivation of microbial vegetative cells and spores. In: Balasubramaniam VM, Barbosa-Cánovas GV, Lelieveld HLM (eds) High pressure processing of food: principles, technology and applications. Springer, New York, pp 271–294

Devi AF, Buckow R, Hemar Y, Kasapis S (2013) Structuring dairy systems through high pressure processing. J Food Eng 114(1):106–122. https://doi.org/10.1016/j.jfoodeng.2012.07.032

Dewettinck K, Rombaut R, Thienpont N, Le TT, Messens K, Van Camp J (2008) Nutritional and technological aspects of milk fat globule membrane material. Int Dairy J 18(5):436–457. https://doi.org/10.1016/j.idairyj.2007.10.014

Dewey KG (2001) Nutrition, growth, and complementary feeding of the brestfed infant. Pediatr Clin North Am 48(1):87–104. https://doi.org/10.1016/S0031-3955(05)70287-X

do Nascimento da Silva E, Leme ABP, Cidade M, Cadore S (2013) Evaluation of the bioaccessible fractions of Fe, Zn, Cu and Mn in baby foods. Talanta 117:184–188. https://doi.org/10.1016/j.talanta.2013.09.008

Donovan SM, Monaco MH, Drnevich J, Kvistgaard AS, Hernell O, Lönnerdal B (2014) Bovine osteopontin modifies the intestinal transcriptome of formula-fed infant rhesus monkeys to be more similar to those that were breastfed. J Nutr 144(12):1910–1919. https://doi.org/10.3945/jn.114.197558

Duncan B, Schifman RB, Corrigan JJ Jr, Schaefer C (1985) Iron and the exclusively breast-fed infant from birth to six months. J Pediatr Gastroenterol Nutr 4(3):421–425. https://doi.org/10.1097/00005176-198506000-00017

E.U. (2013) European Commission, Commission Directive 2013/46/EU amending Directive 2006/141/EC with regard to protein requirements for infant formulae and follow-on formulae. Off J Eur Union 46:16–19

E.U. (2017) Commission implementing regulation (EU) 2017/2470 of 20 December 2017 establishing the Union list of novel foods in accordance with Regulation (EU) 2015/2283 of the European Parliament and of the Council on novel foods. Off J Eur Union 351:72–201

Egbert WR, Payne CT (2009) Plant proteins. In: Tarté R (ed) Ingredients in meat products: properties, functionality and applications. Springer, New York, pp 111–129

Eggersdorfer M, Wyss A (2018) Carotenoids in human nutrition and health. Arch Biochem Biophys 652:18–26. https://doi.org/10.1016/j.abb.2018.06.001

Euromonitor (2016) Market overview: identifying new trends and opportunities in the global infant formula market. Presentation to VitaFoods Europe. https://www.slideshare.net/Euromonitor/market-oveview-identifying-new-trends-and-opportunities-in-the-global-infant-forumla-market66

Evrendilek GA, Baysal T, Icier F, Yildiz H, Demirdoven A, Bozkurt H (2012) Processing of fruits and fruit juices by novel electrotechnologies. Food Eng Rev 4(1):68–87. https://doi.org/10.1007/s12393-011-9045-5

Fiocchi A, Brozek J, Schünemann H et al (2010) World Allergy Organization (WAO) diagnosis and rationale for action against cow's milk allergy (DRACMA) guidelines. World Allergy Organ J 3(4):57–161. https://doi.org/10.1097/WOX.0b013e3181defeb9

Floris R, Lambers T, Alting A, Kiers J (2010) 17—Trends in infant formulas: a dairy perspective. In: Griffiths MW (ed) Improving the safety and quality of milk. Woodhead Publishing, Sawston, pp 454–474

Fontecha J, Brink L, Wu S, Pouliot Y, Visioli F, Jiménez-Flores R (2020) Sources, production, and clinical treatments of milk fat globule membrane for infant nutrition and well-being. Nutrients 12(6):1607

Gibson GR, Roberfroid MB (1995) Dietary modulation of the human colonic microbiota: introducing the concept of prebiotics. J Nutr 125(6):1401–1412. https://doi.org/10.1093/jn/125.6.1401

Golan-Gerstl R, Elbaum Shiff Y, Moshayoff V, Schecter D, Leshkowitz D, Reif S (2017) Characterization and biological function of milk-derived miRNAs. Mol Nutr Food Res 61(10). https://doi.org/10.1002/mnfr.201700009

Goyal A, Sharma V, Upadhyay N, Sihag M, Kaushik R (2013) High pressure processing and its impact on milk proteins: a review. J Dairy Sci Technol 2(1):2319–3409

Guo L, Chen X, Xu L, Xiao M, Lu L (2018) Enzymatic synthesis of 6′-sialyllactose, a dominant sialylated human milk oligosaccharide, by a novel exo-α-sialidase from bacteroides fragilis NCTC9343. Appl Environ Microbiol 84(13):e00071-18. https://doi.org/10.1128/aem.00071-18

Gurnida DA, Rowan AM, Idjradinata P, Muchtadi D, Sekarwana N (2012) Association of complex lipids containing gangliosides with cognitive development of 6-month-old infants. Early Hum Dev 88(8):595–601. https://doi.org/10.1016/j.earlhumdev.2012.01.003

Hammond BR Jr, Wooten BR, Curran-Celentano J (2001) Carotenoids in the retina and lens: possible acute and chronic effects on human visual performance. Arch Biochem Biophys 385(1):41–46. https://doi.org/10.1006/abbi.2000.2184

He S, Tang H, Yi H, Xu W, Ma Y, Wang R (2017) Properties of emulsions from milk fat globule membrane and its components. Int J Food Prop 20(Suppl 2):1342–1353. https://doi.org/10.1080/10942912.2017.1343348

Hernell O, Timby N, Domellöf M, Lönnerdal B (2016) Clinical benefits of milk fat globule membranes for infants and children. J Pediatr 173:S60–S65. https://doi.org/10.1016/j.jpeds.2016.02.077

Hildebrand HV, Arias A, Simons E et al (2021) Adult and pediatric food allergy to chickpea, pea, lentil, and lupine: a scoping review. J Allergy Clin Immunol Pract 9(1):290–301.e2. https://doi.org/10.1016/j.jaip.2020.10.046

Ho SM, Cheong A, Adgent MA et al (2017) Environmental factors, epigenetics, and developmental origin of reproductive disorders. Reprod Toxicol 68:85–104. https://doi.org/10.1016/j.reprotox.2016.07.011

Holma R, Juvonen P, Asmawi MZ, Vapaatalo H, Korpela R (2002) Galacto-oligosaccharides stimulate the growth of bifidobacteria but fail to attenuate inflammation in experimental colitis in rats. Scand J Gastroenterol 37(9):1042–1047. https://doi.org/10.1080/003655202320378239

Hradecky J, Kludska E, Belkova B, Wagner M, Hajslova J (2017) Ohmic heating: a promising technology to reduce furan formation in sterilized vegetable and vegetable/meat baby foods. Innov Food Sci Emerg Technol 43:1–6. https://doi.org/10.1016/j.ifset.2017.07.018

Humphrey J, Iliff P (2001) Is breast not best? Feeding babies born to HIV-positive mothers: bringing balance to a complex issue. Nutr Rev 59(4):119–127. https://doi.org/10.1111/j.1753-4887.2001.tb06999.x

Hussain M, Sun Y, Pan Y, Liu L, Zhang X, Wang Q, Lin S, Qayum A, Hussain K, Li X (2023) Formulation, invitro digestive study, and comparative fatty acid analysis of walnut oil-based infant formula, with human milk, animal milk, and commercial infant formula. Inn Food Sci Emerg Tech 84:103279

Hussain M, Li X, Wang L et al (2021) Recent approaches and methods for the formulation of a risk free infant formula: review. Food Rev Int 39(1):59–72. https://doi.org/10.1080/87559129.2021.1901113

Ikeda K (2002) Buckwheat: composition, chemistry, and processing. Adv Food Nutr Res 44:395–434. https://doi.org/10.1016/s1043-4526(02)44008-9

Jaeger H, Janositz A, Knorr D (2010) The Maillard reaction and its control during food processing. The potential of emerging technologies. Pathol Biol (Paris) 58(3):207–213. https://doi.org/10.1016/j.patbio.2009.09.016

Javed F, Shahbaz HM, Nawaz A et al (2021) Formation of furan in baby food products: identification and technical challenges. Compr Rev Food Sci Food Saf 20(3):2699–2715. https://doi.org/10.1111/1541-4337.12732

Jenness R (1979) The composition of human milk. Semin Perinatol 3(3):225–239

Jewell VC, Mayes C, Tubman TRJ, Northrop-Clewes CA, Thurnham DI (2004) A comparison of lutein and zeaxanthin concentrations in formula and human milk samples from Northern Ireland mothers. Eur J Clin Nutr 58:90–97

Kalach N, Bellaïche M, Elias-Billon I, Dupont C (2019) Family history of atopy in infants with cow's milk protein allergy: a French population-based study. Arch Pediatr 26(4):226–231. https://doi.org/10.1016/j.arcped.2019.02.014

Kelishadi R, Hadi B, Iranpour R et al (2012) A study on lipid content and fatty acid of breast milk and its association with mother's diet composition. J Res Med Sci 17(9):824–827

Koletzko B, Rodriguez-Palmero M, Demmelmair H, Fidler N, Jensen R, Sauerwald T (2001) Physiological aspects of human milk lipids. Early Hum Dev 65(Suppl):S3–S18. https://doi.org/10.1016/s0378-3782(01)00204-3

Koletzko B, Baker S, Cleghorn G et al (2005) Global standard for the composition of infant formula: recommendations of an ESPGHAN coordinated international expert group. J Pediatr Gastroenterol Nutr 41(5):584–599. https://doi.org/10.1097/01.mpg.0000187817.38836.42

Kon IY, Gmoshinskaya MV, Safronova AI, Alarcon P, Vandenplas Y (2014) Growth and tolerance assessment of a lutein-fortified infant formula. Pediatr Gastroenterol Hepatol Nutr 17(2):104–111. https://doi.org/10.5223/pghn.2014.17.2.104

Korpal M, Lee ES, Hu G, Kang Y (2008) The miR-200 family inhibits epithelial-mesenchymal transition and cancer cell migration by direct targeting of E-cadherin transcriptional repressors ZEB1 and ZEB2. J Biol Chem 283(22):14910–14914. https://doi.org/10.1074/jbc.C800074200

Kreiter SR, Schwartz RP, Kirkman HN Jr, Charlton PA, Calikoglu AS, Davenport ML (2000) Nutritional rickets in African American breast-fed infants. J Pediatr 137(2):153–157. https://doi.org/10.1067/mpd.2000.109009

Kultur G, Misra NN, Barba FJ, Koubaa M, Gökmen V, Alpas H (2017) Microbial inactivation and evaluation of furan formation in high hydrostatic pressure (HHP) treated vegetable-based infant food. Food Res Int 101:17–23. https://doi.org/10.1016/j.foodres.2017.07.064

Landrum JT, Bone RA (2001) Lutein, zeaxanthin, and the macular pigment. Arch Biochem Biophys 385(1):28–40. https://doi.org/10.1006/abbi.2000.2171

Le Huërou-Luron I, Blat S, Boudry G (2010) Breast- v. formula-feeding: impacts on the digestive tract and immediate and long-term health effects. Nutr Res Rev 23(1):23–36. https://doi.org/10.1017/S0954422410000065

Le Parc A, Dallas DC, Duaut S, Leonil J, Martin P, Barile D (2014) Characterization of goat milk lactoferrin N-glycans and comparison with the N-glycomes of human and bovine milk. Electrophoresis 35(11):1560–1570. https://doi.org/10.1002/elps.201300619

Le Roux L, Mejean S, Chacon R et al (2020a) Plant proteins partially replacing dairy proteins greatly influence infant formula functionalities. LWT 120:108891. https://doi.org/10.1016/j.lwt.2019.108891

Le Roux L, Ménard O, Chacon R et al (2020b) Are faba bean and pea proteins potential whey protein substitutes in infant formulas? An in vitro dynamic digestion approach. Foods 9(3):362. https://doi.org/10.3390/foods9030362

Lee W-H, Pathanibul P, Quarterman J et al (2012) Whole cell biosynthesis of a functional oligosaccharide, 2′-fucosyllactose, using engineered Escherichia coli. Microb Cell Factories 11(1):48. https://doi.org/10.1186/1475-2859-11-48

Lee H, Padhi E, Hasegawa Y et al (2018) Compositional dynamics of the milk fat globule and its role in infant development. Front Pediatr 6:313. https://doi.org/10.3389/fped.2018.00313

Leong A, Liu Z, Almshawit H et al (2019) Oligosaccharides in goats' milk-based infant formula and their prebiotic and anti-infection properties. Br J Nutr 122(4):441–449. https://doi.org/10.1017/s000711451900134x

Li H, Byers HM, Diaz-Kuan A et al (2018) Acute liver failure in neonates with undiagnosed hereditary fructose intolerance due to exposure from widely available infant formulas. Mol Genet Metab 123(4):428–432. https://doi.org/10.1016/j.ymgme.2018.02.016

Liao Y, Alvarado R, Phinney B, Lönnerdal B (2011) Proteomic characterization of human milk fat globule membrane proteins during a 12 month lactation period. J Proteome Res 10(8):3530–3541. https://doi.org/10.1021/pr200149t

Liu Z, Neuringer M, Erdman JW Jr et al (2019) The effects of breastfeeding versus formula-feeding on cerebral cortex maturation in infant rhesus macaques. Neuroimage 184:372–385. https://doi.org/10.1016/j.neuroimage.2018.09.015

LoCascio RG, Ninonuevo MR, Freeman SL et al (2007) Glycoprofiling of bifidobacterial consumption of human milk oligosaccharides demonstrates strain specific, preferential consumption of small chain glycans secreted in early human lactation. J Agric Food Chem 55(22):8914–8919. https://doi.org/10.1021/jf0710480

Lönnerdal B (2009) Nutritional roles of lactoferrin. Curr Opin Clin Nutr Metab Care 12(3):293–297. https://doi.org/10.1097/MCO.0b013e328328d13e

Lönnerdal B (2014) Infant formula and infant nutrition: bioactive proteins of human milk and implications for composition of infant formulas. Am J Clin Nutr 99(3):712s–717s. https://doi.org/10.3945/ajcn.113.071993

Lönnerdal B, Jiang R, Du X (2011) Bovine lactoferrin can be taken up by the human intestinal lactoferrin receptor and exert bioactivities. J Pediatr Gastroenterol Nutr 53(6):606–614. https://doi.org/10.1097/MPG.0b013e318230a419

Lönnerdal B, Kvistgaard AS, Peerson JM, Donovan SM, Peng YM (2016) Growth, nutrition, and cytokine response of breast-fed infants and infants fed formula with added bovine osteopontin. J Pediatr Gastroenterol Nutr 62(4):650–657. https://doi.org/10.1097/mpg.0000000000001005

Lukasik A, Brzozowska I, Zielenkiewicz U, Zielenkiewicz P (2017) Detection of plant miRNAs abundance in human breast milk. Int J Mol Sci 19(1):37. https://doi.org/10.3390/ijms19010037

Luque V, Closa-Monasterolo R, Escribano J, Ferré N (2015) Early programming by protein intake: the effect of protein on adiposity development and the growth and functionality of vital organs. Nutr Metab Insights 8(Suppl 1):49–56. https://doi.org/10.4137/nmi.S29525

Mackey AD, Albrecht D, Oliver J, Williams T, Long AC, Price PT (2013) Plasma carotenoid concentrations of infants are increased by feeding a milk-based infant formula supplemented with carotenoids. J Sci Food Agric 93(8):1945–1952. https://doi.org/10.1002/jsfa.5996

Malunga LN, Bar-El Dadon S, Zinal E, Berkovich Z, Abbo S, Reifen R (2014) The potential use of chickpeas in development of infant follow-on formula. Nutr J 13:8. https://doi.org/10.1186/1475-2891-13-8

Manoni M, Di Lorenzo C, Ottoboni M, Tretola M, Pinotti L (2020) Comparative proteomics of milk fat globule membrane (MFGM) proteome across species and lactation stages and the potentials of MFGM fractions in infant formula preparation. Foods 9(9):1251. https://doi.org/10.3390/foods9091251

Manzoni P (2016) Clinical benefits of lactoferrin for infants and children. J Pediatr 173(Supp l): S43–S52. https://doi.org/10.1016/j.jpeds.2016.02.075

Marcobal A, Barboza M, Froehlich JW et al (2010) Consumption of human milk oligosaccharides by gut-related microbes. J Agric Food Chem 58(9):5334–5340. https://doi.org/10.1021/jf9044205

Martin CR, Ling PR, Blackburn GL (2016) Review of infant feeding: key features of breast milk and infant formula. Nutrients 8(5):279. https://doi.org/10.3390/nu8050279

Martins CPC, Cavalcanti RN, Couto SM et al (2019) Microwave processing: current background and effects on the physicochemical and microbiological aspects of dairy products. Compr Rev Food Sci Food Saf 18(1):67–83. https://doi.org/10.1111/1541-4337.12409

Masschalck B, Van Houdt R, Michiels CW (2001) High pressure increases bactericidal activity and spectrum of lactoferrin, lactoferricin and nisin. Int J Food Microbiol 64(3):325–332. https://doi.org/10.1016/s0168-1605(00)00485-2

Matthies H, Staak S, Krug M (1996) Fucose and fucosyllactose enhance in-vitro hippocampal long-term potentiation. Brain Res 725(2):276–280. https://doi.org/10.1016/0006-8993(96)00406-4

Mayayo C, Montserrat M, Ramos S et al (2014) Kinetic parameters for high-pressure-induced denaturation of lactoferrin in human milk. Int Dairy J 39(2):246–252

Mazri C, Sánchez L, Ramos SJ, Calvo M, Pérez MD (2012) Effect of high-pressure treatment on denaturation of bovine lactoferrin and lactoperoxidase. J Dairy Sci 95(2):549–557. https://doi.org/10.3168/jds.2011-4665

Mbori-Ngacha D, Nduati R, John G et al (2001) Morbidity and mortality in breastfed and formula-fed infants of HIV-1-infected women: a randomized clinical trial. JAMA 286(19):2413–2420. https://doi.org/10.1001/jama.286.19.2413

Mendonça MA, Araújo WMC, Borgo LA, Alencar ER (2017) Lipid profile of different infant formulas for infants. PLoS One 12(6):e0177812. https://doi.org/10.1371/journal.pone.0177812

Mesías M, Wagner M, George S, Morales FJ (2016) Impact of conventional sterilization and ohmic heating on the amino acid profile in vegetable baby foods. Innov Food Sci Emerg Technol 34: 24–28. https://doi.org/10.1016/j.ifset.2015.12.031

Misselwitz B, Butter M, Verbeke K, Fox MR (2019) Update on lactose malabsorption and intolerance: pathogenesis, diagnosis and clinical management. Gut 68(11):2080–2091

Moro GE, Stahl B, Fanaro S, Jelinek J, Boehm G, Coppa GV (2005) Dietary prebiotic oligosaccharides are detectable in the faeces of formula-fed infants. Acta Paediatr Suppl 94(449):27–30. https://doi.org/10.1111/j.1651-2227.2005.tb02151.x

Mosca F, Giannì ML (2017) Human milk: composition and health benefits. Pediatr Med Chir 39(2): 155. https://doi.org/10.4081/pmc.2017.155

Nieto-Ruiz A, García-Santos JA, Bermúdez MG et al (2019) Cortical visual evoked potentials and growth in infants fed with bioactive compounds-enriched infant formula: results from COGNIS randomized clinical trial. Nutrients 11(10):2456. https://doi.org/10.3390/nu11102456

Niño DF, Sodhi CP, Hackam DJ (2016) Necrotizing enterocolitis: new insights into pathogenesis and mechanisms. Nat Rev Gastroenterol Hepatol 13(10):590–600. https://doi.org/10.1038/nrgastro.2016.119

O'Callaghan DM, O'Mahony JA, Ramanujam KS, Burgher AM (2011) Dehydrated dairy products/infant formulae. In: Fuquay JW (ed) Encyclopedia of dairy sciences (second edition). Academic Press, New York, pp 135–145

O'Kane FE, Vereijken JM, Gruppen H, Van Boekel MAJS (2005) Gelation behavior of protein isolates extracted from 5 cultivars of Pisum sativum L. J Food Sci 70(2):C132–C137. https://doi.org/10.1111/j.1365-2621.2005.tb07073.x

Oddy WH (2017) Breastfeeding, childhood asthma, and allergic disease. Ann Nutr Metab 70(Suppl 2):26–36. https://doi.org/10.1159/000457920

Odom EC, Li R, Scanlon KS, Perrine CG, Grummer-Strawn L (2013) Reasons for earlier than desired cessation of breastfeeding. Pediatrics 131(3):e726–e732. https://doi.org/10.1542/peds.2012-1295

Oozeer R, van Limpt K, Ludwig T et al (2013) Intestinal microbiology in early life: specific prebiotics can have similar functionalities as human-milk oligosaccharides. Am J Clin Nutr 98(2):561s–571s. https://doi.org/10.3945/ajcn.112.038893

Packard VS (1982) 5—Mammary infections, drugs, and environmental pollutants. In: Packard VS (ed) Human milk and infant formula. Academic Press, New York, pp 108–139

Palmano K, Rowan A, Guillermo R, Guan J, McJarrow P (2015) The role of gangliosides in neurodevelopment. Nutrients 7(5):3891–3913. https://doi.org/10.3390/nu7053891

Pediatria SBd (2018) Sociedade Brasileira de Pediatria (SPB) Manual de Alimentação: Orientações para Alimentação do Lactente ao Adolescente, Na Escola, Na Gestante, Na Prevenção de Doenças e Segurança Alimentar. 4nd. SBP, São Paulo. Departamento de Nutrologia, p 32

Pérez-Escalante E, Alatorre-Santamaría S, Castañeda-Ovando A et al (2022) Human milk oligosaccharides as bioactive compounds in infant formula: recent advances and trends in synthetic methods. Crit Rev Food Sci Nutr 62(1):181–214. https://doi.org/10.1080/10408398.2020.1813683

Pires RPS, Cappato LP, Guimarães JT et al (2020) Ohmic heating for infant formula processing: evaluating the effect of different voltage gradient. J Food Eng 280:109989. https://doi.org/10.1016/j.jfoodeng.2020.109989

Pisacane A, De Vizia B, Valiante A et al (1995) Iron status in breast-fed infants. J Pediatr 127(3):429–431. https://doi.org/10.1016/s0022-3476(95)70076-5

Ramiro-Cortijo D, Singh P, Liu Y et al (2020) Breast milk lipids and fatty acids in regulating neonatal intestinal development and protecting against intestinal injury. Nutrients 12(2):534. https://doi.org/10.3390/nu12020534

Rodríguez MH (2001) Alimentación infantil. Ediciones Díaz de Santos

Rollins NC, Bhandari N, Hajeebhoy N et al (2016) Why invest, and what it will take to improve breastfeeding practices? Lancet 387(10017):491–504. https://doi.org/10.1016/s0140-6736(15)01044-2

Rossiter MD, Colapinto CK, Khan MKA et al (2015) Breast, formula and combination feeding in relation to childhood obesity in Nova Scotia, Canada. Matern Child Health J 19:2048–2056

Roux S, Courel M, Birlouez-Aragon I, Municino F, Massa M, Pain J-P (2016) Comparative thermal impact of two UHT technologies, continuous ohmic heating and direct steam injection, on the nutritional properties of liquid infant formula. J Food Eng 179:36–43. https://doi.org/10.1016/j.jfoodeng.2016.02.001

Roux LL, Chacon R, Dupont D, Jeantet R, Deglaire A, Nau F (2020) In vitro static digestion reveals how plant proteins modulate model infant formula digestibility. Food Res Int 130:108917. https://doi.org/10.1016/j.foodres.2019.108917

Rudloff S, Lönnerdal B (1992) Solubility and digestibility of milk proteins in infant formulas exposed to different heat treatments. J Pediatr Gastroenterol Nutr 15(1):25–33. https://doi.org/10.1097/00005176-199207000-00005

Rueda R (2007) The role of dietary gangliosides on immunity and the prevention of infection. Br J Nutr 98(Suppl 1):S68–S73. https://doi.org/10.1017/s0007114507832946
Sakr M, Liu S (2014) A comprehensive review on applications of ohmic heating (OH). Renew Sustain Energy Rev 39:262–269
Salpietro CD, Gangemi S, Briuglia S et al (2005) The almond milk: a new approach to the management of cow-milk allergy/intolerance in infants. Minerva Pediatr 57(4):173–180
Sammut D, Dennison P, Venter C, Kurukulaaratchy RJ (2011) Buckwheat allergy: a potential problem in 21st century Britain. BMJ Case Rep 2011:bcr0920114882. https://doi.org/10.1136/bcr.09.2011.4882
Sandberg AS (2011) Developing functional ingredients: a case study of pea protein. Woodhead Publishing, Cambridge, pp 358–382
Schaafsma G (2008) Lactose and lactose derivatives as bioactive ingredients in human nutrition. Int Dairy J 18:458–465. https://doi.org/10.1016/j.idairyj.2007.11.013
Schack L, Lange A, Kelsen J et al (2009a) Considerable variation in the concentration of osteopontin in human milk, bovine milk, and infant formulas. J Dairy Sci 92(11):5378–5385. https://doi.org/10.3168/jds.2009-2360
Schack L, Stapulionis R, Christensen B et al (2009b) Osteopontin enhances phagocytosis through a novel osteopontin receptor, the alphaXbeta2 integrin. J Immunol 182(11):6943–6950. https://doi.org/10.4049/jimmunol.0900065
Schuh S, Thevenier A, Ran-Ressler R, Rade-Kukic K, Kuslys M (2019) Infant formula for cow's milk protein allergic infants. Google patents
Seoane RG, Garcia-Recio V, Garrosa M et al (2020) Human health effects of lactose consumption as a food and drug ingredient. Curr Pharm Des 26(16):1778–1789. https://doi.org/10.2174/1381612826666200212114843
Setchell KD, Zimmer-Nechemias L, Cai J, Heubi JE (1997) Exposure of infants to phyto-oestrogens from soy-based infant formula. Lancet 350(9070):23–27. https://doi.org/10.1016/s0140-6736(96)09480-9
Skolnick J, Chou C, Miklavcic J (2020) Insights into novel infant milk formula bioactives. Nutr Diet Suppl 12:11–19. https://doi.org/10.2147/NDS.S192099
Sodek J, Ganss B, McKee MD (2000) Osteopontin. Crit Rev Oral Biol Med 11(3):279–303. https://doi.org/10.1177/10454411000110030101
Spitsberg VL (2005) Invited review: bovine milk fat globule membrane as a potential nutraceutical. J Dairy Sci 88(7):2289–2294. https://doi.org/10.3168/jds.S0022-0302(05)72906-4
Takahashi Y, Ichikawa S, Aihara Y, Yokota S (1998) [Buckwheat allergy in 90,000 school children in Yokohama]. Arerugi 47(1):26–33
Tanaka K, Hosozawa M, Kudo N et al (2013) The pilot study: sphingomyelin-fortified milk has a positive association with the neurobehavioural development of very low birth weight infants during infancy, randomized control trial. Brain Dev 35(1):45–52. https://doi.org/10.1016/j.braindev.2012.03.004
Tang M, Krebs NF (2014) High protein intake from meat as complementary food increases growth but not adiposity in breastfed infants: a randomized trial. Am J Clin Nutr 100(5):1322–1328. https://doi.org/10.3945/ajcn.114.088807
Thomson K (2007) The State of Food and Agriculture 2006: Food Aid for Food Security? Food and Agriculture Organization (FAO). FAO Agriculture Series no. 37. xii+168 pp. + mini CD-ROM. Rome: FAO (2006). US $65.00 (Paperback). ISBN 978-92-5-105600-4. J Agric Sci 145(4):415–416
Timby N, Domellöf E, Hernell O, Lönnerdal B, Domellöf M (2014) Neurodevelopment, nutrition, and growth until 12 mo of age in infants fed a low-energy, low-protein formula supplemented with bovine milk fat globule membranes: a randomized controlled trial. Am J Clin Nutr 99(4): 860–868. https://doi.org/10.3945/ajcn.113.064295
Timby N, Hernell O, Vaarala O, Melin M, Lönnerdal B, Domellöf M (2015) Infections in infants fed formula supplemented with bovine milk fat globule membranes. J Pediatr Gastroenterol Nutr 60(3):384–389. https://doi.org/10.1097/mpg.0000000000000624

Timby N, Domellöf M, Lönnerdal B, Hernell O (2017) Supplementation of infant formula with bovine milk fat globule membranes. Adv Nutr 8(2):351–355. https://doi.org/10.3945/an.116.014142

Title AC, Denzler R, Stoffel M (2015) Uptake and function studies of maternal milk-derived microRNAs. J Biol Chem 290(39):23680–23691. https://doi.org/10.1074/jbc.M115.676734

Torres JA, Velazquez G (2005) Commercial opportunities and research challenges in the high pressure processing of foods. J Food Eng 67(1–2):95–112

Triantis V, Bode L, van Neerven RJJ (2018) Immunological effects of human milk oligosaccharides. Front Pediatr 6:190. https://doi.org/10.3389/fped.2018.00190

Turck D, Bresson J-L, Burlingame B et al (2016) Guidance on the preparation and presentation of an application for authorisation of a novel food in the context of Regulation (EU) 2015/2283. EFSA J 14(11):e04594. https://doi.org/10.2903/j.efsa.2016.4594

U.S. (1972) Bureau of Census. Curnegie Commission on HigherEducation, Washington, DC

U.S. (2018) US Department of Health & Human Services, US Food & Drug Administration. GRAS notices. https://www.accessdata.fda.gov/scripts/fdcc/index.cfm?set=GRASNotices

Ulloa JA, Valencia ME, Garcia ZH (1988) Protein concentrate from chickpea: nutritive value of a protein concentrate from chickpea (Cicer arietinum) obtained by ultrafiltration and its potential use in an infant formula. J Food Sci 53(5):1396–1398. https://doi.org/10.1111/j.1365-2621.1988.tb09285.x

Upson K, Sathyanarayana S, Scholes D, Holt VL (2015) Early-life factors and endometriosis risk. Fertil Steril 104(4):964–971.e5. https://doi.org/10.1016/j.fertnstert.2015.06.040

Urashima T, Asakuma S, Kitaoka M, Messer M (2011) Lactose and oligosaccharides | indigenous oligosaccharides in milk. In: Fuquay JW (ed) Encyclopedia of dairy sciences (second edition). Academic Press, San Diego, pp 241–273

Vadivambal R (2010) Non-uniform temperature distribution during microwave heating of food materials—a review. Food Bioprocess Technol 3(2):161–171. https://doi.org/10.1007/s11947-008-0136-0

van de Looij Y, Ginet V, Chatagner A et al (2014) Lactoferrin during lactation protects the immature hypoxic-ischemic rat brain. Ann Clin Transl Neurol 1(12):955–967. https://doi.org/10.1002/acn3.138

Vandenplas Y, Zakharova I, Dmitrieva Y (2015) Oligosaccharides in infant formula: more evidence to validate the role of prebiotics. Br J Nutr 113(9):1339–1344. https://doi.org/10.1017/S0007114515000823

Vandenplas Y, De Mulder N, De Greef E, Huysentruyt K (2021) Plant-based formulas and liquid feedings for infants and toddlers. Nutrients 13(11). https://doi.org/10.3390/nu13114026

Venlet NV, Hettinga KA, Schebesta H, Bernaz N (2021) Perspective: a legal and nutritional perspective on the introduction of quinoa-based infant and follow-on formula in the EU. Adv Nutr 12(4):1100–1107. https://doi.org/10.1093/advances/nmab041

Victora CG, Bahl R, Barros AJ et al (2016) Breastfeeding in the 21st century: epidemiology, mechanisms, and lifelong effect. Lancet 387(10017):475–490. https://doi.org/10.1016/s0140-6736(15)01024-7

Vishwanathan R, Kuchan M, Sen S, Johnson E (2014) Lutein is the predominant carotenoid in infant brain: preterm infants have decreased concentrations of brain carotenoids. J Pediatr Gastroenterol Nutr 59:659–665. https://doi.org/10.1097/MPG.0000000000000389

Wagner CL, Greer FR (2008) Prevention of rickets and vitamin D deficiency in infants, children, and adolescents. Pediatrics 122(5):1142–1152. https://doi.org/10.1542/peds.2008-1862

Wakabayashi H, Yamauchi K, Abe F (2018) Quality control of commercial bovine lactoferrin. Biometals 31(3):313–319. https://doi.org/10.1007/s10534-018-0098-2

Wayler TJ, Kein RS (1965) Applied nutrition. MacMillan, New York

Wazed MA, Farid M (2019) Hypoallergenic and low-protein ready-to-feed (RTF) infant formula by high pressure pasteurization: a novel product. Foods 8(9):408. https://doi.org/10.3390/foods8090408

Weber JA, Baxter DH, Zhang S et al (2010) The microRNA spectrum in 12 body fluids. Clin Chem 56(11):1733–1741. https://doi.org/10.1373/clinchem.2010.147405

Weisselberg B, Dayal Y, Thompson JF, Doyle MS, Senior B, Grand RJ (1996) A lamb-meat-based formula for infants allergic to casein hydrolysate formulas. Clin Pediatr (Phila) 35(10):491–495. https://doi.org/10.1177/000992289603501002

WHO (2010) Guidelines on HIV and infant feeding

World Health Organization, Food, Agriculture Organization of the United Nations (2004) Enterobacter sakazakii and other microorganisms in powdered infant formula: meeting report. World Health Organization, Geneva

Yi DY, Kim SY (2021) Human breast milk composition and function in human health: from nutritional components to microbiome and microRNAs. Nutrients 13(9):3094. https://doi.org/10.3390/nu13093094

Yu S, Liu J-J, Yun EJ, Kwak S, Kim KH, Jin Y-S (2018) Production of a human milk oligosaccharide 2′-fucosyllactose by metabolically engineered Saccharomyces cerevisiae. Microb Cell Factories 17(1):101. https://doi.org/10.1186/s12934-018-0947-2

Zavaleta N, Kvistgaard AS, Graverholt G et al (2011) Efficacy of an MFGM-enriched complementary food in diarrhea, anemia, and micronutrient status in infants. J Pediatr Gastroenterol Nutr 53:561–568

Zempleni J, Aguilar-Lozano A, Sadri M et al (2017) Biological activities of extracellular vesicles and their cargos from bovine and human milk in humans and implications for infants. J Nutr 147(1):3–10. https://doi.org/10.3945/jn.116.238949

Zhang H, Vieira Resende ESB, Cui J (2018) miRDis: a web tool for endogenous and exogenous microRNA discovery based on deep-sequencing data analysis. Brief Bioinform 19(3):415–424. https://doi.org/10.1093/bib/bbw140

Zheng L, Fleith M, Giuffrida F, O'Neill BV, Schneider N (2019) Dietary polar lipids and cognitive development: a narrative review. Adv Nutr 10(6):1163–1176. https://doi.org/10.1093/advances/nmz051

Zielińska-Pukos M, Wesolowska A, Pawlus B, Hamulka J (2017) Health effects of carotenoids during pregnancy and lactation. Nutrients 9:838. https://doi.org/10.3390/nu9080838

Zimmer JP, Hammond BR Jr (2007) Possible influences of lutein and zeaxanthin on the developing retina. Clin Ophthalmol 1(1):25–35

Chapter 20
Role of Probiotics in Gut Micro-flora

Sakshi Rai, Pooja Yadav, Nabendu Debnath, Shalini Arora, and Ashok K. Yadav

Contents

Abstract Human gastro-intestinal tract is inhabited by the most diverse cluster of microorganisms known as gut micro-flora and actively partake in numerous functional activities. With the advancement of recent microbial technologies, the apparent role of gut micro-flora in health and diseases is appreciated in scientific communities. It is helpful to think of the microflora in one's digestive tract, as the "microbiota is comprised of beneficial, harmful, and commensal species," due to the diversity and complexity of its constituent parts. The Human Microbiome Project (HMP) found that many genetic components from the intestinal microbiota are functionally interconnected and complement the genes necessary for important biological processes in the human intestine. Additionally, any disturbances in the

S. Rai · P. Yadav · N. Debnath · A. K. Yadav (✉)
Centre for Molecular Biology, Central University of Jammu, Jammu, Jammu and Kashmir, India
e-mail: akyadav@cujammu.ac.in

S. Arora
Department of Dairy Technology, College of Dairy Science and Technology, Lala Lajpat Rai University of Veterinary and Animal Sciences, Hisar, Haryana, India

J. A. Malik et al. (eds.), *Food Process Engineering and Technology*,
https://doi.org/10.1007/978-981-99-6831-2_20

gut micro-flora can substantially augment the different disease phenotypes. In this regard, probiotics have become an exciting new approach to modulating the gut micro-flora composition, improving gut health. An acceptable dose of "live micro-organisms that bestow a health benefit on the host" is what makes a probiotic effective for human consumption. Although their exact mechanism of action is not fully elucidated yet, probiotics have shown to participate in the production of important biological compounds such as bacteriocin and short chain fatty acid, along with lowering of gut pH, nutrient competition, stimulation of mucosal barrier function and immunomodulation. Apart from these functions, probiotics can also help to restore the gut microbiomes makeup and bring beneficial activities to gut microbial populations, which can help to alleviate or avoid intestinal swelling and other symptoms of intestinal or systemic illness. In this chapter, we'll delve into the ways in which probiotics improve the gut microbiome and the lives of the microorganisms that live there. The discoveries and expanding knowledge about probiotics will also be analysed in this chapter in order to create a novel approach, a probiotic-driven strategy, to deal with gut dysbiosis.

Keywords Gut microbiota · Human microbiome project · Gut-dysbiosis · Immunomodulation

20.1 Introduction

The human gut inhabits trillions of bacteria, approximately 3.8×10^{13}, that account for around 0.2 kg of body weight (Azad et al. 2018) and confer numerous health benefits. Gut microflora consists of a dynamic range of bacterial communities. However, majorly it includes four major bacterial divisions such as *Firmicutes*, *Bacteroidetes*, *Actinobacteria*, and *Proteobacteria* which generally constitute 99% of the gut microbiota, apart from them, *Verrucomicrobia*, *Fusobacteria* and *Cyanobacteria* also exist but in less comparable population size. In a healthy gut, the balance of these microbes is maintained in a way that provides overall beneficial effects to the individual. These bacterial communities show a mutual, symbiotic, and commensal kind of relationship with the gut environment of the host. This balance of the gut microflora is dependent upon the diet being consumed by the individual.

An alternative approach of using probiotics can maintain this healthy balance of the gut microbiome. An acceptable dose of "live microorganisms that bestow a health benefit on the host" is what makes a probiotic effective for human consumption-Food and Agriculture Organization (FAO)/World Health Organization (WHO) (Gogineni et al. 2013). Among all those microbes harboured in our intestine, only a few are considered healthy and can be used as a probiotic. In order to be called a probiotic, there are certain specifications provided by the Indian Council of Medical Research (ICMR) (Gogineni et al. 2013). These parameters include acid tolerance, bile tolerance, lysozyme resistant, colonization ability (hydrophobicity), antibiotic susceptibility, acid and bacteriocin production ability. When a bacterial species possesses these characteristics, it can be considered a probiotic. Different bacterial strain has variable degree of probiotic potential. The first commonly known probiotic Genus was *Lactobacillus* from the group of Lactic Acid Bacteria (LAB) and *Saccharomyces* the yeast. *Lactobacillus* has been reclassified in 25 genera (Zmora et al. 2018). Apart from *Lactobacillus* the second most common probiotic genus is *Bifidobacterium*.

Alteration in these microbial populations leads to various health disorders, e.-g., Inflammatory Bowel Disease (IBD), Colon cancer, Irritable Bowel Syndrome (IBS), Obesity, Metabolic syndrome, atherosclerosis, Type 1 diabetes, autism, allergy, asthma, celiac disease, non-alcoholic steatohepatitis.

With the advent of omics and metagenomic analysis, many undiscovered details of gut microbiome and microbial communities have been unfolded. From the metagenomic analyses, it was discovered that the microbes belonging to similar taxa or even species showed remarkable variation in their abundance amongst different individuals (Lu et al. 2021). These microbial diversities maintain the host homeostasis. The microbiome of the human gut provides a complement to the molecular as well as metabolic functions. After the human genome project, it was uncovered that approximate of 20,000 genes are the protein-coding genes.

Moreover, the additional functions are complemented by the genes of the microbial population. Thus, the human metabolic features come out as a blend of the host and the microbial genome (Tursi et al. 2010). According to the disappearing microbiome hypothesis, it has been stated that the individuals belonging to developed countries possess comparatively less diversity of microbes flourishing in their gut which puts them at higher risk of attaining chronic illnesses than the developing countries (Lu et al. 2021). The HMP specifies certain parameters that are needed to be considered while designing to manipulate the gut microbiome e.g., through probiotics. HMP extended the scope of study from single organism to studying the whole communities. Based upon the 16s rRNA profile of the isolated population, demonstration of the bacterial type could be done effectively. Most of the population is in the colon region dominated by *Bacteroidetes* and *Firmicutes*. A particular type of microbial distribution could lead to several health disorders. Microbial genes carry metabolic conversions that convert certain non-digestible or non-consumable compounds into absorbable and nutritionally essential forms.

The microbiota plays myriad functions including metabolic, immunological and structural functions. Dietary fibres such as lignin, resistant starch, non-starch polysaccharide and oligosaccharides which are not readily digestible by the host's digestive enzymes are digested by the microbial enzymes. Fluctuation in the microbiota could result in leaky-gut conditions or can also lead to disorders ranging from allergy to cancer causation.

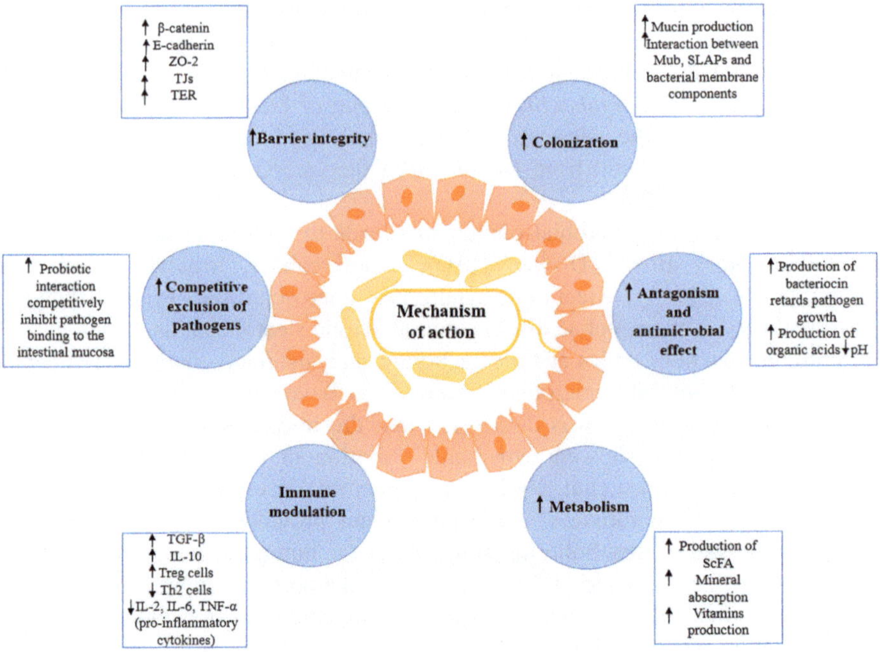

Fig. 20.1 Demonstration of different mechanisms of probiotics action against pathogens; (1) enhancement of barrier integrity, (2) immunomodulation, (3) antagonism and antimicrobial effect, (4) enhanced metabolism, (5) competitive exclusion of pathogens, and (6) colonization

20.2 Gut Dysbiosis and Its Effects on Health

As illustrated in Fig. 20.1, the gut flora impacts various host components such as the intestinal barrier, metabolism, immune system, and so on. A healthy gut is maintained by keeping track of various parameters such as age, environmental conditions, eating habits, antibiotic usage, and so on. Dysbiosis is a condition where harmful microbes' population increases more than healthy microbe population. Due to the breakdown of barrier integrity, this may result in a leaky gut syndrome, exposing the interior of the intestinal layer to opportunistic microorganisms that can be invasive. IBS, IBD, Crohn's disease (CD), Ulcerative colitis (UC), colon cancer, allergic responses, hypersensitivity, diabetes, obesity, and a host of other diseases have all been linked to changes in the microbiome of the gut (Belizário and Faintuch 2018). Gut dysbiosis affects the vaginal gut microbiome, which leads to vaginal dysbiosis leading to reproductive issues. Children born to mothers with a history of antibiotic usage during their pregnancy are more likely to develop long-term postnatal abnormalities such as asthma, allergies, diabetes, enteric disorders, and so on, since early antibiotic exposure may influence new-born gut microbial colonization. Maternal dysbiosis leads to altered microbiota of the infant's gut, for which nowadays, maternal faecal microbiota transplant is performed to attain a balanced gut microbiome. It has also been proposed that altered maternal gut microbiota might lead to neurodevelopmental disorders and behavioural defects (Zhao et al. 2021). It has also been supported through various studies that an altered maternal gut microbiota could act as a trigger for preterm birth. Usually, a healthy vaginal microbiota has *Lactobacillus* dominance, preventing other bacteria from flourishing the vagina. A very recent study has discovered that in the case of approximately 30% of preterm birth there is "preterm prelabour rupture of membrane," and it's possible cause has been linked to an altered vaginal microbial population (Baldassarre et al. 2019). The gut microbiome of an infant is established with birth, which is contributed by the maternal gut microbiome, vaginal microbiome, and the microbiota composed in breast milk; thus, any alteration or dysbiosis in the maternal microbiome could lead to the establishment of a deformed or compromised microbiota in the infant gut.

20.3 Probiotics and Their Mechanism of Action

Probiotics are now being used as an alternative strategy to attain a healthy and disease-free gut. LAB and *Bifidobacterium* are the most frequently used bacterial genera which are considered to deliver probiotic effects to the individual. This can be achieved either by making them a part of functional foods or by their direct consumption. Altered microbial balance could be restored efficiently by such probiotics. The basic mechanisms they employ involve competitive exclusion of pathogens, increasing barrier integrity, promoting mucosal adhesion or colonization with healthy bacteria, promoting bacteriocin and mucin production, and affecting metabolism. By-products of digestion could be carcinogenic, which are required to be metabolized by these microbes residing the gut into absorbable forms. The gut microflora produces biologically active substances that help to confer immune-modulatory effects and reduce inflammation. These substances generally are SCFA (Short-chain Fatty Acids), possibly including acetic acid, butyric acid, and propionic acid.

Probiotics maintain host homeostasis, which eventually deals with plenty of diseases comprising neurological, allergic, inflammatory, metabolic, etc. A healthy balance of gut microbiota resists endogenous disturbances and systemic malfunctions. Multiple studies have stated that these probiotics activate the G-Protein Coupled Receptors (GPCR) and maintain host physiology (Mardini and Grigorian 2014). Antibiotic consumption has been shown to escalate the host susceptibility towards enteric pathogens as it disrupts the microbial barrier that safeguards the gut environment. The mechanisms preferred by these resident bacteria include both direct and indirect pathways as depicted in Fig. 20.1. Role of gut homeostasis is displayed when certain opioids alter and dysregulate the gut microbial balance and contribute to disease progression. This marks the gut microbiota as a potential marker and shows how it leads to disease progression. As per a study, opioids affect the gut's microbial balance and lead to bile dysregulation. Host metabolic homeostasis and the gut microbiome have been supposed to be effectively correlated. The commensals of the gut play their crucial role in transforming the bile into its secondary active forms as deoxycholate and lithocholate by subsequent deconjugation and dihydroxylation events. Now, this secondary active bile compromises the growth of *Clostridium difficile (C. difficile)* and prevents pathogenesis. In this way, altered microbiota disrupts both barrier integrity and bile metabolism and progresses systemic inflammation (Yadav et al. 2020).

20.3.1 Enhancement of Epithelial Barrier

The intestinal mucosal barrier acts as the first line of defense of the GI tract. The pathogen access to the interior is inhibited or checked by attaining and enhancing the barrier integrity. Barrier integrity is sustained by components such as mucus layer, secretory IgA, certain antimicrobial peptides, and the epithelial junction adhesion complex. Compromised barrier integrity makes the GI tract more prone to pathogens which can gain access to the submucosa and elicit an inflammatory response (Bermudez-Brito et al. 2012). Probiotics can be consumed as a single strain or in the form of probiotic preparations, which positively affect gut health by altering the expression of specific genes such as activating p38 and extracellular regulated kinase signaling pathway, increasing the expression of mucin genes such as MUC2, MUC3, MUC5AC. Moreover, the integral protein ZO (Zonula Occludens), JAM (Junctional Adhesion Molecules), trans-membrane protein E-cadherin, and adherents protein β-catenin these all contribute to maintaining the intestinal epithelial barrier functions. *Lactobacillus* strains (*L. acidophilus, L. plantarum, L. reuteri, L. casei, L. rhamnosus, Lactocaseibacillus casei*), *Bifidobacterium* strains (*B. animalis subsp. lactis, B. dentium N8*), *Escherichia coli* Nissle 1917 (EcN), *Bacillus, Clostridium butyricum, Enterococcus, Saccharomyces boulardii*, all these have been reported to affect the barrier integrity by general mechanism of enhancing the expression of TJ (Tight-Junction) proteins such as ZO-1, ZO-2, cingulin, occluding and its associated plaque proteins, tubulin proteasome, Trans-epithelial Electrical Resistance (TER) and down-regulating the activation of inflammatory gene expression (TNF-α, IL-1β, IL-6) (Lesbros-Pantoflickova et al. 2007). The fermented probiotic products also significantly enhance the TER levels which could prevent carcinogenesis (Commane et al. 2005). Moreover, *B. dentium* N8 has been reported to alleviate barrier permeability induced by lipopolysaccharide (LPS) and down-regulate the expression of Toll-Like Receptor-4(TLR4) and pro-inflammatory cytokines (Zheng et al. 2020). EcN, the declared probiotic along with reducing inflammation produces membrane vesicles and certain soluble factors which have been reported to prevent the intestinal epithelial damage brought due to pathogens. Also, TcpC protein synthesised by EcN manages the signalling pathways PKC and ERK1/2 which in turn up-regulates the expression of claudin-14 that forms TJ proteins (Lesbros-Pantoflickova et al. 2007). Up-regulation of genes such as MUC2, MUC3,

MUC5AC promotes mucin production which helps in probiotic colonization and pathogen clearance.

20.3.2 Probiotic Colonization and Competitive Exclusion of Pathogens

Probiotics possess adhesion capabilities with the intestinal mucosa, inhibiting or interfering with the pathogen adhesion to the intestinal epithelial layer. In order to deliver the probiotic effects, these bacteria need to possess the efficient colonization property which is offered due to hydrophobicity possessed by them (Servin and Coconnier 2003). The attachment or adhesion is mediated through determinants that are expressed on their surface and the mechanism involves interactive forces like passive, electrostatic, hydrophobic interactions, steric forces, lipoteichoic acids and an external structure called as appendages covered with lectins (Stier and Bischoff 2016). The lining of the intestinal mucosa is called glycocalyx, which consists of components like glycolipids, glycosylated proteins, i.e., mucins, Immunoglobulins (Igs) and electrolytes (Bron et al. 2012). The bacterial membranes carry specific molecules e.g., Lipoteichoic acid (LPA), Mucin binding proteins (MucBP), and Surface Layer Associated Proteins (SLAPs) through which the bacteria bind to the mucus and its components i.e., to the sugar domain of mucin of the epithelial layer. Particular bacterial type carries specific sort of domain in order to colonize. Initial binding is mediated through non-specific weak-interactions like hydrophobic interactions, further interaction called as the second stage of binding includes the presence of specific components on the bacterial cell wall which bind with the mucin. Two types of protein domains; Mub (exclusive to LAB of GI tract) and MucBP (found in nearly all bacterial types) are involved in bacterial adhesion to mucin (Verdenelli et al. 2016). These domains which are responsible for binding to the mucins are linked to the peptidoglycan of bacteria through the C-terminal motif (Leu-Pro-X-Thr-Gly). Other mode of attachments or adhesion comprise of surface structures such as fimbriae, pili, fibronectin-binding proteins or SLAPs (mediate adhesion to the collagen, laminin as well as self-assembly) (Antikainen et al. 2002). Chemical removal of SLAPs, negatively affects and reduces the adhesion ability (Tang et al. 2010). Attachment of the probiotics with the mucosal epithelium inhibits the pathogen interaction by creating a gut microbial barrier, eventually resulting in competitive exclusion of pathogens. There are possible chances of binding of pathogenic microbes, toxins that already exist in the gut with the intestinal lumen, but probiotics inhibit the pathogen binding to the mucosal surface. As reported through a study, binding of *Enterococcus faecium* is inhibited by structurally similar SpaC protein of the mucus binding pili, or some antibody by the mechanism of competitive exclusion. Valid evidence have been proposed for the successful colonization of the probiotic bacteria in the vagina and urogenital tract (Chapman et al.

2014; Wang and Roy 2017), thus, supplementation with probiotics, can promote healthy bacterial colonization in the gut.

20.3.3 Production of Antimicrobial Agents Against Pathogen

Probiotics like LAB produce specific antimicrobial components, which help ameliorate the microbial population because it slows the development of potentially dangerous organisms and opportunistic bacteria and promotes the growth of healthy communal bacteria. LAB deliver their antagonistic effect by producing organic acids which generally include lactic acid, acetic acid, formic acid, etc. (Turnbaugh et al. 2007). These acidic components lower the pH, thus, inhibiting the pathogenic non-acid-tolerant bacteria from flourishing the gut environment. Moreover, some other primary metabolites like hydrogen peroxide, carbon dioxide, diacetyl, bacteriocins (antimicrobial), SCFA, and bacteriocin-like inhibitory substances are also produced by them, which are collectively called as inhibitory compounds and retard the growth of undesired, pathogenic, opportunistic microbes (Lloyd-Price et al. 2016). Hydrogen peroxide produced by such bacteria shows oxidizing effect and thus acts as a bactericidal component. Antagonism is determined by analyzing the inhibitory zone formation by the bacterial type against specific pathogen being examined. Common pathogens against which the *Lactobacillus* isolates are generally expected to show antagonism are *E. coli, Salmonella typhimurium, Staphylococcus aureus*, etc. Cell-free supernatant (CFs) containing the antimicrobial component of *Lactobacillus* are used to determine such activity. The potential to produce such substance enhances their ability to compete with others and maintain a healthy balance of gut microflora. Productions of compounds like hydrogen peroxide exert a bactericidal impact on most of pathogens. As a product of intestinal fermentation, the lactic acid is metabolized by lactate utilizing bacteria to variable fatty-acids (VFA). Lactic acid and SCFA mediate pathogen clearance by disrupting the outer membrane of the gram-positive pathogenic bacteria, thus retaining only the beneficial ones (Alakomi et al. 2000).

20.3.4 Modulation of Immune System

Probiotics influence both innate and the adaptive immune responses either directly or indirectly. A cross-talk takes place between the probiotic bacteria and the intestinal microbiota (Corthésy et al. 2007). The basic mechanism of their action is, alleviating the release of pro-inflammatory cytokines and influencing the Treg cells which act as the chief player in ameliorating the immune response. They make the mucosal immune system to produce protective immunoglobulins (IgA), bacteriocins as well as defensins. Probiotic supplementation modulates the production of T cells, NK cells (Kaur et al. 2020). Levels of Treg were found to be elevated as per a study contributing toward developing immune-tolerance (Cervantes-Barragan et al. 2017). Numerous strains of *Lactobacillus* have been evaluated to enhance the levels of certain receptors such as TLR2/6 and, inhibition of pro-inflammatory cytokines as IL-2, IL-6, TNF-α and raise the levels of TGF-β and IL-10 (Cristofori et al. 2021). Hydrogen peroxide produced by the probiotic strains as an antagonistic factor induces high levels of CD80 and CD86 the costimulatory molecules which ultimately leads to the clearance of enteropathogenic bacteria from bloodstream (Tallon et al. 2007). Specific strains of probiotics, or mixture of probiotic strains ameliorate allergic response (asthma), atopic dermatitis and rheumatoid arthritis. Studies suggest the protective effects of probiotics against Dinitrobenzene sulphate (DNBS) and Dextran sodium sulphate (DSS) induced colitis. Relation of probiotic strains with candida infections and biofilm production has also been reported which is suppressed by numerous mechanisms comprising caproic acid and hydrogen peroxide production (Lloyd-Price et al. 2016). VSL 3, which is a probiotic mixture considered as a medical food. It comprises eight live bacterial strains out of which four strains are of genus *Lactobacillus* which are *L. acidophilus, L. casei, L. plantarum, L. delbrueckii subspecies bulgaricus*, three strains of *Bifidobacterium* such as *B. bifidum, B. longum, B. infantis*, and one strain of *Streptococcus* wiz., *S. salivaris subspecies thermophilus*. All these strains confer different strain-specific effects on the individual, thus promoting health (Cheng et al. 2020).

20.3.5 Improvement of Intestinal Health

Gut microbiota governs the health-related parameters; probiotics help restore a healthy gut and lead to a healthy metabolism, thus alleviating metabolic disorders including obesity, hyperlipidemia, diabetes, total cholesterol and LDL levels, and lactose tolerance and insulin resistance (Aggarwal et al. 2013). The undigestible forms which are not metabolised by the host enzymes are used as a fermentation substrate for these bacteria including resistant starch, dietary fibres, oligosaccharides, plant-based materials (cell wall) (Cummings and Macfarlane 1991). Phytase, an enzyme possessed by bacteria, plays a crucial role in mineral absorption, it removes the phosphate groups from the phytate which allows easy absorption of the minerals like calcium, phosphorus, amino acids (Tharmaraj and Shah 2009). The carbohydrates are fermented to yield end products in the form of SCFA exclusively, propionate, butyrate and acetate. Butyrate functions as the main fuel for the colonic region's epithelial cells, thus managing their proliferation and differentiation and checking tumorigenic proliferation (Floch 2010).

20.4 Marketed Probiotics Foods

There are plenty of probiotic sources, which include human sources (faecal source, gut), dairy sources (fermented-dairy products), non-dairy sources (fermented foods, vegetables) and animal sources (Yu et al. 2011). Probiotics from LAB have been used for a very long time, for about centuries. A study based on Mongolia's fermented products demonstrates numerous strains involved in the fermentation of food products (Zhang et al. 2021). Characterizing all such bacterial types, gives a standard for basic probiotic characteristics and the types that could be used for industrial fermentation. Moreover, specific characteristics of these bacteria make them suitable for industrial use like their potential to tolerate stress. Probiotics like *L. acidophilus, Bifidobacterium* and *L. casei* are the most prominently used probiotic cultures; other than these certain yeast species, *E. coli*, *Bacillus* are also being used for fermentation of products like milk, meats, vegetables as well as cereals (Caplice and Fitzgerald 1999).

20.5 Probiotics Role in Disease Management

Probiotics have shown promising effects in many diseased conditions where it has been reported to minimize disease-related complications to a great extent and hence could be employed as a therapeutic alternative. As per the studies performed with *Saccharomyces boulardii* in antibiotic-associated diarrhoea, it has been reported that they confer protective effects, also the strain *L. rhamnosus GG* has shown considerable protective impacts (Haller et al. 2000). *S. boulardii* has also been reported to produce a serine protease that directly degrades the toxin A produced by *C. difficile* and its it binding site to brush border receptors (Castagliuolo et al. 1996).

Antimicrobial effects of probiotics have been correlated with *Helicobacter pylori* which is mediated through multiple probable mechanisms as per studied in in-vitro studies. Also, the antimicrobial effect of these probiotics varies from strain to strain. As reported, strains such as *L. acidophilus, L. casei 393 and L. salivarius WB1040, L. casei* strain *Shirota* deliver their inhibitory impacts achieved via manufacture of lactic acid as lactate acts an inhibitory factor to *H. pylori*. Likewise, *L. casei* strain *Shirota* produces heat labile substances, *L. acidophilus LB* and *L. johnsonii La1* produce heat stable proteins and substances, *Weissella confusa* and *L. lactis BH5* produce bacteriocins as inhibitory substances. Mechanisms opted by other strains such as *L. reuteri* include blocking the glycolipid-binding site. *B. subtilis* was shown to produce antibiotic type of substance known as anticoumacin. In all such ways, the inflammation and disease progression is checked and reduced by these probiotics (Li and Gu 2018).

Probiotics have been studied to be the inducer for the remission of ulcerative colitis (UC) to a certain extent; they confer improvement in the clinical disease score (Kocot et al. 2022). Since, no proper evidence has been proposed yet for the cases of the extreme severity of the condition, the suspected remission property of these probiotics needs to be analysed through more studies. VSL 3 has been proposed through a study as a potential adjunctive therapy to treat mild to severe UC (Orel and Trop 2014). Certain studies show that these probiotics, when used in mixture form, can be potentially used to reduce the severity of the condition (Van Tassell and Miller 2011). Also, some studies suggest that IBD, which possibly be caused due to complex immune response of the host's body, can be ameliorated and checked by using probiotic mixtures e.g., VSL3 (Pan et al. 2006). This probiotic mixture named VSL 3 has been reportedly seen in reducing the health risks associated with plenty of more diseases other than the ones mentioned above, such as, hepatic diseases (non-alcoholic fatty liver disease, alcoholic liver disease, cirrhosis, hepatic encephalopathy), diabetes, obesity, allergic diseases, reproductive diseases, preterm birth issues, systemic nervous diseases, bone diseases, atherosclerosis, microscopic colitis, uncomplicated diverticulitis, Antibiotic associated diarrhoea (AAD), *C. difficile* associated diarrhoea (CDAD). Also, in case of chemotherapy-induced diarrhoea (Bowen et al. 2007), rotavirus diarrhoea, and radiation-induced enteritis (Delia et al. 2007) remission has been noted to be associated with the use of VSL 3. In colon cancer, notable effects of VSL 3 have been associated and the mechanism employed is similar as mentioned in Table 20.1 which is a general approach for conferring anti-carcinogenic and anti-inflammatory effects of VSL3. Probable outcomes of probiotics have been noted against the dysbiosis caused due to cancer chemo and radiotherapies. Exposure to these therapies increases the risk of harmful bacteria invading the internal intestinal environment which disrupts the overall gut environment. Supplementation with probiotics helps to check and suppress the probable chances of disease progression (Zmora et al. 2018). Multiple studies have reported that probiotics directly or indirectly cross-communicate with the brain and maintain the bacteria-brain relationship as well.

20.6 Probiotic Forecasts

Various reports have suggested the positive therapeutic impact of probiotics on health issues such as IBS, traveller's diarrhoea, AAD, lactose intolerance, IBD, rheumatoid arthritis, etc. They do so by assisting the naturally flourishing microbiota of the gut. Probiotic supplementation maintains gut dysbiosis induced due to antibiotic consumption. A few more probiotic attributes and potentials have been recently documented, such as anti-carcinogenicity, antigenotoxicity, and antimutagenicity but these further need to be looked forward. Certain studies report that consuming fermented dairy products minimizes the risk of colon cancer as they retard those bacteria from flourishing the gut, possibly converting the pro-carcinogens to carcinogens. The future of these probiotics has been expected to be successful in many aspects, as the nanoencapsulation and microencapsulation techniques could help them to deliver those beneficial effects at the extreme

Table 20.1 List of probiotic foods readily accessible on the market with their manufacturer companies

Brand name	Source	Manufacturer	Country
Activia	*Bifidus actiregularis*	Danone	France
Yakult	*Lactobacillus casei Shirota*	Yakult	Japan
Actimel	*Lactobacillus casei immunitas*	Danone	France
Bacilac	*Lactobacillus acidophilus, Lactobacillus rhamnosus*	THT	Belgium
Jovita probiotisch	*Lactobacillus* strain	H & J Bruggen	Germany
Pro Viva	*Lactobacillus plantarum*	Skanemejerier	Sweden
Provie	*Lactobacillus plantarum*	Skanemejerier	Sweden
Revital active	Probiotics	Olma	Czech republic
Vifit	*Lactobacillus* strain	Campina	The Netherlands
LC1	*Lactobacillus*	Nestle	Switzerland
Symprove	*L. rhamnosus, L. acidophilus, L. plantarum Enterococcus*	Symprove limited	Farnham, UK
Rela	*Lactobacillus reuteri*	Ingman foods	Finland
Bactisubtil	*Bacillus sp.* strain IP5832	Synthelabo	Belgium
Vitamel	*Lactobacillus casei GG, Bifidobacterium bifidum, Lactobacillus acidophilus*	Campina	The Netherlands
Proflora	*Lactobacillus acidophilus, Lactobacillus delbrueckii subsp. bulgaricus, Streptococcus thermophilus, Bifidobacterium*	Chefaro	Belgium
Biola	*Lactobacillus rhamnosus GG*	Orlan	Ukraine
Bioprofit	*Lactobacillus rhamnosus GG, Propionibacterium freudenreichii, Shermanii JS*	Valio Ltd.	Finland
Grainfields wholegrain liquid	*Lactobacillus acidophilus, Lactobacillus delbrueckii, Saccharomyces boulardii, Saccharomyces cerevisiae*	AGM Food Pvt. Ltd.	Australia

conditions of the environment as well because it will elevate their tolerance to more stressful situations.

Moreover, probiotic spores will prevail, which employs a spore germination technique where spores are delivered to the host instead of delivering viable bacteria. The linking of probiotics to gut-brain axis and their possible impact on the brain and liver must also be established by extensive research. Determining their ability to produce extracellular vesicles in order to render their impacts and utilizing them as an alternative approach in medicine and therapeutics is possibly going to flourish in the coming era. Synergistic impacts of pro and prebiotics have been studied, which has expanded the value and demand of symbiotic products in the market.

20.7 Conclusion

Probiotics have shown many promising effects; these are a safe mode for delivering benefits to the host's body. An alternative approach using probiotics has been used to modulate the gut microbiota. It acts as the key player affecting diverse components, including gut health, metabolism, and immune system, making them a potential approach for therapeutic applications. They improve gut health by harboring healthy bacteria and eliminating the ones that could probably be the cause of disease. Probiotics have shown promising effects in diseases like IBD, IBS, Crohn's disease, ulcerative colitis, and colon cancer. Their primary mechanism of action is similar to all the probiotic bacterial types; the remarkable difference arises in their potential, which is strain specific. Healthy gut conditions implicate a healthy response in the brain, which creates a scope for looking for therapeutic effects of probiotics in diseases like Alzheimer's. Fermented foods also dominate the market due to their higher probiotic content, which the consumers are entertained.

References

Aggarwal J, Swami G, Kumar M (2013) Probiotics and their effects on metabolic diseases: an update. J Clin Diagn Res 7(1):173–177

Alakomi HL, Skyttä E, Saarela M, Mattila-Sandholm T, Latva-Kala K, Helander IM (2000) Lactic acid permeabilizes gram-negative bacteria by disrupting the outer membrane. Appl Environ Microbiol 66(5):2001–2005

Antikainen J, Anton L, Sillanpää J, Korhonen TK (2002) Domains in the S-layer protein CbsA of Lactobacillus crispatus involved in adherence to collagens, laminin and lipoteichoic acids and in self-assembly. Mol Microbiol 46(2):381–394

Azad MAK, Sarker M, Li T, Yin J (2018) Probiotic species in the modulation of gut microbiota: an overview. Biomed Res Int 2018:9478630

Baldassarre ME, Di Mauro A, Capozza M, Rizzo V, Schettini F, Panza R, Laforgia N (2019) Dysbiosis and prematurity: is there a role for probiotics? Nutrients 11(6):1273

Belizário JE, Faintuch J (2018) Microbiome and gut dysbiosis. In: Metabolic interaction in infection. Springer International Publishing, Cham, pp 459–476

Bermudez-Brito M, Plaza-Díaz J, Muñoz-Quezada S, Gómez-Llorente C, Gil A (2012) Probiotic mechanisms of action. Ann Nutr Metab 61(2):160–174

Bowen JMC, Stringer AM, Gibson RJ, Yeoh ASJ, Hannam S, Keefe DMK (2007) VSL#3 probiotic treatment reduces chemotherapy-induced diarrhea and weight loss. Cancer Biol Ther 6(9): 1449–1454

Bron PA, van Baarlen P, Kleerebezem M (2012) Emerging molecular insights into the interaction between probiotics and the host intestinal mucosa. Nat Rev Microbiol 10(1):66–78

Caplice E, Fitzgerald GF (1999) Food fermentations: role of microorganisms in food production and preservation. Int J Food Microbiol 50(1):131–149

Castagliuolo I, LaMont JT, Nikulasson ST, Pothoulakis C (1996) Saccharomyces boulardii protease inhibits Clostridium difficile toxin: a effects in the rat ileum. Infect Immun 64(12):5225–5232

Cervantes-Barragan L, Chai JN, Tianero MD, Di Luccia B, Ahern PP, Merriman J, Cortez VS, Caparon MG, Donia MS, Gilfillan S, Cella M, Gordon JI, Hsieh C-S, Colonna M (2017) Lactobacillus reuteri induces gut intraepithelial CD4+CD8αα+ T-cells. Science 357(6353): 806–810

Chapman CMC, Gibson GR, Rowland I (2014) Effects of single- and multi-strain probiotics on biofilm formation and in vitro adhesion to bladder cells by urinary tract pathogens. Anaerobe 27: 71–76

Cheng FS, Pan D, Chang B, Jiang M, Sang LX (2020) Probiotic mixture VSL#3: an overview of basic and clinical studies in chronic diseases. World J Clin Cases 8(8):1361–1384

Commane DM, Shortt CT, Silvi S, Cresci A, Hughes RM, Rowland IR (2005) Effects of fermentation products of pro- and prebiotics on trans-epithelial electrical resistance in an in vitro model of the colon. Nutr Cancer 51(1):102–109

Corthésy B, Gaskins HR, Mercenier A (2007) Cross-talk between probiotic bacteria and the host immune system. J Nutr 137(3):781S–790S

Cristofori F, Dargenio VN, Dargenio C, Miniello VL, Barone M, Francavilla R (2021) Anti-inflammatory and immunomodulatory effects of probiotics in gut inflammation: a door to the body. Front Immunol 12:578386

Cummings JH, Macfarlane GT (1991) The control and consequences of bacterial fermentation in the human colon. J Appl Bacteriol 70(6):443–459

Delia P, Sansotta G, Donato V, Frosina P, Messina G, De Renzis C, Famularo G (2007) Use of probiotics for prevention of radiation-induced diarrhea. World J Gastroenterol 13(6):912–915

Floch MH (2010) The effect of probiotics on host metabolism: the microbiota and fermentation. J Clin Gastroenterol 44(Suppl 1):S19–S21

Gogineni VK, Morrow LE, Malesker MA (2013) Probiotics: mechanisms of action and clinical applications. J Probiotics Health 1(1):1000101

Haller D, Blum S, Bode C, Hammes WP, Schiffrin EJ (2000) Activation of human peripheral blood mononuclear cells by nonpathogenic bacteria in vitro: evidence of NK cells as primary targets. Infect Immun 68(2):752–759

Kaur L, Gordon M, Baines PA, Iheozor-Ejiofor Z, Sinopoulou V, Akobeng AK (2020) Probiotics for induction of remission in ulcerative colitis. Cochrane Database Syst Rev 3:CD005573

Kocot AM, Jarocka-Cyrta E, Drabińska N (2022) Overview of the importance of biotics in gut barrier integrity. Int J Mol Sci 23(5):2896

Lesbros-Pantoflickova D, Corthésy-Theulaz I, Blum AL (2007) Helicobacter pylori and probiotics. J Nutr 137(3):812S–818S

Li P, Gu Q (2018) Antimicrobial effects of probiotics and novel probiotic-based approaches for infectious diseases. In: Probiotics—current knowledge and future prospects. IntechOpen, London, p 178

Lloyd-Price J, Abu-Ali G, Huttenhower C (2016) The healthy human microbiome. Genome Med 8(1):51–57

Lu K, Dong S, Wu X, Jin R, Chen H (2021) Probiotics in cancer. Front Oncol 11:408

Mardini HE, Grigorian AY (2014) Probiotic mix VSL#3 is effective adjunctive therapy for mild to moderately active ulcerative colitis: a meta-analysis. Inflamm Bowel Dis 20(9):1562–1567

Orel R, Trop TK (2014) Intestinal microbiota, probiotics and prebiotics in inflammatory bowel disease. World J Gastroenterol 20(33):11505–11524

Pan WH, Li PL, Liu Z (2006) The correlation between surface hydrophobicity and adherence of Bifidobacterium strains from centenarians' faeces. Anaerobe 12(3):148–152

Servin AL, Coconnier M-H (2003) Adhesion of probiotic strains to the intestinal mucosa and interaction with pathogens. Best Pract Res Clin Gastroenterol 17(5):741–754

Stier H, Bischoff SC (2016) Influence of Saccharomyces boulardii CNCM I-745 on the gut-associated immune system. Clin Exp Gastroenterol 9:269–279

Tallon R, Arias S, Bressollier P, Urdaci MC (2007) Strain- and matrix-dependent adhesion of Lactobacillus plantarum is mediated by proteinaceous bacterial compounds. J Appl Microbiol 102(2):442–451

Tang AL, Wilcox G, Walker KZ, Shah NP, Ashton JF, Stojanovska L (2010) Phytase activity from Lactobacillus spp. in calcium-fortified soymilk. J Food Sci 75(6):M373–M376

Tharmaraj N, Shah NP (2009) Antimicrobial effects of probiotics against selected pathogenic and spoilage bacteria in cheese-based dips. Int Food Res J 16:261–276

Turnbaugh PJ, Ley RE, Hamady M, Fraser-Liggett CM, Knight R, Gordon JI (2007) The human microbiome project. Nature 449(7164):804–810

Tursi A, Brandimarte G, Papa A, Giglio A, Elisei W, Giorgetti GM, Forti G, Morini S, Hassan C, Pistoia MA, Modeo ME, Rodino' S, D'Amico T, Sebkova L, Sacca' N, Di Giulio E, Luzza F, Imeneo M, Larussa T, Di Rosa S, Annese V, Danese S, Gasbarrini A (2010) Treatment of relapsing mild-to-moderate ulcerative colitis with the probiotic VSL#3 as adjunctive to a standard pharmaceutical treatment: a double-blind, randomized, placebo-controlled study. Am J Gastroenterol 105(10):2218–2227

Van Tassell ML, Miller MJ (2011) Lactobacillus adhesion to mucus. Nutrients 3(5):613–636

Verdenelli MC, Cecchini C, Coman MM, Silvi S, Orpianesi C, Coata G, Cresci A, Di Renzo GC (2016) Impact of probiotic SYNBIO® administered by vaginal suppositories in promoting vaginal health of apparently healthy women. Curr Microbiol 73(4):483–490

Wang F, Roy S (2017) Gut homeostasis, microbial dysbiosis, and opioids. Toxicol Pathol 45(1): 150–156

Yadav M, Mandeep, Shukla P (2020) Probiotics of diverse origin and their therapeutic applications: a review. J Am Coll Nutr 39(5):469–479

Yu J, Wang WH, Menghe BLG, Jiri MT, Wang HM, Liu WJ, Bao QH, Lu Q, Zhang JC, Wang F, Xu HY, Sun TS, Zhang HP (2011) Diversity of lactic acid bacteria associated with traditional fermented dairy products in Mongolia. J Dairy Sci 94(7):3229–3241

Zhang Z, Xue C, Ju M, Guo J, Wang M, Yi S, Yi X (2021) Maternal gut dysbiosis alters offspring microbiota and social interactions. Microorganisms 9(8):1742

Zhao L, Xie Q, Etareri Evivie S, Liu D, Dong J, Ping L, Liu F, Li B, Huo G (2021) Bifidobacterium dentium N8 with potential probiotic characteristics prevents LPS-induced intestinal barrier injury by alleviating the inflammatory response and regulating the tight junction in Caco-2 cell monolayers. Food Funct 12(16):7171–7184

Zheng J, Wittouck S, Salvetti E, Franz CMAP, Harris HMB, Mattarelli P, O'toole PW, Pot B, Vandamme P, Walter J, Watanabe K, Wuyts S, Felis GE, Gänzle MG, Lebeer S (2020) A taxonomic note on the genus Lactobacillus: description of 23 novel genera, emended description of the genus Lactobacillus Beijerinck. Int J Syst Evol Microbiol 70(4):2782–2858

Zmora N, Zilberman-Schapira G, Suez J, Mor U, Dori-Bachash M, Bashiardes S, Kotler E, Zur M, Regev-Lehavi D, Brik RB-Z, Federici S, Cohen Y, Linevsky R, Rothschild D, Moor AE, Ben-Moshe S, Harmelin A, Itzkovitz S, Maharshak N, Shibolet O, Shapiro H, Pevsner-Fischer-M, Sharon I, Halpern Z, Segal E, Elinav E (2018) Personalized gut mucosal colonization resistance to empiric probiotics is associated with unique host and microbiome features. Cell 174(6):1388–1405

Chapter 21
Fermented *Vigna mungo* and Carrot Pomace Cookies Using *Lactobacillus casei*

Vibhuti Batra and Abhijit Ganguli

Contents

Abstract This chapter aims to develop an affordable snack using legume (*Vigna mungo*) dough supplemented with carrot pomace under controlled fermentation with *Lactobacillus casei.* The *Vigna mungo*-wheat dough was supplemented with carrot pomace and fermented with *Lactobacillus casei* whose viability was examined in MRS (deMan, Ragosa and Sharpe) with 1%v/v as inoculum size and 4 h of fermentation. Excellent metabolic response observed by *Lactobacillus casei* in the fermentation matrix is evident from survival and growth pattern, lowering of pH and production of organic acids. Bioactives in fermented sourdough were β carotene (376 μg/100 g), EPS (2.5–2.9/100 g), fructo-oligosaccharide (FOS; 2.4–2.7/100 g) and Gamma amino butyric acid. The later was quantified using by high performance liquid chromatography (HPLC) as 748 mg/g. Cookies baked using fermented sourdough were subjected to sensorial analysis based on seven point hedonic scale, scored adequately for general acceptability. Both sensory and nutraceuticals produced in the product were retained for up to 60 days at ambient temperature (28 °C).

V. Batra
Suryan Enterprises, Baddi, Himachal Pradesh, India

A. Ganguli (✉)
SSD Conglomerate Projects DST (Government of India), Science & Technology, Government of Haryana, PHFI, Gurugram, Haryana, India

J. A. Malik et al. (eds.), *Food Process Engineering and Technology*,
https://doi.org/10.1007/978-981-99-6831-2_21

Keywords Carrot pomace · EPS · Fermented cookies · GABA · Lactobacillus casei · Organic acids · Vigna mungo · β carotene

21.1 Introduction

Fruits and vegetables constitute an important part in our diet and human nutrition, and therefore with the increasing world population and change in eating habits their demand have increased significantly. They are fortified with nutritive fibers, vitamins, minerals and bioactive compounds including phytochemicals (Slavin and Lloyd 2012). Lower intake of vegetables and fruits causes cardiac arrest, hypercholesterolemia, osteoporosis, cancers, lung disorders, respiratory diseases and mental instability (Adebawo et al. 2006; Payne et al. 2012). The processing, harvesting, sorting, with shortage of adequate facilities are a major cause of wastage and loss of these required food commodities, as well as their left-over parts, bases and parcel. According to United Nations Food and Agriculture Organization (FAO) estimates, around 1.3 billion metric tons of food is irrecoverable (2014) and lost every year.

Fruits and Vegetable by-products often gets discarded or wasted during stages of food processing. These wastes include peels, shells, pomace and leaves, later being utilized via as cattle feeding, burning and landfills, which leads to loss of significant amounts of nutritional bioactive compounds carbohydrates, phenols, peptides, proteins and lipids which are constituents for preparation of functional foods against acute and chronic diseases. Thus, controlled fermentation of these waste products as a sustainable utilization technique will provide food availability, enhance living and improve the nutrient, diet and overall people across the globe. Controlled fermentation by lactic acid bacteria (LAB), being cheap and efficient will not only improve digestive health, immune response, mental health but aid in greater usability of fruits and vegetables with nutritive and sensory features and less lethality to food commodities (Cagno et al. 2013).

Lactic Acid Bacteria (LAB) has been renowned for its advantageous probiotic properties as it opposes the host gastrointestinal condition, in addition to host intestinal epithelium and prevents the entry of pathogenic bacteria in the animal intestine (Campana et al. 2017) *L. acidophilus, L. plantarum, L. casei* and *L. lactis* are few major species of LAB that are well studied (Sindhu and Khetarpaul 2002). The incorporation of these heath promoting micro-organisms is a favorable choice for delivery of probiotics to humans (Saarela et al. 2000). Controlled fermentation of foods is of great importance as they conserve and improve the texture and aroma of foods to a greater extent. Utilizing probiotic strains in fermented foods and beverage industry such as pickles, sausages, cheese, sourdough breads provides a rich nutritive possibility for ensuring healthy dietary option and health benefits to consumers. LAB (lactic acid bacteria) has also been utilized to generate certain bioactive compounds like such as ϒ-amino butyric acid (GABA) (Batra et al. 2018). GABA is produced from decarboxylation of glutamate via 1-glutamic acid decarboxylase

(GAD) of glutamate and is thus an amino-acid derived neurotransmitter (Shi and Li 2011). Thus; GABA constitutes to be an integral component in pharmaceutical industry, food sector and to regulate sensations of pain, anxiety and lipid metabolism (Miura et al. 2006). Foods rich in GABA prevent Cancer cell proliferation (Park and Oh 2006) and enhanced memory and learning ability skills. However, supplementation of chemically produced GABA through diet is considered precarious and unusual. Thus it is obligatory to come with affordable methods that can produce and increase GABA in foods. EPS (Exopolysaccharides), polymers derived from LAB ameliorates mouth ambience, quality and rheology of constituted fermented foods thus ensuring advantageous physiological effects on human health such as stress tolerance, antitumor action and immunomodulating (Cagginaniello et al. 2016). In addition it has been known to lower cholesterol and show prebiotic effect and modulates the response of immune system. The sourdough is composed of flour and fermented water (by lactic acid bacteria) that evolves acetic acid and lactic acid resulting in pungent pleasant taste of formulated dough (Vuyst and Neysens 2004). Foods such as breads, cakes and crackers work on this principle and ensures beneficial effects such as prolonged shelf life, impeded volume, retarded staling, finished bread aroma and enhanced nutritional value. Increased folate content, decreased tocotrienol and topopherol content (Liukkonen et al. 2003), reduced acute glyceamic and insulinaemic responses (Liljeberg et al. 1995) are a result of lactic acid formed during fermentation. Besides, extra cellular polysaccharides or "exopolysaccharides" improve the appearance, formulation, rheology and mouth feel of food formulations, they confer favourable effects such as stress tolerance, anti tumor activity and immune modulating bioactivity (Cagginaniello et al. 2016) on human health, thus has gained immense commercial value. Organic acids produced by LAB emend bread flavor, aid in gluten swelling and accelerate gases retention, which results in finished products with adequate consistency and rich volume resulting in its application as natural dough conditioner (Arendt et al. 2007).

Among different agricultural products, Carrots (*Daucus carota* L.) are critical root vegetables belonging to Apiaceae family and grown across throughout the world. They are inexpensive and highly nutritious as it contains highest concentration of β carotene and a pectin fiber (calcium pectate) with properties to lower cholesterol, antioxidants and source of Vitamin K, dietary fiber and minerals (Arscott and Tanumihardjo 2010; Nicolle et al. 2004). During initial processing of production, nearly 20–30% gets serated as carrot rejects and pomace. Thus, carrots are gaining much importance as they are rich in nutrients and assure healthy human health. Carrot pomace, wet carrot shavings contains high amounts of residual bioactive (carotenoids, dietary fibre, uronic acid and neutral sugars) (Adeleye et al. 2016) with currently little commercial values. The pomace from carrot juice production could reduce cholesterol levels and can be exploited as a food ingredient with incorporation of lactic acid bacteria to enhance its functionality. On an average during season, 15 kg of carrot pulp are wasted after processing for juice by a single fresh pressed juice vendor and can serve as an excellent raw material for developing functional products. On the other hand, *Vigna mungo* an important legume has been reported (Singh et al. 2012) to have excellent functional properties upon controlled fermentation. In view of this we envisaged a possibility of developing *Vigna mungo* sourdough incorporated with carrot pomace as a simple and affordable cookies as alternate snack for all age groups.

This chapter explores the functional aspects of the cookies developed from the sourdough and other characteristic of the developed product.

21.2 Functional and Proximate Attributes

21.2.1 Microorganism and Culture Medium

The significant technological and probiotic attributes of *L. casei* (Singh et al. 2012) have shown typical growth cycle starting from lag phase to reach exponential phase, starts around 4th to 5th hour where it continuously multiplies and maintains after 18 h. The highest specific growth rate $\mu = 0.27$/h, increase in transmission density shows a high cell density while approaching the exponential phase. Concomitant EPS, GABA (gamma amino butyric acid) production was noted. Decrease of pH and carboxyl acids (Citric acid, lactic acid, and acetic acid) was detected in Fig. 21.1.

21.2.2 Food Formulation and GABA Content

Commercially available wheat flour (*Aashirwad*, ITC, India), dehusked black gram (*Vigna mungo*) was bought from general store in Baddi (H.P). Fresh carrot pomace were obtained from juice bars, washed with hot water thrice, dried in an oven at 50 ± 1 °C, powdered and stored in zip locked food grade containers until use. Pomace and *Vigna mungo* dough's with *Lactobacillus casei* inoculums was kneaded to stiff dough. The dough samples were kept at 35 °C for 10 h for fermentation and 1 g sample was drawn after every hour diluted tenfold with sterilized PBS and spread onto MRS agar. GABA concentration in fermented sourdough escalated with time and attained maximum yield after 18 h in Fig. 21.2.

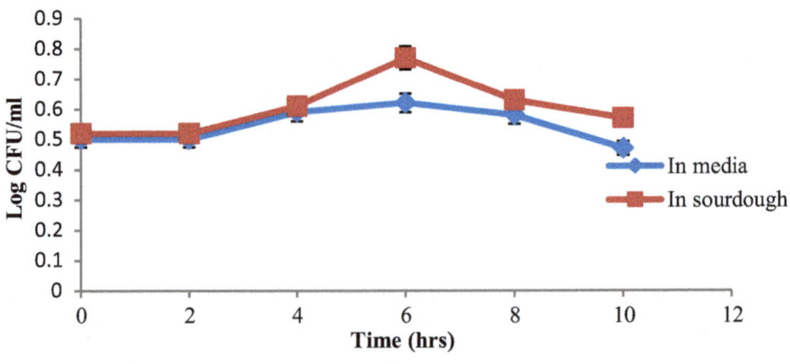

Fig. 21.1 Growth kinetics of *L. casei*

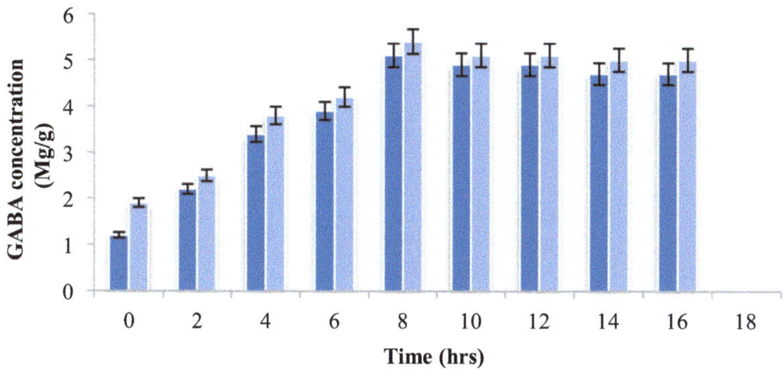

Fig. 21.2 Detection of cell viability (CFU/g) and GABA Conc (mg/g) production by *L. casei* in sourdough

21.2.3 Quantitative Analysis of GABA by HPLC

GABA formation was quantitatively confirmed using HPLC (Agilent 1100 series) to confirm the results obtained by TLC and spectrophotometer (Kook and Cho 2013). A 15 min centrifugation of the sample, with filtration of supernatant using 0.22 μm filter was done. Filtrate and sodium bicarbonate (0.2 mM) at a ratio of 9: 1 and pH = 9.8 was mixed together. Dansyl chloride, a derivatizer (μ = 8 g/L) was added to the sample and later at 30 °C was kept in the dark. HPLC was conducted using GABA standards at 750 mg/L, analyzed at UV detector at 254 nm (Li and Cao 2010). The data of retention time in case of sourdough was 12.287 min and insignificantly different as compared to that of authentic GABA, 12.583 min. The concentration of 748 mg/g was observed, which remained constant for 24 h, with significant ($P > 0.05$) decrease on storage for 2 months (Fig. 21.3).

21.3 Proximate Analysis After Fermentation

21.3.1 Functional Attributes

The pH observed in carrot sourdough was in range 3.7–4.0 and water absorption rate was 62.3/100 mL. The titratable acidity was observed to be 13.7–23.1 mL (Park and Oh 2006). The texture was sticky and granular. Compared to the plain dough, carrot sourdough had a good aroma and a vital appearance. The sourdough (with baking powder, butter, milk powder and kneaded) was rolled to 5 mm thickness, cut into uniform pieces and baked using the process described (Venturi et al. 2016). Gluten content (wet and dry), moisture, protein, fat, fiber and Ash content was described by standard AOAC (2000). The gluten content of the baked cookies were as follows: 9.27 ± 0.02, moisture content: 3.26 ± 0.13, protein content: 6.72 ± 0.07, fat:

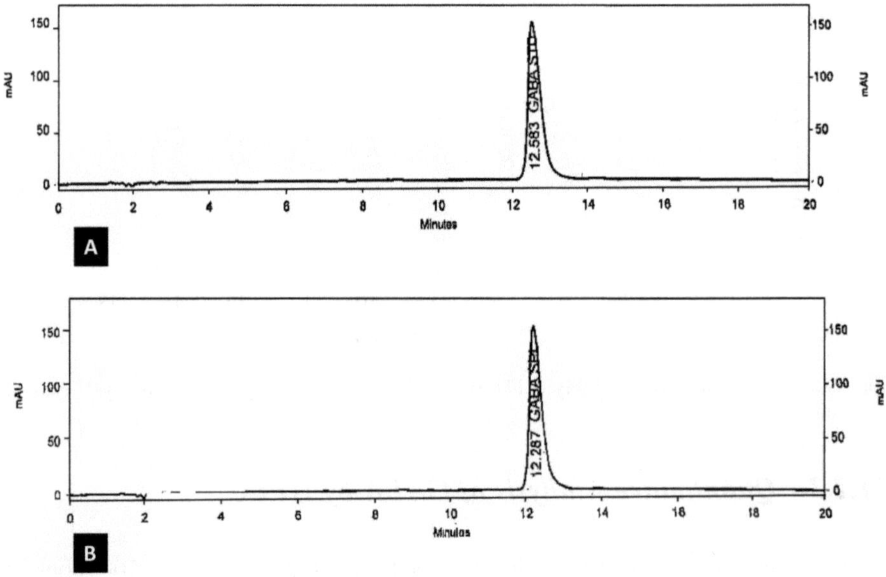

Fig. 21.3 HPLC chromatographs of (**a**) GABA standard and (**b**) optimum sample of 750 ppm and $\lambda = 254$ nm

Table 21.1 Proximate analysis (g/100 g) of fermented *Vigna mungo*-carrot cookies

Parameters	Attributes
Gluten	9.27 ± 0.02
Moisture	3.26 ± 0.13
Crude protein	6.72 ± 0.07
Fat	16.02 ± 0.12
Fiber	5.76 ± 0.02
Ash	0.57 ± 0.01
B carotene (μg)	376 ± 5.52

16.02 ± 0.12, fiber: 5.76 ± 0.02, Ash Content: 0.57 ± 0.01 respectively. The β carotene content was recorded spectrophotometrically using a separation technique of petroleum ether and sulphate at 450 nm as 376 ± 5.52. (RDA 8285 μg/100 g) (Table 21.1).

21.3.2 *Fructo-Oligosaccharides and Exopolysaccharides*

The sourdough (1.7 g ± 0.001) was measured adequately and transferred to a round bottom flask of 50 mL. Adequate conditions for primary extrication was 70–75 °C under air-condenser run time of 30 min with 20 mL of 70% water solution of methanol and 0.1 g of $CaCO_3$ followed by second and third extrication each for

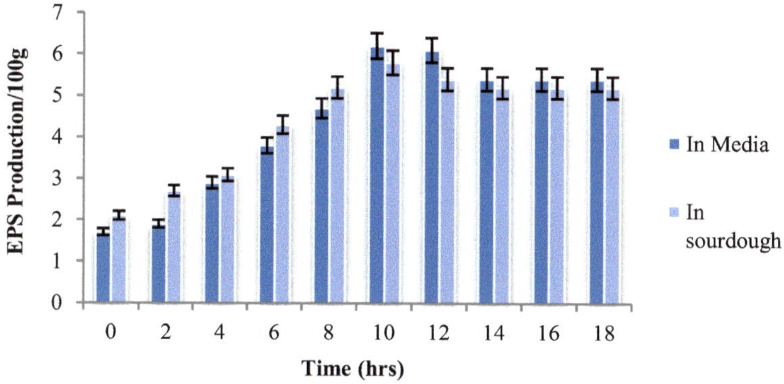

Fig. 21.4 Determination of cell viability versus EPS production in media and sourdough

30 min with 15 mL of 70% methanol. Adequate parameters of HPLC system (Knauer) filled with Euro Chrom 2000 data control system, amino propyl polymer filled shodex NH2P column, RI detector with composition of mobile phase as acetonitrile-water (67–33% v/v) with 0.8 mL/min as flow rate and ambient temperature of 20 °C was used. Fructo-oligosaccharides content determined was 0.3–0.81 g/100 g. The content of fructans reported was near about 2.5–2.9 g/ 100 g, whereas, FOS DP 3–5 was almost 0.7 g/100 g

Amount of EPS produced as a result of chemical extraction of water soluble polysaccharides from flour followed by purification by microbial EPS synthesis, EPS were calculated on the basis of amount of glucose in water soluble polysaccharides. WPS (water soluble polysaccharides) from watery extracts was precipitated using ethanol for 2 days, dialyzed against double distilled water for next 2 days followed by lyophilized hydrolysis of dried polysaccharide using incubation by sulphuric acid (4 mol/L) for 2 h at 80 °C followed by neutralization (4 mol/L), KOH samples were analyzed by HPLC technique with a polysphere OAKC column. The amount of exopolysaccharides at pH 5, in the dough sample was 251.2 mg/g (Fig. 21.4).

21.3.3 Carboxyl Acids

The presence of carboxyl acids like acetic acid, lactic acid and citric acid were shown as peaks obtained by HPLC analysis (Mueen-ud-Din et al. 2009). The thorough sour dough bread made have lowering pH and inflated mass of acetic acid with greater volume and lesser speed of degradation. The results depicts that the content of acetic acid is far greater in acidified thorough dough. L-Lactic acid was produced by the strain; whereas both L-lactic acid and D-lactic was formed by other cultures. Evaluation of carboxyl acids did not affect the baking characteristics and aroma of sourdough (Fig. 21.5).

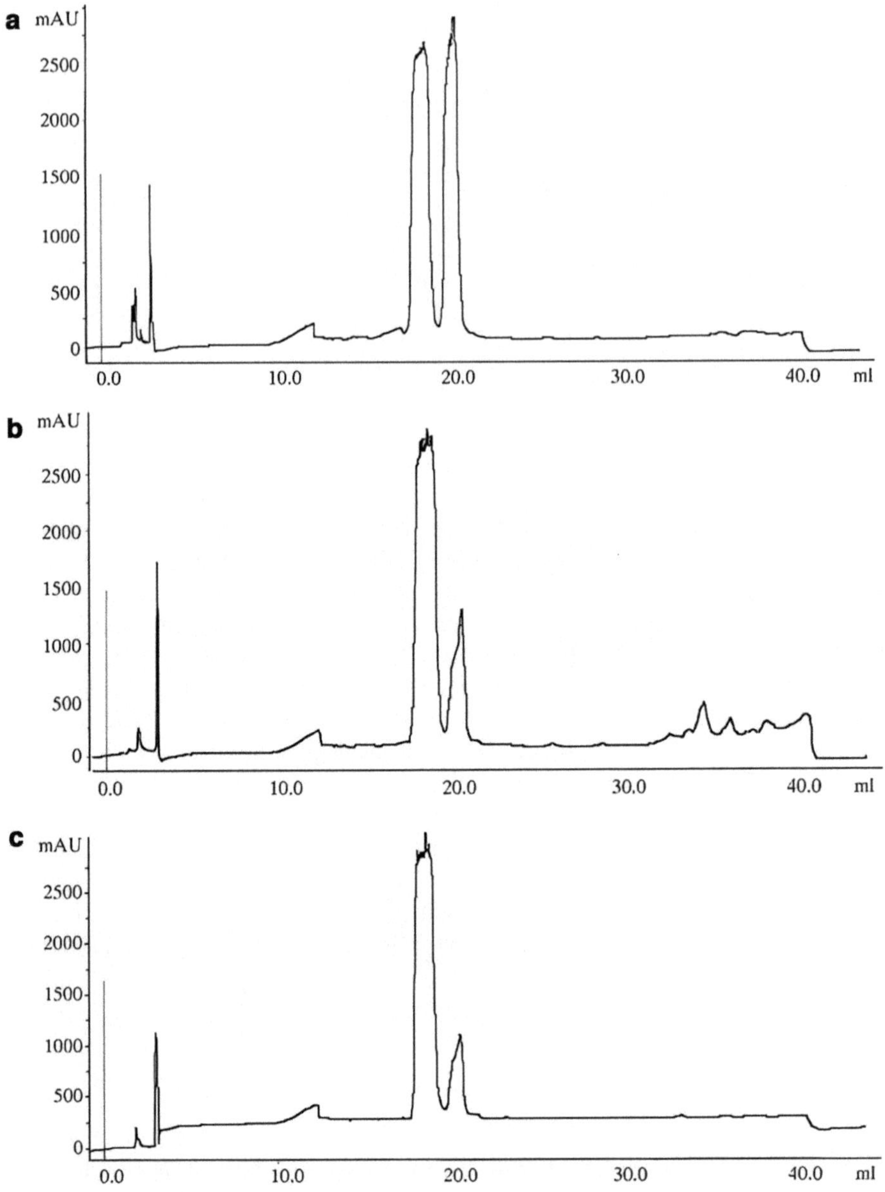

Fig. 21.5 HPLC profiles of carboxyl acids (**a**) acetic acid, (**b**) lactic acid and (**c**) citric acid in fermented Vigna mungo-carrot cookies

Fermented carrot cookies were coded and subjected to sensory analysis with parameters as aroma, consistency, color, texture and acceptability by team of judges using hedonic seven point scale. The sensory scores reveal overall acceptability remained favourable for 2 months after storage with variable deterioration in color and presentation (Figs. 21.6 and 21.7, Table 21.2).

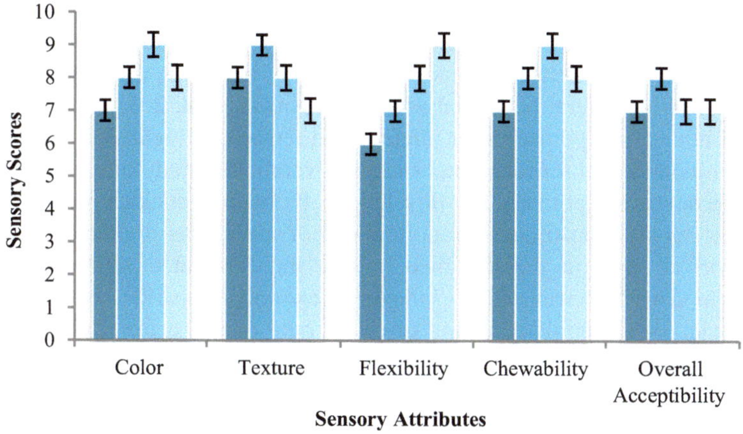

Fig. 21.6 Comparative sensory attributes of fermented *Vigna mungo* carrot cookies (freshly prepared, 15 days, 1 month and 2 month)

Fig. 21.7 Fermented *Vigna mungo*-carrot cookies

Table 21.2 Sensory evaluations of freshly baked fermented *Vigna mungo*-carrot pomace cookies

Attributes	Freshly prepared
Color (7)	5.12. ± 0.02
Consistency (7)	4.62 ± 0.12
Texture (7)	5.09 ± 0.02
Color and appearance (7)	5.17 ± 0.17
Overall acceptability (7)	5.92 ± 0.02

21.4 Conclusion

Utilizing microbial cultures in food matrices using traditional and molecular approaches have been used in designing fermented foods has gained considerable momentum due to enhanced functional properties. Fermented foods using LAB have not only known to add aroma and flavor, it serves to improve nutritive value and its digestibility. Fermentation of *Vigna mungo* (flour) as expounded in our previous studies reduces non/anti-nutrients substantially favoring digestibility besides endowing it with several beneficial aspects (Bhanwar et al. 2013). Thus, a combination of both carrot pomace after juicing with fermented *Vigna mungo* could be a novel combination apt for functional snacks. Overall, this study offered a novel way to repurpose carrot pomace for developing fermented cookies with functional properties; cookies constitute an important part of snack. The strain of *Lactobacillus casei* survived well in the designed matrix and fermented sourdough mixture resulting in GABA concentration (748 mg/g), exopolysaccharides in the sourdough and organic acids. Supplementation of carrot pulp enriched the sourdough with β carotene and EPS improved the dough texture notably without altering its baking characteristics.

Acknowledgments We wish to acknowledge S.E Enterprises, Baddi (H.P) for providing financial assistance to carry out this work for larger public benefit.

References

Adebawo O, Salau B, Ezima E, Oyefuga O, Ajani E, Idowu G, Famodu A, Osilesi O (2006) Fruits and vegetables moderate lipid cardiovascular risk factor in hypertensive patients. Lipids Health Dis 5(5):14–20

Adeleye AS, Oyerinde AS, Bratte AG (2016) Comparison of nutritional and colour properties of fresh and dried carrot (Daucus carota L.) slices and carrot pomace. J Multidiscip Eng Sci Technol 3(8):5366–5370

AOAC (2000) Association of official analytical chemists. Official methods of analysis. Vol. II, 17th Edition, AOAC, Washington DC

Arendt EK, Ryan LAM, Bello FD (2007) Impact of sourdough on the texture of bread. Food Microbiol 24(2):165–174

Arscott SA, Tanumihardjo SA (2010) Carrots of many colors provide basic nutrition and bioavailable phytochemicals acting as a functional food. Compr Rev Food Sci Food Saf 9(2):223–239

Batra V, Lomash H, Ganguli A (2018) Fermentation of groundnut brittle by Lactococcus lactis produces γ-aminobutyric acid and enhances nutritional quality and safety. Food Qual Saf 2(2): 83–87

Bhanwar S, Bamnia M, Ghosh M (2013) Use of Lactococcus lactis to enrich sourdough bread with γ-amino butyric acid. Int J Food Sci Nutr 64(1):77–81

Cagginaniello G, Kleerebezem M, Spano G (2016) Exopolysaccharides produced by lactic acid bacteria: from health-promoting benefits to stress tolerance mechanisms. Appl Microbiol Biotechnol 100(9):3877–3886

Cagno RN, Coda R, Angelis MD, Gobbetti E (2013) Exploitation of vegetables and fruits through lactic acid fermentation. Food Microbiol 33(1):1–10

Campana R, Hemert SV, Baffone W (2017) Strain-specific probiotic properties of lactic acid bacteria and their interference with human intestinal pathogens invasion. Gut Pathog 9:12–16

Kook MC, Cho SC (2013) Production of GABA (gamma amino butyric acid) by lactic acid bacteria. Korean J Food Sci Anim Resour 33(3):377–389

Li H, Cao Y (2010) Lactic acid bacterial cell factories for gamma-amino butyric acid (GABA). Amino Acids 39(5):1107–1116

Liljeberg HG, Lonner CH, Bjorck IM (1995) Sourdough fermentation or addition of organic acids or corresponding salts to bread improves nutritional properties of starch in healthy humans. J Nutr 125(6):1503–1511

Liukkonen KH, Katina K, Wilhelmsson A, Myllymaki O, Lampi AM, Kariluoto S, Piironen V, Heinonen SM, Nurmi T, Aldercreuttz H, Peltoketo A, Pihlava JM, Hietaniemi V, Poutanen K (2003) Process-induced changes on bioactive compounds in whole grain rye. Proc Nutr Soc 62(1):117–122

Miura E, Fukaya M, Sato T, Sugihara K, Asano M, Yoshioka K, Watanabe M (2006) Expression and distribution of JNK/SAPK-associated scaffold protein JSAP1 in developing and adult mouse brain. J Neurochem 97(5):1431–1446

Mueen-ud-Din G, Rehman SU, Anjum FM, Nawaz H (2009) Studies on organic acids and minerals content of sourdough naans made from different extraction rate wheat flour and starter cultures. Pak J Nutr 8(6):877–881

Nicolle C, Simon G, Rock E, Amouroux P, Rémésy C (2004) Genetic variability influences carotenoid, vitamin, phenolic, and mineral content in white, yellow, purple, orange, and dark-orange carrot cultivars. J Am Soc Hortic Sci 129(4):523–529

Park KB, Oh SH (2006) Production of yogurt with enhanced levels of gamma-amino butyric acid and valuable nutrients using lactic acid bacteria and germinated soybean extract. Bioresour Technol 98(8):1675–1679

Payne ME, Steck SE, George RR, Steffens DC (2012) Fruit, vegetable, and antioxidant in-takes are lower in older adults with depression. J Acad Nutr Diet 112(12):2022–2027

Saarela M, Mogensen G, Fonden R, Matto J, Mattila-Sandholm T (2000) Probiotic bacteria: safety, functional and technological properties. J Biotechnol 84(3):197–215

Shi F, Li Y (2011) Synthesis of γ-aminobutyric acid by expressing Lactobacillus brevis-derived glutamate decarboxylase in the Corynebacterium glutamicum strain ATCC 13032. Biotechnol Lett 33(12):2469–2474

Sindhu SC, Khetarpaul N (2002) Effect of probiotic fermentation on anti-nutrients and in vitro protein and starch digestibilities of indigenously developed RWGT food mixture. Nutr Health 16(3):173–181

Singh MK, Singla R, Singh A, Ghosh M, Ganguli A (2012) Survival of probiotic strains in non-dairy Indian spice condiment exhibiting cholesterol reducing properties. Food Sci Biotechnol 21(5):1309–1315

Slavin JL, Lloyd B (2012) Health benefits of fruits and vegetables. Adv Nutr 3(4):506–516

Venturi F, Sanmartin C, Tagliere I, Nari A, Andrich G (2016) Effect of the baking process on artisanal sourdough bread-making: a technological and sensory evaluation. Agrochim Pisa 60(3):222–234

Vuyst LD, Neysens P (2004) The sourdough microflora: biodiversity and metabolic interactions. Trends Food Sci Technol 16(1–3):43–56

Chapter 22
Metabolomics Applications in Food Science and Nutritional Research

Lakshita Rao, Deepika Yadav, Neha Rai, and Pawan Jalwal

Contents

Abstract The high-throughput study of small molecule metabolites in biological matrices is the focus of metabolomics, a rapidly expanding field of omics research. As a result, metabolomics is well situated to be employed in food nutrition and food science research. Technology and databases are the two factors that now limit the potential impact of metabolomics. For metabolite detection and quantification systems to become more accurate, automated, and comprehensive, significant equipment changes are obviously needed. Food quality, monitoring of food intake, food component analysis, physiological supervision in food intercession, food authenticity assessment, and diet challenge research are imperative fields of food science and technology. The study's main emphasis is on metabolomics' present potential future usage and trends.

L. Rao (✉)
Department of Pharmaceutical Sciences, Gurugram University, Gurugram, Haryana, India

D. Yadav
Aravali College of Pharmacy, Rewari, Haryana, India

N. Rai
Department of Pharmaceutical Sciences, Kurukshetra University, Kurukshetra, Haryana, India

P. Jalwal
Faculty of Pharmacy, Baba Mastnath University, Rohtak, Haryana, India

J. A. Malik et al. (eds.), *Food Process Engineering and Technology*,
https://doi.org/10.1007/978-981-99-6831-2_22

Keywords Analysis · Metabolomics · Food nutrition · Food quality · Food technology

22.1 Introduction

The high-throughput quantification/identification of metabolites (small molecules) in the metabolome is the focus of the relatively new branch of omics research known as "metabolomics." The collection of all small molecule compounds or metabolites that may be discovered in a cell, organ, or organism is known as the "metabolome," according to formal definitions. These molecules have diverse exogenous and endogenous components, including carbohydrates, amino acids, peptides, alkaloids, vitamins, organic acids, polyphenols, minerals, and nucleic acids pretty much any other substance that can be utilized, ingested, or produced by a specific cell or organism (Cevallos-Cevallos and Reyes-de-Corcuera 2012).

Metabolomics, also known as metabonomics or metabolic profiling, was only made practical recently by advances in the technology of small molecule isolation and identification. These include reliable, high-resolution mass spectrometry devices for accurate mass determination, high-resolution, high-throughput NMR spectrometers, capillary electrophoresis, and ultra-high pressure liquid chromatography systems for quick compound separation, as well as new software tools for quickly processing spectral or chromatographic patterns. The creation of electronic databases with spectral and descriptive data on the chemicals present in various metabolomes has been crucial to the advancement of metabolomics. These advances in technology and software have made it feasible to quickly identify and analyze dozens of small molecule metabolites rather than just one or two at a time (Fraga-Corral et al. 2022).

Similar to other emerging "omics" domains, there is sometimes a propensity for quick technological advancement, which is then followed by a protracted phase of consolidation and methodological development. No exclusion pertains to metabolics. There are several metabolomics technologies available, and numerous analytical methods can be applied, each with unique benefits and drawbacks (Table 22.1). In reality, for metabolomic data analyzing and processing, two different schools of thought have evolved in recent years. In one form, just the spectral patterns and intensities of chemical compounds are recorded, statistically compared, and utilized to determine the pertinent spectral properties that separate sample classes. Unsupervised clustering or supervised categorization are frequently used in these statistical analyses and feature identification methods (Cevallos-Cevallos et al. 2009).

PCA is formally a dimensional reduction approach that makes it simple to plot, display, and cluster various metabolomic data sets based on linear combinations of their comparable assets. PCA is massively pertained as clustering approach to determine how one sample differs from another, which factors are mainly responsible for this difference, and whether those variables are conscientious for the

Table 22.1 Comparison of different metabolomics technologies

Technology	Advantages	Disadvantages
Liquid Chromatography-Mass Spectroscopy (LC-MS)	Quantitative non-destructive fast (2–3 min/sample); necessitates neither derivatization nor separation; identifies every organic class; ID of new compounds is permitted; dependable, advanced technology; large amount of databases/software and for metabolite ID; and metabolite imaging (fMRI)	Lacking in sensitivity; high-cost instrumentation substantial instrument footprint; unable to identify salts and inorganic ions; non-protonated substances cannot be detected; and bigger (0.5 mL) samples are necessary
Gas Chromatography-Mass Spectroscopy (GC-MS)	Robust, established technology, reasonably priced, quantitative (with calibration), small sample size need, and high sensitivity; extensive databases and software for metabolite identification; most organic and some inorganic compounds are detected, and the separation repeatability is excellent	Sample not recoverable; sample derivatization is necessary; needs to be separated; (20–30 min per sample); not suitable for imaging; and novel compound ID is difficult
Nuclear Magnetic Resonance Spectroscopy (NMR)	Excellent sensitivity; highly adaptable technology; most organic and some inorganic compounds are detected; minimum sample size necessary metabolite imaging (MALDI) is a possible use; direct injection, which can be performed without separation; and has the capacity to identify the majority of the metabolome	Unrecoverable sample not very numerical; high-cost instrumentation slow (20–30 min/sample); poor resolution and repeatability of the separation (compared to GC); less durable equipment than NMR or GC-MS; databases and software for metabolite identification are few; novel compound ID is challenging

difference jointly or separately. By turning the PCA components so that utmost severance among classes is attained, PLS-DA, in contrast to PCA, is a supervised classification method that applies to improvement of groups of data separation. PLS-DA uses a second piece of information, the labeled set of class identities, but its fundamental concepts are comparable to those of PCA. Even though chemometric techniques like PCA and PLS-DA alone do not allow for the direct identification or quantification of chemicals, they do allow for an objective, chemically thorough comparison between various samples (Pinu 2016).

The goal of the other metabolomics technique which is also known as quantitative metabolomics or focused profiling, is to identify and/or quantify as many molecules in the sample as is practicable. This is often accomplished by contrasting the NMR or MS spectrum of the sample with a spectral reference library compiled from pure substances. Following the quantification/identification of the constituent molecules, the data are statistically validated in estimating the most significant biomarkers or instructive metabolic pathways. Quantitative metabolomics can be either focused or comprehensive, based on the goals and tool capabilities. Given the significance linked towards discovering bioactives, quantitative metabolomics is increasingly preferred in many fields of food nutrition and food science research. The quantitative metabolomics and chemometrics offer benefits and limitations. The majority of metabolomics research, also known as metabonomics, has, up until recently, been largely focused on clinical or pharmaceutical applications, such as drug development, drug evaluation, clinical toxicology, and clinical chemistry. But during the past few years, metabolomics has also grown to be a subject that interests food and nutrition specialists more and more (Ibáñez et al. 2013).

Identification of several compounds simultaneously in sample specimens is made possible by metabolomics, which gives nutrition and food explorers a fantastic opportunity to build up a much more detailed and comprehensive molecular picture of food composition, food consumption, and the metabolomic effects of various diets. To put it another way, metabolomics essentially opens the door to studying many facets of "molecular nutrition," such as: (1) food component analysis; (2) food quality/authenticity sensing; (3) food intake monitoring; and (4) work in science in food interference or diet conundrum research.

22.2 Metabolomics Applications in FCA (Food Component Analysis)

Dietary constituent analysis is traditionally used to identify and categorize food components into relatively broad categories such proteins, lipids, carbs, fiber, vitamins, trace elements, solids, and/or ash. Conversely, as metabolomics has advanced, meals and drinks are now being examined with a great deal more molecular detail, and in certain foods, different chemical identities are being discovered and/or recognized. Food chemists have a unique chance to comprehend the molecular specifics of what gives some meals and drinks their distinctive flavor, texture, scent, or color because of the possibility to chemically "deconstruct" foods and beverages into their chemical components. Nutritional scientists have an equally compelling opportunity to accurately pinpoint bioactive dietary components and get a deeper understanding of their possible advantages (Li et al. 2021).

Most meals may be conceptualized from the perspective of a metabolomics researcher as complex chemical mixes made up of assorted metabolites and chemical additives in a solid, semi-solid, or liquid matrix. Some "produced" foods only include 10–20 distinct chemicals, whereas others may have hundreds or even thousands (Consonni and Cagliani 2019).

A common name for this collection is "Everything" Added to Food in the United States. The nutrition metabolome or simply the food metabolome is the term used to describe this collection of 15,000 food components, which includes both artificial and natural nutrients as well as artificial and natural non-nutrients. The food metabolome exhibits significant changes in chemical abundance in addition to high chemical variety. In reality, ingredients concentration can range for any particular food item from femtomolar to millimolar. It is a tremendous technological and computational challenge to detect, identify, and measure so many distinct substances across such wide concentration ranges. Nevertheless, metabolomics has the prospective to significantly advance towards mechanistic comprehending the underlying health and illness with novel technology development and enlarged spectrum libraries (Chin and Slupsky 2013).

Particularly, a remarkable variety of natural foods, spices, and drinks—including milk, grapes, tomatoes, tomato juice, rhubarb, beer, celery seeds, coriander, and several other herbs and spices—have previously been the subject of thorough metabolomic-based component analyses. To identify up to 100 distinct phytochemicals or 200 different carbohydrates in chosen fruit, vegetable, and beverage samples, these investigations combined NMR, GC-MS, LC-MS, and CE methods. A far wider variety of metabolites may often be detected when various separation and detection methods are used, with NMR being especially helpful for high abundance polar metabolite identification and GC-MS or LCMS being ideal for low abundance non-polar or semi-polar metabolite detection. Also noteworthy found that bulk of food-based metabolomics studies that has been published to far have analyzed liquids or liquid extracts rather than food solids. This is typically owing to the alleviation of most nutrients that haul out into liquids and mainstream of chemical detection devices are designed to function with liquids (McGhie and Rowan 2012).

Recent analyses of milk's metabolome have proved very instructive. These investigations have demonstrated the capability of high resolution Fourier Transform mass spectrometry to characterize more than 200 distinct oligosaccharides. In order to follow the characteristics of the herd, the feeding practices, and the quality of the milk, LC-MS techniques have also been used to profile around 15 different vitamins, terpenoids, and other phytochemicals in the milk of cows. These MS-based investigations are not the only ones that have demonstrated that NMR approaches may be used to quickly and nondestructively identify and quantify over a dozen distinct milk fatty acids, free sugars, acetylcarbohydrates, and other small molecule metabolites. NMR spectroscopy may soon take the place of the more conventional and time-consuming techniques employed to exemplify milk, given the speed and accuracy (Putri et al. 2013).

NMR and LC-MS studies on tomatoes and tomato juices have demonstrated how these two techniques may offer complimentary details about various tomato ingredients. Many low abundance and non-polar secondary metabolites, including polyphenols, flavo-glucosides, and alkaloids, may be quickly identified using LC-MS techniques in conjunction with the right reference databases. On the other hand, it was discovered that NMR was very helpful in the quantification/identification of the lycopene and carotenoid isomers that LC or LC-MS approaches were unable to discriminate. Numerous of the more prevalent tomato chemicals, such as sugars, flavonoid glycosides, amino acids, and organic acids, have been discovered and measured in other NMR investigations. The slight differences in tomato tastes, smells, ripening phases, and colors are mostly caused by variances in certain of these tomato-specific phytochemicals. They are also to blame for any putative anti-prostate cancer effects. The more pronounced tomato flavor differences are caused by variances in the more prevalent, polar metabolites (Herrero et al. 2012).

With more thorough metabolomic analyses being performed pharmaceutically or economically more significant fruits and vegetables, grains, nuts, and fruits and vegetables, food composition research is probably going to spread like wildfire in the subsequent period. Additionally, it's likely that metabolomic-based food constituent analysis will facilitate researchers to correctly scrutinize the outcomes of various food preparation and preservation methods, such as freezing, drying, smoking, and refrigerating, on essential food components. Examples of these processes include frying, baking, steaming, and boiling. Furthermore, these researches will aid food scientists in the selection, breeding, and modification of better crops, the raising and feeding of fish and cattle, and the production of numerous processed foods (Putri et al. 2022).

22.3 Metabolomics in Food Quality/Authenticity Assessment

Metabolic profiling's ability to detect food elements exploited to evaluate food quality and food adulteration. In particular, the detection of tainted or adulterated food items takes use of the fact that specific chemicals or chemical concentrations are particularly typical of particular types of juices, extracts, and oils. Food items with desired qualities that cannot be recognized by flavor, fragrance, or color can be distinguished from those with less desirable attributes using similar chemical composition criteria. Food quality control is impacted by food quality evaluation as well. In reality, batch-to-batch product reproducibility monitoring or quality control monitoring may be where metabolomic approaches in the food business are most useful (López-Ruiz et al. 2019).

The majority of metabolomic research on food quality and food adulteration that have been published so far have centered on describing oils or drinks such vegetable oils, fish oils, fruit juices, wines, and beers. Because it is so challenging to tell if fruit juice is adulterated by flavor or color, it happens rather frequently. For instance, orange juice and less expensive grapefruit juice can be mixed without any discernible flavor or color changes. However, those who are on specific drugs may experience substantial side effects if grapefruit juice is present in an ostensibly pure orange juice product. Numerous coumarin-like flavanoids and other potent CYP450 inhibitors can be found in grapefruit juice. Therefore, it is particularly crucial to identify and stop this sort of adulteration. Traditional methods for detecting this form of juice mixing involve time-consuming HPLC procedures. Recently, an NMR-based chemometric method utilizing Independent Component Analysis (ICA), a PCA version, has been created. The main idea behind this strategy is to examine certain areas of the ^{1}H-NMR spectrum that are known to have distinctive flavanoid glycoside signals. Ninety-two juices comprising of 59 orange, 23 grape, and 10 combinations, were used to evaluate the approach. It was discovered to be highly accurate, properly classifying 98% of the samples. Additionally, the time required by this NMR technique was less than half that of conventional chromatographic methods (Hu and Xu 2013).

The difference between juices made from pulp washes and juices that have just been squeezed is another frequent problem with juice production. Similar to how pulp wash is used to cut expenses while producing grapefruit juice, pulp might adjoined to freshly squeeze orange juice without leaving a noticeable taste, texture, or visual difference. Scientists employed ^{1}H-NMR and principal component analysis to separate more than 300 samples of freshly squeezed and pulp-washed orange juice in order to solve this problem. The accuracy of the approach was found to be close to 95%. Furthermore, compared to freshly squeezed juice, pulp-wash juice was found to contain considerably greater amounts of the particular biomarker component dimethylproline. The creation of a test to identify this changed amino acid might make the detection of pulp-wash relatively straightforward and low-cost (O'Gorman and Brennan 2015).

Additionally, NMR has been used successfully to track variations in beer production sites and batch-to-batch quality. Large multinational or pannational brewers are required to produce their brews across a wide geographic range. It is crucial to create methods for quality control that can be performed at a more precise molecular level rather than merely at the level of color, alcohol content, or specific gravity given the need to make beers that taste and are of comparable quality at these various locations. The study found that various manufacturing locations, likely producing the same beer, could be easily recognized based on their concentration of lactic acid, pyruvic acid, dextrans, adenosine, inosine, uridine, tyrosine, and 2-phenylethanol. By quantifying these chemicals, the researchers are familiar with one production site that had far poorer quality control than the other production sites and shown considerably more variability in these compounds. Other international firms might use these techniques to enhance quality control in the manufacture of juice, soft drinks, and vegetable oils. The quality control of functional foods and nutraceuticals—food extracts with favorable therapeutic effects—harvested from various geographical regions may also be monitored using comparable techniques (Sotelo and Slupsky 2013).

In order to detect and categorize samples, the majority of published food/beverage adulteration studies to date have used NMR in conjunction with PCA, artificial neural networks, or other multivariate statistical approaches. Mass spectrometry may also be used to look for food adulteration, though. In example, beet sugar detection and exogenous citric acid to fruit juices has long been done using isotope ratio mass spectrometry, as well as the adulteration of wines and olive oils. The comparative loads of isotopes or isotope fractionation in a particular sample is measured using the mass spectrometry instrument in IRMS, a specialized variant of the technique. Isotope abundances can alter based on material's age, physical source, geographic location, and metabolic response. As a result, isotope abundances can be utilized to determine the biological or geographical origins of numerous substances when compared to a known internal isotope reference (Li et al. 2016).

It is expected that stricter and more accurate food monitoring systems will need to be put in place over the next years due to the international nature of food production and processing as well as the rising concerns about food safety and food quality. Metabolomics may soon play a significant role in many aspects of food quality assessment and quality control, as evidenced by the proven ability of metabolomic techniques to quickly and affordably recognize adulterated oils and beverages as well as their capacity to precisely scrutinize previously undetectable quality control issues (Cubero-Leon et al. 2014).

22.4 Metabolomics Applications in FCM (Food Consumption Monitoring)

For nutritional experts, one of the most crucial and difficult duties is to monitor or measure food consumption. Nutrition specialists would like to be able to precisely manage what a person eats or keep track of everything they ingest, similar to how we can do with the meals of caged lab animals. Using this method would undoubtedly enable significant diet response discoveries. Unfortunately, people are not lab rats, and conducting closed-facility food treatments on people is quite expensive. The majority of research findings on the benefits or drawbacks of certain foods or diets come from epidemiological studies of sizable population cohorts that complete food frequency questionnaires. Sadly, FFQ recall for retrospective diet research and FFQ compliance for prospective diet studies haven't been all that great. The plasma levels of micronutrients or nutrients, which might vary from person to person by factors of 10 or more, cannot be directly measured with these dietary frequency questionnaires (Kim 2012).

In light of FFQs' drawbacks, metabolomics is being used to determine food consumption. In this method, volunteers in diet intervention studies' blood, urine, or saliva are analyzed using the information being gained in food component analysis and food quality testing. Using metabolomics for pharmacological safety or effectiveness assessment is quite comparable to this, both in terms of design and idea. On the other hand, in the context of nutritional monitoring, the "drugs" are rather ill-defined chemical admixtures of nutrients and non-nutrients (González-Peña and Brennan 2019).

The value of metabolomics-based consumption of food supervision comes from a better read-out of eating habits as well as a more accurate assessment of the physiological consequences of a particular diet. The aptitude for differentiating spectral changes in the urine or blood resulting from diet-specific non-nutrient "spill over" and those spectral changes resulting from diet-induced physiological changes, like peaks from amend metabolites (endogenous). Any diet intervention trial will be gravely misinterpreted if these distinctions are not recognized. Indeed, if the diet-derived signals are not properly accounted for, all that one may learn from an extensive metabolomic study is the fact that two groups are different obvious fact that needs no secondary confirmation (Lee 2009).

Urine is often the best biofluid to analyze nutritional consumption or to find biomarkers unique to certain foods. The urinary system, and any nutrient or non-nutrient that is present in excess or is not required will end up in the urine. Indeed, the proportion of metabolites to non-metabolites is a significant distinction between urine and blood. In practically every instance that has been documented to date, urine has a disproportionately larger amount of non-metabolites than blood. Such as biomarkers study of *Allium* ingestion discovered that only 5 lower abundance quercitin metabolites could be identified in blood, but 11 very high abundance quercitin metabolites were detectable in urine (Hall et al. 2008).

Sometimes the existence of the chemical alone signals intake of a certain meal, and other times its concentration is noticeably high for a while. The majority of food-specific biomarkers stay in the blood or urine for 5–10 h, with others staying there for up to 48. Numerous of these substances are phytochemicals or unmetabolized non-nutrients. Others are secondary metabolites that were identified as conjugates that needed to be eliminated because they were glycosylated, glucuronidated, or sulfated. The approach the body responds to these substances is often quite similar to how it reacts to medications or other xenobiotics. Consequently, monitoring of the food utilization may be done using the same principles that underlie drug testing and drug monitoring. The distinction is that medications often consist of relatively pure chemicals with minimal metabolites. Foods include many more components and metabolites since they are not pure chemicals (Munekata et al. 2022).

Additionally, it is probable that food consumption biomarkers will change from comprising just of one or two somewhat generic molecules to a much more precise "profile" made up numerous components. It should also be possible to identify individual variations in metabolite processing or absorption by using concentration data in conjunction with information on the time since, or amount of, ingestion. Few food intake surveys (short-term) employed in food interference or eventual diet studies may eventually reinstated by metabolomic methods (Table 22.2) (Munekata et al. 2022).

22.5 Metabolomics in Physiological Monitoring of Diet and Nutrition Studies

The identification of the crucial nutrients required for human growth and survival has advanced significantly over the past century thanks to the work of food scientists and biochemists. Most nutritional "deficiencies" have been successfully treated with improved dietary recommendations and required food supplements of critical minerals and vitamins. Today's nutritional experts are faced with the task of developing novel methods for treating or preventing illnesses including obesity, diabetes, chronic inflammation, and cardiovascular disease that are brought on by dietary "oversufficiencies." Additionally, they are expected to name any bioactive dietary ingredients that may lengthen life, lower weight, improve physical or mental

Table 22.2 Summary of diet biomarkers

Food	Biomarkers
Alcohol	Ethyl glucuronide
Apple	Phloretin
Black tea	Kaempferol
Carbohydrate rich diet	Lactate
Citrus fruit	Naringenin
Cocoa	Naringenin
Coffee	Caffeic acid
Cooked onions	Quercetin glucuronides
Cooked vegetables	Alpha-carotene
Fish	Trimethylamine
Garlic	Allylmercapturic acid
Grapefruit	Naringenin
Meat-rich diet	Creatine
Milk	Hippuric acid
Orange	Hesperetin
Pomegranate	Ellagic acid
Vegetarian diet	Salicyluric acid
Wine	Gallic acid

performance, and shield against ailments including atherosclerosis, heart disease, cancer, and arthritis (Brown 2010).

In order to investigate how diet or bioactive food ingredients affect human health, nutrition experts might choose one of two approaches. One way is through retrospective epidemiological studies that aim to link a person's long-term food intake to their health or susceptibility to disease. The other way is by conducting research with shorter-term prospective dietary interventions or diet challenges. Epidemiological studies designated several positive connections between diet and better health status, however these correlations do not always imply causality. Therefore, diet-challenge studies that concentrate on quantitatively detecting molecular biomarkers and physiological impacts have emerged as the ideal method for examining the mechanisms behind the unfavorable or favorable effects of important nutrients (Adebo et al. 2017).

In many diet-challenge or food intervention investigations, genomic, proteomic, and most recently metabolomic approaches are increasingly often employed. Small molecule indicators of inflammation, redox potential, anti-oxidant activity, oxidative stress, and cardiovascular disease risk may all be quantified using metabolomic tests, based upon the physiological reaction under examination. The benefit of using small molecule biomarkers is that the same small molecule monitoring techniques can be used to look at the short-term concentration changes of diet-related chemicals or metabolites. Although proteomic and transcriptomic markers are also available for tracking these responses, small molecule biomarkers have this advantage over them. To measure dietary intakes and physiological outputs, just one technological platform is required (Sébédio 2017).

The molecules that are utilized as biomarkers of physiological response to meals are usually non-polar and most are detected in blood, in contrast to the urine polar compounds that can be employed as biomarkers of food ingestion. This is due to the fact that blood is often a more consistent and trustworthy indication of physiological processes. Additionally, compared to urine, blood contains a lot more non-polar molecules. NMR or MS approaches haven't historically been used to find all of the chemicals. In reality, some are often evaluated using centrifugation techniques or identified using immunoassays. However, improvements in MS-based lipidomic technologies and lipoprotein characterisation now make it possible to identify and/or quantify these compounds using relatively conventional, high-throughput MS or NMR methods (Ahad and Nissar 2017).

One of the initial investigations on the effects of soy-derived proteins and soy isoflavones in a diet-intervention experiment used metabolomic methods. Pre-menopausal women on a multi-week soy diet had their blood plasma alterations examined using ^{1}H-NMR and chemometric methods. The findings demonstrated that distinct variations in lipoprotein, amino acid, and carbohydrate profiles associated to the metabolic pathways governing osmolyte fluctuation and energy consumption could be observed. The effects of whole-grain and wheat flour diets on rats were investigated in a related study utilizing both NMR-based chemometric techniques. These experiments demonstrated that the rats given whole grains had increased urine levels of hippuric acid, aromatic acids, and Kreb's cycle intermediates. Both a change in the animal's basal metabolic rate and a change in the metabolism of the gut microbiota may be indicated by these metabolites. Increases in glutathione and betaine, two markers of a healthy redox state and lessened oxidative stress, were seen in NMR tests on plasma and liver (Brennan 2017).

Many of the more recent studies have concentrated on employing more sensitive MS techniques or immunoassays since many of the effects of food modification are likely to be more subtly expressed and the biomarkers of interest will be at lower concentrations. The impact of concentrated red grape juice on oxidative stress, lipidemic, and inflammatory markers in blood, the impact of extra virgin olive oil on plasma inflammatory and oxidative stress markers, the effects of macadamia nut consumption on markers for cardiovascular disease, and the impact of gazpacho soup consumption on plasma oxidative biomarkers are a few of the studies included in this list. The findings of prior retrospective epidemiological research were essentially confirmed by all of these investigations, all of which demonstrated some statistically significant, favorable effect. Many of these researches measured the physiological effects of interest using both small molecule and large molecule protein biomarkers (Puiggròs et al. 2015).

Future investigations on the physiological effects of foods or diet-intervention trials are expected to utilize less varied food sources or specific bioactive components as diet supplements as a result of advances in perceptive of bioactives present in many foods. The physiological elucidations are frequently simplified, and the outcomes are frequently more repeatable, when pure or partly purified bioactive components are used. Studies of this nature are already being conducted. For instance, researchers looked at how plasma homocysteine levels were affected by two distinct compounds from food, betaine and choline. A known risk factor for heart disease and several other chronic illnesses is homocysteine. Similar research on people, animals, and other cell systems have been done with other pure bioactives, such as folate, omega-3 polyunsaturated fatty acids, and a range of purified fruit polyphenols (Utpott et al. 2022).

While long-term food exposure eventually affects human health and illness, the most efficient way to employ metabolomics to investigate diet-health interactions is probably to use short-term diet challenges to assess their expected consequences on long-term health outcomes. Sensitive metabolomic techniques combined with experiments using glucose, fructose, fat, and high-calorie diet challenges are likely to yield some important insights into the temporal progression of chronic diseases like obesity, diabetes, cardiovascular disease, and metabolic syndrome. Additionally, studies using fish/olive oil, flaxseed, cruciferous vegetables, polyphenol-rich fruit, and low-calorie diet challenges may provide similarly significant insights into their effectiveness for treating metabolic syndrome, preventing cardiovascular disease, and prolonging life (Van der Werf et al. 2001).

Diet-challenge studies will aid in revealing the range of individual responses to these various dietary perturbations if they are adequately and thoroughly tracked by the more sensitive metabolomic approaches. Certain disease-risk biomarkers or metabolites will undoubtedly exhibit far bigger or longer-lasting disturbances in certain "high-risk" individuals, whilst other "low-risk" persons would show essentially none. The assimilated utilization of transcriptomic, SNP-profiling, copy-number variation, gut metagenomic, and proteomic/enzymatic studies in these diet-challenge experiments will also allow for a much more thorough understanding of the inter-individual differences or proclivities due to epigenetic and genetic effects, albeit metabolomics undertake to endow with a much more chemically and quantitative comprehensive perceptive of phenotypic alterations (Kim et al. 2016).

As was previously said, there many different kinds of actively responding coordination to nutrition in addition to a wide range of dietary preferences among people. Certainly, this variation creates profiling of metabolites and metabolite "standardization" challenging. Gender, diurnal influences, and specifically age also affect the concentrations and even the existence of several metabolites, complicating consistencies, and inter-subject comparability. This might be substantially aided by the development of specialized metabolite or metabolomic databases that incorporate these gender-specific ranges, age-specific variances, or temporal impacts. This kind of database building would undoubtedly be made much simpler by the establishment and exchange of extensive, longitudinal metabolomic data sets gathered from well-designed and well-documented diet-challenge trials. Additionally, it could strengthen group comparisons and metabolite calibrations, making their results considerably more reliable (Capozzi and Trimigno 2014).

22.6 Summary

The creation of equipment with extremely low limits of detection and high sample throughput, as well as sample production and manipulation, has frequently been the bottleneck despite enormous advancements in analytical methodologies. It is obvious that metabolomics participate in a critical function in plant improvement, whether conventional hybridization or transgenic plants are utilized, from the standpoint of producing plant meals with improved quality. However, it is very challenging to demonstrate cause-and-effect since there are several metabolic pathways and outside influences. Metabolomics plays an even more important function in nutrition, however it is outside the purview of this book chapter. In fact, it doesn't matter how many more vitamins, antioxidants, or probiotics may be added to a meal or food composition; what matters is whether human absorption improves consumer health. Similar to this, describing not only the metabolome of a specific plant, animal, or microbe but also the interaction that takes place when the metabolomes interact, such as in plant and animal infection, may prove to be the greatest difficulty in the future. Large interdisciplinary teams are required to tackle these problems and put the "omics" puzzle together logically.

22.7 Conclusion

In several fields of food nutrition and food science research, such as food intake monitoring, food quality, food component analysis, food authenticity evaluation, and physiological monitoring in food intervention studies, metabolomics is well-positioned to be applied. Metabolomics' potential influence is currently constrained by two elements, namely technology and databases. It is obvious that considerable apparatus upgrades are required to make metabolite detection and quantification technologies more reliable, automated, and thorough. Even though noteworthy advancement has been achieved, current methods can only identify maybe one-tenth of the important metabolome. In research pertaining to food and nutrition, this broadened range and depth of coverage are especially crucial. It is necessary to create and make available many more reference spectral or chromatographic databases on metabolites, dietary components, and phytochemicals. Only with the assistance of databases can nutritionally important chemicals be regularly discovered or measured. In fact, a thorough endeavor should be channelized to finish and annotate the "food metabolome," comparable to the work required to document the human metabolome. Similar efforts must be made to develop openly accessible, comprehensive databases of nutritional phenotypes that contain quantitative metabolomic information obtained from diet-challenge or intrusion studies. It is anticipated that the food science community would find a manner of coordinating its actions to carry out these initiatives, even if they may take years to accomplish and cost millions of dollars. It is true that having access to resources like databases pertaining to nutritional phenotype or food metabolome that might be just as beneficial to the explorers (molecular biologists) (Balkir et al. 2021).

References

Adebo OA, Njobeh PB, Adebiyi JA, Gbashi S, Kayitesi E (2017) Food metabolomics: a new frontier in food analysis and its application to understanding fermented foods. In: Hueda MC (ed) Functional food: improve health through adequate food. IntechOpen, London, pp 211–234

Ahad T, Nissar J (2017) Metabolomic and its approach in food science. Pharm Innov J 6(11): 745–756

Balkir P, Kemahlioglu K, Yucel U (2021) Foodomics: a new approach in food quality and safety. Trends Food Sci Technol 108:49–57

Brennan L (2017) Metabolomics: a tool to aid dietary assessment in nutrition. Curr Opin Food Sci 16:96–99

Brown R (2010) Metabolomics applications for food. Food Sci Technol 24(2):34–35

Capozzi F, Trimigno A (2014) Using metabolomics to describe food in detail. In: Sebedio JL, Brennan L (eds) Metabolomics as a tool in nutrition research. Woodhead Publishing, Cambridge, pp 203–229

Cevallos-Cevallos JM, Reyes-de-Corcuera JI (2012) Metabolomics in food science. Adv Food Nutr Res 67:1–24

Cevallos-Cevallos JM, Reyes-De-Corcuera JI, Etxeberria E, Danyluk MD, Rodrick GE (2009) Metabolomic analysis in food science: a review. Trends Food Sci Technol 20(11–12):557–566

Chin E, Slupsky CM (2013) Applications of metabolomics in food science: food composition and quality, sensory and nutritional attributes. In: Weimer BC, Slupsky C (eds) Metabolomics in food and nutrition. Woodhead Publishing, Cambridge, pp 217–230

Consonni R, Cagliani LR (2019) The potentiality of NMR-based metabolomics in food science and food authentication assessment. Magn Reson Chem 57(9):558–578

Cubero-Leon E, Peñalver R, Maquet A (2014) Review on metabolomics for food authentication. Food Res Int 60:95–107

Fraga-Corral M, Carpena M, Garcia-Oliveira P, Pereira AG, Prieto MA, Simal-Gandara J (2022) Analytical metabolomics and applications in health, environmental and food science. Crit Rev Anal Chem 52(4):712–734

González-Peña D, Brennan L (2019) Recent advances in the application of metabolomics for nutrition and health. Annu Rev Food Sci Technol 10:479–519

Hall RD, Brouwer ID, Fitzgerald MA (2008) Plant metabolomics and its potential application for human nutrition. Physiol Plant 132(2):162–175

Herrero M, Simó C, García-Cañas V, Ibáñez E, Cifuentes A (2012) Foodomics: MS-based strategies in modern food nutrition and food science. Mass Spectrom Rev 31(1):49–59

Hu C, Xu G (2013) Mass-spectrometry-based metabolomics analysis for foodomics. Trends Anal Chem 52:36–46

Ibáñez C, García-Cañas V, Valdés A, Simó C (2013) Novel MS-based approaches and applications in food metabolomics. Trends Anal Chem 52:100–111

Kim HJ (2012) Application of metabolomics in food science. Food Sci Ind 45(2):36–44

Kim S, Kim J, Yun EJ, Kim KH (2016) Food metabolomics: from farm to human. Curr Opin Biotechnol 37:16–23

Lee SH (2009) New technology-metabolomics: applications to food nutrition and food science research. Bull Food Technol 22(1):69–77

Li B, Dunham SJ, Dong Y, Yoon S, Zeng M, Sweedler JV (2016) Analytical capabilities of mass spectrometry imaging and its potential applications in food science. Trends Food Sci Technol 47:50–63

Li S, Tian Y, Jiang P, Lin Y, Liu X, Yang H (2021) Recent advances in the application of metabolomics for food safety control and food quality analyses. Crit Rev Food Sci Nutr 61(9):1448–1469

López-Ruiz R, Romero-González R, Frenich AG (2019) Metabolomics approaches for the determination of multiple contaminants in food. Curr Opin Food Sci 28:49–57

McGhie TK, Rowan DD (2012) Metabolomics for measuring phytochemicals, and assessing human and animal responses to phytochemicals, in food science. Mol Nutr Food Res 56(1): 147–158

Munekata PE, Pateiro M, Rocchetti G, Dominguez R, Rocha JM, Lorenzo JM (2022) Application of metabolomics to decipher the role of bioactive compounds in plant and animal foods. Curr Opin Food Sci 46:100851

O'Gorman A, Brennan L (2015) Metabolomic applications in nutritional research: a perspective. J Sci Food Agric 95(13):2567–2570

Pinu FR (2016) Metabolomics: applications to food safety and quality research. In: Beale DJ, Kouremenos KA, Palombo EA (eds) Microbial metabolomics. Springer International Publishing, Cham, pp 225–259

Puiggròs F, Canela N, Arola L (2015) Metabolome responses to physiological and nutritional challenges. Curr Opin Food Sci 4:111–115

Putri SP, Nakayama Y, Matsuda F, Uchikata T, Kobayashi S, Matsubara A, Fukusaki E (2013) Current metabolomics: practical applications. J Biosci Bioeng 115(6):579–589

Putri SP, Ikram MMM, Sato A, Dahlan HA, Rahmawati D, Ohto Y, Fukusaki E (2022) Application of gas chromatography-mass spectrometry-based metabolomics in food science and technology. J Biosci Bioeng 133(5):425–435

Sébédio JL (2017) Metabolomics, nutrition, and potential biomarkers of food quality, intake, and health status. Adv Food Nutr Res 82:83–116

Sotelo J, Slupsky CM (2013) Metabolomics using nuclear magnetic resonance. In: Weimer BC, Slupsky CM (eds) Metabolomics in food and nutrition. Woodhead Publishing, Cambridge, pp 29–43

Utpott M, Rodrigues E, de Oliveira RA, Mercali GD, Flôres SH (2022) Metabolomics: an analytical technique for food processing evaluation. Food Chem 366:130685

Van der Werf MJ, Schuren FHJ, Bijlsma S, Tas AC, Ommen BV (2001) Nutrigenomics: application of genomics technologies in nutritional sciences and food technology. J Food Sci 66(6): 772–780

Chapter 23
Microbial Exopolysaccharides: Production, Properties, and Food Applications

Ramesh Sharma, Pinku Chandra Nath, Biswanath Bhunia, and Tarun Kanti Bandyopadhyay

Contents

R. Sharma · P. C. Nath (✉) · B. Bhunia
Department of Applied Biology, School of Biological Sciences, University of Science and Technology Meghalaya, Ri-Bhoi, Techno City, Killing Road, Baridua, Meghalaya, India

T. K. Bandyopadhyay
Department of Chemical Engineering, National Institute of Technology Agartala, Jirania, Tripura, India

J. A. Malik et al. (eds.), *Food Process Engineering and Technology*,
https://doi.org/10.1007/978-981-99-6831-2_23

Abstract The obtained polysaccharides for food application are mainly from sources like plants, animals, and algae. The exopolysaccharides (EPSs) production from lactic acid bacteria (LAB) added to the significant importance of the application of bio-thickener and thickeners in recent years. Several studies conducted under in vitro conditions to study the health-associated benefits of EPS produced from LAB. These biopolymers are synthesized in lesser quantities which added obstacles to large-scale production to its restriction on the commercial production of these polymers. In this chapter, emphasis is given to the EPS application for food and health applications.

Keywords Biopolymer · Exopolysaccharides · Food application · Genetic engineering · LAB

23.1 Introduction

Polysaccharides added significant functions in food processing as viscous agents, stabilizing, emulsifying, gelling agents. The hydrocolloids in the global market derived from algal polysaccharides and plant sources including starch, galactomannan, pectin, carrageenan, and alginate have flourished (Freitas et al. 2011). In 2008, the commercial value of more than four million US$ for xanthan gum was obtained based on bacterial EPS reported for 4% of the overall commercial value (Imeson 2011). However, the derived polysaccharides were obtained after chemical or enzymatic hydrolysis on starch, pectin, carrageenan, alginates, and cellulose and hence, their use was prohibited in food products. Hence, the other effective source of EPS is microbes. The EPS obtained from microbes added enhanced rheological properties due to its improved formations of viscous solutions at a lower dose with a behaving pseudoplastic nature (Becker et al. 1998).

EPS derived from microbes are xanthan and gellan using the species including *Pseudomonas elodea* and *Xanthomonas campestris* respectively. The products added with these EPS needed to be labeled since these bacteria are pathogenic to plants. Particularly, LAB are present in fermented food items like yoghurt, cheese, sour cream, buttermilk, and curd where they added taste, flavor and extended the fermented food shelf life (Shah and Prajapati 2014). Since LAB are considered GRAS microbes and can be added to food products without avoiding details on the labels. The attention towards bacterial strains of LAB for producing the EPS is oriented due to adding texture and rheology in LAB-containing food products (Cerning and Marshall 1999). The EPS producing LAB are grouped to *Pediococcus, Lactococcus, Lactobacillus* and *Leuconostoc* species. Similarly, EPS production is also seen from the non-starter culture of LAB including *Bifidobacterium*. LAB synthesis the polysaccharides in two forms as per cell location which include intracellular and extracellular polysaccharides (Degeest et al. 2001). The bacteria may produce the EPS in capsular form or slime form and some produced in both EPS form (Broadbent et al. 2003; Yang et al. 2010). The LAB bacteria can survive higher processing stress conditions (Stack et al. 2010); and therefore, can easily pass through the gastrointestinal tract (GI) when compared to other bacteria that do not produce anything. In addition, EPS induced a constructive role as well as cholesterol-reducing properties (Amatayakul et al. 2006; Li and Wang 2012) a reduction over pathogenic biofilms production (Kim et al. 2000) adhesion modulation in epithelial cells (Ruas-Madiedo et al. 2006) with the improvement of the bifidobacterial population with adding prebiotic features (Degeest et al. 2001; Hongpattarakere et al. 2012). Therefore, the selection of a starter culture with EPS-producing potential added significant results over the non-producing culture.

23.2 Probiotics for EPS Secretion and Other Sources

The probiotic are remoted from dairy and non-dairy assets. Dairy derived included milk and milk-based items while non-dairy products included grains, vegetables, and fruits. Even though dairy products added health benefits but are not suitable for people having problems with lactose intolerant, allergy to milk proteins, and retaining higher cholesterol people (Shori 2016). The food source obtained from non-dairy products including cereals fruits and vegetables, enriched with vitamin, dietary fibers, antioxidants, minerals, vitamins, and other therapeutic characteristics facilitates the growth of probiotics (Homayouni et al. 2012). The food products retaining fermentable sugar promote the probiotics growth. In past years, the novelty of probiotics LAB was tested which possesses the features like improved textured and flavour, enhanced shelf life of obtained products (Patel et al. 2012a, b, c). Similarly, the probiotic microbes are also isolated from meat and the products. Mainly fermented food is used for their production in the industry. The EPS isolated from fermented food added considerable benefits by adding probiotics and functional properties.

23.3 EPS Types and Composition

EPS are long-chain sugar molecules with sugar residues and are branched with lengthy chains. Molecular weight, chemical structure, functions, and chemical bonding classified bacterial polysaccharides. Mainly, EPS are classified as per their chemical compositions into homopolysaccharides (HOPS) and heteropolysaccharides (HEPS). HOPS occurred with molecular weight of 107 Da. HOPS are further divided as per the availability of linkage bonds and monomeric units into: (1) α-D-glucan (2) β-D-glucan (3) Polygalactan (4) Fructan. Additionally, HEPS consist of D-glucose, D-galactose, and L-rhamnose and contain *N*-acetylglucosamine, glucuronic acid or *N*-acetyl galactosamine, and molecular mass of HEPS with ranging 104–106 Da (Nguyen et al. 2020). The phosphate, acetyl, and glycerol are present as non-carbohydrates groups which are attached to it. The monomeric units are attached with 1,4- or 1,3-linkages and 1,2- or 1,6-linkages.

The HOPS are distinguished not only by linkage bond and chemical nature but also get distinguished on synthesis site and enzyme used. The repeating unit in the heteropolysaccharide is formed intracellular region and glycosyl lipids are associated to translocate the precursors in the membrane site whereas homopolysaccharides sucrose is required as a particular substrate. The EPS are produced from gram-positive and gram-negative bacteria (Nguyen et al. 2020).

23.4 Physical Properties of EPS

EPS originated from LAB has improved properties like mouth-feel, enhances, and are stable food products that ultimately improve the production capacity of novel food items. The EPS based on microbes added the functional properties in food products with function of viscous, bio-thickening, emulsifying and stabilizing effects and which are altered by pH, temperature, and ionic strength (Abid et al. 2018; Patten and Laws 2015). The stability of EPS under thermal treatment is important in dairy processing retaining emulsifying properties and flavor in milk-based products. The EPS were purified with the following microorganism including *Leuconostoc citreum* (sausages prepared from bovine meat), *Leuconostoc mesenteroides* (sausages prepared from turkey meat), *Pediococcus pentosaceus* (extracted using sap from date palm) and *Leuconostoc pseudomesenteroides* (extracted from cow milk) and were thermally stable with higher melting points of 224 °C with showing their higher potential of use in thermally processed food products (Abid et al. 2018). The EPS obtained using *Streptococcus thermophilus* microbes added beneficial features of gastro-intestinal protection, therefore could be a dairy industry as the additives to improve oil and water and holding capacity, improving solubility under aqueous conditions, emulsification, and flocculation properties (Lobo et al. 2019).

The EPS produced from *Leuconostoc citreum* was added with higher solubility in water, higher emulsification activity, and greater thermal stability, therefore, could be a good substrate as an food additives (Feng et al. 2018). Similarly, the novel probiotic strain called *Enterococcus faecium* has been exploited for its application in the field of food processing based on emulsifying characteristics under thermal processing conditions with improving other physio-chemical properties (Bhat and Bajaj 2018). The EPS produced from *Lactobacillus rhamnosus* had shown promising results under in-vitro conditions by improving flocculation, emulsification, and solubility, with other functional features by adding use in a commercial application (Raji et al. 2014). The dextran produced from *Leuconostoc pseudomesenteroides* in wine has improved viscosity and signified the ideal thickening agent. Besides that, it has improved water-holding capacity and its solubility added a convincing result for its use as hydrocolloid and stabilizing agent (Zhou et al. 2018).

23.5 Biosynthesis of Bacterial EPS

Mostly EPS are produced during the growth period of bacterial development whereas in some case it is synthesized during the stationary or late log phase. The imbalance of nutrients due to higher carbon-to-nitrogen (C: N) ratios, and temperature at sub-optimal incubation are factors that accelerate to produces of these polymers (Sutherland 1977). Polysaccharides are considered the main source of EPS production (Gupta and Diwan 2017). In bacteria, EPS is a synthesis with two approaches (1) intracellular and (2) extracellular regions. The synthesis of homopolysaccharides occurred under the extracellular region whereas the heteropolysaccharides are intracellularly synthesized with their transfer to the extracellular region. The various enzymes group are presently performance a key function in the synthesis of EPS. As per the phase involvement in EPS production, the enzymes involved are classified into four groups (Badel et al. 2011).

- **Group 1:** These groups are present in intracellular regions. Hexokinase is a primary enzyme in this group. Hexokinase phosphorylases facilitate the transfer of glucose into glucose-6-phosphate (Glu-6-pho). Phosphoglucomutase is the other enzyme playing a role to convert (Glu-6-pho) to glucose-1-phosphate.
- **Group 2:** The second enzyme groups are present in intracellular regions which play vital function in the production of sugar nucleotides which might be EPS simple unit. UDP-GPPR is a major enzyme in this group. The UDP-Glu catalyzed with the transforming of glucose-1-phosphate to UDP-Glu. With the presence of UDP-Gal4-epimerase, a nucleotide is also interconverted to other sugar molecules like UDP-Glu to UDP-Gal.
- **Group 3:** The enzyme called GTFs, is available in the periplasmic membrane of the cell and is considered in these groups. GTFs facilitate the over transport of sugar nucleotide (UDP-Glu or UDP-Gal) by forming a repeating unit with glycosyl linkage after involving with a carrier agent like lipid, isoprenoid alcohol (Panosyan et al. 2018).
- **Group 4:** Polysaccharides get modified with the involvement of enzyme activity like acetylation, acylation, sulphation, and methylation which mainly occurred in extracellular regions with the use of hydrophobic enzymes including permease, flippase, ABC transporters during the mechanism of production under fourth stage (Panosyan et al. 2018).

There are four stages of EPS Biosynthesis in Bacterial Cells.

23.5.1 Pathway Dependent on Wzx/Wzy

Firstly, sugar nucleotides are synthesized, assembling the repeating unit, polymerization, and exported to all route parts (Schmid et al. 2015). The activated sugar is formed and gets transferred with connecting carriers like lipid and isoprenoid alcohol at the earlier first stage. GTFs promote the joining of sugar molecules with forming repeating units. These units are transported through a cytoplasmic membrane with Wzx flippase (De Vuyst and Degeest 1999; Islam and Lam 2014; Rehm 2010; Schmid et al. 2015). Oligosaccharides get translocated by adding several enzymatic changes including methylation, and acylation before their polymerization by Wzy protein (Islam and Lam 2014). ABC transporters facilities the transport of obtained polysaccharides into the cell surface (De Vuyst and Degeest 1999).

23.5.2 Pathway Dependent on ABC Transporters

The members are involved in ABC Transporter which included the Periplasmic protein and polysaccharide co-polymerase (Cuthbertson et al. 2009). Glycosyltransferase facilitates induced polysaccharides before it is exported through efflux pump complex. This pathway is deployed in capsular polysaccharides (Cuthbertson et al. 2009).

23.5.3 Pathway Dependent on Synthase

The obtained products after following the synthase pathway are mainly homopolysaccharides like cellulose (Ua-Arak et al. 2017; Wang et al. 2016). This pathway may be operated with or without the availability of a lipid acceptor compound. A single synthase protein and membrane-localized GTFs also operated with the polymerization of EPS precursor and eventually transported the molecules with this pathway (Whitney and Howell 2013). While compared to another pathway, this pathway's metabolism is not dependent on central carbon. The biosynthesis in the extracellular region by sucrase protein in which sucrase facilitates the transformation of sucrose into monomers unit (Ua-Arak et al. 2017). The EPS including Dextran, alternant, and Levan are obtained under extracellular conditions. The dextran is produced from *Leuconostoc mesenteroides* with a molecular weight of 15–20,000 kDa. As in the reaction below, glycosyl transferase dextransucrase facilitates the transfers of glucose from sucrose molecules at the reducing end with developing the dextran chain.

$$Sucrose \rightarrow Dextran + D - fructose$$

The EPS biosynthesis in probiotics occurred with two sites mainly in intra and extracellular regions. HOPS are produced from biosynthetic pathways under extracellular conditions whereas HEPS may be produced by extracellular pathways. In biosynthesis under the extracellular pathway, the HOPS are induced by extracellular fructosyl transferases and glycosyl transferases during polymerization (Li and Wang 2012; Yother 2011). The residues of sugar molecules underwent cleavage into monomeric units external cellular region and get associated in extracellular conditions and facilitate to polymerized by glycosyl transferases.

Similarly, sucrose is used as a ingredient and gets cleavage to form fructose and glucose and polymerization is taken place with an enzyme called fructan sucrase and glucan sucrase forming fructan and glucan. Glucan sucrases are categorized into mutan sucrase, reuteran sucrase, alternan sucrase and dextran sucrases with producing EPS called mutan, reuteran, alternan dextran, and respectively and similarly fructan sucrases are classified to inulin sucrase and levan sucrases with producing inulin and levan respectively (Patel et al. 2012a, b, c). The polymerized form of these HOPS is released into the external environment. In the case of the intracellular bio-synthesis pathway, the residues of sugar are transferred to the cell which is converted to monomeric units attached to an isoprenoid lipid carrier along with its partial polymerization (Madhuri and Prabhakar 2014). In this stage, the required modification in the polymer is achieved and transported to the extracellular region and assembled into polysaccharides.

The synthesis of HEPS with an intracellular pathway is relatively challenging due to the involvement of various factors like different enzymes, involvement of transporters and carriers' proteins which are encoded by genes of plasmids or chromosomes (Pattanaik et al. 2020). The HEPS synthesis was performed with Wzx/Wzy pathway whereas besides polymerization all the processes occurred in cytoplasm. Transport of sugar, its nucleotide synthesis, synthesis of repeating units, and repeating unit's polymerizations in the cytoplasm are the major steps involved in HEPS synthesis. The substrate selection depends on the types of sugar used and mainly occurred through active or passive transport system (Badel et al. 2011; Freitas et al. 2011; Werning et al. 2012).

The precursors for the synthesis of bacterial EPS are obtained from intracellular regions with the intermediates formed from the metabolism of central carbon. The donor monomers and precursors involved with bio-synthesis of repeating units mainly the sugar nucleotides like nucleoside di-phosphate (NDP) sugars, sugar acids, sugar derivatives including glucose, mannuronic acid, and sugar derivatives (Nath et al. 2021).

With the pathway following Wzx/Wzy, the mentioned sugar nucleotides are regarded as particular repeating units that are associated with undecaprenyl di-phosphate anchor and found in the inner region of membrane and linked by glycosyl transferases and translocation occurred throughout the membrane of cytoplasm by a flippase which is called as Wzx protein (Welman and Maddox 2003). In the second stage, the polysaccharides are modified with enzyme activity like sulphation, acetylation and methylation. Therefore, these polymers obtained after modification are released to the bacterial cell wall and considered as CPS or EPS and underwent with the enzyme action like flippase and APC transporters (De Vuyst and Degeest 1999; Sørensen et al. 2022).

23.6 Physiological Functions of EPS

Besides the EPS application in industrial areas, it is used to add functions to the microbes cell. However, the EPS role in the physiology of bacteria is not studied which is complex and diverse. The major EPS role is clear in ecological niches, which depend on the associated environment of microorganisms. The EPS present in microbes protected the microbes' cells from adverse environmental conditions like antibiotics, osmotic stress desiccation, or toxic substances including, metal ions, ethanol and sulfur dioxide, protozoan's predations, phage attack, and phagocytosis. The surrounding microorganisms with hydrated exopolysaccharides induced protection against protozoan predation and desiccations (Suresh Kumar et al. 2007).

Additionally, the availability of gelled polysaccharide layer in the surrounding cells affects over-diffusing characteristics of to inward and outward of the cell (Sutherland 1998). The researchers claimed with the fact that EPS-associated cells promote with reduction of sensitivity toward lysozyme and bacteriophage, by protecting the enzyme and phage target to the cell. The higher EPS-producing bacterial strains were relatively insensitive towards phages and nisin incorporation relatively added physiological functions in the environment. The β-glucan-produced from *L. paracasei* was used to analyze under gastrointestinal and technological stress conditions (Stack et al. 2010).

Heat stress assays were performed and the presence of polysaccharides was associated with improving protection under thermal stress conditions (60-fold), stress under acidic conditions (20-fold), and stimulated stress by gastric juice (15-fold). Similarly, the assays for bile stress revealed modest results with a 5.5-fold increment in surviving the β-glucan-releasing strains after comparing the control strain. Therefore, the β-glucan releasing strains may be used as probiotics with greater protection during growing and processing conditions. The in vitro experiment for EPS-releasing strains called *L. rhamnosus* was performed with a protective effect on the intestine innate immune system (Lebeer et al. 2011).

The mutant strain not producing the EPS functions with higher sensitivity towards host innate defense substances including complement factors and LL-37 antimicrobial peptide. This was revealed by the results that EPS added the protective function for LGG in GI tract. EPS production in capsular form is well-known in pathogenic microbes, and the organism's pathogenicity is dependent on the synthesis rate of EPS. The attachment of capsules to the pathogen allows for phagocytosis. The CPS are devoid to activate immune system due to the symmetrical nature of chemical structure with components of the host cell surface. Besides the protection by EPS produced from LAB, it also exhibits a health-promoting effect through the process called immunomodulation (Lin and Chien 2007; Rodríguez et al. 2009; Suresh Kumar et al. 2007).

23.7 Prebiotic Functions

The EPS based on fructan produced by *L. sanfranciscensis* has been used as prebiotic mentioned (Korakli et al. 2001). The EPS type called levan which was released by the same strains of other species had shown bifidogenic effect (Dal Bello et al. 2001). Another experiment carried out by Salazar et al. (2008) mentioned that EPS produced from intestinal *Bifidobacterium* was fermented with microbes by gut bacteria with interaction modifications among intestinal populations.

23.8 Scope of EPS in Food Applications

Dextran is commercially produced from LAB called *Leu. mesenteroides.* The discovery was done in 1880 in beet syrups or sugar cane and was playing the role of thickening and gelation properties in syrups (Crescenzi 1995). As per the difference in structure, dextran may be insoluble and soluble in water. Similarly, dextran has been used in confectionary for improving the retention of moisture, and viscosity retentions and for inhibiting the crystallization of sugar. It is used in jam and jellies as the gelling agent. It is used as a crystallization inhibitor in ice cream and added the mouth feel and body texture during pudding mixing (Whistler and Daniel 1990).

Additionally, it is used for blood plasma extender purposes (Kulicke and Heinze 2005) and is considered a basic ingredient in chromatography as the stationary phase. Dextran is produced by *Leu. mesenteroides* protected the organisms during the starvation period and added resistance to survival under acidic and alkaline environments (Kim and Kim 2009).

Similarly, Xanthan gum is produced from a pathogenic plant called *Xanthomonas campestris* which is a EPS approved for its use in food items in 1969. Xanthan is considered a "benchmark" items depends on the potential application in food and non-food items (Sutherland 1990) and food items including pet food, dressing, syrups, confectionary, pet food, bakery, dairy products. The processing cost of xanthan production is lower due to the higher conversion rate to polymer (60–70%) (Sutherland 1990).

The biopolymers liberated higher viscosity at the lower concentration showing strong pseudo-plasticity and are mostly stable with temperature, wider pH, and ionic strength (Qin et al. 2007). The polysaccharides produced from microbes retain the rheological properties which are demanded by industries and can be produced in higher quantity at higher purity level. Since the time of 1940, levan and dextran were used in the food and pharmaceutical sectors. Fructose-oligosaccharides (FOS) have been used for food products as it is lower sweater than sucrose, considered calorie-free with retaining non-carcinogenic properties (Yun 1996). The inulin and FOS are used in the food sector is due to their prebiotic function in the body. Fortunately, a polymer based on fructose has a higher ability to get fermented by microflora in the gut which ultimately improves the intestinal flora by improving the mineral uptake capacity (Xu et al. 2006; Yamamoto et al. 1999).

The microorganism *L. sanfranciscensis* LTH 2590 is used to produce levan which has also demonstrated prebiotic functions under in vitro experimental conditions (Dal Bello et al. 2001; Korakli et al. 2001). The stress tolerance in plants by stabilizing the surface of the membrane is mentioned (Oliver et al. 2001). The exceptional property of EPS and FOS with protecting the bacteria under stressful conditions promote their application for survival and promote the activity of LAB for storage and processing of milk based fermented products. EPS derived from LAB, has commonly used for enhancing the texture and body of yogurt and Dahi (Broadbent et al. 2003; Faber et al. 2001).

Dahi contains 3.5–8% fat, like yogurt, it is been widely consumed in every house (Nair and Prajapati 2003). The milk fat contributed to texture, flavor, and body in dairy products, and removing the fats resulted in functional and textural defects in fermented milk products. For yogurt and dahi with a lower fat range, weak body, texture, and flavor shortage are the major problem associated with it (Guven et al. 2005). The EPS produced from LAB are considered bio thickeners which are more acceptable and preferred additives in food products. These cultures fulfill the consumer requirements by developing the products by lowering the use of chemical ingredients (De Vuyst and Degeest 1999; Jolly et al. 2002), lowering solid concentration without altering the textural parameters (De Vuyst and Degeest 1999; Wacher-Rodarte et al. 1993) as well as improving sensory features (Folkenberg et al. 2006).

Dahi with lower fat content prepared from EPS releasing culture of *S. thermophilus* and *L. lactis* were reported with higher acceptability in terms of body, flavour, and texture after comparing with dahi prepared with other culture of *L. lactis* (EPS not present) (Behare et al. 2009). The preparation of dahi from EPS releasing strains had higher percentage of pores and open structure as well as casein matrix were present discontinuously present than compared to controlled dahi sample which was having compact linear structure (Praveen 2000).

For defect in low fat containing cheese, the producers added the texture promoting or ropy culture especially to the products where stabilizer addition is restricted. There ropy strains promote the flavour intensity in fermented milk due to action of carbohydrates over masking the flavor, imparting mouth feel and other attributes. The skim milk gel is prepared with ropy culture including *Lc. lactis* subsp. Lactis had higher apparent viscosity after comparing to non-ropy cultures (Behare et al. 2009; Dabour et al. 2005).

The rheology of goat cheese (Florencia 2013) prepared from lactic cultures with various EPS producing capacities was studied and revealed that EPS-releasing strains have lower storage module, loss module and obtained viscosity over wider ranges with lower critic stress than compared to cheese produced from non-EPS releasing strains. The culturing with *L. rhamnosus* JAAS8 (EPS-producing strains) after combining with non-EPS-producer yogurt species could enhanced the capacity of water-binding of fermented food than non-EPS culture-produced yogurt products (Yang et al. 1999). The replacement of non-EPS-producing strains under the total or partial condition to *L. bulgaricus* with EPS-releasing *L. rhamnosus* could result to improve apparent viscosity by 16% and 21%, respectively, with a 2% improvement in WHC in fermented food items. To achieve reduction over added milk solids, enhance yogurt viscosity, improve mouth feel and texture at the fermentation time or during storage, EPS producing strains added the benefits.

Preparation of yogurt with EPS releasing strains retains higher water binding action with decreasing the products enhances over syneresis conditions (Amatayakul et al. 2006; Hassan et al. 1995, 2003). It is always helpful to utilize the grouping of ropy with non-ropy starter culture (Laws and Marshall 2001). The EPS producing culture effect over product rheology in yoghurt, cheese and dahi, the other fermented milk products beside these the effect over rheology were seen in kefir, sour cream, and other dairy products. The culturing with slime producers strain called *S. thermophilus* could improve the rheological characteristics in cream turo, cultured milk and cream items (Obert 1984).

Kefir is acidic in nature and are fermented milk-based products originating from Eastern Europe (Shah and Prajapati 2014; Tamime et al. 2005). Kefir grains are involved to prepared kefir using acetic acid releasing microbes, yeast and LAB. The cells are attached with kefiran, which is a slimy polysaccharide, which is associated with releasing the texture in kefir (Duboc and Mollet 2001). Additionally, the incorporated LAB strains also release the EPS with in the matrix and function as viscosifying agent. Similarly, dextran produced from *L. mesenteroides* were commercially applied as baking improver.

The evidence was added with improvement in dough rheology and quality of bread (Brandt et al. 2003). The EPS released under in situ conditions was more efficient than EPS (Levan) and the 1% sucrose addition to wheat dough induced with formation of polymers by *Lactobacilli.* The EPS formation under in-situ conditions also further promotes with formation of metabolite like mannitol, glucose, and acetate with imparting the bread quality (Korakli et al. 2002). The *Weissella* sp. producing the EPS, increases the textural features and bread quality (Di Cagno et al. 2006; Galle et al. 2010). Polymers released from the culture of *Lactobacilli* added benefits to bread and dough including (a) dough water absorption capacity (b) rheological dough properties (c) stability of dough under freezing conditions (d) volume of loaf and (e) staling in bread.

The longer duration of fermentation in idly batters resulted in collapsing and separations of whey which makes the idli hard in texture. Adding the hydrocolloids to batter may promote the texture of processed products (Nisha et al. 2005). Similarly, Xanthan additions at 0.1% imparted textural characteristics which improve the sensory scores compared to idli with avoided additives (Thakur et al. 1995).

The over addition of hydrocolloids to idli batter resulted with improvement over viscosity (Nisha et al. 2005). The additions of agar and guar gum to the idli improve the sensory properties with improving the texture and mouth feel. The EPS leasing microbes is used to impart stabilizing function under in-situ conditions. Similarly, the EPS may be used as thickening agent with improving viscosity and reducing the syneresis effect (De Vuyst et al. 2001) thus improving the product quality without incorporation of additives.

23.9 Health Potential of EPS

The EPS biological capacity is affected by the fermentation parameters while cultivating the probiotics (De Vuyst et al. 2001; Salazar et al. 2008). EPS with various sources are affected by glycosidic linkages, compositions of monosaccharides, chemical modifications that are held within specific conformations, expanded chains, and molecular weight (Crescenzi 1995). The availability of these components added the EPS health-promoting properties. The EPS released from probiotics added much importance for therapeutic use (Oliver et al. 2001; Xu et al. 2006) and different application are described below:

23.9.1 Antimicrobial Activity

The LAB strains released the antimicrobial components like diacetyl, bacteriocins, organic acid, and CO_2 with other molecules including reutericyclin, reuterin, and peptides (Li et al. 2014). Besides this, the EPS released from LAB performed an antagonistic effect against pathogenic bacteria. The EPS released from *Lactobacillus rhamnosus* from human milk functions with antibacterial function against *E. coli* and *S. typhimurium* (Raji et al. 2014). EPS released from *Bifidobacterium longum* functions with impairment of cell divisions rather than inhibitions of bacteria (Wu et al. 2010). The HEPS produced from *Lactobacillus gasseri* were used to investigate the antibacterial activity and it could successfully inhibit the *Listeria monocytogenes* (Rani et al. 2018). Similarly, EPS obtained from *Lactobacillus* sp. retarded the proliferation of *Salmonella enterica* and *Micrococcus luteus* with an inhibition zone of 14 and 10 mm, respectively. EPS(C-70) produced from milk of camel performed with 2–3 log food pathogen population reduced from initial populations of 9 log (CFU/mL) and maximum inhibitions were seen to *E. coli* and *S. aureus* (Ayyash et al. 2020).

23.9.2 Activity of the Immuno-modulatory System

The different polysaccharides derived from LAB modulated the host immunity by improving the T/B lymphocytes proliferation functioning as natural killer as anti-tumoricidal activity, improving the phagocytosis of the mononuclear cell, mitogenic activity, induce the functions of cytokine with elevating the host immune response towards pathogen (Shao et al. 2014). The immunomodulatory functions of EPS released from LAB are mainly forbidden by their physico-chemical characteristics like molecular weight, compositions of monosaccharide, electric charges, its solubility towards the water, and stereochemistry (Lopez et al. 2012). The conclusions are made by some researchers that the EPS with lower molecular weight or negatively charged molecules functions with strong immunomodulating activities whereas the neutral polymers with higher molecular weight function with weak immunomodulating activity and act as immunosuppressive agents. Therefore, EPS is applied to regulate the adaptive and innate immune systems in the body (Notararigo et al. 2014). They also reduced the chances of infections, prevent cancer in the GI tract, and prevent inflammatory bowel disease (Raji et al. 2014).

Similarly, the cytokines are modulated by these immunomodulators which ultimately change the immune system's regulatory behavior (Singh and Saini 2017). The EPS produced from *Lactobacillus johnsonnii* isolated from mice intestines and under experimental condition induced the IBD with promoting the dendritic cell differentiations, with triggering the cytokines productions mainly the Th-2 immune response (Górska et al. 2016). *Lactobacillus plantarum* produced the EPS by stimulating the immunomodulating activity in macrophage cells by inducing cyclophosphamide in immunosuppressed mice (Wang et al. 2016).

The experiment was performed under *in-vivo* conditions in Swiss albino mice which revealed the fact that the EPS released by *Weissella confusa* could stimulate the synthesis of antibodies called IgM and IgG while the mutant strain produces the higher antibodies called IgA (Adebayo-Tayo et al. 2018). The EPS produced from *Lactobacillus casei* was experimented in mice and reported with improvement in cellular and humoral immune response with elevation of serum antibodies, proliferation of T cells, better expression of cytokines with up-regulating the maturation of dendritic cell (Xiu et al. 2018).

23.9.3 Anti-inflammatory Activity

Inflammation phenomenon where normal tissue gets restored with its functions responding to infection and injuries in the host organism. The inflammation for prolonged period leads to cancer and inflammatory diseases (Smiderle et al. 2014). The initiation of inflammations by microorganisms occurred through endotoxin lipopolysaccharide (LPS) by targeting the receptor-4 (TLR-4) by promoting the genes for inflammatory expression (Toshchakov et al. 2002). LPS facilities to produce cytokines and mediators including nitric oxide and other cytokines like IL-1β, IL-6, and TNF-α with suppressing the anti-inflammatory cytokines including IL-10 (Bhatia and Moochhala 2004).

The excess in pro-inflammatory cytokines production through activation of NF-κB resulted in chronic and acute inflammations leading to inflammatory diseases like asthma, inflammatory bowel diseases, tissue injuries, and rheumatoid arthritis. The biomolecules from microbes activated the macrophages thereby killing the bacteria by releasing the cytokines, regulating the immunity, stimulating the phagocytosis, and functions with the availability of bacterial antigens to help the T cells (Serbina et al. 2008).

EPS released from probiotics functions with an anti-inflammatory mechanism in the immune cell (Azad et al. 2018). The *in-vitro* experiment was performed on macrophage cells and several researchers evaluated over EPS induced innate immune response. The purified form of EPS produced from *Lactobacillus pentosus* functioned with immune-stimulatory activity by enhancing the macrophage cells viability, improving phagocytosis, an increasing over activation of macrophage and enhancing the synthesis of TNF-α, Il-1β, IL-6 and IL-10, NO (You et al. 2020). EPS produced from *Lactobacilli* after isolating from a vagina of human body demonstrated the anti-inflammatory functions increasing the synthesis of IL-10 and TNF-α in HeLa cells (Sungur et al. 2017). The polymer derived from *B. longum* functions with immunomodulatory activities towards J77A.1 macrophage with improving of IL-10 secretion and reducing TNF-α level (Wu et al. 2010).

Similarly, under *in-vitro* conditions, the EPS released from *Lactobacillus rhamnosus* stimulate to produce cytokines for pro-inflammatory and anti-inflammatory conditions in mouse peritoneal macrophages (Ciszek-Lenda et al. 2011). The *L. paraplantarum* was used as probiotic strains and experienced in peritonitis rat model and carrageenan was use as inducing agent for decreasing the pro-inflammatory mediators, IL-1β, TNF-α and iNOS, with enhancing the IL-10 secretion (Dinić et al. 2018).

The *B. licheniformis* and *L. mesenteroides* were used to produce the EPS and inhibit secrete the IL-6 in LPS stimulated mouse macrophages by enhancing the secretion of IL-10. Therefore, this feature added the potential for EPS application in the pharmaceutical, food, and cosmetic industries (Kook et al. 2019). The EPS obtained from *B. animalis* were reported to stimulate the TLR-4 in the mouse colitis model (Castro-Bravo et al. 2019). The EPS produced from the *L. reuteri* (Mh-001) containing galactose influences anti-inflammatory functions on the macrophages (Chen et al. 2019).

The EPS produced from *L. plantarum* reduced the excess release of IL-6, TNF-α, NO, and prostaglandin with suppressing NF-κB activation by inhibiting the Iκ-Bα phosphorylation in macrophages by activation using LPS (Wang et al. 2010). Similarly, the EPS produced from *L. fermentum* after combining with yogurt was fed to mice which revealed enhancement in the production of butyrate and acetate (Ale et al. 2019). These produced fatty acids are volatile compound that is released in the gut by microbiota and perform anti-inflammatory functions. The bifidobacteria are protective against enteropathogenic infections with producing the acetate molecules (Fukuda et al. 2011). Similarly, butyrate involved to mobilize the colon, reduction of inflammations, visceral irrigation improvement, inducing apoptosis, inhibition in progress of tumour cell with playing role to prevent colon cancer (Canani et al. 2011).

23.9.4 Antioxidant Activity

The ROS including hydroxyl radicals, superoxide, and hydrogen peroxide is released after cellular metabolism responds with ill effects on human health leading to atherosclerosis, neurodegenerative diseases, cancer and rheumatoid arthritis. The Superoxide radicals are the most destructive among ROS, served with a precursor to other ROS including H_2O_2, OH radical, and singlet oxygen. The phenomena of oxidative stress have occurred with these ROS leading to damage to the proteins, DNA mutations, membrane lipid oxidations, and low-density lipoproteins modifications leading to damage of tissues and death of the cell (Kogani et al. 2008; Yin et al. 2010). Antioxidants are compound functions with retardations and hinder the process of oxidations. Even though synthetic antioxidants are available including BHA, BHT and PG to scavenge these radicals but their functions fail with complete eliminations leading to carcinogenesis and damage of liver cells (Liu et al. 2011).

Many reports are available on probiotic-based EPS with functions of antioxidant activity. Generally, LAB stains scavenge the free radicals through the cellular antioxidant defense with the release of an enzyme called superoxide dismutase assisting the antioxidant production non-enzymatically with glutathione productions. The exopolysaccharides produced by probiotics were used to evaluate their antioxidant activity with the degradation of hydrogen peroxide and superoxide ions, reducing the metal chelating properties and ROS with their desirability of applications as supplementary items or in the formulation of drugs (Liu and Pan 2010).

The purified form of EPS released from *L. plantarum* performed the antioxidant activity with scavenging activity towards ROS, up-regulating the enzymatic and non-enzymatic antioxidant characteristics. Additionally, it was mentioned that the probiotics with secondary metabolites were efficient in lowering the accumulations of ROS by degrading the H_2O_2 and superoxide anions (Seo et al. 2015). Similarly, the EPS produced from *L. delbrueckii* derived with fermented milk performed the functions of antioxidant activity (Tang et al. 2017).

The *Lactobacillus plantarum* with its produced EPS had shown improved results of antioxidant activity after the process of sulfonation (Zhang et al. 2016). The EPS showing DPPH scavenging activity produced with the culture of *L. gasseri* after isolating from native chicken showed its activity at a absorption of 4 mg/mL concentration with 75.95% results after comparing to ascorbic acid results of 92.94% as standard (Rani et al. 2018).

Glucan obtained from idli at the concentration of 500 μg/mL performed the antioxidant activity with DPPH reagent by 74% and 97.8% by OH radical antioxidant assay. This reveals the promising application of glucan in the food and medical industries (Saravanan et al. 2019). Two EPS from acidic heteropolysaccharides enriched with galactose and mannose produced by *W. cibaria* performed strong anti-oxidant properties (Zhu et al. 2018).

23.9.5 Hypo-cholestrolemic and Antidiabetic Activity

Cholesterol accumulation leads to cardiovascular disorders in the human body (Majeed et al. 2019). The higher level of serum cholesterol level is manageable through drugs but added a side effect to the body. The cholesterol can be reduced by probiotic bacteria with the mentioned mechanisms (a) cholesterol assimilation by growing bacteria, (b) involving the bile salt hydrolase to de-conjugate bile salt (c) Involvement of de-conjugated bile to precipitate the cholesterol (d) involvement of cell for binding the cholesterol.

Limited research is available on the use of EPS to lower body cholesterol level in recent studies. The *Lactobacillus paracasei* based EPS was reported with reducing the cholesterol level from 30 μg at 25 °C for 20 min to 70.79% under *in vitro* conditions with the EPS dose of 0.1% (Bhat and Bajaj 2019). Similarly, the *Lactobacillus plantarum*-based EPS was efficient in lowering the cholesterol to 42.25% (Dilna et al. 2015). The EPS derived from a species called *E. faecium* after isolating from milk based items and could lower the cholesterol level after incorporation was functional food (Bhat and Bajaj 2018). Metabolism of carbohydrate is crucial metabolic process in the body and a disorder in metabolism leads to diabetes mellitus (Type-2). The drugs with antidiabetic functions including the voglibose, miglitol and acarbose cause bowel disruption, and diarrhea, which are not suitable for people suffering with GI disorders (Wang et al. 2018).

The most appropriate remedy with anti-diabetic functions is EPS based on LAB which was identified in the year 2020. The HEPS produced from *L. plantarum* H31 lowered the activity of α-amylase by increasing the gene expression in GLUT-4, AKT-2 and AMPK towards insulin-resistant HepG2 cell line under *in vitro* experimental parameters (Huang et al. 2020). These genes promote by improving the uptake of glucose towards resistant cells of insulin with a pathway called insulin signaling. Similarly, the usual insulin concentration inefficiently transfers the glucose to the cell.

The T2DM occurred is due to insulin resistivity, which is due to the failure in translocation of GLUT-4 vesicles to plasma membrane with inhibiting the glucose consumptions. AMPK in the liver play the master role with switching to regulate the glucose and metabolism of lipids (Meng et al. 2015). EPS H31 under medium dose treating with insulin-resistant cell called HepG2 improves the expressions of GLUT-4, AMPK and AKT-2 indicted with improvement in glucose consumption. Therefore, this revealed with the results that EPS is used to treat hyperglycemia with vigorous in-vitro studies on the reduction of blood glucose levels (Huang et al. 2020).

23.9.6 Anti-biofilm Activity

The exogenous stress induced by environmental conditions on the microorganism, some groups of microorganisms forced them to form the biofilm and remained attached to get protection from the antagonistic activity of the host organism (Merghni et al. 2017). These biofilm-producing groups of pathogenic organisms added antibiotic resistance, and recurrent and chronic infections due to their residing features on the surface of medicines and towards dwelling devices. Therefore, the biofilms added for improving safety of food by building resistance to traditional therapy (Sarikaya et al. 2017). The EPS produced from LAB is used for lowering the concentration of biofilm produced by the pathogenic organism to prevent infectious diseases. The disruption of bio-film or biofilm inhibitions was seen by EPS-BMS which is released by *Leuconostoccitreum* after isolating from meat sausages of meat (Abid et al. 2018).

The supernatant collected under cell-free condition for *Lactobacillus acidophilus* A4 prohibited the creation of biofilms in entero-haemorrhagic *E. coli* O157:H7 strain (Kim and Thomas 2000). The *Lactobacillus fermentum* based EPS after isolating from children's faces inhibited the biofilm production in *Bacillus cereus* (Sarikaya et al. 2017). The culturing with *Weissella confusa* and after isolating dextran from Romanian yogurt performed with anti-biofilm activity to 70% against *Candia albicans* SC5314 strain (Rosca et al. 2018).

23.9.7 Antiviral Activity

In general cases, viral diseases are prevented with chemotherapy, chemopreservation, and vaccination. Apart from the mentioned methods, the probiotics microbes with their metabolic products favor against countering the virus. LAB performed anti-viral activities with the following mechanism interacting with viruses, producing the inhibitory substances of viruses, or stimulating the immune system. The sulfated polysaccharides are highly defensive against the virus in crude form (Liu et al. 2011). Immunobiotics are supplementary diets that function with the combined effect of prebiotics, probiotics, and immunoglobulin with boosting immunity and promoting intestinal health. Immunobiotics protected against the virus by building the innate and adaptive immune system thereafter reducing the duration of infections and viral shedding. EPS origin from LAB communicated the immunobiotics to the host organism with interaction by pattern recognition receptors after expressing in non-immune and immune cells. The response for antiviral immune systems is stimulated using polyinosinic-polycytidylic acid, which is an analog for viral ds RNA and is regarded the basic tool to carryout research in the immunity. EPS originated from *Lactobacillus delbruecki* function with antiviral function in porcine epithelial cell of intestine induced with poly (I:C), adding significant results for expression of MxA, IFN-β, IFN-α, and RNase L (Kanmani et al. 2018).

The EPS obtained from probiotics including *Lactobacillus, Leuconostoc,* and *Pediococcus* functioned with antagonistic function against adenovirus in the human body. The EPS originated from *Lactobacillus plantarum* LRCC5310 functions by reducing the diarrhea durations, reducing the damaged epithelial regions, reducing the replication rate of rotavirus, and reducing the recovery time in suckling mice. The EPS based on *Lactobacillus plantarum* performed with inhibitory functions on the proliferation of gastroenteritis corona virus.

23.9.8 Anti-gastritis, Antiulcer, and Cholesterol-Reducing Activities

The EPS produced from *S. thermophilus* effectively prevented chronic gastritis (Rodríguez et al. 2009). The interaction of EPS-protein responded to gastroprotective functions, and the interaction was affected by the processing conditions. The EPS function of anti-ulcer properties from the producer strains including *bifidobacterial, lactobacilli,* and *streptococci* strains was mentioned (Nagaoka et al. 1994). The EPS function over the protection of gastric epithelium has not been described. The EPS of the fermented milk obtained after culturing with *Lc. lactis* ssp. *cremoris* added the cholesterol-reducing functions whereas the action of mechanism has not been defined yet (Nakajima et al. 1992).

The research was carried out with cholesterol adsorption by polysaccharides and determined under in-vitro conditions with enzymatic reactions following the procedure of polysaccharides precipitation (Soh et al. 2003). They reported with the results that 0.1% of polysaccharides additions after dissolving in distilled H_2O and absorption capacity were measured for alginate, pectins, gellan gum, xanthan gum, and zooglan with values of 2.9, 2.88, 2.5, 2.9, and 2.4 mg/dL, respectively. The zooglan at the 0.2% concentration adsorbed the cholesterol completely whereas dextran was inefficient over it. Tok and Aslim (2010) selected the five strains of *L. delbrueckii* subsp. *bulgaricus*, and isolated from yoghurt, three strains were selected for higher productivity of EPS with removing the cholesterol from selected medium after comparing the low EPS releasing strains. They also revealed that immobilized cells efficiently adsorbed than free cells.

23.9.9 Anti-mutagenic Properties

The attachment of EPS with the strains called *L. plantarum* function with the binding ability with mutagen including heterocyclic amines thereby inactivating the bind mutagen (Tsuda et al. 2008). There is limited research on physiological or immunological effects in animals feeding doses of EPS. The fluctuation of serum components and blood pressure in animal tissues was studied in SHRSP/HOS rats with kefiran doses of 100 and 300 mg/kg of rats (Maeda et al. 2004). With the duration of 30 days, there was a suppression of blood pressure. The capsular polysaccharide promoted for adherence of bacteria towards biological surfaces promoting the colonisations with various ecological niches. Similarly, the EPS was used as an adherent in biofilms (Reddy et al. 1996; Suresh Kumar et al. 2007) with EPS functioning for initial adhesion and permanent adhesion components (Allison and Sutherland 1987).

Additionally, the EPS performed adhesive agent with facilitating the association between plants and bacteria, i.e., the production of levan from sugar cane root with invasions of *G. diazotrophicus* (Hernandez et al. 1995). Homopolysaccharides including glucans and fructan produced from oral *streptococci* added a role in forming dental plaque. They facilitate the bacteria adherence among each other on the tooth surface, controlling diffusions through plaque and serving as energy reserves (Russell 1990). Additionally, the fructan promotes bacterial cell adhesions with the surface of the plant cell (Hernandez et al. 1995). Although biopolymer towards bacterial adhesions with epithelial cells in intestines has not been reported under in-vivo conditions whereas some in-vitro studies suggested the EPS from LAB promoted the adhesions towards epithelial layers in GI (Fernandez de Palencia et al. 2009).

The EPS originated from *Streptococcus mutans* added a significant role in bacterium adhesion in tooth surfaces by forming dental plaque (Klein et al. 2009) by promoting the colonization of bacteria by protecting them from the hostile organism. Several studies mentioned that EPS originated from LAB function with cellular recognitions, bio-film formations, and adhesions Klein et al. 2009). The adhesions of probiotics bacteria including *Lactobacilli* were seen towards to HT29 cells as suggested by Kodaikkal et al. (2012) and noticed by flow cytometry. These strains are familiar with the EPS producing capacity.

23.9.10 Antitumor Properties

The EPS with an antitumor activity was revealed which was produced from *L. helveticus* sp. *jugurti* (Lee et al. 2008). The EPS activity trial with Sarcoma-180 after injecting the prepared EPS intraperitoneally. The EPS treated with mice at the concentration of 20 mg kg^{-1} for 9 days recover the shelf-life by 144%, and the values were greater by 233% at a concentration of 40/80 mg kg^{-1} of concentration. The authors also revealed with the results that EPS functions as an anti-tumor activity with host-mediated actions. The EPS antitumor action was studied with an understanding of the immune system. The surface material of cells mainly lipoteichoic acids, in *Lc. lactis* sp. *cremoris* released the mitogenic activity of T-cells in the lymphocytes of humans (Forsén et al. 1987). The other researchers also mentioned anti-tumour and immunomodulating activities of EPS (Chabot et al. 2001; Kitazawa et al. 1992).

The slime producing strain called *B. adolescentis* added the immunomodifier activity on mouse splenocytes (Gomez et al. 1988). The slime producing strain called *Lc. Lactis* sp. *cremoris* which had an antitumor mechanism, and slime was present with B-cell-reliant mitogenic substances (Kitazawa et al. 1998). In the recent studies, conducted by Wang et al. (2020) the EPS produced from *L. reuteri* inhibited haemagglutination induced by ETEC in porcine erythrocytes representing the EPS application by incorporating probiotics for inhibitions of pathogenicity in ETEC. However, the effect was not seen on dextran produced from *Weissella cibaria* as well as with oligo- and polysaccharides. The applications of EPS like compound called heparin which are low molecular weight compound isolated from *Alteromonas infernus* which exhibited anti-coagulant activity and collected from deep-sea hydrothermal vents (Jouault et al. 2001). The polysaccharides containing Clavan, an L-fructose has been potentially applied in preventing colonization of tumor cell in the lung with control over white blood cell formations during the treatment of rheumatoid arthritis, with synthesizing antigen for producing antibody (Vanhooren and Vandamme 2000).

The EPS released from *L. paracasei subsp. paracasei* and *L. plantarum* function with anti-oxidant activity including scavenging the radicals like 1,1-diphenyl-2-picrylhydrazyl, ferrous ions chelation and peroxidation inhibitions in linoleic acid (Liu et al. 2009). EPS functions with stimulation in cell proliferation and immune-modulating activities in macrophage with induction of cytokine productivity including, TNF-α, IL-6, and IL-1β at the concentration range of 5–500 μg mL^{-1}. Additionally, studies by Martin et al. (1984) revealed that bacterial EPS in conjugation with or single substance is considered as potent somnogen with inducing sleepiness without adding side effect and eliminating the xenobiotics for inducing the sleepiness.

23.10 Prospects for Bacterial EPS

The EPS applications under commercial conditions are limited due to their higher production cost. Some of the species like *streptococci* produced EPS at lower amounts whereas other strains like *Lactobacilli* reported a production capacity of 1–10 g of EPS/L of growing media (Ahuja et al. 2023). The *Weissella* sp. reported the highest production with 18 g/L of EPS under maximized conditions (Adebayo-Tayo et al. 2018). Therefore, such strains are an important source to explore for solving the problems. Similarly, the *Pediococc*i groups of organisms produced in milligrams per liter of EPS under favorable conditions as mentioned by Kumar et al. (2020).

Even though the EPS application extended to industry, health, and environment the application in the field of flocculation significant role in health promotions and eco-friendly nature added it application wastewater and municipalities. The flocculation for the wastewater treatment is carried out using inorganic salts including $AI_2(SO_4)_3$ and poly-$AICL_3$ and under organic groups, the synthetic derivatives include polyacrylamide and polyethyleneimine. These flocculants function by adding adverse effects on human health causing neurotoxicity, carcinogenicity, and Alzheimer's disease (Rudén 2004).

Therefore, the EPS from microbes is considered a potential safe source for a flocculating agent. Many researchers have claimed biopolymers regarding flocculating efficiency produced from *Bacillus* sp., and *Artrobacter* sp. (Cosa et al. 2011; Mabinya et al. 2012). These results concluded that bacterial EPS is an efficient flocculant to be used on an industrial scale for wastewater and water treatment plants. More studies must be performed on the feasibility of EPS production by LAB under a sustainable approach. EPS also reduced the oxygen tension by involving up taking of metal ions (Patel et al. 2012a, b, c). The EPS were produced from *Pseudoalteromonas* sp. functions by improving the cold-adopted protease stability after secreting from the same strains with the prevention of autolysis. Similarly, the EPS can bind with metal ions, such as: Fe^{2+}, Cu^{2+}, Fe^{2+}, Zn^{2+}, and Co^{2+} (Prasanna et al. 2012).

23.11 Conclusion

Bacterial EPS are available with wider diversity. Besides, considering the natural protective agent produced by LAB, it also functions to protect the cell from technological stress induced during the processing of fermented milk-based products. Under recent conditions, the applications of EPS for human use have been diversified including medical, cosmetics, pharmaceutical, dairy products, industrial and environmental sectors. However, the limitation of use occurred due to its higher production cost at the industrial level. The finding of strains with higher productivity of EPS is an ongoing process to date, whereas the estimation of optimum conditions for fermentations, applications of biological tools for metabolic and genetic engineering, and selection of cheap substrate for industrial production is another alternative approach to improve the commercial EPS production.

References

Abid Y, Casillo A, Gharsallah H, Joulak I, Lanzetta R, Corsaro MM, Attia H, Azabou S (2018) Production and structural characterization of exopolysaccharides from newly isolated probiotic lactic acid bacteria. Int J Biol Macromol 108:719–728

Adebayo-Tayo B, Ishola R, Oyewunmi T (2018) Characterization, antioxidant and immunomodulatory potential on exopolysaccharide produced by wild type and mutant Weissella confusa strains. Biotechnol Rep 19:e00271

Ahuja V, Bhatt AK, Banu J. R, Kumar V, Kumar G, Yang YH, Bhatia SK (2023) Microbial Exopolysaccharide Composites in Biomedicine and Healthcare: Trends and Advances. Polym 15(7):1801

Ale EC, Bourin MJ-B, Peralta GH, Burns PG, Ávila OB, Contini L, Reinheimer J, Binetti AG (2019) Functional properties of exopolysaccharide (EPS) extract from Lactobacillus fermentum Lf2 and its impact when combined with Bifidobacterium animalis INL1 in yoghurt. Int Dairy J 96:114–125

Allison DG, Sutherland IW (1987) The role of exopolysaccharides in adhesion of freshwater bacteria. Microbiology 133:1319–1327

Amatayakul T, Halmos A, Sherkat F, Shah N (2006) Physical characteristics of yoghurts made using exopolysaccharide-producing starter cultures and varying casein to whey protein ratios. Int Dairy J 16:40–51

Ayyash M, Abu-Jdayil B, Itsaranuwat P, Galiwango E, Tamiello-Rosa C, Abdullah H, Esposito G, Hunashal Y, Obaid RS, Hamed F (2020) Characterization, bioactivities, and rheological properties of exopolysaccharide produced by novel probiotic Lactobacillus plantarum C70 isolated from camel milk. Int J Biol Macromol 144:938–946

Azad M, Kalam A, Sarker M, Wan D (2018) Immunomodulatory effects of probiotics on cytokine profiles. Biomed Res Int 2018:8063647

Badel S, Bernardi T, Michaud P (2011) New perspectives for Lactobacilli exopolysaccharides. Biotechnol Adv 29:54–66

Becker A, Katzen F, Pühler A, Ielpi L (1998) Xanthan gum biosynthesis and application: a biochemical/genetic perspective. Appl Microbiol Biotechnol 50:145–152

Behare P, Singh R, Singh RP (2009) Exopolysaccharide-producing mesophilic lactic cultures for preparation of fat-free Dahi—an Indian fermented milk. J Dairy Res 76:90–97

Bhat B, Bajaj BK (2018) Hypocholesterolemic and bioactive potential of exopolysaccharide from a probiotic Enterococcus faecium K1 isolated from kalarei. Bioresour Technol 254:264–267

Bhat B, Bajaj BK (2019) Hypocholesterolemic potential and bioactivity spectrum of an exopolysaccharide from a probiotic isolate Lactobacillus paracasei M7. Bioactive Carbohydr Diet Fibre 19:100191

Bhatia M, Moochhala S (2004) Role of inflammatory mediators in the pathophysiology of acute respiratory distress syndrome. J Pathol 202:145–156

Brandt M, Roth K, Hammes W (2003) Effect of an exopolysaccharide produced by Lactobacillus sanfranciscensis LTH1729 on dough and bread quality. In: Sourdough from fundamentals to applications. Vrije Universiteit Brussel, Brussel, p 80

Broadbent JR, McMahon DJ, Welker DL, Oberg CJ, Moineau S (2003) Biochemistry, genetics, and applications of exopolysaccharide production in Streptococcus thermophilus: a review. J Dairy Sci 86:407–423

Canani RB, Di Costanzo M, Leone L, Pedata M, Meli R, Calignano A (2011) Potential beneficial effects of butyrate in intestinal and extraintestinal diseases. World J Gastroenterol 17:1519

Castro-Bravo N, Margolles A, Wells JM, Ruas-Madiedo P (2019) Exopolysaccharides synthesized by Bifidobacterium animalis subsp. lactis interact with TLR4 in intestinal epithelial cells. Anaerobe 56:98–101

Cerning J, Marshall V (1999) Exopolysaccharides produced by the dairy lactic acid bacteria. Rec Res Dev 3:195–209

Chabot S, Yu H-L, De Léséleuc L, Cloutier D, Van Calsteren M-R, Lessard M, Roy D, Lacroix M, Oth D (2001) Exopolysaccharides from Lactobacillus rhamnosus RW-9595M stimulate TNF, IL-6 and IL-12 in human and mouse cultured immunocompetent cells, and IFN-γ in mouse splenocytes. Le Lait 81:683–697

Chen YC, Wu YJ, Hu CY (2019) Monosaccharide composition influence and immunomodulatory effects of probiotic exopolysaccharides. Int J Biol Macromol 133:575–582

Ciszek-Lenda M, Nowak B, Śróttek M, Gamian A, Marcinkiewicz J (2011) Immunoregulatory potential of exopolysaccharide from Lactobacillus rhamnosus KL37. Effects on the production of inflammatory mediators by mouse macrophages. Int J Exp Pathol 92:382–391

Cosa S, Mabinya LV, Olaniran AO, Okoh OO, Bernard K, Deyzel S, Okoh AI (2011) Bioflocculant production by Virgibacillus sp. Rob isolated from the bottom sediment of Algoa Bay in the Eastern Cape, South Africa. Molecules 16:2431–2442

Crescenzi V (1995) Microbial polysaccharides of applied interest: ongoing research activities in Europe. Biotechnol Prog 11:251–259

Cuthbertson L, Mainprize IL, Naismith JH, Whitfield C (2009) Pivotal roles of the outer membrane polysaccharide export and polysaccharide copolymerase protein families in export of extracellular polysaccharides in gram-negative bacteria. Microbiol Mol Biol Rev 73:155–177

Dabour N, Kheadr E, Fliss I, LaPointe G (2005) Impact of ropy and capsular exopolysaccharide-producing strains of Lactococcus lactis subsp. cremoris on reduced-fat Cheddar cheese production and whey composition. Int Dairy J 15:459–471

Dal Bello F, Walter J, Hertel C, Hammes WP (2001) In vitro study of prebiotic properties of levan-type exopolysaccharides from lactobacilli and non-digestible carbohydrates using denaturing gradient gel electrophoresis. Syst Appl Microbiol 24:232–237

De Vuyst L, De Vin F, Vaningelgem F, Degeest B (2001) Recent developments in the biosynthesis and applications of heteropolysaccharides from lactic acid bacteria. Int Dairy J 11:687–707

De Vuyst L, Degeest B (1999) Heteropolysaccharides from lactic acid bacteria. FEMS Microbiol Rev 23:153–177

Degeest B, Vaningelgem F, De Vuyst L (2001) Microbial physiology, fermentation kinetics, and process engineering of heteropolysaccharide production by lactic acid bacteria. Int Dairy J 11: 747–757

Di Cagno R, De Angelis M, Limitone A, Minervini F, Carnevali P, Corsetti A, Gaenzle M, Ciati R, Gobbetti M (2006) Glucan and fructan production by sourdough Weissella cibaria and Lactobacillus plantarum. J Agric Food Chem 54:9873–9881

Dilna SV, Surya H, Aswathy RG, Varsha KK, Sakthikumar DN, Pandey A, Nampoothiri KM (2015) Characterization of an exopolysaccharide with potential health-benefit properties from a probiotic Lactobacillus plantarum RJF4. LWT-Food Sci Technol 64:1179–1186

Dinić M, Pecikoza U, Djokić J, Stepanović-Petrović R, Milenković M, Stevanović M, Filipović N, Begović J, Golić N, Lukić J (2018) Exopolysaccharide produced by probiotic strain Lactobacillus paraplantarum BGCG11 reduces inflammatory hyperalgesia in rats. Front Pharmacol 9:1

Duboc P, Mollet B (2001) Applications of exopolysaccharides in the dairy industry. Int Dairy J 11: 759–768

Faber EJ, Kamerling JP, Vliegenthart JF (2001) Structure of the extracellular polysaccharide produced by Lactobacillus delbrueckii subsp. bulgaricus 291. Carbohydr Res 331:183–194

Feng F, Zhou Q, Yang Y, Zhao F, Du R, Han Y, Xiao H, Zhou Z (2018) Characterization of highly branched dextran produced by Leuconostoc citreum B-2 from pineapple fermented product. Int J Biol Macromol 113:45–50

Fernandez de Palencia P, Werning ML, Sierra-Filardi E, Dueñas MT, Irastorza A, Corbí AL, López P (2009) Probiotic properties of the 2-substituted (1,3)-β-D-glucan-producing bacterium Pediococcus parvulus 2.6. Appl Environ Microbiol 75:4887–4891

Florencia FS (2013) Rheology of spreadable goat cheese made with autochthonous lactic cultures differing in their ability to produce exopolysaccharides. Food Sci Technol 33:233–238

Folkenberg DM, Dejmek P, Skriver A, Guldager HS, Ipsen R (2006) Sensory and rheological screening of exopolysaccharide producing strains of bacterial yoghurt cultures. Int Dairy J 16: 111–118

Forsén R, Heiska E, Herva E, Arvilommi H (1987) Immunobiological effects of Streptococcus cremoris from cultured milk 'viili'; application of human lymphocyte culture techniques. Int J Food Microbiol 5:41–47

Freitas F, Alves VD, Reis MA (2011) Advances in bacterial exopolysaccharides: from production to biotechnological applications. Trends Biotechnol 29:388–398

Fukuda S, Toh H, Hase K, Oshima K, Nakanishi Y, Yoshimura K, Tobe T, Clarke JM, Topping DL, Suzuki T (2011) Bifidobacteria can protect from enteropathogenic infection through production of acetate. Nature 469:543–547

Galle S, Schwab C, Arendt E, Gänzle M (2010) Exopolysaccharide-forming Weissella strains as starter cultures for sorghum and wheat sourdoughs. J Agric Food Chem 58:5834–5841

Gomez E, Melgar M, Silva GP, Portoles A, Gil I (1988) Exocellular products from Bifidobacterium adolescentis as immunomodifiers in the lymphoproliferative responses of mouse splenocytes. FEMS Microbiol Lett 56:47–52

Górska S, Sandström C, Wojas-Turek J, Rossowska J, Pajtasz-Piasecka E, Brzozowska E, Gamian A (2016) Structural and immunomodulatory differences among lactobacilli exopolysaccharides isolated from intestines of mice with experimentally induced inflammatory bowel disease. Sci Rep 6:1–16

Gupta P, Diwan B (2017) Bacterial exopolysaccharide mediated heavy metal removal: a review on biosynthesis, mechanism and remediation strategies. Biotechnol Rep 13:58–71

Guven M, Yasar K, Karaca O, Hayaloglu A (2005) The effect of inulin as a fat replacer on the quality of set-type low-fat yogurt manufacture. Int J Dairy Technol 58:180–184

Hassan A, Frank J, Farmer M, Schmidt K, Shalabi S (1995) Observation of encapsulated lactic acid bacteria using confocal scanning laser microscopy. J Dairy Sci 78:2624–2628

Hassan A, Ipsen R, Janzen T, Qvist K (2003) Microstructure and rheology of yogurt made with cultures differing only in their ability to produce exopolysaccharides. J Dairy Sci 86:1632–1638

Hernandez L, Arrieta J, Menendez C, Vazquez R, Coego A, Suarez V, Selman G, Petit-Glatron M, Chambert R (1995) Isolation and enzymic properties of levansucrase secreted by Acetobacter diazotrophicus SRT4, a bacterium associated with sugar cane. Biochem J 309:113–118

Homayouni A, Payahoo L, Azizi A (2012) Effects of probiotics on lipid profile: a review. Am J Food Technol 7:251–265

Hongpattarakere T, Cherntong N, Wichienchot S, Kolida S, Rastall RA (2012) In vitro prebiotic evaluation of exopolysaccharides produced by marine isolated lactic acid bacteria. Carbohydr Polym 87:846–852

Huang Z, Lin F, Zhu X, Zhang C, Jiang M, Lu Z (2020) An exopolysaccharide from Lactobacillus plantarum H31 in pickled cabbage inhibits pancreas α-amylase and regulates metabolic markers in HepG2 cells by the AMPK/PI3K/Akt pathway. Int J Biol Macromol 143:775–784

Imeson A (2011) Food stabilisers, thickeners and gelling agents. Wiley, New York, p 368

Islam ST, Lam JS (2014) Synthesis of bacterial polysaccharides via the Wzx/Wzy-dependent pathway. Can J Microbiol 60:697–716

Jolly L, Vincent SJ, Duboc P, Neeser J-R (2002) Exploiting exopolysaccharides from lactic acid bacteria. In: Lactic acid bacteria: genetics, metabolism and applications. Oxford University Press, Oxford, pp 367–374

Jouault SC, Chevolot L, Helley D, Ratiskol J, Bros A, Sinquin C, Roger O, Fischer A-M (2001) Characterization, chemical modifications and in vitro anticoagulant properties of an exopolysaccharide produced by Alteromonas infernus. Biochim Biophys Acta 1528:141–151

Kanmani P, Albarracin L, Kobayashi H, Iida H, Komatsu R, Kober AH, Ikeda-Ohtsubo W, Suda Y, Aso H, Makino S (2018) Exopolysaccharides from Lactobacillus delbrueckii OLL1073R-1 modulate innate antiviral immune response in porcine intestinal epithelial cells. Mol Immunol 93:253–265

Kim D-S, Thomas S, Fogler HS (2000) Effects of pH and trace minerals on long-term starvation of Leuconostoc mesenteroides. Appl Environ Microbiol 66:976–981

Kim Y, Kim SH (2009) Released exopolysaccharide (r-EPS) produced from probiotic bacteria reduce biofilm formation of enterohemorrhagic Escherichia coli O157: H7. Biochem Biophys Res Commun 379:324–329

Kitazawa H, Yamaguchi T, Itoh T (1992) B-cell mitogenic activity of slime products produced from slime-forming, encapsulated Lactococcus lactis ssp. cremoris. J Dairy Sci 75:2946–2951

Kitazawa H, Harata T, Uemura J, Saito T, Kaneko T, Itoh T (1998) Phosphate group requirement for mitogenic activation of lymphocytes by an extracellular phosphopolysaccharide from Lactobacillus delbrueckii ssp. bulgaricus. Int J Food Microbiol 40:169–175

Klein M, Duarte S, Xiao J, Mitra S, Foster T, Koo H (2009) Structural and molecular basis of the role of starch and sucrose in Streptococcus mutans biofilm development. Appl Environ Microbiol 75:837–841

Kodaikkal V, Prajapati J, Ljungh A (2012) Evaluation of adhesion of Lactobacillus strains to HT-29 cells by a flow cytometric assay. Int J Appl Anim Sci 1:1–7

Kogani G, Pajtinka M, Babincova M, Miadokova E, Rauko P, Slamenova D, Korolenko T (2008) Yeast cell wall polysaccharides as antioxidants and antimutagens: can they fight cancer? Minireview. Neoplasma 55:387

Kook S-Y, Lee Y, Jeong E-C, Kim S (2019) Immunomodulatory effects of exopolysaccharides produced by Bacillus licheniformis and Leuconostoc mesenteroides isolated from Korean kimchi. J Funct Foods 54:211–219

Korakli M, Rossmann A, Gänzle MG, Vogel RF (2001) Sucrose metabolism and exopolysaccharide production in wheat and rye sourdoughs by Lactobacillus sanfranciscensis. J Agric Food Chem 49:5194–5200

Korakli M, Gänzle MG, Vogel RF (2002) Metabolism by bifidobacteria and lactic acid bacteria of polysaccharides from wheat and rye, and exopolysaccharides produced by Lactobacillus sanfranciscensis. J Appl Microbiol 92:958–965

Kulicke WM, Heinze T (2005) Improvements in polysaccharides for use as blood plasma expanders. Macromol Symp 231:49–59

Kumar R, Bansal P, Singh J, Dhanda S (2020) Purification, partial structural characterization and health benefits of exopolysaccharides from potential probiotic Pediococcus acidilactici NCDC 252. Process Biochem 99:79–86.

Laws AP, Marshall VM (2001) The relevance of exopolysaccharides to the rheological properties in milk fermented with ropy strains of lactic acid bacteria. Int Dairy J 11:709–721

Lebeer S, Claes IJ, Verhoeven TL, Vanderleyden J, De Keersmaecker SC (2011) Exopolysaccharides of Lactobacillus rhamnosus GG form a protective shield against innate immune factors in the intestine. Microb Biotechnol 4:368–374

Lee DK, Jang S, Kim MJ, Kim JH, Chung MJ, Kim KJ, Ha NJ (2008) Anti-proliferative effects of Bifidobacterium adolescentis SPM0212 extract on human colon cancer cell lines. BMC Cancer 8:1–8

Li J, Wang N (2012) The gpsX gene encoding a glycosyltransferase is important for polysaccharide production and required for full virulence in Xanthomonas citri subsp. citri. BMC Microbiol 12: 1–16

Li S, Huang R, Shah NP, Tao X, Xiong Y, Wei H (2014) Antioxidant and antibacterial activities of exopolysaccharides from Bifidobacterium bifidum WBIN03 and Lactobacillus plantarum R315. J Dairy Sci 97:7334–7343

Lin T, Chien M-FC (2007) Exopolysaccharides production as affected by lactic acid bacteria and fermentation time. Food Chem 100:1419–1423

Liu C-F, Pan T-M (2010) In vitro effects of lactic acid bacteria on cancer cell viability and antioxidant activity. J Food Drug Anal 18:8

Liu J, Luo J, Ye H, Sun Y, Lu Z, Zeng X (2009) Production, characterization and antioxidant activities in vitro of exopolysaccharides from endophytic bacterium Paenibacillus polymyxa EJS-3. Carbohydr Polym 78:275–281

Liu C-F, Tseng K-C, Chiang S-S, Lee BH, Hsu WH, Pan TM (2011) Immunomodulatory and antioxidant potential of Lactobacillus exopolysaccharides. J Sci Food Agric 91:2284–2291

Lobo RE, Gómez MI, de Valdez GF, Torino MI (2019) Physicochemical and antioxidant properties of a gastroprotective exopolysaccharide produced by Streptococcus thermophilus CRL1190. Food Hydrocoll 96:625–633

Lopez P, Monteserin DC, Gueimonde M, Clara G, Margolles A, Suarez A, Ruas-Madiedo P (2012) Exopolysaccharide-producing Bifidobacterium strains elicit different in vitro responses upon interaction with human cells. Food Res Int 46:99–107

Mabinya LV, Cosa S, Nwodo U, Okoh AI (2012) Studies on bioflocculant production by Arthrobacter sp. Raats, a freshwater bacterium isolated from Tyume River, South Africa. Int J Mol Sci 13:1054–1065

Madhuri K, Prabhakar KV (2014) Microbial exopolysaccharides: biosynthesis and potential applications. Oriental J Chem 30:1401

Maeda H, Zhu X, Suzuki S, Suzuki K, Kitamura S (2004) Structural characterization and biological activities of an exopolysaccharide kefiran produced by Lactobacillus kefiranofaciens WT-2BT. J Agric Food Chem 52:5533–5538

Majeed M, Majeed S, Nagabhushanam K, Arumugam S, Beede K, Ali F (2019) Evaluation of the in vitro cholesterol-lowering activity of the probiotic strain Bacillus coagulans MTCC 5856. Int J Food Sci Technol 54:212–220

Martin SA, Karnovsky M, Krueger J, Pappenheimer J, Biemann K (1984) Peptidoglycans as promoters of slow-wave sleep. I. Structure of the sleep-promoting factor isolated from human urine. J Biol Chem 259:12652–12658

Meng F, Ning H, Sun Z, Huang F, Li Y, Chu X, Lu H, Sun C, Li S (2015) Ursolic acid protects hepatocytes against lipotoxicity through activating autophagy via an AMPK pathway. J Funct Foods 17:172–182

Merghni A, Dallel I, Noumi E, Kadmi Y, Hentati H, Tobji S, Amor AB, Mastouri M (2017) Antioxidant and antiproliferative potential of biosurfactants isolated from Lactobacillus casei and their anti-biofilm effect in oral Staphylococcus aureus strains. Microb Pathog 104:84–89

Nagaoka M, Hashimoto S, Watanabe T, Yokokura T, Mori Y (1994) Anti-ulcer effects of lactic acid bacteria and their cell wall polysaccharides. Biol Pharm Bull 17:1012–1017

Nair B, Prajapati J (2003) History of fermented foods. CRC Press, Boca Raton, pp 17–42

Nakajima H, Hirota T, Toba T, Itoh T, Adachi S (1992) Structure of the extracellular polysaccharide from slime-forming Lactococcus lactis subsp. cremoris SBT 0495. Carbohydr Res 224:245–253

Nath PC, Tiwari ON, Devi I, Bandyopadhyay TK, Bhunia B (2021) Biochemical and morphological fingerprints of isolated Anabaena sp.: a precious feedstock for food additives. Biomass Convers Biorefin 11:2723–2733

Nguyen P-T, Nguyen T-T, Bui D-C, Hong P-T, Hoang Q-K, Nguyen H-T (2020) Exopolysaccharide production by lactic acid bacteria: the manipulation of environmental stresses for industrial applications. AIMS Microbiol 6:451

Nisha P, Ananthanarayan L, Singhal RS (2005) Effect of stabilizers on stabilization of idli (traditional south Indian food) batter during storage. Food Hydrocoll 19:179–186

Notararigo S, de Las Casas-Engel M, de Palencia PF, Corbí AL, López P (2014) Immunomodulation of human macrophages and myeloid cells by 2-substituted (1–3)-β-d-glucan from P. parvulus 2.6. Carbohydr Polym 112:109–113

Obert G (1984) Manufacturing of technology of cream turo with extended shelf life & isolation of a special strain of lactic acid bacteria which improves product consistency. Tejipar 33:47–48

Oliver AE, Leprince O, Wolkers WF, Hincha DK, Heyer AG, Crowe JH (2001) Non-disaccharide-based mechanisms of protection during drying. Cryobiology 43:151–167

Panosyan H, Di Donato P, Poli A, Nicolaus B (2018) Production and characterization of exopolysaccharides by Geobacillus thermodenitrificans ArzA-6 and Geobacillus toebii ArzA-8 strains isolated from an Armenian geothermal spring. Extremophiles 22:725–737

Patel A, Lindström C, Patel A, Prajapati J, Holst O (2012a) Probiotic properties of exopolysaccharide producing lactic acid bacteria isolated from vegetables and traditional Indian fermented foods. Int J Fermented Foods 1:87–102

Patel A, Lindström C, Patel A, Prajapati J, Holst O (2012b) Screening and isolation of exopolysaccharide producing lactic acid bacteria from vegetables and indigenous fermented foods of Gujarat, India. Int J Fermented Foods 1:77–86

Patel S, Majumder A, Goyal A (2012c) Potentials of exopolysaccharides from lactic acid bacteria. Indian J Microbiol 52:3–12

Pattanaik A, Samal D, Sukla LB, Pradhan D (2020) Advancements and use of OMIC technologies in the field of bioleaching: a review. Biointerface Res Appl Chem 11:10185–10204

Patten D, Laws AP (2015) Lactobacillus-produced exopolysaccharides and their potential health benefits: a review. Benef Microbes 6:457–471

Prasanna P, Bell A, Grandison AS, Charalampopoulos D (2012) Emulsifying, rheological and physicochemical properties of exopolysaccharide produced by Bifidobacterium longum subsp. infantis CCUG 52486 and Bifidobacterium infantis NCIMB 702205. Carbohydr Polym 90:533–540

Praveen K (2000) Physico-chemical and microstructural properties of dahi using EPS producing strains. Doctoral dissertation, NDRI, India

Qin G, Zhu L, Chen X, Wang PG, Zhang Y (2007) Structural characterization and ecological roles of a novel exopolysaccharide from the deep-sea psychrotolerant bacterium Pseudoalteromonas sp. SM9913. Microbiology 153:1566–1572

Raji HA, Moayed RZ, Nozari M (2014) The viscosity of xanthan gum grout with different pH and ionic strength. Int J Geotech Geol Eng 15:38–43

Rani RP, Anandharaj M, Ravindran AD (2018) Characterization of a novel exopolysaccharide produced by Lactobacillus gasseri FR4 and demonstration of its in vitro biological properties. Int J Biol Macromol 109:772–783

Reddy K, Soper B, Tang J, Bradley R (1996) Phenotypic variation in exopolysaccharide production in the marine, aerobic nitrogen-fixing unicellular cyanobacterium Cyanothece sp. World J Microbiol Biotechnol 12:311–318

Rehm BH (2010) Bacterial polymers: biosynthesis, modifications and applications. Nat Rev Microbiol 8:578–592

Rodríguez C, Medici M, Rodriguez AV, Mozzi F, de Valdez GF (2009) Prevention of chronic gastritis by fermented milks made with exopolysaccharide-producing Streptococcus thermophilus strains. J Dairy Sci 92:2423–2434

Rosca I, Petrovici AR, Peptanariu D, Nicolescu A, Dodi G, Avadanei M, Ivanov IC, Bostanaru AC, Mares M, Ciolacu D (2018) Biosynthesis of dextran by Weissella confusa and its in vitro functional characteristics. Int J Biol Macromol 107:1765–1772

Ruas-Madiedo P, Gueimonde M, Margolles A, de los Reyes-Gavilán CG, Salminen S (2006) Exopolysaccharides produced by probiotic strains modify the adhesion of probiotics and enteropathogens to human intestinal mucus. J Food Prot 69:2011–2015

Rudén C (2004) Acrylamide and cancer risk—expert risk assessments and the public debate. Food Chem Toxicol 42:335–349

Russell R (1990) Molecular genetics of glucan metabolism in oral streptococci. Arch Oral Biol 35: S53–S58

Salazar N, Gueimonde M, Hernández-Barranco AM, Ruas-Madiedo P, de los Reyes-Gavilán CG (2008) Exopolysaccharides produced by intestinal Bifidobacterium strains act as fermentable substrates for human intestinal bacteria. Appl Environ Microbiol 74:4737–4745

Saravanan C, Kavitake D, Kandasamy S, Devi PB, Shetty PH (2019) Production, partial characterization and antioxidant properties of exopolysaccharide α-d-glucan produced by Leuconostoc lactis KC117496 isolated from an idli batter. J Food Sci Technol 56:159–166

Sarikaya H, Aslim B, Yuksekdag Z (2017) Assessment of anti-biofilm activity and bifidogenic growth stimulator (BGS) effect of lyophilized exopolysaccharides (l-EPSs) from Lactobacilli strains. Int J Food Prop 20:362–371

Schmid J, Sieber V, Rehm B (2015) Bacterial exopolysaccharides: biosynthesis pathways and engineering strategies. Front Microbiol 6:496

Seo BJ, Bajpai VK, Rather IA, Park YH (2015) Partially purified exopolysaccharide from Lactobacillus plantarum YML009 with total phenolic content, antioxidant and free radical scavenging efficacy. Indian J Pharm Educ Res 49:282–292

Serbina NV, Jia T, Hohl TM, Pamer EG (2008) Monocyte-mediated defense against microbial pathogens. Ann Rev Immunol 26:421

Shah N, Prajapati J (2014) Effect of carbon dioxide on sensory attributes, physico-chemical parameters and viability of probiotic L. helveticus MTCC 5463 in fermented milk. J Food Sci Technol 51:3886–3893

Shao L, Wu Z, Zhang H, Chen W, Ai L, Guo B (2014) Partial characterization and immunostimulatory activity of exopolysaccharides from Lactobacillus rhamnosus KF5. Carbohydr Polym 107:51–56

Shori AB (2016) Influence of food matrix on the viability of probiotic bacteria: a review based on dairy and non-dairy beverages. Food Biosci 13:1–8

Singh P, Saini P (2017) Food and health potentials of exopolysaccharides derived from Lactobacilli. Microbiol Res J Int 22:1–14

Smiderle FR, Baggio CH, Borato DG, Santana-Filho AP, Sassaki GL, Iacomini M, Van Griensven LJ (2014) Anti-inflammatory properties of the medicinal mushroom Cordyceps militaris might be related to its linear (1→3)-β-D-glucan. PLoS One 9:e110266

Soh H-S, Kim C-S, Lee S-P (2003) A new in vitro assay of cholesterol adsorption by food and microbial polysaccharides. J Med Food 6:225–230

Sørensen HM, Rochfort KD, Maye S, MacLeod G, Brabazon D, Loscher C, Freeland B (2022) Exopolysaccharides of lactic acid bacteria: production, purification and health benefits towards functional food. Nutrients 14:2938

Stack HM, Kearney N, Stanton C, Fitzgerald GF, Ross RP (2010) Association of beta-glucan endogenous production with increased stress tolerance of intestinal lactobacilli. Appl Environ Microbiol 76:500–507

Sungur T, Aslim B, Karaaslan C, Aktas B (2017) Impact of exopolysaccharides (EPSs) of Lactobacillus gasseri strains isolated from human vagina on cervical tumor cells (HeLa). Anaerobe 47:137–144

Suresh Kumar A, Mody K, Jha B (2007) Bacterial exopolysaccharides—a perception. J Basic Microbiol 47:103–117

Sutherland IW (1977) Surface carbohydrates of the prokaryotic cell. Academic, New York, p 472

Sutherland IW (1990) Biotechnology of microbial exopolysaccharides. Cambridge University Press, Cambridge, p 163

Sutherland IW (1998) Novel and established applications of microbial polysaccharides. Trends Biotechnol 16:41–46

Tamime A, Saarela M, Sondergaard AK, Mistry V, Shah N (2005) Production and maintenance of viability of probiotic microorganisms in dairy products. In: Probiotic dairy products. Wiley, New York, pp 39–63

Tang W, Dong M, Wang W, Han S, Rui X, Chen X, Jiang M, Zhang Q, Wu J, Li W (2017) Structural characterization and antioxidant property of released exopolysaccharides from Lactobacillus delbrueckii ssp. bulgaricus SRFM-1. Carbohydr Polym 173:654–664

Thakur S, Prasad M, Rastogi N (1995) Effect of xanthan on textural properties of idli (traditional south Indian food). Food Hydrocoll 9:141–145

Tok E, Aslim B (2010) Cholesterol removal by some lactic acid bacteria that can be used as probiotic. Microbiol Immunol 54:257–264

Toshchakov V, Jones BW, Perera P-Y, Thomas K, Cody MJ, Zhang S, Williams BR, Major J, Hamilton TA, Fenton MJ (2002) TLR4, but not TLR2, mediates IFN-β-induced STAT1α/β-dependent gene expression in macrophages. Nat Immunol 3:392–398

Tsuda H, Hara K, Miyamoto T (2008) Binding of mutagens to exopolysaccharide produced by Lactobacillus plantarum mutant strain 301102S. J Dairy Sci 91:2960–2966

Ua-Arak T, Jakob F, Vogel R (2017) Fermentation pH modulates the size distributions and functional properties of Gluconobacter albidus TMW 2.1191 levan. Front Microbiol 8:807

Vanhooren P, Vandamme E (2000) Microbial production of clavan, an L-fucose rich exopolysaccharide. Prog Biotechnol 17:109–114

Wacher-Rodarte C, Galvan MV, Farres A, Gallardo F, Marshall VM, Garcia-Garibay M (1993) Yogurt production from reconstituted skim milk powders using different polymer and non-polymer forming starter cultures. J Dairy Res 60:247–254

Wang Y, Gänzle MG, Schwab C (2010) Exopolysaccharide synthesized by Lactobacillus reuteri decreases the ability of enterotoxigenic Escherichia coli to bind to porcine erythrocytes. Appl Environ Microbiol 76:4863–4866

Wang P-C, Zhao S, Yang B-Y, Wang Q-H, Kuang H-X (2016) Anti-diabetic polysaccharides from natural sources: a review. Carbohydr Polym 148:86–97

Wang J, Wu T, Fang X, Min W, Yang Z (2018) Characterization and immunomodulatory activity of an exopolysaccharide produced by Lactobacillus plantarum JLK0142 isolated from fermented dairy tofu. Int J Biol Macromol 115:985–993

Wang J, Fang X, Wu T, Fang L, Liu C, Min W (2020) In vitro immunomodulatory effects of acidic exopolysaccharide produced by Lactobacillus planetarium JLAU103 on RAW264.7 macrophages. Int J Biol Macromol 156:1308–1315

Welman AD, Maddox IS (2003) Exopolysaccharides from lactic acid bacteria: perspectives and challenges. Trends Biotechnol 21:269–274

Werning ML, Notararigo S, Nácher M, Fernández de Palencia P, Aznar R, López P (2012) Biosynthesis, purification and biotechnological use of exopolysaccharides produced by lactic acid bacteria. In: Food additives. IntechOpen, London, pp 83–114

Whistler R, Daniel J (1990) Functions of polysaccharides in foods. In: Food additives. IntechOpen, London, pp 395–424

Whitney J, Howell P (2013) Synthase-dependent exopolysaccharide secretion in Gram-negative bacteria. Trends Microbiol 21:63–72

Wu M-H, Pan T-M, Wu Y-J, Chang S-J, Chang M-S, Hu C-Y (2010) Exopolysaccharide activities from probiotic bifidobacterium: immunomodulatory effects (on J774A.1 macrophages) and antimicrobial properties. Int J Food Microbiol 144:104–110

Xiu L, Zhang H, Hu Z, Liang Y, Guo S, Yang M, Du R, Wang X (2018) Immunostimulatory activity of exopolysaccharides from probiotic Lactobacillus casei WXD030 strain as a novel adjuvant in vitro and in vivo. Food Agric Immunol 29:1086–1105

Xu Q, Yajima T, Li W, Saito K, Ohshima Y, Yoshikai Y (2006) Levan (β-2,6-fructan), a major fraction of fermented soybean mucilage, displays immunostimulating properties via Toll-like receptor 4 signalling: induction of interleukin-12 production and suppression of T-helper type 2 response and immunoglobulin E production. Clin Exp Allergy 36:94–101

Yamamoto Y, Takahashi Y, Kawano M, Iizuka M, Matsumoto T, Saeki S, Yamaguchi H (1999) In vitro digestibility and fermentability of levan and its hypocholesterolemic effects in rats. J Nutr Biochem 10:13–18

Yang Z, Staaf M, Widmalm G, Tenhu H (1999) Separation, purification and characterisation of extracellular polysaccharides produced by slime-forming Lactococcus lactis ssp. cremoris strains. Int Dairy J 9:631–638

Yang Z, Li S, Zhang X, Zeng X, Li D, Zhao Y, Zhang J (2010) Capsular and slime-polysaccharide production by Lactobacillus rhamnosus JAAS8 isolated from Chinese sauerkraut: potential application in fermented milk products. J Biosci Bioeng 110:53–57

Yin JY, Nie SP, Zhou C, Wan Y, Xie MY (2010) Chemical characteristics and antioxidant activities of polysaccharide purified from the seeds of Plantago asiatica L. J Sci Food Agric 90:210–217

Yother J (2011) Capsules of Streptococcus pneumoniae and other bacteria: paradigms for polysaccharide biosynthesis and regulation. Annu Rev Microbiol 65:563–581

You X, Yang L, Zhao X, Ma K, Chen X, Zhang C, Wang G, Dong M, Rui X, Zhang Q (2020) Isolation, purification, characterization and immunostimulatory activity of an exopolysaccharide produced by Lactobacillus pentosus LZ-R-17 isolated from Tibetan kefir. Int J Biol Macromol 158:408–419

Yun JW (1996) Fructooligosaccharides—occurrence, preparation, and application. Enzyme Microb Technol 19:107–117

Zhang Z, Liu Z, Tao X, Wei H (2016) Characterization and sulfated modification of an exopolysaccharide from Lactobacillus plantarum ZDY2013 and its biological activities. Carbohydr Polym 153:25–33

Zhou Q, Feng F, Yang Y, Zhao F, Du R, Zhou Z, Han Y (2018) Characterization of a dextran produced by Leuconostoc pseudomesenteroides XG5 from homemade wine. Int J Biol Macromol 107:2234–2241

Zhu Y, Wang C, Jia S, Wang B, Zhou K, Chen S, Yang Y, Liu S (2018) Purification, characterization and antioxidant activity of the exopolysaccharide from Weissella cibaria SJ14 isolated from Sichuan paocai. Int J Biol Macromol 115:820–828

Chapter 24
Microbial Production of Vitamin B12 Using Food Matrices

Ramesh Sharma, Amiya Ojha, and Biswanath Bhunia

Contents

Abstract Vitamin B12 (Vit B12) is a commonly consumed ingredient in food and feed, health care products and is produced by fermentation due to its complexity for synthesis. Therefore, the selection of better microbial strains to produce vit B12 with optimum bioprocessing parameters has been targeted in past decades. This chapter emphasizes the biosynthesis characteristics of vitamin B12 from selected strains for their industrial applications, including *Propionibacterium freudenreichii* (PF) and *Pseudomonas denitrificans*.

Keywords Vit B12 · Fermentation · Biosynthesis · Industrial application

R. Sharma (✉) · A. Ojha · B. Bhunia
Bioengineering Department, National Institute of Technology Agartala, Jirania, Tripura, India

J. A. Malik et al. (eds.), *Food Process Engineering and Technology*,
https://doi.org/10.1007/978-981-99-6831-2_24

24.1 Introduction

Vitamin B12 (vit B12) is mainly available in animal-based food products and it can be produced through fermentation in food matrix or fortified products (Watanabe et al. 2013), the deficiency of vit B12 and folate are associated with health issues among the public. The deficiency of vit B12 is mainly spread over developing countries due to lower eating habits to animal-based food products (Allen 2010; Marsh et al. 2012). Similarly, in developed countries, Ontarians with elder people are highly affected due to a deficiency of vit B12 (Allen 2009; Pawlak et al. 2014). In the current scenario, the substation of animal protein with plant proteins is a better option for reducing the dietary intake of vit B12 (Elmadfa and Singer 2009; Marsh et al. 2012). Therefore, finding a sustainable approach to fulfill the demand for vit B12 deficiency is achieved with the fortification of plant-based materials with the natural process including fortification with fermentation with association micro-organism microorganisms. Efforts were made to obtain plant-derived products including tempeh vegetable-based table-based fermented products (Babuchowski et al. 1999) incorporated with vit B12.

The content of vit B12 varies in the products ranging from 1.6 to 80 μg/kg of fresh sample weight and is dependent on the used microorganism associated with the fermentation process (Mo et al. 2013; Watanabe et al. 2013). In the case of fermentation in tempeh, with the production of vit B12, the contaminants organism including *Citrobacter freundii* and *Klebsiella pneumoniae* are involved (Watanabe et al. 2013). Similarly, the content of vit B12 was found at 180 μg/L in yogurt produced from soy using *Lactobacillus reuteri* ZJ03 (Gu et al. 2015). The fermentation of soymilk was performed using *L. reuteri* 1098 and was fed to pregnant mice and offspring and the vit B12 deficiency was overcome (Molina et al. 2012). The content of vit B12 was analyzed using MBA, which was performed with lacking specificity for distinction in the presence of an actively present state of vit B12 with attached DMBI as a lower ligand forming the inactive analogs for human consumption (Watanabe et al. 2013). The characterization of vit B12 is equally important in fermented products because different microorganisms are involved with the synthesis of vit B12 analogs which are a biologically inactive form of vit B12 to human health (Watanabe et al. 2013). Similarly, *L. reuteri* was known for producing pseudo-vit B12 with the binding of adenine towards the lower ligand (Crofts et al. 2013; Santos et al. 2007) which is inactive in the human body (Watanabe et al. 2013).

The application of liquid chromatography is more reliable for the estimation of the active form of vit B12 in food products mainly in samples retaining the analogs form of vit B12 for which the MBA method is unreliable (Selva Kumar et al. 2010). In a research, a selective and sensitive approach were applied to determine vit B12 fermented food samples using UHPLC after purification of vit B12 with immunoaffinity columns (Chamlagain et al. 2015).

PF is the most used bacterium and is considered the GRAS product for the synthesis of vit B12 (Thierry et al. 2011). Similarly, other groups of microorganisms

Table 24.1 Microbial strain under different growth conditions to produce vitamin B12

Microorganism/ strain	Media components	Pathway *De novo*	Pathway Salvage	Production (mg/L)	References
Bacillus megaterium DSM 319	Terrific broth media	Yes	–	0.21	Mohammed et al. (2014)
Escherichia coli	ALA	–	Yes	0.00065	Ko et al. (2014)
Lactobacillus reuteri ZJ03	Soy milk	–	Yes	0.204	Gu et al. 2015)
P. denitrificans	Beet molasses, sucrose, betaine	Yes	Yes	214.13	Li et al. (2008a, b, c)
P. denitrificans	Glucose, corn steep liquor, betaine	Yes	Yes	188	Wang et al. (2010)
P. denitrificans	Maltose syrup, corn steep liquor, betaine	Yes	Yes	198.27	Xia et al. (2015a, b)
P. denitrificans SC510	Sucrose, betaine, DMB	–	Yes	214	Li et al. (2008a, b, c)
P. freudenreichii	Glucose, DMB, corn steep liquor, cobalt	Yes	Yes	59.5	Wang et al. (2015)
P. freudenreichii CICC 10019	Glucose, corn extract, DMB	–	Yes	58.8	Wang et al. (2012a, b)
P. freudenreichii DF13	Supplemented whey permeates	–	Yes	0.75	Hugenschmidt et al. (2011)
P. freudenreichii DSM 20271	Wheat bran dough and water	–	Yes		Xie et al. (2019)
P. freudenreichii PTCC 1674	Yeast extract, tryptone, various source of carbon	–	Yes	2.74	Hajfarajollah et al. (2015)
Propionibacterium freudenreichii IFO12426	Glucose; ALA	–	Yes	1.46	Piao et al. (2004a, b)
Pseudomonas denitrificans	Maltose, peptone, betaine	Yes	Yes	54.7	Cheng et al. (2014)
Pseudomonas sp. PCSIR-B-99	Methanol, DMB	Yes	Yes	3.5	Riaz et al. (2007)
S. meliloti CGMCC 9638	Sucrose, glycine betaine, corn liquor, DMB	–	Yes	60–170	Balabanova et al. (2021)
S. olivaceus NRRL B-1125	Glucose, DMB	–	Yes	1–3.3	Hall et al. (1953)
Sinorhizobium meliloti MC5-2	Sucrose, DMB	–	Yes	156	Cai et al. (2018)

are also involved in the synthesis of vit B12 under different conditions of growth as presented in Table 24.1.

The application of PF for fermentation in food matrix including bread is limited for shelf-life extension (Tinzl-Malang et al. 2015). However, other studies mentioned the ability of organisms for producing active vit B12 in other cereal-based food matrices (Chamlagain et al. 2015; Edelmann et al. 2016). The cobalt availability and DMBI (lower ligand of vit B12) in PF improved the production of vit B12 (Hugenschmidt et al. 2011). The demerit of DMBI is that it is not applied to food products. However, DMBI is supplied for the commercial production of vit B12 (Martens et al. 2002). The experiment targeted improving the production of vit B12 by PF in a whey media was used and substitution of DMBI was done with RF and NAM and considered as co-supplement ingredients (Chamlagain et al. 2016). Therefore, the fortification needs to be carried out without using DMBI as supplementary ingredients for improving the production of vit B12 in a food-based matrix with substation by RF and NAM. In addition to it, the lower content of cobalt in cereal matrix (10–60 μg kg^{-1}) may restrict the productivity of vit B12 in cereals-based matrices (Ekholm et al. 2007).

24.2 Sources of Vit B12

Vitamin B12 is synthesized only with some group of bacteria human body mainly collect this vitamin through animal food. The vit B12 is derived in the form of adenosylcobalamin (Adnlcbl), hydroxy cobalamin (OHcbl), methylcobalamin (Methcbl), and cyanocobalamin (CN cbl), where cyanocobalamin is used in dietary supplements and fortified food items. The optimum content of vit B12 is available in the kidney and liver, which is estimated to be more than 14 and 24 μg per 100 g, respectively. The other sources include fish, eggs, muscles of meats, and shellfish. Milk is considered a major source of vit B12 for vegetarian groups of people and an important nutrient for infant children (Souci et al. 2000).

Algal supplements are being pushed more and more as vegetarians' alternative vit B12 sources. *Spirulina* sp. has been the major source of vit B12 supplements. These microorganisms produce a B12 analog that has no biological function and may even obstruct the digestion of vit B12 (Marie Sych et al. 2016). Recent studies mentioned edible algae called *Porphyra* sp., as the active source to produce vit B12 (Watanabe et al. 2014).

The synthesis of vitamins for the host is one of the gut microbiota's primary nutritional effects. Early research emphasized the need for more dietary vit B12 to maintain animal health which is devoid of germs. However, the vitamins are not fully sufficient to contribute to gut microbiota in the human body (Degnan et al. 2014). Vit B12 is a necessary nutrient, with the RDA for adults in the majority of nations ranging from 2.0 to 3.0 g. There is no maximum limit for vit B12 because excessive intake has not yet been linked to any negative consequences (Flynn et al. 2003).

24.3 Biosynthesis of Vitamin B12

24.3.1 The Pathway Followed Under Aerobic and Anaerobic Conditions

The knowledge of biology and the structural role of vit B12 has added interest for the researcher to understand the biosynthetic pathway in Cbl-releasing organisms. The structure of vit B12 is shown in Fig. 24.1.

The complexity of structure due to the involvement of large and complicated biosynthesis activity involving more than 30 genes with various enzymatic steps occurred for molecule synthesis under the "de novo" pathway. This pathway is followed by groups of bacteria and archaea because the evidence is not available for its synthesis in a eukaryotic organism with the production of Cbl (Martens et al. 2002; Roth et al. 1996). The intermediates compound was isolated in earlier studies (Leeper 1989), till the 1990s, the pathway for biosynthesis was unknown but after that, the pathway was revealed for *Pseudomonas denitrificans* (Kräutler et al. 1998). The involvement of genes in Cbl synthesis was mentioned with the cob prefix and the referred letter signifies the gene positions in the operon. The isolation and characterization of cob enzymes and intermediary of cbl from *P. denitrificans* was performed by a French-based company (Crouzet et al. 1990). In the later period, the

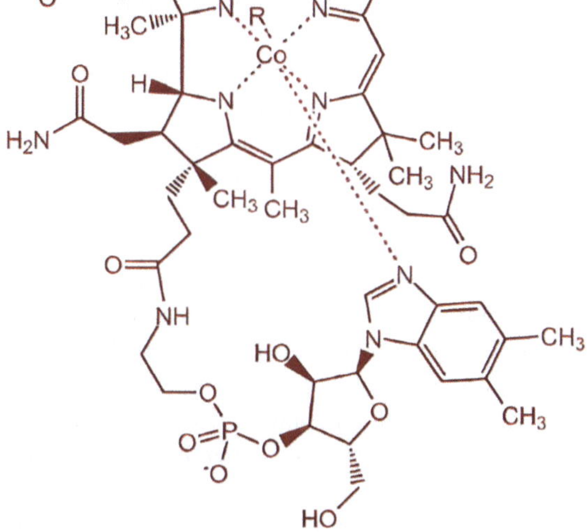

Fig. 24.1 Structure of vitamin B12

characterization of the involvement of genes for cbl biosynthesis was performed in *Bacillus megaterium*, *Salmonella enterica*, and PF.

The pathway involved in synthesis in the later organism was identical to that found in *P. denitrificans* but was varied under genetic conditions. The difference was seen with lacking the monooxygenase and with various cobaltochelatase. With this consideration, two pathways were discovered for cbl synthesis: (a) an aerobic pathway which is seen in *P. denitrificans* (Stamford et al. 1996), and this pathway also seen in other organisms including *Ensifer casida* and *Sinorhizobium meliloti*; and (b) and another is an anaerobic (also called pathway of early cobalt insertion) pathway which is seen in PF, *B. megaterium* and *S. enterica*. The pathway for aerobic and anaerobic conditions is shown in Fig. 24.2.

Without being dependent on the biosynthetic pathway, the synthesis of tetrapyrrole occurred with the synthesis of ALA. After that, the ALA is converted to a tetrapyrrolic macrocycle structure, which involved the reaction with enzymatic conversion. Firstly, an ALA dehydratase, which depends on the Zn^{2+} and Mg^{2+} enzyme, catalyzes with the condensation reaction occurring between two molecules of ALA with forming porphobilinogen (PBG) (Jaffe 2000). The PBG deaminase is involved in the polymerization of PBG by forming linear tetrapyrrole molecules. Under final conditions, uroporphyrinogen III synthase is involved to convert to the final unit of pyrrole and get linked with the first unit of pyrrole with the formation of uroporphyrinogen III, which is considered an unsymmetrical isomer of hexahydro porphyrin (Jordan 1991). The formed molecules are the last intermediates that are attached to prosthetic groups including chlorophyll and heme groups (Raux and Schubert 2000). The uroporphyrinogen III again gets transferred to precorrin-2 which is catalyzed with an enzyme called uroporphyrinogen III methyltransferase, which occurred in presence of SAM and is considered a methyl donor.

Under specific conditions, methylation by enzymes occurred at C-2 for the uroporphyrinogen III with the creation of precorrin I, and tautomerization occurred under phototropic conditions with the involvement of the same enzymes for methylation at C-7, resulting in obtained precorrin-2 (Raux and Schubert 2000; Warren et al. 2002).

The difference between aerobic pathways and anaerobic pathway mainly lies in the contraction of rings and steps involved in cobalt chelation. For the case of the aerobic pathway, contraction of the ring is performed with the requirement of O_2 molecules with monooxygenase (CobG) for the formation of precorrin-3B, an intermediate called a hydroxylated γ-lactone which undergoes the arrangement with masked pinacol concept through contraction of the ring, during the contraction of the ring, thrusting the position of methylated C20 occurs. The contraction of the ring occurs and acetic acid gets liberated from the process (Estevinho et al. 2019). Similarly, for the anaerobic pathway, ring contraction occurred in a later period at the time in which cobalt is already inserted with molecules. This step is performed with catalyzation with an enzyme that is codified with the cbiH gene, without the requirement of oxygen. Thereafter, the methylation takes place which is SAM-dependent at C17, for the extrusion promotion whose methylation is done at the C20 position with the formation of a δ-lactone ring (Moore and Warren 2012).

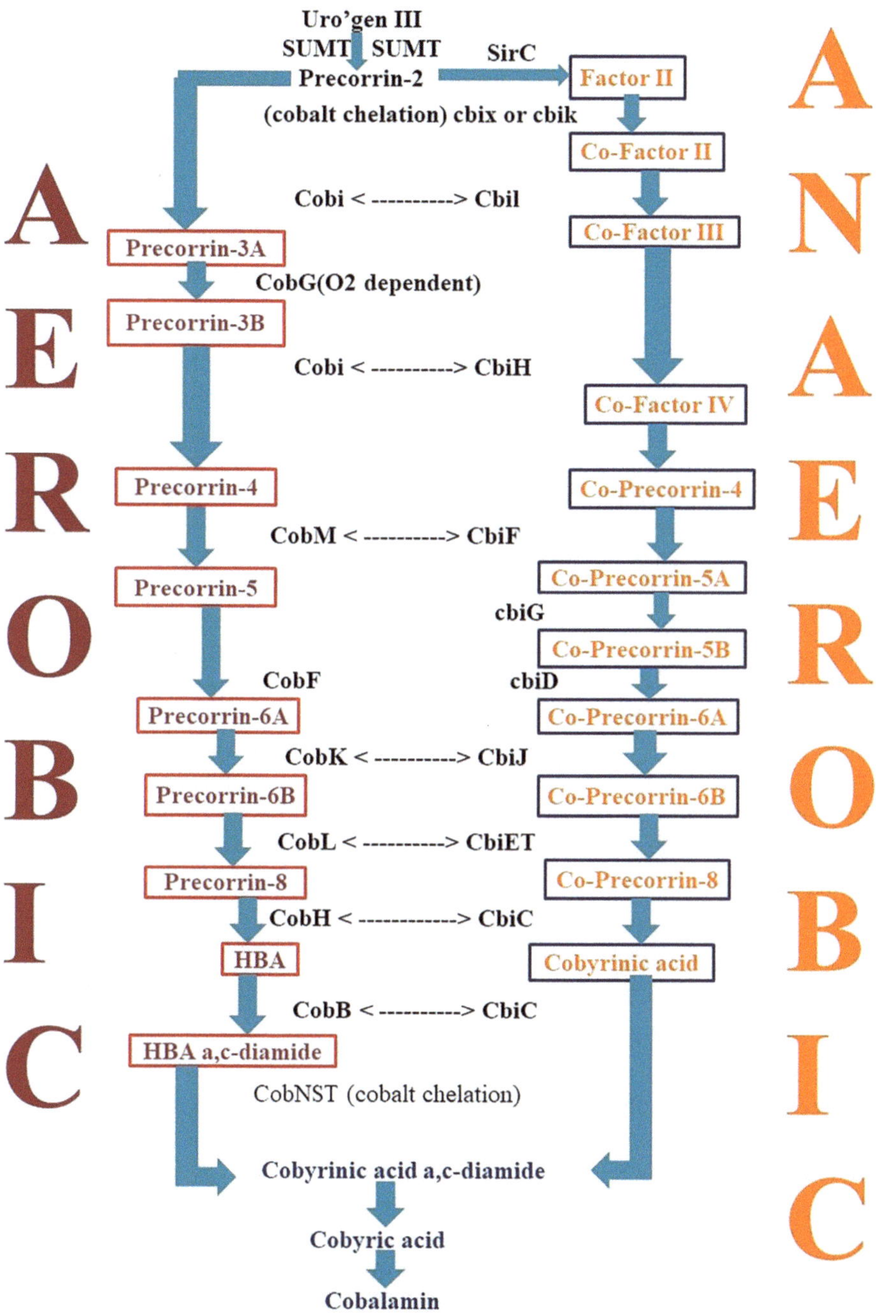

Fig. 24.2 Aerobic and anaerobic pathways for the synthesis of vitamin B12

The chelation with cobalt is for both pathways. For the case of the aerobic pathway, the Co atom late insertion is seen after the full contraction of the ring. The catalyzation of insertion is occurred by ATP-associated multienzyme complex with the occurrence of Mg element (Moore and Warren 2012). For the case of the anaerobic pathway, these steps are involved at the earlier point of the route at the time when either cbiX or cbiK genes encoded the Co chelatase, or canalization of Co insertion occurs with the ring at a non-contracted state (Moore and Warren 2012). With the independence of the biosynthetic pathway in the corrin ring, the produced DMBI is attached to the configuration at α-axis. The synthesis of the lower ligand is described by presenting transform routes including the aerobic and the anaerobic pathway and based on the requirement of O_2.

Under aerobic conditions, the DMBI synthesis is catalyzed by DMBI synthase BluB, which functions with fragmenting and contracting the bound form of FMN cofactor with breakage of ribityl tail to produce DMBI and D-erythrose 4-phosphate with the availability of O_2. Whereas in later conditions, the phosphoribosyl transferase is involved in introducing the DMBI via a reaction involving the substitution of nucleophiles (Mathur et al. 2020). This mentioned pathway was first revealed by *S. meliloti* (Campbell et al. 2006) and later it was discovered in bacteria producing Cbl (6), which included PF (Deptula et al. 2015) and *P. denitrificans* (Nguyen-Vo et al. 2018). These results revealed the fact of incapability of PF for producing Cbl under anaerobic conditions by avoiding the addition of DMBI.

Similarly, under the anaerobic biosynthesis for the DMBI formation underwent catalysis by the gene bzaA-bzaB-cobT-bzaC-bzaD-bzaE operon, which facilitates with the production of DMBI together with 5-hydroxybenzimidazole, and 5-methoxy-6 methyl benzimidazole that are formed as intermediate products. The mentioned route was also described for obligate anaerobic bacteria which included *Eubacterium limosum* (Hazra et al. 2015) and *Acetobacterium woodii* (Mathur et al. 2020).

24.4 Optimization of Bioprocess Conditions for Synthesis of Cobalamin

The requirement for cbl in the field of dietary, nutraceutical food, and beverage industries have been increasing significantly due to consumer health awareness with the gaining popularity of substitution diets for vegan and vegetarian groups of people. Considering this, various efforts are made for strain selection and optimization of processes for cbl synthesis (Fang et al. 2017; Martens et al. 2002). The various types of cbl including the adenosyl Adocbl, Metcbl, CNcbl, and various other carbamides with various lower axial ligands are available and considered as key enzyme cofactors that depend on corrinoids mainly important for gut microbiota (Hardlei et al. 2013; Sobczyńska-Malefora et al. 2021). For considering various cbl, only cyanocobalamin is reviewed, which gets absorbed by the human body easily

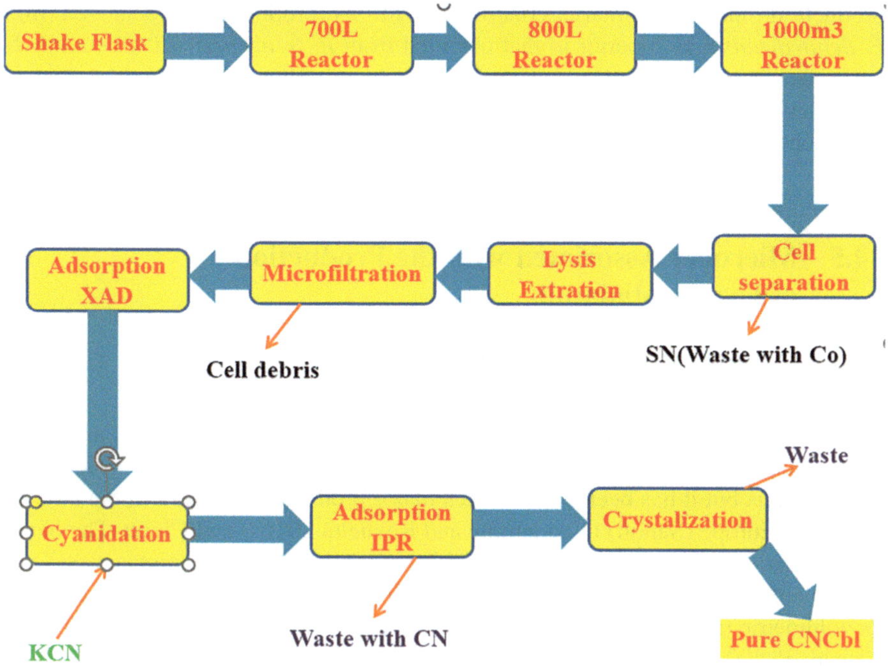

Fig. 24.3 The pure form of cbl extraction process

and is industrially produced the pure form of the cbl extraction process that is shown in Fig. 24.3.

In the present situation, the production of cbl is seen industrially with a selection of productive strains which included the strains of PF and *P. denitrificans*, with additional related strains including *Pseudomonas nitroreducens* and *E. casida* (Fang et al. 2017; Martens et al. 2002). The common strategies applied for improving the productivity of strains included the application of random mutagenesis techniques with increasing the yield of B12 or with developing resistance to the formation of intermediates of toxic compounds in media (Kojima et al. 1985). The genes with overexpression for cbl bio-synthesis (Piao et al. 2004b), foreign genes expression under heterologous manner (Piao et al. 2004a), and down regulation (Biedendieck et al. 2010) with considering various genes reported producing better-producing strains. It is worth with development of productive strains with good results which included *B. megaterium* (Mohammed et al. 2014) and *Acetobacter pasteurianus* (Bernhardt et al. 2019), and the biosynthesis pathway expression in a heterologous manner is a well-known platform for cells mainly in *E. coli* (Fang et al. 2017).

The extensive work was reviewed on biosynthetic regulation and genetics and used the genetic tool for the improvement of cobalamin production in cell factories. On this approach, frequent research is carried out for vit B12 production with optimization of bioprocess parameters. The studies were mainly carried out for improving production with genetic engineering for the selected strains. The production under volumetric conditions with its productivities achieved from *P. denitrificans* was superior to that obtained from *P. freudenreichii* which may be varied as per strain differences and conditions of culturing. For the later strains, the strategies for decreasing the inhibitory effect of propionic acid seem an effective approach for improved production.

24.5 Microbes Associated with the Production of Cyanocobalamin

24.5.1 Culturing with Pseudomonas denitrificans

The *P. denitrificans* belongs to gram-negative bacterial groups and used an aerobic biosynthetic pathway to release vit B12. This strain has not been considered as GRAS status but it has been considered a major producer of vit B12 in the industry mainly in Europe (Sanofi) (Vandamme and Revuelta 2016a) and China. The Sanofi strain was obtained with a combination of the process of mutagenesis and implementing biological methods but no information is available officially regarding its volumetric production with the consideration of other aerobic strains produced at a yield of 200–300 mg/L (Vandamme and Revuelta 2016b). The strain with the optimum result was developed which is called MB-580, and get patented in 1962 (US3018225A). In the past few years, the French company developed with amplification of cob genes used for vit B12 biosynthesis, and some higher-producing strains were developed which are called SBL27 and, finally developed called SC510 (Crouzet et al. 1990). The vit B12 is produced dominantly by different China companies with an approximate quantity of 31.41 ton in the year 2020 with a commercial value of 339.8 million USD (Calvillo et al. 2022). The originality of the strain used for commercial production is not known precisely but the assumption was made as an aerobic strain due to the availability of various publications from different groups of research as mentioned by Huarong Pharmaceutical Company (Li et al. 2008a, b, c, 2012; Wang et al. 2010), and as per the latest patent on optimization of bioprocess using *P. denitrificans* reported with volumetric production capacity reaching 281 mg/L. Besides considering the genetic modifications for improving the productivity of vit B12 using *P. denitrificans,* the productivity can be improved with optimization of media and with changing the bioprocessing conditions. The various processing factors including media trace elements, control over dissolved oxygen, and pH with the addition of supplements were used for testing. The addition of Zn^{2+} resulted with giving significant results in synthesizing ALA and PBG, which are the main precursor for the synthesis of cobalamin while supplementing with Co^{2+} and DMBI, considered as the based ingredient for incorporation in nucleotide loop, which ultimately improved the production (Li et al. 2008a, b, c). The optimization of the initial quantity of the compound with the design of the experiment improved the cobalamin production by 13% (Li et al. 2008a, b, c).

Similarly, the composition of media affected the stability of culture pH giving significant results on cobalamin production. For obtaining better control over pH, the feed was controlled with supplementing glucose as a source of carbon, and betaine was used to donate the methyl group for optimum production of vit B12 in a bioreactor (Li et al. 2008a, b, c, 2012; Wang et al. 2010). Mainly betaine is used to donate methyl groups for the biosynthesis of vit B12 (Kusel et al. 1984) and functions in forming intermediates including glutamate, ALA, methionine, and glycine (Xia et al. 2015b), and at higher betaine concentrations may restrict cell growth (Li et al. 2008b; Xia et al. 2015b). Therefore, there was a need for optimization of its concentration to reduce the negative influence over cell growth and for improving the production of cbl, which was successful for industrial production (Li et al. 2008b). Oxygen transfer rate (OTR) is an important bioprocessing parameter for optimization in *P. denitrificans*. The OTRs with higher value at initial culturing conditions facilitates cell growth whereas the lower OTRs at later culturing conditions were adding critical results for improving productivity (Wang et al. 2010). In the later studies, it is mentioned that the production was improved at lower oxygen concentrations which are related to altering the morphology of the cell, stimulation of changes occurred at the growth phase of the cell to the elongation state with improving the production of vit B12 (Wang et al. 2016). Considering these issues, the various dissolved oxygen control approach under multi-step were created by reducing the agitation and aeration till achieving the dissolved oxygen to 2%, which improved the production by 17% (Li et al. 2012; Wang et al. 2010). Additionally, the incorporation of respiratory chain inhibitors, including rotenone promoted the production of vitamins but added a harmful effect on cell growth (Cheng et al. 2014). The other nutrient source including maltose syrup, glucose, corn steep liquor, and beet molasses was used as an alternative and considered as a cheap source in place of expensive components including refined sucrose and glucose. Some components performed with adding negative effect on pH stability and finally altered the vitamin productivity (Li et al. 2008a). Whereas the blend of corn steep liquor, maltose syrup, and betaine is mentioned as a cheaper substitution for media compared to traditionally prepared media (Xia et al. 2015a).

24.5.2 Culturing with Propionibacterium freudenreichii

The PF strains belonged to gram-positive rod-shaped in morphology and are capable to synthesize the propionic acid to larger volume through the Wood–Workman pathway. With consideration of the aerobic pathway for synthesizing B12, PF has additional advantages of GRAS status which is approved by the FDA, and QPS status approved by EFSA. The approach with genetic engineering was applied in PF for obtaining higher production of vit B12. The main genes were overexpressed for the synthesis of cobalamin (Piao et al. 2004b) and the phenomena of genome shuffling (Zhang et al. 2010) improved the production of cbl. However, for industrial production of strain, random mutagenesis with the application of mutagen including

UV light, or maybe chemical compounds is applied to induce the mutation to produce cobalamin. PF is considered the higher yield-producing strain and showed higher resistance and tolerance for the accumulation of propionic acid (Kojima et al. 1985). Similarly, PF is considered facultative anaerobic strains that follows an anaerobic pathway for the synthesis of cobalamin. The higher yield of cobalamin is produced at lower oxygen concentrations, but oxygen requires for the synthesis of DMBI and its attachment to the corrin ring. Therefore, two stages are followed for culturing (1) firstly the culturing of cells performed under the absence of O_2, and (2) secondly the cell is cultured under aerobic conditions usually after 72–96 h of culturing period (Chamlagain et al. 2016; Gray and Escalante-Semerena 2007; Pillai et al. 2018) which is performed by providing the gentle micro aeration for the synthesis of DMBI and production of cobalamin (Vandamme and Revuelta 2016a). The GRAS status of Vit B12 assures its use for the development of fortified food items. With this consideration, food fortification under in situ conditions with the incorporation of PF is tested in food matrices including cheese with selecting like propionic medium or liquid medium based on whey protein (Chamlagain et al. 2016; Deptula et al. 2017), cereal matrices (Chamlagain et al. 2018) and in tempeh for in situ fortifications (Signorini et al. 2018). Although the production levels and cell density under food fortification were lower than compared to traditional media the requirement content of cobalamin is at a higher amount in non-traditional sources fulfilling the recommended daily requirement of vit B12 with consuming a lower amount of fermented food products (Deptula et al. 2017). As this group of bacteria has the ability the production of propionic acid in higher quantity which eventually lead to toxicity and inhibited the growth of cell (Wang et al. 2012a). Hence, different approaches for optimization to decrease the accumulation of propionic acid are applied. The various product removal techniques under in situ conditions are applied with simultaneous production of vit B12 and propionic acid. Among various techniques, the application of EBABs with the high attachment of biocompatible resins (ZGA330), was successfully employed for volumetric production of vit B12 ranging from 40 to 60 mg/L (Wang et al. 2012a). In EBABs, during column expansion, absorption is occurred and allowed the passage of culture from the chromatographic column with avoiding clogging, with the retention of propionic acid in the resin (Wang et al. 2014). The various conditions for culturing (Wang et al. 2014) including sources of nitrogen and carbon (Wang et al. 2020) with adding the supplement of media including DMBI (Wang et al. 2015), were tested for continuous improvement of B12 and propionic acid production. EBAB system with supplementation to glycerol and glucose (Wang et al. 2014) with corn stalk hydrosylates (Wang et al. 2020) have been considered as an effective source of carbon with volumetric production of CNCbl levels at 43.2 and 47.6 mg/L respectively. The other steps to reduce the formation of concentration of propionic acid is using the co-fermentation process of PF by considering other microbes that could metabolize the propionic acid. The co-culturing of PF with *Ralstonia eutropha* had shown improvement in cbl production improving from 6.73 to 19 mg/L (Miyano et al. 2000). The co-fermentation was fruitfully implemented not only for reducing the content of propionic acid but also useful for the production of more than one type of product simultaneously or used for the fortification of cell culture. The B12 and folate were simultaneously produced with co-fermentation of PF with *Lactobacillus plantarum* (Hugenschmidt et al. 2011; Zheng et al. 2020) and the strain called *Basidiomycota* and PF were applied for co-fermentation produced vit B12 and vitamin D and has been patented (Frettlöh et al. 2022). The experiment was performed with food fortification under in situ conditions for B12 production with selecting bread dough as the food matrix and whey-based media was selected for co-cultivation of PF with *Lactobacillus brevis* (Zheng et al. 2020) for achieving stability and microbial safety (Xie et al. 2019). The supplementation of nutrients with the precursor of cobalamin is important to improve production. The incorporation of precursors with an additional compound including ALA and Co^{2+}, benefited in improving the production of vitamins (Martens et al. 2002). Mostly all the strains of PF function with their own synthesis of DMBI whereas the base biosynthesis is typically low. Additionally, the formation of DMBI is restricted under anaerobic conditions as it required oxygen for the synthesis (Deptula et al. 2015). With DMBI unavailability, the inactive form of B12 is formed with the cell, starting the formation of the incomplete form of vitamin including cobamamide or pseudo vit B12. Therefore, the addition of DMBI or maybe its precursor is needed which includes Riboflavin or Nicotinamide, which has constantly added the positive factor for the synthesis of cobalamin (Chamlagain et al. 2016; Deptula et al. 2017; Fang et al. 2017; Martens et al. 2002; Wang et al. 2012b, 2015). Among the other groups of vit B12 the addition of its analogues reduced the feedback inhibition with increasing the production of cobalamin (Quesada-Chanto et al. 1994; Thirupathaiah et al. 2012). The strain of PF can be used for culturing for industrial purpose with considering the wide source of nitrogen and carbon, may be spent and waste media including (Quesada-Chanto et al. 1994), glycerol (Kośmider et al. 2012), waste obtained after frying sunflower oil (Hajfarajollah et al. 2015), pomace derived from tomato (Haddadin et al. 2001), soybean-based protein residue (Assis et al. 2020) and use spent media based on vegetable juice (Gardner and Champagne 2005).

24.6 Downstream Processing of Vit B12

Strategies for vit B12 recovery is important to the process that has been applied to industrial scale for the past few decades. The culture broth underwent different steps including filtration, adsorption, and extraction which affected over feasibility and yield of products. The downstream processing is initialized with the conversion of biomass by reducing its volume mainly performed with the centrifugation process. The bioprocess can be selected because the Cbl may be available under the extracellular region therefore the purification may be started from the selection of whole broth. Mostly all the corrinoid species are used to extract with thermal treatment at 80–120 °C with maintaining the pH ranging from 6.5 to 8.5 for an extraction time of 10–30 min. Cyanidation is followed at the time of extraction which, maybe after the initial process of filtration and extraction (Calvillo et al. 2022; Vandamme and Revuelta 2016a). With this involved process, the corrinoids are converted into CNCbl by adding thiocyanate or potassium cyanide. The process is carried out with a supply of sodium nitrate in presence of heat (Gray and Escalante-Semerena 2007). In the later process clarification of CNCbl is involved with applying a filtration process including microfiltration and/or nanofiltration and the adsorption process is followed using XAD resin. Before adding the obtained Cbl to the animal feed the solution of vitamin underwent treatment with zinc chloride and precipitation is performed with organic solvent including acetone, for obtaining the final process products (Survase et al. 2006). For obtaining the higher purity vitamins for use in pharmacy the adsorption steps are followed using different resign including IRA and Alumina and are often used to achieve higher purity products. After obtaining the purified form of vit B12 it undergoes the post-modification steps for use in food supplements with improving the bioavailability in the human body. The protection of these compounds is crucial for application in the non-functional intrinsic factor where the bioavailability of Cbl is very low (Berlin et al. 1968; Nielsen et al. 2012). Therefore, several technologies are applied for the protection of oral supplements with the environmental condition of the gastrointestinal region (Gamboa and Leong 2013). Considering various technology, the microencapsulation, the method is mostly used in pharmacy and cosmetic industries (Dhakal and He 2020), where improvement in the stability of vitamin B12 is seen mainly using food grade W1/O/W2 emulsions system (Matos et al. 2015; Ramalho et al. 2021), liposomes (Bochicchio et al. 2016) or with considering various encapsulating agents based on food grade, such as Arabic gum, xanthan, modified starch, maltodextrin, chitosan, sodium alginate, carrageenan, cyanobacterial extracellular polymeric pectin (Carlan et al. 2017; Estevinho et al. 2019). Additionally, Fidaleo et al., reviewed over application of nanocarrier which improves vit B12 therapies by reducing the side effect under effective cost with improving the patient's life quality (Fidaleo et al. 2021).

24.7 The Commercial Significance of Vit B12

The market for vit B12 in our food and feed industry and its efficient use and security purpose have been verified (Additives et al. 2018), additionally, its use is extended for use in pharmacy and supplementary diets. For food industry and feed purposes, the CNCbl is incorporated with pigs, calf feed and poultry feeds at a concentration ranging from 10 to 30 mg/ton which is followed in Europe and USA (Vandamme and Revuelta 2016a). It is used for the additive purposes in various food items including cereal, where its function is for the retention of organoleptic features with retention of chemical properties including odor, taste, and water solubility, which added the advantages for food fortifications. The presence of brighter red color restricted its application for use in white bread. Regarding its supplementation use, it is gaining popularity in past decades, especially in the diet related to vegetarian people (Rizzo et al. 2016). The CNCbl form is mostly used in the food industry due to retention of higher stability, lower price, and safety issues (Obeid et al. 2015) and its efficiency of performance is similar to other forms (Obeid et al. 2015; Rizzo et al. 2016). Similarly, the dried algae were also present with a sufficient amount of B12 (Watanabe et al. 2013), whereas Cbl is not present in vegetables (Rizzo et al. 2016), and the obtained dietary source are from animal-based items. Although the content of B12 is available in eggs and dairy products mainly in a vegetarian diet which is quantitively low when compared to other options. The content of B12 with 0.40 μg/100 g, 9.40 μg/100 g, 8.9 μg/100 g, and 52.4 μg/100 g is present in milk, eggs, meats, and shellfish respectively (Watanabe et al. 2013). The Cbl derived from food sources works with 50% bioavailability and their loss mainly occurred during the processing like cooking, light exposure, pasteurization, etc. (Rizzo et al. 2016), making the difficult fulfillment of a daily intake of 2.40 μg with consumption of a vegetarian diet with avoiding supplements of vit B12 (Rizzo et al. 2016). Deficiency of vit B12 is mainly seen among middle-income people with consumption of Vitamin with consumption of plant-based food products with lower consumption of meat (Titcomb and Tanumihardjo 2019). In considering some high-income countries also the deficiency of vit B12 is seen. This mainly affects the elder groups with 20% of the population and mostly affected the above 60 age group of people in USA and UK as per NIH (Oh et al. 2021). In the case of elder groups, the deficiencies mainly occurred due to lower intake and phenomena of malabsorption based on food-bound which is mainly caused due to age-related problems like gastric atrophy and lower IF levels (Rusher and Pawlak 2013; Sobczyńska-Malefora et al. 2021). The other people with higher risk are lactating women, and people suffering from autoimmune diseases causing gastric infection including atrophic gastritis (Watanabe et al. 2013). To overcome the issues a higher intake of vit B12 is necessary through the process of supplementation. Mostly the supplementation is based on CNcbl source but with a cause like trafficking in cellular region and alteration of protein processing which are due to genetic diseases (Watkins and Rosenblatt 2011), therefore supplementing with another form including MetCbl or OHCbl, is required. Similarly, the supplementation with CNCbl is not favorable for the smoker populations (Linnell et al. 1968; Shepherd and Velez 2008). It is also mentioned that supplementation of B12 should be done individually because its incorporation with multivitamin tablets may degrade to an inactive form due to the availability of copper and vitamin C (Rizzo et al. 2016). Besides using for supplementation, it can be used to fortify food products. Among the Cbl, the CNCbl is the most stable form due to its higher stability in undercooked and processed conditions (Watanabe et al. 2013). Fortified product based on B12 is mostly found in the US and in other countries including milk and fortified cereals which significantly meet the requirement of daily intake of Cbl (Watanabe et al. 2013). Similarly, other products like flour fortification were also reported (Rizzo et al. 2016). A review of in situ fortifications is collected (Chamlagain et al. 2018; Xie et al. 2019) which may be the appropriate alternative for the future. Vit B12 is used in the pharmacological sector which mainly used CNCbl, with also considering other forms for use including OHCbl, AdoCbl, and MetCbl, because of higher uptake capacity with producing sustainable serum levels (Vandamme and Revuelta 2016a). Pharmacological vit B12 based on pharmacology is available in various forms including nasal sprays, oral and sublingual products, and injecting may treat pernicious anaemia, deficiency of vit B12, poisoning associated with cyanide and lower the levels of homocysteine. It is also claimed that the vit B12 added the positive effect to the patient associated with Alzheimer's disease and help to stimulate the immune system for adding more evidence additional experiment needs to be performed (Hamel 2011; Vandamme and Revuelta 2016a). Considering the various application and uses of vit B12 its demand and consumption is increasing worldwide although worldwide prediction is difficult due to the scarcity of information. However, in the past few decades, production has improved to a greater extent. In the year of 1989, the production improved to 3 ton in a year (Vandamme and Revuelta 2016a) and with reaching to 2005, it improved to 10 ton with commercial value of 77 million EUR (Moine et al. 2000). In the year 2020, China produced a capacity of 31.41 ton with a marketing potential of a USD 339.48 million (Li et al. 2008b), while some surveys mentioned commercial value of 410 million USD by the year 2027 (Calvillo et al. 2022). The rise in elder group populations, searching the alternative diet for vegan and vegetarian people due to avoiding or scarcity of animal food has led to an enhance in the production of vit B12 in the market.

24.8 Conclusion

From the published paper it is familiar that vit B12 production from PF strains faced several challenges and failures which are needed to be surpassed by competing for aerobic strains. The anaerobic strains reported with highest production capacity of 76 mg/L (IN201827044769 A), though the result is inferiority with a production capacity of 250–280 mg/L for aerobic strains under studies (CN108949866 A) (Calvillo et al. 2022). Considering the desired function of selected the PF is also useful for the synthesis of propionic acid and trehalose besides vit B12 which added industrial value. In the current situation, the production of propionic acid is mostly associated with petrochemical processes, and various studies are carried out for bio-based alternative sources which are to be named "natural preservatives" (Zhou and Li 2021). Therefore, the application of PF for the biorefinery purpose would simultaneously improve the productivity and extraction yield of vit B12 and propionic acid with improving the commercial feasibility and industrial productivity.

References

Additives EPO, Feed SUIA, Rychen G, Aquilina G, Azimonti G, Bampidis V, Bastos ML, Bories G, Chesson A, Cocconcelli PS, Flachowsky G (2018) Safety and efficacy of vitamin B12 produced by Ensifer spp. as a feed additive for all animal species based on a dossier submitted by VITAC EEIG. EFSA J 16(7):e05336

Allen LH (2009) How common is vitamin B-12 deficiency? Am J Clin Nutr 89(2):693S–696S

Allen LH (2010) Bioavailability of vitamin B12. Int J Vitam Nutr Res 80(4–5):330–335

Assis DA, Matte C, Aschidamini B, Rodrigues E, Záchia Ayub MA (2020) Biosynthesis of vitamin B12 by Propionibacterium freudenreichii subsp. shermanii ATCC 13673 using liquid acid protein residue of soybean as culture medium. Biotechnol Prog 36(5):e3011

Babuchowski A, Laniewska-Moroz L, Warminska-Radyko I (1999) Propionibacteria in fermented vegetables. Lait 79(1):113–124

Balabanova L, Averianova L, Marchenok M, Son O, Tekutyeva L (2021) Microbial and genetic resources for cobalamin (vitamin B12) biosynthesis: from ecosystems to industrial biotechnology. Int J Mol Sci 22(9):4522

Berlin H, Berlin R, Brante G (1968) Oral treatment of pernicious anemia with high doses of vitamin B12 without intrinsic factor. Acta Med Scand 184(4):247–258

Bernhardt C, Zhu X, Schütz D, Fischer M, Bisping B (2019) Cobalamin is produced by Acetobacter pasteurianus DSM 3509. Appl Microbiol Biotechnol 103(9):3875–3885

Biedendieck R, Malten M, Barg H, Bunk B, Martens JH, Deery E, Leech H, Warren MJ, Jahn D (2010) Metabolic engineering of cobalamin (vitamin B12) production in Bacillus megaterium. Microb Biotechnol 3(1):24–37

Bochicchio S, Barba AA, Grassi G, Lamberti G (2016) Vitamin delivery: carriers based on nanoliposomes produced via ultrasonic irradiation. LWT Food Sci Technol 69:9–16

Cai Y, Xia M, Dong H, Qian Y, Zhang T, Zhu B, Wu J, Zhang D (2018) Engineering a vitamin B12 high-throughput screening system by riboswitch sensor in Sinorhizobium meliloti. BMC Biotechnol 18(1):27

Calvillo Á, Pellicer T, Carnicer M, Planas A (2022) Bioprocess strategies for vitamin B12 production by microbial fermentation and its market applications. Bioengineering (Basel) 9(8):365

Campbell GR, Taga ME, Mistry K, Lloret J, Anderson PJ, Roth JR, Walker GC (2006) Sinorhizobium meliloti bluB is necessary for production of 5,6-dimethylbenzimidazole, the lower ligand of B12. Proc Natl Acad Sci U S A 103(12):4634–4639

Carlan IC, Estevinho BN, Rocha F (2017) Study of microencapsulation and controlled release of modified chitosan microparticles containing vitamin B12. Powder Technol 318:162–169

Chamlagain B, Edelmann M, Kariluoto S, Ollilainen V, Piironen V (2015) Ultra-high performance liquid chromatographic and mass spectrometric analysis of active vitamin B12 in cells of Propionibacterium and fermented cereal matrices. Food Chem 166:630–638

Chamlagain B, Deptula P, Edelmann M, Kariluoto S, Grattepanche F, Lacroix C, Varmanen P, Piironen V (2016) Effect of the lower ligand precursors on vitamin B12 production by food-grade Propionibacteria. LWT Food Sci Technol 72:117–124

Chamlagain B, Sugito TA, Deptula P, Edelmann M, Kariluoto S, Varmanen P, Piironen V (2018) In situ production of active vitamin B12 in cereal matrices using Propionibacterium freudenreichii. Food Sci Nutr 6(1):67–76

Cheng X, Chen W, Peng WF, Li KT (2014) Improved vitamin B12 fermentation process by adding rotenone to regulate the metabolism of Pseudomonas denitrificans. Appl Biochem Biotechnol 173(3):673–681

Crofts TS, Seth EC, Hazra AB, Taga ME (2013) Cobamide structure depends on both lower ligand availability and CobT substrate specificity. Chem Biol 20(10):1265–1274

Crouzet J, Cauchois L, Blanche F, Debussche L, Thibaut D, Rouyez MC, Rigault S, Mayaux JF, Cameron B (1990) Nucleotide sequence of a Pseudomonas denitrificans 5.4-kilobase DNA fragment containing five cob genes and identification of structural genes encoding S-adenosyl-L-methionine: uroporphyrinogen III methyltransferase and cobyrinic acid a,c-diamide synthase. J Bacteriol 172(10):5968–5979

Degnan PH, Barry NA, Mok KC, Taga ME, Goodman AL (2014) Human gut microbes use multiple transporters to distinguish vitamin B12 analogs and compete in the gut. Cell Host Microbe 15(1):47–57

Deptula P, Kylli P, Chamlagain B, Holm L, Kostiainen R, Piironen V, Savijoki K, Varmanen P (2015) BluB/CobT2 fusion enzyme activity reveals mechanisms responsible for production of active form of vitamin B12 by Propionibacterium freudenreichii. Microb Cell Factories 14(1): 1–12

Deptula P, Chamlagain B, Edelmann M, Sangsuwan P, Nyman TA, Savijoki K, Piironen V, Varmanen P (2017) Food-like growth conditions support production of active vitamin B12 by Propionibacterium freudenreichii 2067 without DMBI, the lower ligand base, or cobalt supplementation. Front Microbiol 8:368

Dhakal SP, He J (2020) Microencapsulation of vitamins in food applications to prevent losses in processing and storage: a review. Food Res Int 137:109326

Edelmann M, Chamlagain B, Santin M, Kariluoto S, Piironen V (2016) Stability of added and in situ-produced vitamin B12 in breadmaking. Food Chem 204:21–28

Ekholm P, Reinivuo H, Mattila P, Pakkala H, Koponen J, Happonen A, Hellström J, Ovaskainen M-L (2007) Changes in the mineral and trace element contents of cereals, fruits and vegetables in Finland. J Food Compos Anal 20(6):487–495

Elmadfa I, Singer I (2009) Vitamin B-12 and homocysteine status among vegetarians: a global perspective. Am J Clin Nutr 89(5):1693S–1698S

Estevinho BN, Mota R, Leite JP, Tamagnini P, Gales L, Rocha F (2019) Application of a cyanobacterial extracellular polymeric substance in the microencapsulation of vitamin B12. Powder Technol 343:644–651

Fang H, Kang J, Zhang D (2017) Microbial production of vitamin B12: a review and future perspectives. Microb Cell Factories 16(1):15

Fidaleo M, Tacconi S, Sbarigia C, Passeri D, Rossi M, Tata AM, Dini L (2021) Current nanocarrier strategies improve vitamin B12 pharmacokinetics, ameliorate patients' lives, and reduce costs. Nanomaterials (Basel) 11(3):743

Flynn A, Moreiras O, Stehle P, Fletcher RJ, Müller DJ, Rolland V (2003) Vitamins and minerals: a model for safe addition to foods. Eur J Nutr 42(2):118–130

Frettlöh M, Haag T, Zorn H, Zajul M, Ahlborn J (2022) Sequential co-culturing method for producing a vitamin- and protein-rich food product. Google patents

Gamboa JM, Leong KW (2013) In vitro and in vivo models for the study of oral delivery of nanoparticles. Adv Drug Deliv Rev 65(6):800–810

Gardner N, Champagne CP (2005) Production of Propionibacterium shermanii biomass and vitamin B12 on spent media. J Appl Microbiol 99(5):1236–1245

Gray MJ, Escalante-Semerena JC (2007) Single-enzyme conversion of FMNH2 to 5, 6-dimethylbenzimidazole, the lower ligand of B12. Proc Natl Acad Sci U S A 104(8): 2921–2926

Gu Q, Zhang C, Song D, Li P, Zhu X (2015) Enhancing vitamin B12 content in soy-yogurt by Lactobacillus reuteri. Int J Food Microbiol 206:56–59

Haddadin MS, Abu-Reesh IM, Haddadin FA, Robinson RK (2001) Utilisation of tomato pomace as a substrate for the production of vitamin B12—a preliminary appraisal. Bioresour Technol 78(3):225–230

Hajfarajollah H, Mokhtarani B, Mortaheb H, Afaghi A (2015) Vitamin B12 biosynthesis over waste frying sunflower oil as a cost effective and renewable substrate. J Food Sci Technol 52(6): 3273–3282

Hall HH, Benedict RG, Wiesen CF, Smith CE, Jackson RW (1953) Studies on vitamin B12 production with Streptomyces olivaceus. Appl Microbiol 1(3):124–129

Hamel J (2011) A review of acute cyanide poisoning with a treatment update. Crit Care Nurse 31(1):72–81; quiz 82

Hardlei TF, Obeid R, Herrmann W, Nexo E (2013) Cobalamin analogues in humans: a study on maternal and cord blood. PLoS One 8(4):e61194

Hazra AB, Han AW, Mehta AP, Mok KC, Osadchiy V, Begley TP, Taga ME (2015) Anaerobic biosynthesis of the lower ligand of vitamin B12. Proc Natl Acad Sci U S A 112(34): 10792–10797

Hugenschmidt S, Schwenninger SM, Lacroix C (2011) Concurrent high production of natural folate and vitamin B12 using a co-culture process with Lactobacillus plantarum SM39 and Propionibacterium freudenreichii DF13. Process Biochem 46(5):1063–1070

Jaffe EK (2000) The porphobilinogen synthase family of metalloenzymes. Acta Crystallogr D Biol Crystallogr 56(2):115–128

Jordan PM (1991) The biosynthesis of 5-aminolaevulinic acid and its transformation into uroporphyrinogen III. In: New comprehensive biochemistry, vol 19. Elsevier, Amsterdam

Ko Y, Ashok S, Ainala SK, Sankaranarayanan M, Chun AY, Jung GY, Park S (2014) Coenzyme B12 can be produced by engineered Escherichia coli under both anaerobic and aerobic conditions. Biotechnol J 9(12):1526–1535

Kojima I, Komiya K, Sato H, Oguchi Y (1985) Process for producing vitamin B12 by the fermentation technique, and vitamin B12-producing microorganism. Google patents

Kośmider A, Białas W, Kubiak P, Drożdżyńska A, Czaczyk K (2012) Vitamin B12 production from crude glycerol by Propionibacterium freudenreichii ssp. shermanii: optimization of medium composition through statistical experimental designs. Bioresour Technol 105:128–133

Kräutler B, Arigoni D, Golding BT (1998) Vitamin B12 and B12-proteins. In: European symposium on vitamin B12 and B12-proteins 1996, Innsbruck, Austria

Kusel JP, Fa YH, Demain AL (1984) Betaine stimulation of vitamin B12 biosynthesis in Pseudomonas denitrificans may be mediated by an increase in activity of δ-aminolaevulinic acid synthase. Microbiology 130(4):835–841

Leeper FJ (1989) The biosynthesis of porphyrins, chlorophylls, and vitamin B12. Nat Prod Rep 6(2):171–203

Li KT, Liu DH, Chu J, Wang YH, Zhuang YP, Zhang SL (2008a) An effective and simplified pH-stat control strategy for the industrial fermentation of vitamin B12 by Pseudomonas denitrificans. Bioprocess Biosyst Eng 31(6):605–610

Li KT, Liu DH, Li YL, Chu J, Wang YH, Zhuang YP, Zhang SL (2008b) Improved large-scale production of vitamin B12 by Pseudomonas denitrificans with betaine feeding. Bioresour Technol 99(17):8516–8520

Li KT, Liu DH, Zhuang Y-P, Wang YH, Chu J, Zhang SL (2008c) Influence of Zn2+, Co2+ and dimethylbenzimidazole on vitamin B12 biosynthesis by Pseudomonas denitrificans. World J Microbiol Biotechnol 24(11):2525–2530

Li KT, Zhou J, Cheng X, Wei S (2012) Study on the dissolved oxygen control strategy in large-scale vitamin B12 fermentation by Pseudomonas denitrificans. J Chem Technol Biotechnol 87(12):1648–1653

Linnell JC, Smith AD, Smith CL, Wilson J, Matthews DM (1968) Effects of smoking on metabolism and excretion of vitamin B12. Br Med J 2(5599):215–216

Marie Sych J, Lacroix C, Stevens MJ (2016) Vitamin B12-physiology, production and application. In: Industrial biotechnology of vitamins, biopigments, and antioxidants. Wiley, New York, pp 129–159

Marsh K, Zeuschner C, Saunders A (2012) Health implications of a vegetarian diet: a review. Am J Lifestyle Med 6(3):250–267

Martens JH, Barg H, Warren MJ, Jahn D (2002) Microbial production of vitamin B12. Appl Microbiol Biotechnol 58(3):275–285

Mathur Y, Sreyas S, Datar PM, Sathian MB, Hazra AB (2020) CobT and BzaC catalyze the regiospecific activation and methylation of the 5-hydroxybenzimidazole lower ligand in anaerobic cobamide biosynthesis. J Biol Chem 295(31):10522–10534

Matos M, Gutiérrez G, Iglesias O, Coca J, Pazos C (2015) Enhancing encapsulation efficiency of food-grade double emulsions containing resveratrol or vitamin B12 by membrane emulsification. J Food Eng 166:212–220

Miyano K-i, Ye K, Shimizu K (2000) Improvement of vitamin B12 fermentation by reducing the inhibitory metabolites by cell recycle system and a mixed culture. Biochem Eng J 6(3):207–214

Mo H, Kariluoto S, Piironen V, Zhu Y, Sanders MG, Vincken JP, Wolkers-Rooijackers J, Nout MJ (2013) Effect of soybean processing on content and bioaccessibility of folate, vitamin B12 and isoflavones in tofu and tempe. Food Chem 141(3):2418–2425

Mohammed Y, Lee B, Kang Z, Du G (2014) Development of a two-step cultivation strategy for the production of vitamin B12 by Bacillus megaterium. Microb Cell Factories 13(1):102

Moine G, Hohmann HP, Kurth R, Paust J, Hähnlein W, Pauling H, Weimann BJ, Kaesler B (2000) Vitamins, 6. B vitamins. In: Ullmann's encyclopedia of industrial chemistry. Wiley, New York

Molina V, Médici M, Font de Valdez GF, Taranto MP (2012) Soybean-based functional food with vitamin B12-producing lactic acid bacteria. J Funct Foods 4(4):831–836

Moore SJ, Warren MJ (2012) The anaerobic biosynthesis of vitamin B12. Biochem Soc Trans 40(3):581–586

Nguyen-Vo TP, Ainala SK, Kim JR, Park S (2018) Analysis and characterization of coenzyme B12 biosynthetic gene clusters and improvement of B12 biosynthesis in Pseudomonas denitrificans ATCC 13867. FEMS Microbiol Lett 365(21):fny211

Nielsen MJ, Rasmussen MR, Andersen CB, Nexø E, Moestrup SK (2012) Vitamin B12 transport from food to the body's cells—a sophisticated, multistep pathway. Nat Rev Gastroenterol Hepatol 9(6):345–354

Obeid R, Fedosov SN, Nexo E (2015) Cobalamin coenzyme forms are not likely to be superior to cyano- and hydroxyl-cobalamin in prevention or treatment of cobalamin deficiency. Mol Nutr Food Res 59(7):1364–1372

Oh S, Cave G, Lu C (2021) Vitamin B12 (cobalamin) and micronutrient fortification in food crops using nanoparticle technology. Front Plant Sci 12:668819

Pawlak R, Lester SE, Babatunde T (2014) The prevalence of cobalamin deficiency among vegetarians assessed by serum vitamin B12: a review of literature. Eur J Clin Nutr 68(5):541–548

Piao Y, Kiatpapan P, Yamashita M, Murooka Y (2004a) Effects of expression of hemA and hemB genes on production of porphyrin in Propionibacterium freudenreichii. Appl Environ Microbiol 70(12):7561–7566

Piao Y, Yamashita M, Kawaraichi N, Asegawa R, Ono H, Murooka Y (2004b) Production of vitamin B12 in genetically engineered Propionibacterium freudenreichii. J Biosci Bioeng 98(3): 167–173

Pillai VV, Prakash G, Lali AM (2018) Growth engineering of Propionibacterium freudenreichii shermanii for organic acids and other value-added products formation. Prep Biochem Biotechnol 48(1):6–12

Quesada-Chanto A, Afschar AS, Wagner F (1994) Microbial production of propionic acid and vitamin B12 using molasses or sugar. Appl Microbiol Biotechnol 41(4):378–383

Ramalho MJ, Loureiro JA, Pereira MC (2021) Poly(lactic-co-glycolic acid) nanoparticles for the encapsulation and gastrointestinal release of vitamin B9 and vitamin B12. ACS Appl Nano Mater 4(7):6881–6892

Raux E, Schubert H (2000) Biosynthesis of cobalamin (vitamin B12): a bacterial conundrum. Cell Mol Life Sci CMLS 57(13):1880–1893

Riaz M, Ansari ZA, Iqbal F, Akram M (2007) Microbial production of vitamin B12 by methanol utilizing strain of Pseudomonas specie. Pak Biochem Mol Biol 40:5–10

Rizzo G, Laganà AS, Rapisarda AMC, La Ferrera GMG, Buscema M, Rossetti P, Nigro A, Muscia V, Valenti G et al (2016) Vitamin B12 among vegetarians: status, assessment and supplementation. Nutrients 8(12):767

Roth JR, Lawrence JG, Bobik TA (1996) Cobalamin (coenzyme B12): synthesis and biological significance. Annu Rev Microbiol 50(1):137–181

Rusher DR, Pawlak R (2013) A review of 89 published case studies of vitamin B12 deficiency. J Hum Nutr Food Sci 1(2):1008

Santos F, Vera JL, Lamosa P, de Valdez GF, de Vos WM, Santos H, Sesma F, Hugenholtz J (2007) Pseudovitamin is the corrinoid produced by Lactobacillus reuteri CRL1098 under anaerobic conditions. FEBS Lett 581(25):4865–4870

Selva Kumar S, Chouhan RS, Thakur MS (2010) Trends in analysis of vitamin B12. Anal Biochem 398:139–149

Shepherd G, Velez LI (2008) Role of hydroxocobalamin in acute cyanide poisoning. Ann Pharmacother 42(5):661–669

Signorini C, Carpen A, Coletto L, Borgonovo G, Galanti E, Capraro J, Magni C, Abate A, Johnson SK et al (2018) Enhanced vitamin B12 production in an innovative lupin tempeh is due to synergic effects of Rhizopus and Propionibacterium in cofermentation. Int J Food Sci Nutr 69(4):451–457

Sobczyńska-Malefora A, Delvin E, McCaddon A, Ahmadi KR, Harrington DJ (2021) Vitamin B12 status in health and disease: a critical review. Diagnosis of deficiency and insufficiency–clinical and laboratory pitfalls. Crit Rev Clin Lab Sci 58(6):399–429

Souci SW, Fachmann W, Kraut H (2000) Food composition and nutrition tables. Medpharm GmbH Scientific Publishers, Stuttgart

Stamford NP, Crouzet J, Cameron B, Alanine AI, Pitt AR, Yeliseev AA, Battersby AR (1996) Biosynthesis of vitamin B12: the preparative multi-enzyme synthesis of precorrin-3A and 20-methylsirohydrochlorin (a 2,7,20-trimethylisobacteriochlorin). Biochem J 313(1):335–342

Survase SA, Bajaj IB, Singhal RS (2006) Biotechnological production of vitamins. Food Technol Biotechnol 44(3):381–396

Thierry A, Deutsch SM, Falentin H, Dalmasso M, Cousin FJ, Jan G (2011) New insights into physiology and metabolism of Propionibacterium freudenreichii. Int J Food Microbiol 149(1): 19–27

Thirupathaiah Y, Swarupa Rani C, Sudhakara Reddy M, Venkateswar Rao L (2012) Effect of chemical and microbial vitamin B12 analogues on production of vitamin B12. World J Microbiol Biotechnol 28(5):2267–2271

Tinzl-Malang SK, Rast P, Grattepanche F, Sych J, Lacroix C (2015) Exopolysaccharides from co-cultures of Weissella confusa 11GU-1 and Propionibacterium freudenreichii JS15 act synergistically on wheat dough and bread texture. Int J Food Microbiol 214:91–101

Titcomb TJ, Tanumihardjo SA (2019) Global concerns with B vitamin statuses: biofortification, fortification, hidden hunger, interactions, and toxicity. Compr Rev Food Sci Food Saf 18(6): 1968–1984

Vandamme EJ, Revuelta JL (2016a) Industrial biotechnology of vitamins, biopigments, and antioxidants. Wiley, New York

Vandamme EJ, Revuelta JL (2016b) Vitamins, biopigments, antioxidants and related compounds: a historical, physiological and (bio)technological perspective. In: Industrial biotechnology of vitamins, biopigments, and antioxidants. Wiley, New York, pp 1–14

Wang ZJ, Wang HY, Li YL, Chu J, Huang MZ, Zhuang YP, Zhang SL (2010) Improved vitamin B12 production by step-wise reduction of oxygen uptake rate under dissolved oxygen limiting level during fermentation process. Bioresour Technol 101(8):2845–2852

Wang P, Wang Y, Liu Y, Shi H, Su Z (2012a) Novel in situ product removal technique for simultaneous production of propionic acid and vitamin B12 by expanded bed adsorption bioreactor. Bioresour Technol 104:652–659

Wang P, Wang Y, Su Z (2012b) Improvement of adenosylcobalamin production by metabolic control strategy in Propionibacterium freudenreichii. Appl Biochem Biotechnol 167(1):62–72

Wang P, Jiao Y, Liu S (2014) Novel fermentation process strengthening strategy for production of propionic acid and vitamin B12 by Propionibacterium freudenreichii. J Ind Microbiol Biotechnol 41(12):1811–1815

Wang P, Zhang Z, Jiao Y, Liu S, Wang Y (2015) Improved propionic acid and 5,6-dimethylbenzimidazole control strategy for vitamin B12 fermentation by Propionibacterium freudenreichii. J Biotechnol 193:123–129

Wang ZJ, Shi HL, Wang P (2016) The online morphology control and dynamic studies on improving vitamin B12 production by Pseudomonas denitrificans with online capacitance and specific oxygen consumption rate. Appl Biochem Biotechnol 179(6):1115–1127

Wang P, Shen C, Li L, Guo J, Cong Q, Lu J (2020) Simultaneous production of propionic acid and vitamin B12 from corn stalk hydrolysates by Propionibacterium freudenreichii in an expanded bed adsorption bioreactor. Prep Biochem Biotechnol 50(8):763–767

Warren MJ, Raux E, Schubert HL, Escalante-Semerena JC (2002) The biosynthesis of adenosylcobalamin (vitamin B12). Nat Prod Rep 19(4):390–412

Watanabe F, Yabuta Y, Tanioka Y, Bito T (2013) Biologically active vitamin B12 compounds in foods for preventing deficiency among vegetarians and elderly subjects. J Agric Food Chem 61(28):6769–6775

Watanabe F, Yabuta Y, Bito T, Teng F (2014) Vitamin B12-containing plant food sources for vegetarians. Nutrients 6(5):1861–1873

Watkins D, Rosenblatt DS (2011) Inborn errors of cobalamin absorption and metabolism. Am J Med Genet C Semin Med Genet 157C(1):33–44

Xia W, Chen W, Peng WF, Li KT (2015a) Industrial vitamin B12 production by Pseudomonas denitrificans using maltose syrup and corn steep liquor as the cost-effective fermentation substrates. Bioprocess Biosyst Eng 38(6):1065–1073

Xia W, Peng WF, Chen W, Li KT (2015b) Interactive performances of betaine on the metabolic processes of Pseudomonas denitrificans. J Ind Microbiol Biotechnol 42(2):273–278

Xie C, Coda R, Chamlagain B, Varmanen P, Piironen V, Katina K (2019) Co-fermentation of Propionibacterium freudenreichii and Lactobacillus brevis in wheat bran for in situ production of vitamin B12. Front Microbiol 10:1541

Zhang Y, Liu JZ, Huang JS, Mao ZW (2010) Genome shuffling of Propionibacterium shermanii for improving vitamin B12 production and comparative proteome analysis. J Biotechnol 148(2–3): 139–143

Zheng J, Wittouck S, Salvetti E, Franz CMAP, Harris HMB, Mattarelli P, O'Toole PW, Pot B, Vandamme P et al (2020) A taxonomic note on the genus Lactobacillus: description of 23 novel genera, emended description of the genus Lactobacillus Beijerinck 1901, and union of Lactobacillaceae and Leuconostocaceae. Int J Syst Evol Microbiol 70(4):2782–2858

Zhou X, Li Y (2021) Atlas of oral microbiology: from healthy microflora to disease. Springer Nature, Berlin

Chapter 25
The Role of Dietary Fiber in Promoting Health: A Review of Choice and Outcomes

Faisal Sualeh Hayyat, Sartaj Ahmad Allayie, Junaid Ahmad Malik, and Sabzar Ahmad Dar

Contents

F. S. Hayyat
Department of Physical Education and Sports, Government Degree College, Pulwama, Pulwama, Jammu and Kashmir, India

S. A. Allayie · S. A. Dar
Zoology Research Laboratory, Department of Zoology, Government Degree College, Anantnag, Anantnag, Jammu and Kashmir, India

J. A. Malik (✉)
Department of Zoology, Government Degree College Kulgam, Kulgam, Jammu and Kashmir, India

J. A. Malik et al. (eds.), *Food Process Engineering and Technology*,
https://doi.org/10.1007/978-981-99-6831-2_25

Abstract Nutrition plays a vital role in overall health, and fiber has emerged as a key component in promoting well-being. Recent research highlights its significance in maintaining gut health, aiding weight management, and benefiting cardiovascular health. Soluble fiber supports beneficial gut bacteria, aids digestion, and prevents constipation. High-fiber diets contribute to weight control by promoting satiety and stabilizing blood sugar levels. Additionally, soluble fiber helps lower cholesterol, reducing the risk of heart disease. Further studies on specific fiber types can optimize dietary recommendations for various health conditions. Encouraging fiber-rich diets empowers individuals to make informed choices and improve their quality of life. This chapter deals with the comprehensive analysis of the current literature on the health effects of fiber-rich diets and their potential implications for healthcare professionals and policymakers.

Keywords Fiber-rich diets · Health outcomes · Digestive health · Weight management · Chronic diseases

25.1 Introduction

Nobody can undervalue the importance of a nutritious diet in improving overall health and wellbeing. Dietary habits are of great importance in shaping one's physical, mental as well as emotional health, impacting the prevention of chronic diseases, cognitive function, mood regulation, and even life expectancy. As researchers, healthcare professionals, and policymakers continue to recognize the critical link between nutrition and health, scientific proof supports the significance of adopting and maintaining a healthy diet.

A balanced diet is characterized by a well-balanced intake of essential nutrients, vitamins, minerals, and other bioactive compounds that the body requires to function optimally. While restricting consumption of processed foods, sugary beverages, and excessive levels of saturated and trans fats, it places an emphasis on the consumption of a wide variety of nutrient-dense foods, such as fruits, vegetables, whole grains, lean proteins, and healthy fats. Adopting a nutritious diet has been linked to many beneficial health outcomes and a lower chance of developing chronic diseases.

A balanced diet is essential for both preventing and treating a number of chronic diseases, according to research. For instance, cardiovascular disorders continue to be the world's top cause of death. By lowering blood pressure, enhancing blood lipid profiles, and supporting cardiovascular health, studies have demonstrated that a diet high in fruits, vegetables, whole grains, and unsaturated fats can considerably lessen the risk of heart disease (Estruch et al. 2018; Mente et al. 2017). A healthy diet has also been associated with a lower incidence of type 2 diabetes, obesity, certain malignancies, and metabolic disorders (Hu 2002; Schwingshackl et al. 2017).

Beyond physical health, a healthy diet has profound implications for mental well-being and cognitive function. According to recent studies, eating habits high in antioxidants, omega-3 fatty acids, and specific vitamins and minerals may improve cognitive function and lower the risk of age-related cognitive decline and neurodegenerative disorders like Alzheimer's (Morris et al. 2015; Parletta et al. 2017). Additionally, research shows a strong link between diet and mental health, with diets rich in nutrient-dense foods being associated with a lower incidence of depression and anxiety (Opie et al. 2017; Sarris et al. 2015).

The significance of a balanced diet extends beyond the individual level to influence broader societal and economic aspects. The prevalence of diet-related chronic diseases places a substantial burden on healthcare systems and economies worldwide. As a result, promoting healthier dietary habits has become a public health priority, with governments and organizations implementing various initiatives to raise awareness and improve nutrition education predominantly fiber rich nutrients. Encouraging the adoption of a healthy diet also offers significant potential for improving workforce productivity, reducing absenteeism, and enhancing overall quality of life.

Despite the abundant evidence supporting the significance of a healthy diet, numerous challenges persist in encouraging widespread dietary behaviour change. Socioeconomic factors, cultural influences, marketing of unhealthy foods, and lack of access to nutritious options in certain regions pose barriers to adopting healthier diets. In order to influence evidence-based interventions and policies, it is crucial for researchers to keep researching the intricate relationships between nutrition, lifestyle, and health outcomes. This chapter aims to explore the scientific evidence regarding the impact of fiber-rich diets on health outcomes and their potential therapeutic applications.

25.2 Understanding Fiber

The indigestible parts of plant-based meals, such as the cell walls of fruits, vegetables, grains, legumes, and nuts, are referred to as dietary fiber. In contrast to other carbohydrates, fiber is resistant to being broken down by human digestive enzymes, allowing it to travel through the gastrointestinal tract with little or no absorption and offers a host of health advantages. Dietary fiber is an essential part of a balanced diet because of this special quality that has beneficial health effects.

A balanced diet must contain fiber since it has a substantial impact on one's overall health and wellbeing. Since fiber is a complex carbohydrate found in plant-based meals, it is different from other carbohydrates in that the small intestine is not where it is digested or absorbed. Instead, it enters the digestive tract almost undisturbed and has special health advantages. To acquire a thorough grasp of fiber's role in supporting health, we have evaluated many forms of fiber, their purposes, health implications, and recommended intake in this chapter (Cronin et al. 2021).

Based on how easily it dissolves in water, fiber can be divided into two basic categories: soluble fiber and insoluble fiber. Oats, legumes, and fruits all contain soluble fiber, which gels up in the intestines to slow down digestion and increase feelings of fullness. Whole grains, nuts, and vegetables all include insoluble fiber, which gives the stool volume and promotes regular bowel motions (Lattimer and Haub 2010).

The main purposes of fiber are related to digestive health. Both forms of fiber serve as prebiotics, feeding good gut bacteria and promoting a healthy gut microbiota. Soluble fiber lowers cholesterol levels by attaching to cholesterol in the intestines. Diets high in fiber have been linked to a lower risk of developing chronic conditions like type 2 diabetes and cardiovascular illnesses. Intake of fiber has also been connected to satiety and weight management.

Depending on one's age, gender, and stage of life, a certain amount of fiber per day is advised. The recommended daily consumption for people is between 21 and 38 g. To avoid gastrointestinal discomfort, it is recommended to gradually increase fiber intake. Constipation can also be avoided by drinking enough of water (Dhingra et al. 2012). Understanding fiber and its multifaceted health benefits can empower individuals to make informed dietary choices, optimizing their nutritional intake and improving overall well-being (Cronin et al. 2021).

25.2.1 *Types of Dietary Fiber*

Based on how easily it dissolves in water, dietary fibre can be divided into two primary categories: soluble fibre and insoluble fibre.

25.2.1.1 Soluble Fiber

Fiber that is soluble in water turns into a gel-like material in the intestines. Oats, barley, beans, lentils, apples, and citrus fruits are a few examples of foods that contain it. Soluble fiber slows down digestion and nutrient absorption when it is ingested by binding with water and other substances. This trait aids in blood sugar stabilization and leads to sensations of fullness. Additionally, soluble fiber has been linked to lowering blood cholesterol levels through its ability to bind to cholesterol and aid in the body's elimination of it, supporting cardiovascular health (Ioniță-Mîndrican et al. 2022).

25.2.1.2 Insoluble Fiber

The bulky component of the stool is insoluble fiber, which does not dissolve in water. Whole grains, wheat bran, nuts, seeds, the skins of fruits and vegetables, and wheat bran are all sources of this type of fiber. By facilitating the faster passage of food through the intestines, insoluble fiber promotes regular bowel motions and digestive health. Constipation and maintaining regularity are also helped by its bulking effect (Ioniță-Mîndrican et al. 2022).

A balanced diet should include a variety of plant-based foods that are high in dietary fiber because they all have important health advantages. Understanding the differences between soluble and insoluble dietary fiber enables people to make educated dietary decisions to maximize their nutrient intake and enhance their general health. A fuller understanding of fiber's significance is likely to influence dietary recommendations and public health programmes, promoting healthier lifestyles and well-being, as research continues to elucidate the complexities of fiber's effect on human health (Anderson et al. 2009).

25.3 Sources of Dietary Fiber

A number of plant-based foods are rich sources of dietary fiber, which is essential to a balanced diet (Dhingra et al. 2012). In order to assist people in making educated decisions to satisfy their daily fiber needs, the various dietary fiber sources that one must include in his daily food chart are as under:

25.3.1 Fruits and Vegetables

Fruits and vegetables provide a variety of vitamins, minerals, and phyto-nutrients and are excellent sources of dietary fiber. Bananas, apples, pears, berries, oranges, and orange juice are a few fruits high in fiber. Broccoli, carrots, spinach, sweet potatoes, and Brussels sprouts stand out as superior providers of fiber in the vegetable group. A diet rich in a rainbow of fruits and vegetables increases fiber consumption while also supplying vital elements for good health (Stewart and Schroeder 2013).

25.3.2 Whole Grains

Integral to a diet high in fiber are whole grains. Whole grains are higher in dietary fiber than refined grains because they preserve their bran, germ, and endosperm. Due to their delayed digestion, foods including oats, quinoa, brown rice, barley, and whole wheat pasta are great sources of fiber and give you continuous energy. A quick and easy strategy to improve fiber consumption is to substitute whole grains for processed grains (Barber et al. 2020).

25.3.3 Legumes and Pulses

Legumes, including beans, lentils, chickpeas, and peas, are nutritional powerhouses, rich in both protein and dietary fiber. These plant-based protein sources are versatile and can be added to soups, salads, stews, and curries, increasing the fiber content of various dishes. Incorporating legumes into the diet is an excellent strategy to boost fiber intake while also supporting heart health and weight management (Ioniță-Mîndrican et al. 2022).

25.3.4 Nuts and Seeds

In addition to offering good lipids and protein, nuts and seeds also help with dietary fiber. Among the best sources of fiber in this group are almonds, chia seeds, flaxseeds, and sunflower seeds. A handful of nuts or seeds can increase fiber intake and add a pleasing crunch to meals or snacks. It's crucial to consume the necessary amount of dietary fiber each day to maintain general health and fend off a variety of chronic diseases. A range of plant-based foods can be incorporated into regular meals to create a diet high in fiber. It is simpler for people to consume enough fiber while savouring a variety of mouth-watering meals because to the wide diversity of flavours, textures, and nutritional advantages offered by fruits, vegetables, whole grains, legumes, nuts, and seeds (Dodevska et al. 2022).

25.4 Role of Fiber Rich Foods in Prompting Health

A balanced and healthy diet must include foods high in fibre. They offer a host of health advantages, including as better digestion, control over weight and blood sugar, and a lower risk of developing chronic illnesses. Recent research has highlighted that fiber-rich diets offer numerous health benefits and should be an integral part of a balanced and nutritious eating pattern. Consuming a variety of fiber sources, such as fruits, vegetables, whole grains, legumes, and nuts, supports digestive health, aids in weight management, and lowers the risk of chronic diseases (Barber et al. 2020). Promoting the adoption of fiber-rich diets should be a public health priority to improve overall well-being and reduce the burden of diet-related health issues.

25.4.1 Improved Digestion

In order to keep the digestive system functioning properly, fiber is essential. Dietary fiber gives the stool bulk, which encourages regular bowel motions and wards off constipation. Additionally, it functions as a prebiotic, feeding the good bacteria in the gut, promoting overall gut health, and lowering the risk of gastrointestinal illnesses (Slavin 2013).

25.4.2 Weight Management

Foods high in fibre can help you manage your weight. Meals that are high in fiber usually fill you up longer and can help you consume fewer calories overall (Slavin 2015). Additionally, fibre promotes satiety and better blood sugar regulation by slowing the absorption of nutrients, including glucose. Obesity is a complex disorder that is influenced by a number of variables, such as genetics, environment, nutrition, physical activity, and socioeconomic position. In addition to obesogenic surroundings with easy access to high-calorie foods and few opportunities for physical activity, genetic predisposition can make some people more prone to weight gain (Malik et al. 2013). Additionally, the obesity pandemic is made worse by the proliferation of "food deserts," which are places with little access to cheap and wholesome food (Larson et al. 2009).

25.4.2.1 Challenges in Weight Management

Many people find it difficult to manage their weight and lose weight permanently. The human body has many physiological defences against weight reduction, which frequently result in weight increase following initial achievements (Mantzoros et al. 2019). Additionally, people find it challenging to follow calorie-restricted diets due to the present obesogenic environment, which is full of extremely appetising and energy-dense items (Lowe and Butryn 2021). Various psychological and emotional issues can also make it difficult to maintain your weight, including stress, depression, and emotional eating (Arends et al. 2020).

25.4.2.2 Implications for Public Health

Obesity poses significant implications for public health, exerting a considerable economic burden on healthcare systems worldwide. The direct costs associated with obesity-related medical care and indirect costs due to productivity loss are substantial (Wang et al. 2011). Additionally, the obesity epidemic raises people's risk of developing a number of chronic diseases, which lowers their quality of life and shortens their life expectancy (Prospective Studies Collaboration 2009). Effective weight management interventions, including lifestyle modifications, behavioural counselling, and bariatric surgery, are essential for reducing obesity-related health risks and improving overall public health outcomes.

Obesity presents a multifaceted challenge with implications for weight management and public health. Addressing obesity requires a comprehensive approach that acknowledges the influence of genetics, environment, and individual behaviour. In order to battle obesity and enhance general wellbeing, it is essential to implement evidence-based interventions, promote healthy lifestyle choices, and fight for laws that support access to wholesome meals and opportunities for physical activity.

Obesity-related problems can be prevented and treated with dietary fiber. Reduced absorption of macronutrients and increased satiety are two mechanisms behind this defense. High-fiber diets appear to protect against obesity and play a function in the treatment of obesity. Emerging theories to explain this include changes in the gut flora and short-chain fatty acids (Dayib et al. 2020).

25.4.3 Blood Sugar Control

Foods high in fiber, particularly soluble fiber, can help control blood sugar levels. In order to avoid blood sugar increases following meals, soluble fiber slows down the absorption of glucose (Jovanovski et al. 2017). For people with diabetes or at risk of developing diabetes, this is very advantageous and must be included in diet to keep the sugar under control.

25.4.4 Reduced Risk of Chronic Diseases

A high-fiber diet has been associated with a lower risk of chronic illnesses, such as heart disease, Type-2 diabetes, and several cancers. According to studies, every additional 10 g of dietary fiber consumed each day reduces the risk of coronary heart disease by 10% (Threapleton et al. 2013a, b).

25.5 Impact on Gut Microbiome and Digestive Health

A vital part of preserving digestive health and general wellbeing is the gut microbiome, a complex ecology of microorganisms living in the gastrointestinal tract. The immune system, pathogen defence, and nutritional absorption all depend on the delicate balance of the gut microbiota. Recent studies have shown the numerous influences on the gut microbiome and their potential effects on digestive health.

Diet has a substantial impact on the composition and diversity of the gut microbiome, which has an impact on digestive health. A diet high in fibre, prebiotics, and probiotics encourages the development of good bacteria, creating a favourable environment for gut health (Smith et al. 2018). On the other hand, eating a diet rich in processed foods, added sweets, and bad fats might cause dysbiosis and raise your chance of developing digestive problems (Smith et al. 2018). Consequently, maintaining the health of the gut microbiota and achieving optimal digestive function require adopting a balanced and nutrient-rich diet.

The gut microbiome can be harmed by the use of some medicines. For instance, antibiotics may indiscriminately attack both dangerous and helpful bacteria, which may result in dysbiosis (Francino 2016). Proton pump inhibitors (PPIs) and non-steroidal anti-inflammatory drugs (NSAIDs) are two other pharmaceuticals that have been linked to altered gut microbiomes and digestive issues (Jackson et al. 2018). To reduce their impact on the gut flora and avoid digestive disorders, health professionals should use prudence when administering drugs, further using fiber rich diet may neutralize the harm they have on microbiomes.

The gut-brain axis, a two-way communication mechanism that links the central nervous system and the gut, can be dramatically affected by stress and anxiety. Chronic stress can impair immunological, motility, and secretory processes in the gut, which affects the gut microbiome's composition (Mayer et al. 2015). According to Faresjo et al. (2019), these disturbances can cause the onset or worsening of digestive illnesses such irritable bowel syndrome (IBS). Consequently, stress reduction methods and fiber rich diets may be able to support a healthy gut flora and enhance digestive health.

25.6 Recommended Daily Intake and Dietary Guidelines from Health Organisations

It is impossible to emphasise the value of a healthy, balanced diet in fostering overall wellbeing. To ensure proper nutrient consumption, lower the risk of chronic diseases, and maintain a healthy body weight, health organisations all over the world offer dietary advice and guidelines for the general public.

To help people fulfil their nutritional needs, health organisations offer recommended daily intakes (RDIs) of important nutrients, vitamins, and minerals. In order to account for characteristics like growth, pregnancy, and lactation, these RDIs are often stated as reference values based on age, sex, and life stage (Institute of Medicine 2005). For instance, the recommended nutritional Intakes (RNIs) in Europe and the Recommended Dietary Allowances (RDAs) in the United States describe the typical daily nutritional intake sufficient to meet the needs of almost all healthy people. For proper health Women should attempt to consume at least 21–25 g of fiber daily, and males should aim for 30–38 g (Larson 2015).

Health organisations including World Health Organisation (2018) also provide detailed dietary recommendations that go beyond RDIs and provide broader suggestions for food selections and eating habits to support general health. The necessity of eating a range of nutrient-dense foods, such as fruits, vegetables, whole grains, lean meats, and healthy fats is emphasised by these recommendations (U.-S. Department of Health and Human Services and U.S. Department of Agriculture 2020). In order to lower the risk of developing chronic diseases including obesity, cardiovascular disease, and type 2 diabetes, the guidelines frequently recommend limiting the consumption of added sugars, saturated fats, and sodium.

The dietary recommendations published by many health organisations share a few fundamental themes despite differences in terminology and particular values. All of the recommendations place equal emphasis on eating a variety of nutrient-rich foods, such as fruits, vegetables, and whole grains (WHO 2018). Similar to this, all recommendations emphasise limiting the consumption of processed, unhealthy meals that are heavy in added sugars and saturated fats.

However, differences in dietary guidelines also exist, reflecting the varying cultural and regional dietary habits. Some organizations may prioritize specific nutrients or foods based on regional dietary patterns and nutritional research. For instance, owing to the health advantages of olive oil and seafood, the dietary recommendations of Mediterranean countries may place a strong emphasis on these foods (Ministry of Health – Italy 2020).

To maintain good health and fend against many chronic diseases, a well-balanced diet that contains enough fibre and nutrient-rich foods is crucial. Plant-based foods contain fibre, which is essential for maintaining digestive health, controlling weight, and lowering the risk of cardiovascular illnesses as mentioned above in this chapter.

25.7 Benefits of Fiber in the Diet

The body cannot digest fiber, a type of carbohydrate present in meals including fruits, vegetables, whole grains, legumes, and nuts. Dietary fiber has several health advantages, but it especially encourages regular bowel movements and guards against constipation (Slavin 2013). Additionally, a diet high in fiber is linked to lower cholesterol levels and a lower chance of developing heart disease (Threapleton et al. 2013a, b). Consuming meals high in fiber also aids in controlling blood sugar levels and may reduce the chance of developing Type-2 diabetes (Chandalia et al. 2000). Additionally, fibre promotes satiety, which can help with weight management and weight loss (Bolton et al. 1981).

Including nutrient-rich foods is key for ensuring that the body gets the needed vitamins, minerals, and other necessary elements, in addition to fibre. Variety of colourful fruits and vegetables, which offer antioxidants and phytochemicals that boost the immune system and lower the risk of chronic diseases, should be a part of a nutrient-rich diet (Liu 2013). Complex carbs, fibre, and important vitamins and minerals like iron and B vitamins can all be found in abundance in whole grains like brown rice, quinoa, and oats. Muscle maintenance and regeneration depend on lean protein sources such poultry, fish, beans, and lentils (Phillips 2012). Furthermore, it is essential for brain function and the absorption of fat-soluble vitamins to consume healthy fats from foods like avocados, nuts, seeds, and olive oil (Tomaino et al. 2010).

25.8 Practical Strategies for Dietary Incorporation

Simple yet efficient methods can be used to include foods high in fiber and nutrients in the diet. To give the gut time to adjust, begin by gradually increasing your fiber consumption. To do this, increase the amount of vegetables and whole grains in your diet, add beans or lentils to your soups and salads, and go for whole fruit instead of fruit juice (Dahl and Stewart 2015). The availability of a variety of nutrient-dense foods can also be guaranteed by meal planning and preparation. Furthermore, mindful eating, which entails savouring each bite and doing it deliberately, can aid in portion control and food enjoyment.

25.9 Fiber Rich Recipes and Food Preparation Ideas

For the purpose of supporting digestive health, controlling weight, and lowering the risk of chronic diseases, fiber-rich foods must be a part of our meals. However, finding creative and delicious ways to include these foods in our meals can be challenging. Following are collection of recipes and food preparation ideas that are rich in fiber, making it easier for individuals to adopt a fiber-friendly diet and reap its numerous health benefits.

25.9.1 Recipe 1: Quinoa and Roasted Vegetable Salad

This vibrant salad combines fiber-rich quinoa with a variety of roasted vegetables. Cook quinoa according to package instructions and let it cool. While waiting, roast a variety of colorful veggies, including cherry tomatoes, zucchini, and bell peppers, with garlic, olive oil, and your preferred herbs. Toss the cooled quinoa and roasted vegetables together, and dress the salad with light lemon vinaigrette for a refreshing and fiber-packed meal (Ioniță-Mîndrican et al. 2022).

25.9.2 Recipe 2: Chickpea and Vegetable Curry

This hearty and flavorful curry is loaded with fiber from chickpeas and a medley of vegetables. In a large pot, sauté onions, garlic, and ginger until fragrant. Add chopped vegetables like carrots, cauliflower, and spinach, and cook until tender. Stir in canned chickpeas and coconut milk. Season the curry with curry powder, turmeric, cumin, and chili powder. Serve the mixture over brown rice after simmering until the flavours have combined to provide a fulfilling and high-fiber meal (Jideani et al. 2021).

25.9.3 Recipe 3: Black Bean and Sweet Potato Tacos

These tasty tacos are an excellent way to enjoy the fiber goodness of black beans and sweet potatoes. Roast diced sweet potatoes with olive oil, paprika, and cumin until they are caramelized and tender. In a separate pan, sauté black beans with onions, garlic, and diced tomatoes. Warm up corn tortillas and fill them with the black bean mixture, roasted sweet potatoes, and your favorite taco toppings, like avocado, salsa, and cilantro (Slavin 2013).

25.9.4 Recipe 4: Whole Grain Berry Parfait

Prepare a whole grain berry parfait. Layer cooked quinoa or rolled oats with a mixture of fresh berries and plain Greek yogurt. Drizzle honey and sprinkle some chopped nuts or seeds for added crunch and flavor. This parfait is a wonderful way to start the day with a fibre boost in addition to being a delightful treat (Dhingra et al. 2012).

25.9.5 Recipe 5: Lentil and Vegetable Soup

This comforting soup combines lentils with a variety of vegetables for a fiber-packed meal. In a large pot, sauté onions, carrots, and celery until softened. Add dried lentils, vegetable broth, and diced tomatoes. Simmer until the lentils are tender and the flavors meld together. Season the soup with thyme, bay leaves, and a pinch of red pepper flakes for a warming and nutritious bowl of goodness (Anderson et al. 2009).

25.9.6 Future Research and Advancement in the Field

The importance of a fiber-rich diet in promoting overall health and well-being has been well-established. However, as research continues to advance, there is a need to delve deeper into this area to better understand its mechanisms of action and potential applications in various health conditions. While the benefits of dietary fiber are well-known, future research should aim to elucidate the specific types of fiber and their unique health effects. Investigating the roles of soluble and insoluble fibers, prebiotics, and resistant starch in gut health, metabolic regulation, and immune function could provide valuable insights for tailoring dietary recommendations to specific health conditions (Holscher 2017). Additionally, individualized nutrition and disease prevention show tremendous potential when it comes to studying the relationship between dietary fiber and the gut flora and its consequences for diverse health outcomes (Sonnenburg and Bäckhed 2016).

Moreover, research should explore the potential synergistic effects of combining fiber-rich diets with other nutrients and bioactive compounds. For instance, investigating the impact of fiber in combination with antioxidants, polyphenols, or probiotics could provide a more comprehensive understanding of their collective health benefits (Lamuela-Raventós and Reguant 2018). Further studies should also explore the potential use of novel fiber sources, such as seaweed-derived fibers or insect-based proteins, to expand dietary options and sustainable nutrition practices (Ramos-Romero et al. 2018).

25.10 Conclusion

The research summarized in this chapter highlights the significant health benefits of fiber consumption. Incorporating fiber-rich foods into the diet can contribute to improved cardiovascular health, better diabetes management, enhanced digestive function, effective weight management, and potentially reduced cancer risk. Fiber is an easily obtained and reasonably priced dietary component that provides a potent tool for enhancing general health and lowering the risk of chronic diseases. Public health initiatives and personalized dietary recommendations can capitalize on these findings to encourage increased fiber intake and support better health outcomes for individuals across diverse populations.

References

Anderson JW, Baird P, Davis RH, Ferreri S, Knudtson M, Koraym A, Waters V, Williams CL (2009) Health benefits of dietary fiber. Nutr Rev 67(4):188–205

Arends I, van der Heijden AA, van Dijk AM (2020) Emotional eating and food intake after sadness and joy. Appetite 155:104803

Barber TM, Kabisch S, Pfeiffer AFH, Weickert MO (2020) The health benefits of dietary fibre. Nutrients 12(10):3209

Bolton RP, Heaton KW, Burroughs LF (1981) The role of dietary fiber in satiety, glucose, and insulin: studies with fruit and fruit juice. Am J Clin Nutr 34(2):211–217

Chandalia M, Garg A, Lutjohann D, Bergmann K, Grundy SM, Brinkley LJ (2000) Beneficial effects of high dietary fiber intake in patients with type 2 diabetes mellitus. N Engl J Med 342(19):1392–1398

Cronin P, Joyce SA, O'Toole PW, O'Connor EM (2021) Dietary fibre modulates the gut microbiota. Nutrients 13(5):1655

Dahl WJ, Stewart ML (2015) Position of the Academy of Nutrition and Dietetics: health implications of dietary fiber. J Acad Nutr Diet 115(11):1861–1870

Dayib M, Larson J, Slavin J (2020) Dietary fibers reduce obesity-related disorders: mechanisms of action. Curr Opin Clin Nutr Metab Care 23(6):445–450

Dhingra D, Michael M, Rajput H, Patil RT (2012) Dietary fibre in foods: a review. J Food Sci Technol 49(3):255–266

Dodevska M, Kukic Markovic J, Sofrenic I, Tesevic V, Jankovic M, Djordjevic B, Ivanovic ND (2022) Similarities and differences in the nutritional composition of nuts and seeds in Serbia. Front Nutr 16(9):1003125

Estruch R, Ros E, Salas-Salvado J, Covas MI, Corella D, Aros F, Fiol M (2018) Primary prevention of cardiovascular disease with a Mediterranean diet supplemented with extra-virgin olive oil or nuts. N Engl J Med 378(25):34

Faresjo A, Grodzinsky E, Johansson S (2019) A modified Mediterranean diet with limited alcohol consumption decreases the risk for symptoms of depression and anxiety in subjects with the metabolic syndrome: a randomized controlled trial in primary care. Ann Gen Psychiatry 18(1): 1–10

Francino MP (2016) Antibiotics and the human gut microbiome: dysbioses and accumulation of resistances. Front Microbiol 6:1543

Holscher HD (2017) Dietary fiber and prebiotics and the gastrointestinal microbiota. Gut Microbes 8(2):172–184

Hu FB (2002) Dietary pattern analysis: a new direction in nutritional epidemiology. Curr Opin Lipidol 13(1):3–9

Institute of Medicine (2005) Dietary reference intakes for energy, carbohydrate, fiber, fat, fatty acids, cholesterol, protein, and amino acids. The National Academies Press, Washington, DC

Ioniță-Mîndrican CB, Ziani K, Mititelu M, Oprea E, Neacşu SM, Moroşan E, Dumitrescu DE, Roşca AC, Drăgănescu D, Negrei C (2022) Therapeutic benefits and dietary restrictions of fiber intake: a state of the art review. Nutrients 14(13):2641

Jackson MA, Goodrich JK, Maxan ME (2018) Proton pump inhibitors alter the composition of the gut microbiota. Gut 65(5):749–756

Jideani AIO, Silungwe H, Takalani T, Omolola AO, Udeh HO, Anyasi TA (2021) Antioxidant-rich natural fruit and vegetable products and human health. Int J Food Prop 24(1):41–67

Jovanovski E, Marques AC, Vieira RM, Khan TA, Jenkins AL (2017) Effect of added viscosity in beverages on postprandial glycemia: a systematic review and meta-analysis. J Clin Nutr 42(7): 672–683

Lamuela-Raventós RM, Reguant J (2018) Health effects of wine and alcohol from a dietetic and nutritional perspective. In: Bioactive foods in promoting health. Academic, New York, pp 217–230

Larson H (2015) Easy ways to boost fiber in your daily diet. Eatright.org. Academy of Nutrition and Dietetics, Chicago, p 6

Larson NI, Story MT, Nelson MC (2009) Neighbourhood environments: disparities in access to healthy foods in the US. Am J Prev Med 36(1):74–81

Lattimer JM, Haub MD (2010) Effects of dietary fiber and its components on metabolic health. Nutrients 2(12):1266–1289

Liu RH (2013) Health-promoting components of fruits and vegetables in the diet. Adv Nutr 4(3): 384S–392S
Lowe MR, Butryn ML (2021) Hedonic hunger: a new dimension of appetite? Physiol Behav 228: 113173
Malik VS, Willett WC, Hu FB (2013) Global obesity: trends, risk factors, and policy implications. Nat Rev Endocrinol 9(1):13–27
Mantzoros CS, Magkos F, Brinkoetter M (2019) Leptin in human physiology and pathophysiology. Am J Physiol Endocrinol Metab 301(4):E567–E584
Mayer EA, Knight R, Mazmanian SK (2015) Gut microbes and the brain: paradigm shift in neuroscience. J Neurosci 34(46):15490–15496
Mente A, O'Donnell M, Rangarajan S, Dagenais G, Lear S, McQueen M, Teo K (2017) Association of dietary nutrients with blood lipids and blood pressure in 18 countries: a cross-sectional analysis from the PURE study. Lancet Diabetes Endocrinol 5(10):774–787
Ministry of Health – Italy (2020) Guidelines for a healthy diet. https://www.salute.gov.it/imgs/C_17_pubblicazioni_2851_allegato.pdf. Accessed 4 Aug 2023
Morris MC, Tangney CC, Wang Y, Sacks FM, Bennett DA, Aggarwal NT (2015) MIND diet associated with reduced incidence of Alzheimer's disease. Alzheimers Dement 11(9): 1007–1014
Opie RS, O'Neil A, Itsiopoulos C, Jacka FN, Cotton S (2017) The impact of whole-of-diet interventions on depression and anxiety: a systematic review of randomised controlled trials. Public Health Nutr 20(11):2074–2093
Parletta N, Zarnowiecki D, Cho J, Wilson A, Bogomolova S, Villani A, O'Dea K (2017) A Mediterranean-style dietary intervention supplemented with fish oil improves diet quality and mental health in people with depression: a randomized controlled trial (HELFIMED). Nutr Neurosci 20(3):151–162
Phillips SM (2012) Dietary protein requirements and adaptive advantages in athletes. Br J Nutr 108 (S2):S158–S167
Prospective Studies Collaboration (2009) Body-mass index and cause-specific mortality in 900,000 adults: collaborative analyses of 57 prospective studies. Lancet 373(9669):1083–1096
Ramos-Romero S, González-Aguilar GA, Ayala-Zavala JF (2018) Edible coatings for fresh-cut fruits and vegetables. In: Edible coatings and films to improve food quality, 2nd edn. Woodhead Publishing, Cambridge, pp 369–384
Sarris J, Logan AC, Akbaraly TN, Amminger GP, Balanzá-Martínez V, Freeman MP, Parker G (2015) Nutritional medicine as mainstream in psychiatry. Lancet Psychiatry 2(3):271–274
Schwingshackl L, Schwedhelm C, Hoffmann G, Lampousi AM, Knüppel S, Iqbal K, Boeing H (2017) Food groups and risk of type 2 diabetes mellitus: a systematic review and meta-analysis of prospective studies. Eur J Epidemiol 32(5):363–375
Slavin JL (2013) Fiber and prebiotics: mechanisms and health benefits. Nutrients 5(4):1417–1435
Slavin J (2015) Fiber and prebiotics: mechanisms and health benefits. J Am Assoc Nurse Pract 27(8):452–456
Smith MI, Yatsunenko T, Manary MJ (2018) Gut microbiomes of Malawian twin pairs discordant for kwashiorkor. Science 339(6119):548–554
Sonnenburg JL, Bäckhed F (2016) Diet-microbiota interactions as moderators of human metabolism. Nature 535(7610):56–64
Stewart ML, Schroeder NM (2013) Dietary treatments for childhood constipation: efficacy of dietary fiber and whole grains. Nutr Rev 71(2):98–109
Threapleton DE, Greenwood DC, Evans CEL, Cleghorn CL, Nykjaer C, Woodhead C, Burley VJ (2013a) Dietary fibre intake and risk of cardiovascular disease: systematic review and meta-analysis. BMJ 347:6879
Threapleton DE, Greenwood DC, Evans CEL, Cleghorn CL, Nykjaer C, Woodhead C, Burley VJ (2013b) Dietary fiber intake and risk of first stroke: a systematic review and meta-analysis. Stroke 44(5):1360–1368

Tomaino A, Martorana M, Arcoraci T, Monteleone D, Giovinazzo C, Saija A (2010) Antioxidant activity and phenolic profile of pistachio (Pistaciavera L., variety Bronte) seeds and skins. Biochimie 92(9):1115–1122

U.S. Department of Health and Human Services, U.S. Department of Agriculture (2020) Dietary guidelines for Americans 2020-2025, 9th ed

Wang YC, McPherson K, Marsh T (2011) Health and economic burden of the projected obesity trends in the USA and the UK. Lancet 378(9793):815–825

World Health Organization (WHO) (2018) Healthy diet. https://www.who.int/news-room/fact-sheets/detail/healthy-diet. Accessed 4 Aug 2023

Index

J. A. Malik et al. (eds.), *Food Process Engineering and Technology*,
https://doi.org/10.1007/978-981-99-6831-2

MIX
Papier aus verantwortungsvollen Quellen
Paper from responsible sources
FSC® C105338

If you have any concerns about our products,
you can contact us on
ProductSafety@springernature.com

In case Publisher is established outside the EU,
the EU authorized representative is:
Springer Nature Customer Service Center GmbH
Europaplatz 3, 69115 Heidelberg, Germany

Printed by Libri Plureos GmbH
in Hamburg, Germany